S0-AOJ-741

VOLUME FIVE HUNDRED AND SEVENTY SIX

METHODS IN ENZYMOLOGY

Synthetic Biology and Metabolic Engineering in Plants and Microbes Part B: Metabolism in Plants

METHODS IN ENZYMOLOGY

VOLUME FIVE HUNDRED AND SEVENTY SIX

METHODS IN ENZYMOLOGY

Synthetic Biology and Metabolic Engineering in Plants and Microbes Part B: Metabolism in Plants

Edited by

SARAH E. O'CONNOR
The John Innes Centre
Department of Biological Chemistry
Norwich, NR4 7UH, UK

AMSTERDAM • BOSTON • HEIDELBERG • LONDON
NEW YORK • OXFORD • PARIS • SAN DIEGO
SAN FRANCISCO • SINGAPORE • SYDNEY • TOKYO
Academic Press is an imprint of Elsevier

AP

Academic Press is an imprint of Elsevier
50 Hampshire Street, 5th Floor, Cambridge, MA 02139, United States
525 B Street, Suite 1800, San Diego, CA 92101–4495, United States
The Boulevard, Langford Lane, Kidlington, Oxford OX5 1GB, United Kingdom
125 London Wall, London, EC2Y 5AS, United Kingdom

First edition 2016

ISBN: 978-0-12-804539-8
ISSN: 0076-6879

For information on all Academic Press publications
visit our website at https://www.elsevier.com/

Publisher: Zoe Kruze
Acquisition Editor: Zoe Kruze
Editorial Project Manager: Sarah Lay
Production Project Manager: Magesh Kumar Mahalingam
Cover Designer: Maria Ines Cruz

Typeset by SPi Global, India

CONTENTS

Contributors *xi*
Preface *xvii*

1. Comparative Biochemistry and In Vitro Pathway Reconstruction as Powerful Partners in Studies of Metabolic Diversity **1**
P. Fan, G.D. Moghe, and R.L. Last

1. Introduction 2
2. Characterization of Specialized Metabolite Diversity Using LC/MS and NMR 4
3. Phylogeny-Driven Analysis of the Biochemical Basis of Specialized Metabolite Diversity 8
4. Conclusions 15

Acknowledgments 16
References 16

2. De Novo Deep Transcriptome Analysis of Medicinal Plants for Gene Discovery in Biosynthesis of Plant Natural Products **19**
R. Han, A. Rai, M. Nakamura, H. Suzuki, H. Takahashi, M. Yamazaki, and K. Saito

1. Methods for Transcriptomic Study Prior to High-Throughput Sequencing 20
2. Deep Transcriptome Outline 22
3. Preparation of the Plant Materials and RNA Extraction 24
4. cDNA Library Construction 27
5. High-Throughput Sequencing 30
6. Data Interpretation 31
7. Application of RNA-Seq on Medicinal Plants and Perspectives 38

Acknowledgments 39
References 39

3. Genomics-Based Discovery of Plant Genes for Synthetic Biology of Terpenoid Fragrances: A Case Study in Sandalwood oil Biosynthesis **47**
J.M. Celedon and J. Bohlmann

1. Introduction 48
2. Prior Knowledge of Sandalwood TPSs and P450s and Development of a Hypothesis 50

3. Replication, Sampling, and Statistical Design 51
4. Defining Temporal and Spatial Variables for Tissue Sampling 52
5. Tissue Sampling 53
6. Metabolite Profiling 53
7. Isolation of High-Quality RNA from Recalcitrant Tissues 55
8. Transcriptome Sequencing and De Novo Assembly 56
9. Transcriptome Mining and Annotation 57
10. Expression Analysis and Candidate Gene Selection 58
11. Functional Characterization of Candidate Genes 59
12. Product Identification 63
Acknowledgments 65
References 65

4. A Workflow for Studying Specialized Metabolism in Nonmodel Eukaryotic Organisms 69

M.P. Torrens-Spence, T.R. Fallon, and J.K. Weng

1. Introduction 70
2. "Omics"-Based Novel Specialized Metabolic Pathway Discovery 72
3. Structure–Function Analysis of Specialized Metabolic Enzymes 80
4. Reconstitution of Specialized Metabolic Pathways in Heterologous Systems 87
5. Summary 90
Acknowledgments 91
References 91

5. Gene Discovery for Synthetic Biology: Exploring the Novel Natural Product Biosynthetic Capacity of Eukaryotic Microalgae 99

E.C. O'Neill, G. Saalbach, and R.A. Field

1. Introduction 100
2. Natural Product Synthases 102
3. Genome Mining for the Identification of Natural Products 106
4. Natural Product Discovery 113
5. Conclusions 115
Acknowledgments 116
References 116

6. *cis*-Prenyltransferase and Polymer Analysis from a Natural Rubber Perspective 121

M. Kwon, E.-J.G. Kwon, and D.K. Ro

1. Introduction 122
2. Rationale: Observation of Revertants from *rer2* Mutant 125

3. Generation of *rer2* and *srt1* Double Knockout Yeast Strain 128
4. Complementation of *rer2Δ srt1Δ* with *CPT* and *CBP* 131
5. CPT Biochemical Assay Using Yeast Microsomes 135
6. General Discussion 142
Acknowledgments 143
References 144

7. Generation and Functional Evaluation of Designer Monoterpene Synthases 147

N. Srividya, I. Lange, and B.M. Lange

1. Introduction 148
2. Equipment 150
3. Materials 150
4. Step 1—Generation of Expression Constructs 153
5. Step 2—Production of Purified, Recombinant Target Enzyme 158
6. Step 3—Functional Evaluation of Recombinant Monoterpene Synthases 161
7. Conclusions 164
Acknowledgments 164
References 164

8. Prequels to Synthetic Biology: From Candidate Gene Identification and Validation to Enzyme Subcellular Localization in Plant and Yeast Cells 167

E. Foureau, I. Carqueijeiro, T. Dugé de Bernonville, C. Melin, F. Lafontaine, S. Besseau, A. Lanoue, N. Papon, A. Oudin, G. Glévarec, M. Clastre, B. St-Pierre, N. Giglioli-Guivarc'h, and V. Courdavault

1. Introduction 168
2. Identification of Candidate Genes Through Transcriptomic Data Mining and Analysis 171
3. Validation of Candidate Gene Function by Biolistic-Mediated VIGS 187
4. Studying the Subcellular Localization of Biosynthetic Pathway Enzymes in Plant and Yeast Cells to Alleviate Bottlenecks in Bioengineering Approaches 191
5. Concluding Remarks 202
Acknowledgments 203
References 203

9. Functional Expression and Characterization of Plant ABC Transporters in *Xenopus laevis* Oocytes for Transport Engineering Purposes **207**

D. Xu, D. Veres, Z.M. Belew, C.E. Olsen, H.H. Nour-Eldin, and B.A. Halkier

1. Introduction 208
2. Preparation of cDNA of Plant ABC Transporter Genes by In Planta "Exon Engineering" 211
3. ABC Transporter Expression in *Xenopus* Oocytes 215
4. Optimization of Transport Assay for Diffusible ABA in *Xenopus* Oocytes 217
5. Case Study: Characterization of the ABA Exporter at ABCG25 in *Xenopus* Oocytes 219
6. Conclusions 220

Acknowledgments 221
References 221

10. Quantifying the Metabolites of the Methylerythritol 4-Phosphate (MEP) Pathway in Plants and Bacteria by Liquid Chromatography–Triple Quadrupole Mass Spectrometry **225**

D. González-Cabanelas, A. Hammerbacher, B. Raguschke, J. Gershenzon, and L.P. Wright

1. Introduction 226
2. Preparation of Stable Isotope-Labeled Internal Standards 229
3. Extraction of Methylerythritol Phosphate Pathway Intermediates from Biological Sources 234
4. Analysis of Methylerythritol Phosphate Pathway Metabolites by LC–MS/MS 236
5. Discussion and Summary 244

References 245

11. Establishing the Architecture of Plant Gene Regulatory Networks **251**

F. Yang, W.Z. Ouma, W. Li, A.I. Doseff, and E. Grotewold

1. Introduction 252
2. The *cis*-Regulatory Apparatus 254
3. The *Trans*-Acting Factors 260
4. Transcription Factor Centered Approaches 266
5. Gene-Centered Approaches 274
6. Resources for Studying Plant GRNs 281
7. Conclusions 286

Acknowledgment 287
References 287

12. Engineering of Tomato Glandular Trichomes for the Production of Specialized Metabolites **305**

R.W.J. Kortbeek, J. Xu, A. Ramirez, E. Spyropoulou, P. Diergaarde, I. Otten-Bruggeman, M. de Both, R. Nagel, A. Schmidt, R.C. Schuurink, and P.M. Bleeker

1. Introduction 306
2. Materials and Technology 310
3. Proof of Concept: Targeted Expression of a Terpene Precursor Gene in Tomato Glandular Trichomes 318
4. Summary 325

Acknowledgments 326
References 327

13. Tomato Fruits—A Platform for Metabolic Engineering of Terpenes **333**

M. Gutensohn and N. Dudareva

1. Introduction 334
2. Terpenoid Formation in Tomato Fruits 338
3. Transgene Expression in Ripening Tomato Fruits 340
4. Overexpression of Terpene Biosynthetic Genes in Tomato Fruits 343
5. Analysis of Terpenes in Tomato Fruits 349
6. Conclusions 350

Acknowledgments 352
References 352

14. Libraries of Synthetic TALE-Activated Promoters: Methods and Applications **361**

T. Schreiber and A. Tissier

1. Introduction 361
2. Construction of Libraries of Synthetic Promoters Using Golden Gate Cloning 366
3. Analyzing Promoter Activity in Transient Assays 371
4. Conclusion 373

References 375

Author Index *379*
Subject Index *419*

CONTRIBUTORS

Z.M. Belew
DynaMo Center, Faculty of Science, University of Copenhagen, Frederiksberg C, Denmark

S. Besseau
Université François-Rabelais de Tours, EA2106 "Biomolécules et Biotechnologies Végétales", Tours, France

P.M. Bleeker
Swammerdam Institute for Life Sciences, University of Amsterdam, Amsterdam, The Netherlands

J. Bohlmann
Michael Smith Laboratories, University of British Columbia, Vancouver, BC, Canada

I. Carqueijeiro
Université François-Rabelais de Tours, EA2106 "Biomolécules et Biotechnologies Végétales", Tours, France

J.M. Celedon
Michael Smith Laboratories, University of British Columbia, Vancouver, BC, Canada

M. Clastre
Université François-Rabelais de Tours, EA2106 "Biomolécules et Biotechnologies Végétales", Tours, France

V. Courdavault
Université François-Rabelais de Tours, EA2106 "Biomolécules et Biotechnologies Végétales", Tours, France

M. de Both
Keygene N.V., Wageningen, The Netherlands

P. Diergaarde
Keygene N.V., Wageningen, The Netherlands

A.I. Doseff
Heart and Lung Research Institute, The Ohio State University, Columbus, OH, United States

N. Dudareva
Purdue University, West Lafayette, IN, United States

T. Dugé de Bernonville
Université François-Rabelais de Tours, EA2106 "Biomolécules et Biotechnologies Végétales", Tours, France

T.R. Fallon
Whitehead Institute for Biomedical Research; Massachusetts Institute of Technology, Cambridge, MA, United States

P. Fan
Michigan State University, East Lansing, MI, United States

R.A. Field
John Innes Centre, Norwich, United Kingdom

E. Foureau
Université François-Rabelais de Tours, EA2106 "Biomolécules et Biotechnologies Végétales", Tours, France

J. Gershenzon
Max Planck Institute for Chemical Ecology, Jena, Germany

N. Giglioli-Guivarc'h
Université François-Rabelais de Tours, EA2106 "Biomolécules et Biotechnologies Végétales", Tours, France

G. Glévarec
Université François-Rabelais de Tours, EA2106 "Biomolécules et Biotechnologies Végétales", Tours, France

D. González-Cabanelas
Max Planck Institute for Chemical Ecology, Jena, Germany

E. Grotewold
Center for Applied Sciences (CAPS), The Ohio State University, Columbus, OH, United States

M. Gutensohn
Davis College of Agriculture, Natural Resources and Design, West Virginia University, Morgantown, WV, United States

B.A. Halkier
DynaMo Center, Faculty of Science, University of Copenhagen, Frederiksberg C, Denmark

A. Hammerbacher
Forestry and Agricultural Biotechnology Institute (FABI), University of Pretoria, Pretoria, South Africa

R. Han
Graduate School of Pharmaceutical Sciences, Chiba University, Chiba, Japan

R.W.J. Kortbeek
Swammerdam Institute for Life Sciences, University of Amsterdam, Amsterdam, The Netherlands

E.-J.G. Kwon
University of Calgary, Calgary, AB, Canada

M. Kwon
University of Calgary, Calgary, AB, Canada

F. Lafontaine
Université François-Rabelais de Tours, EA2106 "Biomolécules et Biotechnologies Végétales", Tours, France

B.M. Lange
Institute of Biological Chemistry, M. J. Murdock Metabolomics Laboratory, Washington State University, Pullman, WA, United States

I. Lange
Institute of Biological Chemistry, M. J. Murdock Metabolomics Laboratory, Washington State University, Pullman, WA, United States

A. Lanoue
Université François-Rabelais de Tours, EA2106 "Biomolécules et Biotechnologies Végétales", Tours, France

R.L. Last
Michigan State University, East Lansing, MI, United States

W. Li
Heart and Lung Research Institute, The Ohio State University, Columbus, OH, United States

C. Melin
Université François-Rabelais de Tours, EA2106 "Biomolécules et Biotechnologies Végétales", Tours, France

G.D. Moghe
Michigan State University, East Lansing, MI, United States

R. Nagel
Max Planck Institute for Chemical Ecology, Jena, Germany

M. Nakamura
Graduate School of Pharmaceutical Sciences, Chiba University, Chiba, Japan

H.H. Nour-Eldin
DynaMo Center, Faculty of Science, University of Copenhagen, Frederiksberg C, Denmark

C.E. Olsen
Faculty of Science, University of Copenhagen, Frederiksberg C, Denmark

E.C. O'Neill
University of Oxford, Oxford, United Kingdom

I. Otten-Bruggeman
Keygene N.V., Wageningen, The Netherlands

A. Oudin
Université François-Rabelais de Tours, EA2106 "Biomolécules et Biotechnologies Végétales", Tours, France

W.Z. Ouma
Center for Applied Sciences (CAPS), The Ohio State University, Columbus, OH, United States

N. Papon
Université d'Angers, Groupe d'Etude des Interactions Hôte-Pathogène, UPRES EA 3142, Angers, France

B. Raguschke
Max Planck Institute for Chemical Ecology, Jena, Germany

A. Rai
Graduate School of Pharmaceutical Sciences, Chiba University, Chiba, Japan

A. Ramirez
Swammerdam Institute for Life Sciences, University of Amsterdam, Amsterdam, The Netherlands

D.K. Ro
University of Calgary, Calgary, AB, Canada

G. Saalbach
John Innes Centre, Norwich, United Kingdom

K. Saito
Graduate School of Pharmaceutical Sciences, Chiba University, Chiba; RIKEN Center for Sustainable Resource Science, Yokohama, Japan

A. Schmidt
Max Planck Institute for Chemical Ecology, Jena, Germany

T. Schreiber
Leibniz Institute of Plant Biochemistry, Halle (Saale), Germany

R.C. Schuurink
Swammerdam Institute for Life Sciences, University of Amsterdam, Amsterdam, The Netherlands

E. Spyropoulou
Swammerdam Institute for Life Sciences, University of Amsterdam, Amsterdam, The Netherlands

N. Srividya
Institute of Biological Chemistry, M. J. Murdock Metabolomics Laboratory, Washington State University, Pullman, WA, United States

B. St-Pierre
Université François-Rabelais de Tours, EA2106 "Biomolécules et Biotechnologies Végétales", Tours, France

H. Suzuki
Kazusa DNA Research Institute, Kisarazu, Japan

H. Takahashi
Medical Mycology Research Center, Chiba University, Chiba, Japan

A. Tissier
Leibniz Institute of Plant Biochemistry, Halle (Saale), Germany

M.P. Torrens-Spence
Whitehead Institute for Biomedical Research, Cambridge, MA, United States

D. Veres
DynaMo Center, Faculty of Science, University of Copenhagen, Frederiksberg C, Denmark

J.K. Weng
Whitehead Institute for Biomedical Research; Massachusetts Institute of Technology, Cambridge, MA, United States

L.P. Wright
Max Planck Institute for Chemical Ecology, Jena, Germany

D. Xu
DynaMo Center, Faculty of Science, University of Copenhagen, Frederiksberg C, Denmark

J. Xu
Swammerdam Institute for Life Sciences, University of Amsterdam, Amsterdam, The Netherlands

M. Yamazaki
Graduate School of Pharmaceutical Sciences, Chiba University, Chiba, Japan

F. Yang
Center for Applied Sciences (CAPS), The Ohio State University, Columbus, OH, United States

PREFACE

Advances in sequencing, bioinformatics, and genome editing now enable us to access the rich chemistry encoded within the metabolic pathways of plants and microbes. A major focus in metabolism is the secondary, or specialized, pathways that produce small, biologically active molecules with applications in pharmaceutical, agrochemical, or other biotechnological sectors. In recent years, metabolic engineering/synthetic biology approaches have shown remarkable promise for the exploitation of these pathways for human use. In these two volumes, we highlight some of the most important approaches that have been used to harness microbial and plant metabolic pathways.

In Volume 1, we focus on advances that have been made in microbial-based systems. The discovery some three decades ago that bacterial specialized metabolic pathways are clustered on the genome has greatly facilitated the identification and characterization of these pathway genes. Coupled with the fact that bacterial genomes can now be sequenced rapidly and inexpensively, the last decade has seen a staggering increase in our knowledge of bacterial specialized metabolism. Additionally, it is now known that fungal specialized pathways also cluster on the genome. While fungal genomes are larger than those from bacteria, these genomes can still can be easily sequenced, and substantial advances in elucidating fungal metabolism have been made. Consequently, a wealth of new opportunities in metabolic engineering have been opened. In this volume, we highlight how better production of these compounds can be achieved, and how these biosynthetic enzymes can be engineered to generate new biocatalysts and new products. We also discuss how microbial species can be manipulated to serve as a host for reconstitution of plant pathways. The volume concludes with several representative examples of new tools that allow us to rapidly manipulate the genetic material of the microbial host.

Volume 2 focuses on the metabolism of plants. Historically, elucidating plant metabolism has been challenging due to the lack of tightly genome-clustered pathways that are observed in microbial systems, along with the large size of plant genomes and transcriptomes. The first set of articles in this volume describe a variety of strategies to elucidate plant-specialized metabolism. Notably, the specialized metabolism of plants is controlled by complex regulatory processes. Furthermore, plant biosynthetic processes are also complicated by the fact that the metabolic reactions occur in a variety of different cell types and subcellular compartments. Therefore, several articles in this volume also describe efforts to control the regulatory networks that maintain the levels of metabolism production in plants, along with methods to understand the mechanisms of transport and localization of specialized metabolic intermediates. Finally, we highlight emerging tools to harness plant metabolism: new plant-based expression platforms and expression tools for production of metabolites are discussed.

Metabolic engineering has progressed rapidly in the last several years. The advent of genome editing, the ability to sequence complex genomes quickly and inexpensively, and the successful manipulation of plant and microbial hosts for more effective pathway reconstitution have collectively demonstrated that metabolic engineering holds substantial promise for improving our access to the end products of specialized metabolism. I note that these two volumes scratch the surface of this field, providing only a survey of some of the efforts being made in this area. I am deeply indebted to all of the contributors to this volume who graciously provided their time and effort to make a contribution to this work.

S.E. O'Connor
The John Innes Centre
Department of Biological Chemistry

CHAPTER ONE

Comparative Biochemistry and In Vitro Pathway Reconstruction as Powerful Partners in Studies of Metabolic Diversity

P. Fan, G.D. Moghe, R.L. Last[1]

Michigan State University, East Lansing, MI, United States

[1]Corresponding author: e-mail address: lastr@msu.edu

Contents

1. Introduction 2
2. Characterization of Specialized Metabolite Diversity Using LC/MS and NMR 4
   2.1 Metabolite Structure Elucidation by LC/MS with Multiplexed CID 4
   2.2 Sample Purification for NMR Spectroscopy 7
3. Phylogeny-Driven Analysis of the Biochemical Basis of Specialized Metabolite Diversity 8
   3.1 Phylogeny-Based Screening of Metabolic Phenotypes 8
   3.2 The Biochemical Basis for Differences in Furanose Ring Acylation 10
   3.3 Identifying a Key Polymorphism Driving Enzyme Function Divergence 12
4. Conclusions 15

Acknowledgments 16

References 16

Abstract

There are estimated to be >300,000 plant species, producing >200,000 metabolites. Many of these metabolites are restricted to specific plant lineages and are referred to as "specialized" metabolites. These serve varied functions in plants including defense against biotic and abiotic stresses, plant–plant and plant–microbe communication, and pollinator attraction. These compounds also have important applications in agriculture, medicine, skin care, and in diverse aspects of human culture. The specialized metabolic repertoire of plants can vary even within and between closely related species, in terms of the number and classes of specialized metabolites as well as their structural variants. This phenotypic variation can be exploited to discover the underlying variation in the metabolic enzymes. We describe approaches for using the diversity of specialized metabolites and variation in enzyme structure and function to identify novel enzymatic activities and understand the structural basis for these differences. The knowledge obtained from these studies will provide new modules for the synthetic biology toolbox.

Methods in Enzymology, Volume 576
ISSN 0076-6879
http://dx.doi.org/10.1016/bs.mie.2016.02.023

1. INTRODUCTION

The staggering diversity of specialized metabolites (historically referred to as secondary metabolites) in the plant kingdom is a seemingly limitless source of new compounds for medicinal, cosmetic, culinary, agricultural, and religious use. Much effort has been made to understand the biosynthetic basis of these specialized metabolites in specific species. However, advances in small molecule mass spectrometry, DNA sequencing, and computational methods allow exploration of specialized metabolic diversity using natural genetic variation. Exploration of plant natural variation provides new strategies for discovery of novel biosynthetic activities, enables synthetic pathway reconstruction to produce commercially important plant specialized metabolites, and allows precise alteration of existing pathways. For example, *Escherichia coli* was recently engineered to produce the terpenoid limonene and its derivative perillyl alcohol using enzymes from bacteria, yeast, the gymnosperm plant grand fir (*Abies grandis*), and the dicot spearmint (*Mentha spicata*) (Alonso-Gutierrez et al., 2013). Another study focused on the commercially important labdane terpenoids, using natural and synthetically optimized enzyme activities from several plant species to develop a yeast "plug-and-play" system for the production of diverse compounds of interest (Ignea et al., 2015). These and other examples demonstrate that continued exploration of the full range of plant metabolic diversity and its biosynthetic machinery can put us in a strong position to synthesize more plant natural products (Lau & Sattely, 2015).

Secreting glandular trichomes are epidermal structures that produce specialized metabolites in a wide variety of plant species (Schilmiller, Last, & Pichersky, 2008; Schilmiller et al., 2010; Tissier, 2012; Wagner, 1991) to attract pollinators or deter herbivores (Pichersky & Gershenzon, 2002; Shepherd, Bass, Houtz, & Wagner, 2005; Simmons & Gurr, 2005; Weinhold & Baldwin, 2011). Acylsugars are trichome-produced bioactive compounds with roles in defense against insect herbivores (Hawthorne, Shapiro, Tingey, & Mutschler, 1992; Weinhold & Baldwin, 2011) (Fig. 1). Acylsugars produced by type I trichomes on aerial tissues of cultivated tomato *Solanum lycopersicum* have a sucrose core with fatty acid esters of carbon chains ranging from C2–C12 in length (Ghosh, Westbrook, & Jones, 2014; Schilmiller, Charbonneau, & Last, 2012), with their nomenclature described in Fig. 1. Published work from our lab led to identification of four cultivated tomato trichome-expressed acylsugar acyltransferases: Sl-ASAT1, Sl-ASAT2,

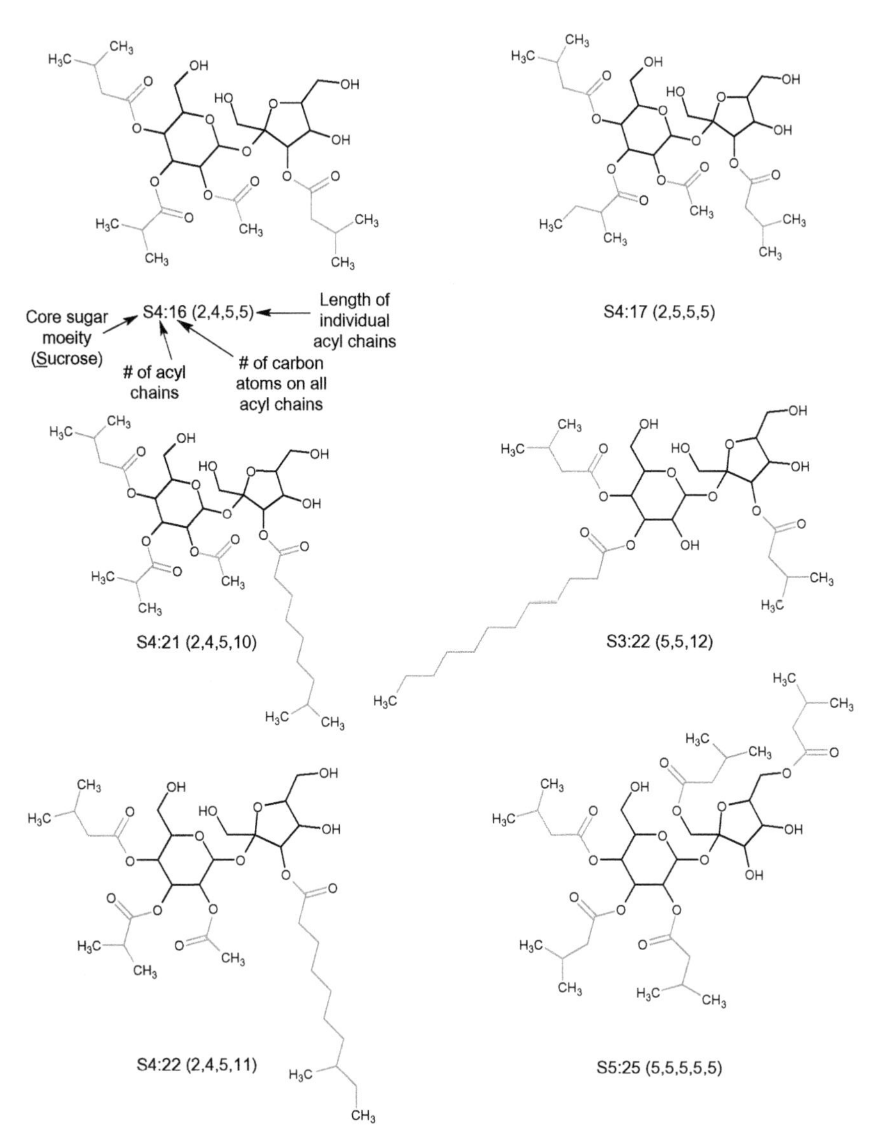

Fig. 1 Representative acylsugar structures identified from *S. lycopersicum* and *S. habrochaites*. The acyl chains are highlighted in *blue* (*light gray* in the print version) and the sugar core is in *black*. The acylsugars are named using the scheme described for the first structure.

Sl-ASAT3, and Sl-ASAT4 (Fan et al., 2016; Schilmiller et al., 2012, 2015). These enzymes were demonstrated to sequentially add acyl chains to sucrose using acyl-CoA donor substrates, and together are sufficient to reconstruct the acylsucrose biosynthetic pathway in vitro (Fan et al., 2016).

ASATs show preference for acyl-CoA donor chain length as well as specificity for the position on the sucrose ring that is acylated. For example, the related wild tomato species *Solanum habrochaites* accumulates a larger number and more diverse group of detectable acylsugars than cultivated tomato (Ghosh et al., 2014; Kim et al., 2012), and prior research from our lab has associated this variation to differences in activities of Sh-ASAT2, Sh-ASAT3 (Fan et al., 2016; Schilmiller et al., 2015), and Sh-ASAT4 (Kim et al., 2012). In this chapter, we illustrate how metabolite diversity can guide discovery of novel enzyme activities, using *S. lycopersicum* and *S. habrochaites* ASAT3 as a case study. We also discuss ways in which this strategy can be used in service of synthetic biology. These diversity-based approaches are generally applicable to pathway engineering in plants and microbes.

2. CHARACTERIZATION OF SPECIALIZED METABOLITE DIVERSITY USING LC/MS AND NMR

2.1 Metabolite Structure Elucidation by LC/MS with Multiplexed CID

Elucidating the structures of important metabolites is a prerequisite toward understanding the biosynthetic pathways that make the compounds. The first step is to understand the number and types of metabolites extracted. Reverse-phase LC separation coupled to time-of-flight (ToF) mass spectrometry (MS) with multiplexed collision-induced dissociation (CID) is a robust approach to perform efficient nontargeted metabolite profiling (Gu, Jones, & Last, 2010). This method generates energy dependent ion fragmentation patterns by rapidly switching among different CID Aperture 1 voltages, and the products are recorded by the ToF-MS. Using both negative- and positive-ion mode electrospray ionization (ESI) provides more information by inducing different types of fragmentation. Accurate ion mass measurement under lower energy nonfragmenting and higher energy fragmenting conditions provides structural information for each compound eluting from the LC. For acylsucroses, spectral information is obtained using three parallel Aperture 1 potentials 10, 40, and 80 V. At the lowest collision voltage (10 V), the nonfragmented acylsugar formate adducts in ESI^- and the ammonium adducts in ESI^+ are the most abundant ions. Fragment ions are detected at higher collision energies. ESI^- fragmentation generates successive neutral loss of acyl groups as ketenes, providing information about the masses of the acyl groups attached to the sucrose core (Fig. 2, third row). The complementary ESI^+ method—which breaks the glycosidic bond between the pyranose and furanose rings of the sucrose moiety—provides complementary information on the masses of the component rings (Fig. 2, fourth row) and allows discrimination of isomers with different ring acylations. Interpreting the ESI^- and ESI^+ results together allows deduction of the acylation patterns of each ring.

2.1.1 Example of Acylsugar Structure Profiling by LC/MS

Here we present an example of the analysis of two acylsugar positional isomers in the tomato relatives *S. lycopersicum* M82 and *S. habrochaites* LA1777 trichomes. The acylsugars extracted from leaf trichomes of the two tomatoes were used for metabolite profiling with LC/MS. The trichome metabolite total ion chromatograms (Fig. 2, first row) revealed a more complex mixture of metabolites in the LA1777 trichome extracts than M82. Extracted ion chromatograms led to identification of three closely eluting peaks with m/z 751.4 (Fig. 2, second row). The coeluting highlighted peak ESI^- CID spectra are consistent with the hypothesis that each includes a C2, two C5, and a C10 chain (Fig. 2, the third row). However, more information is revealed by examining the ESI^+ fragmentation patterns of the two acylsugars (Fig. 2, fourth row): the most abundant M82 acylsugar ion fragment corresponds to a single C5 acyl chain attached to one ring. In contrast, LA1777 has a single C10 acyl chain on one ring. These results indicate that the C10 chain is on different rings on the two acylsugar isomers.

2.1.2 Protocol for Plant Trichome Acylsugar Extraction and LC/MS

1. Seeds are germinated on filter paper soaked with water for 1 week, then transferred to soil and grown for another 2 weeks in a growth chamber under a photoperiod of 16 h light (28°C; cool white fluorescent lights of 300 $\mu E\ m^{-2}\ s^{-1}$) and 8 h (20°C) dark before acylsugar extraction. Note that times and conditions will vary for different species and genotypes; seed treatment with 0.05–0.5 m*M* gibberellin A3 can be useful in promoting germination of recalcitrant plants.
2. Leaf surface metabolites are extracted by placing the leaflet from the next to the youngest leaf in a 1.5 mL microcentrifuge tube containing 1 mL of isopropanol:acetonitrile:water (3:3:2, v/v/v) with 0.1% formic acid as well as 10 μ*M* propyl-4-hydroxybenzoate as internal standard. The tubes are gently agitated on a rotary platform shaker for 2 min, and the solvent moved to a glass LC vial, which can be stored at −20°C until further analysis.

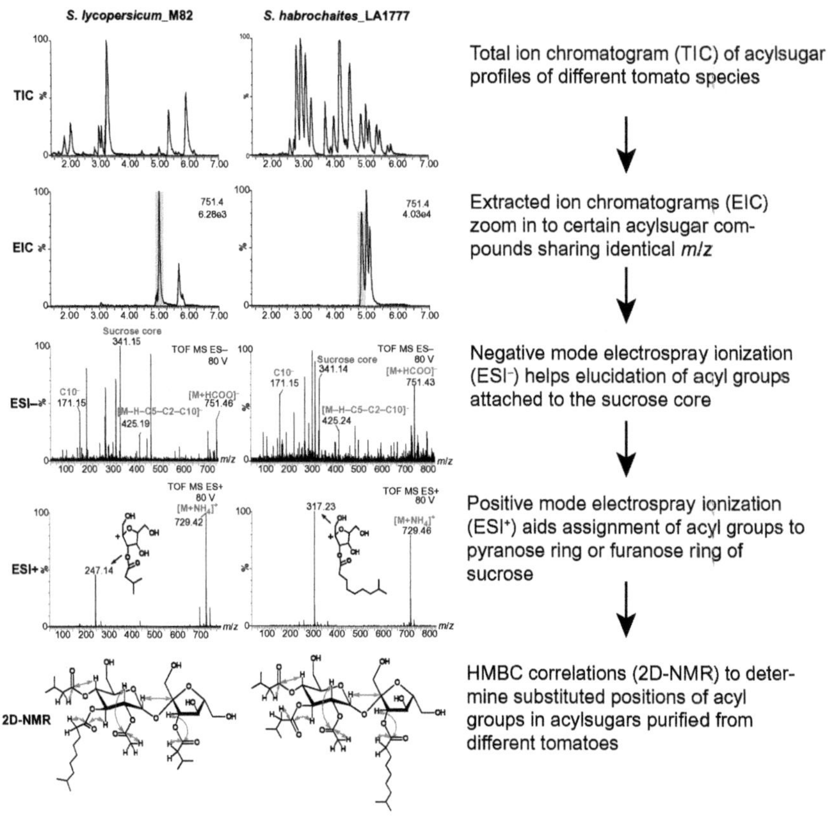

Fig. 2 The workflow used to characterize structures of two acylsugar positional isomers extracted from *S. habrochaites* LA1777 and *S. lycopersicum* M82. The *top row* shows the total ion chromatograms of acylsugars extracted from the two species. The *second row* displays the extracted ion chromatograms of the acylsugar isomers with *m/z* of 751.4. The *yellow* highlighted peaks are selected for further ESI^- and ESI^+ analysis. The *third row* shows the mass spectra of the selected *yellow* highlighted peaks obtained using ESI^- with the CID Aperture 1 voltage of 80 V. $[M+HCOO]^-$ is the acylsugar formate adduct. $[M-H-C5-C2-C10]^-$ is the ion fragment that lost a proton and three acyl chains. The *fourth row* shows the mass spectra of the selected peaks obtained using ESI^+ with the CID Aperture 1 voltage at 80 V. The most abundant ion $[M+NH_4]^+$ is the acylsugar ammonium adduct. The ion fragments released from the acylsugars after breaking the glycosidic bond are also shown (*m/z* of 247.14 and 317.46). The *last row* shows the ^1H and ^{13}C correlations determined by 2D-NMR (HMBC) to elucidate the acyl groups substituted positions. *This figure is adapted from Ghosh, B., Westbrook, T.C., & Jones, A.D. (2014). Comparative structural profiling of trichome specialized metabolites in tomato (*Solanum lycopersicum*) and* S. habrochaites*: Acylsugar profiles revealed by UHPLC/MS and NMR.* Metabolomics, 10, *496–507.* (See the color plate.)

3. The trichome acylsugar samples are analyzed by a Shimadzu LC-20AD HPLC system (Shimadzu) connected to a Waters LCT Premier ToF-MS (Waters). Ten microliters of the extracts are injected into a fused core Ascentis Express C18 column (2.1 mm × 10 cm, 2.7 μm particle size; Sigma Aldrich, catalog number 53823-U) for reverse phase separation using a gradient of 40–100% acetonitrile in 0.15% formic acid–water for 7 min, with a flow rate at 0.4 mL/min. The HPLC is coupled with MS for mass spectra acquisition using the parameters as follows: electrospray is performed in both negative and positive mode; the capillary voltage is set to 3000 V; the desolvation and source temperatures are 300°C and 90°C, respectively; multiplexed CID is performed by switching the Aperture 1 voltage between 10, 40, and 80 V. Each scan time is 0.4 s with an interscan delay of 0.1 s.

2.2 Sample Purification for NMR Spectroscopy

In the example above, LC/MS data revealed that both S4:22 (2,5,5,10) have C10 acylations on different rings, but the exact C10 acylation positions, as well as the positions and branching patterns of the C5 chains (eg, n-pentanoic acid, nC5/2-methylbutanoic acid, aiC5/3-methylbutanoic acid, iC5) could not be determined.

Nuclear magnetic resonance (NMR) spectroscopy is an excellent approach to fully elucidate the structure of a compound. However, typically hundreds of micrograms to milligrams are needed for NMR analysis, which poses a challenge starting with complex mixtures of a biological sample. Here we describe collection of large amounts of acylsugars from tomato trichomes and purification of two acylsugar S4:22 (2,5,5,10) isomers from the mixture. 2D-NMR (for example, heteronuclear multiple bond correlation spectra, HMBC), which shows correlations of ^{1}H and ^{13}C nuclei separated by 2–4 bonds (Fig. 2, the last row), is used to assign the acyl chain substitution positions. It was found that the C10 acyl chain is on the furanose ring of the LA1777 acylsugar, while for the M82 acylsugar, C10 acyl chain is on the pyranose ring and the furanose ring position is occupied by C5 (Ghosh et al., 2014).

2.2.1 Protocol for Purifying Individual Acylsugars from Trichomes for NMR Analysis

1. To collect enough plant acylsugars for NMR analysis, a large amount of the plant tissue (about 150 young leaflets) is harvested from the plants and placed in a 2 L beaker containing 1 L of 100% methanol, followed by

2 min gentle shaking by hand. The extraction solvent is dried down under water aspirator vacuum using a rotary evaporator at room temperature. The purified compound is dissolved in 3 mL of acetonitrile:water (4:1, v/v) in a glass tube, sonicated in a water bath at room temperature for 10 min before being split into three 1.5 mL centrifuge tubes and subjected to centrifugation at 3000 × *g* for 5 min at room temperature to remove any insoluble material.

2. To purify the individual acylsugars from the trichome metabolite mixture, the 3 mL of clarified supernatant is split into fifteen 200 μL aliquots in HPLC vials with 250 μL vial inserts. The acylsugar purification is performed by semipreparative LC using a Waters 2795 HPLC system coupled with a LKB BROMMA 221 Superrac fraction collector. Two-hundred microliter of the extract is injected into a Thermo Scientific Acclaim 120 C18 HPLC column (4.6 × 150 mm, 5 μm particle size, Thermo Scientific, catalog number 059148) with column temperature set at 40°C, using a gradient of 58–65% acetonitrile in 0.15% formic acid in water for 55 min, with flow rate of 1.5 mL/min. The eluted fractions are collected 1 min/fraction in 20 mL glass tubes.
3. The fractions containing the target metabolite are identified by analytical LC/MS, pooled and dried down using a centrifugal evaporator at room temperature. The dry samples are dissolved in 250 μL of $CDCl_3$ and NMR spectra obtained using a Bruker Avance 900 NMR spectrometer (Bruker).

3. PHYLOGENY-DRIVEN ANALYSIS OF THE BIOCHEMICAL BASIS OF SPECIALIZED METABOLITE DIVERSITY

3.1 Phylogeny-Based Screening of Metabolic Phenotypes

The metabolite diversity observed in the two species raises questions about its underlying biochemical basis. One of the first steps in utilizing this type of natural variation-driven strategy involves judicious choice of species or accessions to study. Selecting distantly related species can drown out a potential correlation signal between genotype and phenotype, while selecting very closely related species/accessions may reveal no differences in metabolites and enzymes. By choosing individuals that are as closely related as possible—yet still have phenotypic differences—it is possible to identify differences in enzyme structure and function.

We hypothesized that characterizing species closely related to both *S. lycopersicum* and *S. habrochaites* can provide further insights into the observed acylsugar diversity between the two species. A likely possibility was that the shift in enzymatic activity from furanose ring C10 acylation to pyranose ring C10 acylation/furanose ring C5 acylation occurred sometime since the two species diverged (<7 million years ago), and we expected to see phenotypic differences in the closely related species.

At least two accessions from each of eight *Solanum* species (*S. neorickii, S. arcanum, S. huaylasense, S. peruvianum, S. corneliomulleri, S. chilense, S. habrochaites,* and *S. pennellii*) that cover the entire tomato clade (Fig. 3) were analyzed. LC coupled with ESI^+ MS (see Section 2.1) revealed that six species accumulate acylsucroses with a single short acyl chain (C4 or C5) on the furanose ring, similar to M82. In contrast, *S. pennellii* produces acylsucroses lacking furanose ring acylation (Fig. 3). Accessions of *S. habrochaites* showed the most diversity in furanose ring acylation profiles; these range from none (as seen for *S. pennellii*) to single chains of C4, C5, C10, C11, or C12. These results led us to focus our analysis on the acylsugar diversity in *S. habrochaites*.

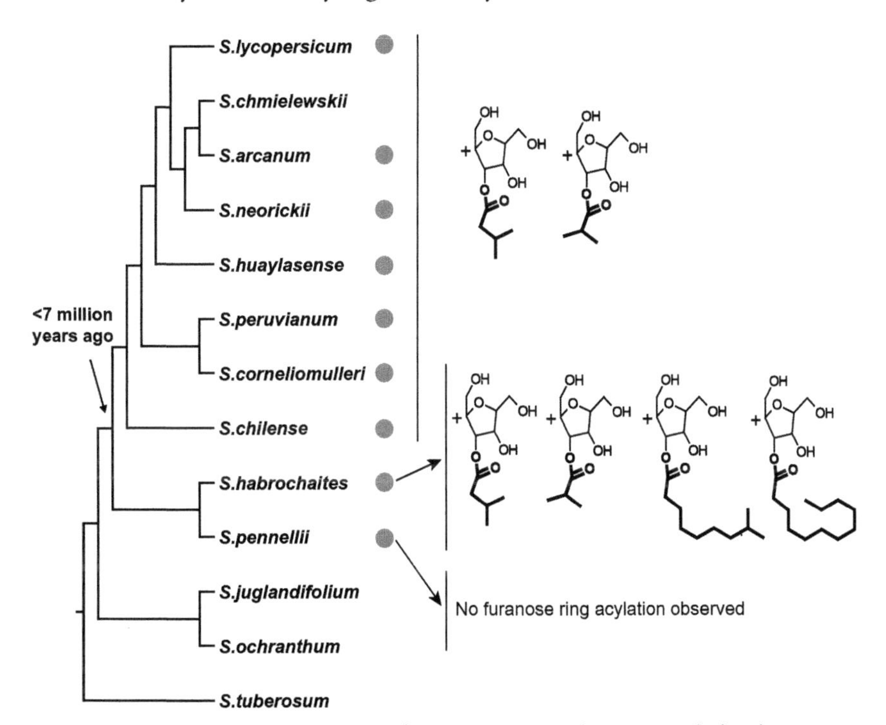

Fig. 3 Comparison of furanose ring acylation patterns in the tomato clade. The *gray circles* identify the species used for characterization of the furanose ring acylation patterns. *The cladogram showing the species relationships is adapted from Rodriguez, F., Wu, F., Ané, C., Tanksley, S., & Spooner, D. M. (2009). Do potatoes and tomatoes have a single evolutionary history, and what proportion of the genome supports this history?* BMC Evolutionary Biology, *9, 191. http://dx.doi.org/10.1186/1471-2148-9-191.*

3.2 The Biochemical Basis for Differences in Furanose Ring Acylation

Once metabolite diversity is identified, the next step is to look for its biochemical basis. Frequently, knowledge about the chemical structures and understanding the details of the pathway in other species can act as a guide to select candidate enzymes for hypothesis testing. *S. lycopersicum* ASAT3 (Sl-ASAT3) was known to catalyze furanose ring acylation using only short-chain CoA esters (Schilmiller et al., 2015) (Fig. 4A). As some of the acylsugars made in *S. habrochaites* have structures similar to those in *S. lycopersicum* M82, the acylsugar biosynthetic pathway in cultivated tomato might also be shared by *S. habrochaites*. We hypothesized that the furanose ring acylation diversity is due to the functional variation in the *S. habrochaites* ASAT3 (Sh-ASAT3) activity—which has a broader substrate specificity and uses different lengths of acyl-CoAs as substrates—in comparison to Sl-ASAT3, which uses only short-chain acyl-CoAs. To test this hypothesis, ASAT3 variants from different *S. habrochaites* accessions were cloned and the in vitro activities characterized (Schilmiller et al., 2015). As predicted, Sh-ASAT3 added both short and long acyl chain to the acylsucrose furanose ring, in contrast to Sl-ASAT3, which only adds short acyl chains.

3.2.1 Protocol to Clone and Characterize S. habrochaites *ASAT3 Isoforms*

1. To collect trichome tissue of selected *S. habrochaites* accessions for RNA extraction, stem, and petiole tissues from 3- to 5-week-old plants are frozen in liquid nitrogen, and the trichomes gently scraped off the frozen tissue into a prechilled 1.5 mL centrifuge tube on dry ice using a liquid nitrogen prechilled plastic stick.
2. Total RNA (0.25 μg) extracted from the trichome samples is used as template for first-strand cDNA synthesis by reverse transcriptase. *ASAT3* isoform sequences are amplified using forward and reverse primers that have *Nhe*I and *Xho*I restriction sites at either end, and the products are subcloned into pET28b with the coding sequence in frame and a His tag at the N-terminus. The *ASAT3* isoform nucleic acid sequences are also determined by sequencing the individual pET28b vector *ASAT3* clones.
3. To express the ASAT3 proteins for in vitro assay, the plasmids are transformed into BL21 Rosetta cells. *E. coli* cultures are grown in 100 mL LB medium in 250 mL glass flasks in a shaker at 37°C at 250 rpm until the

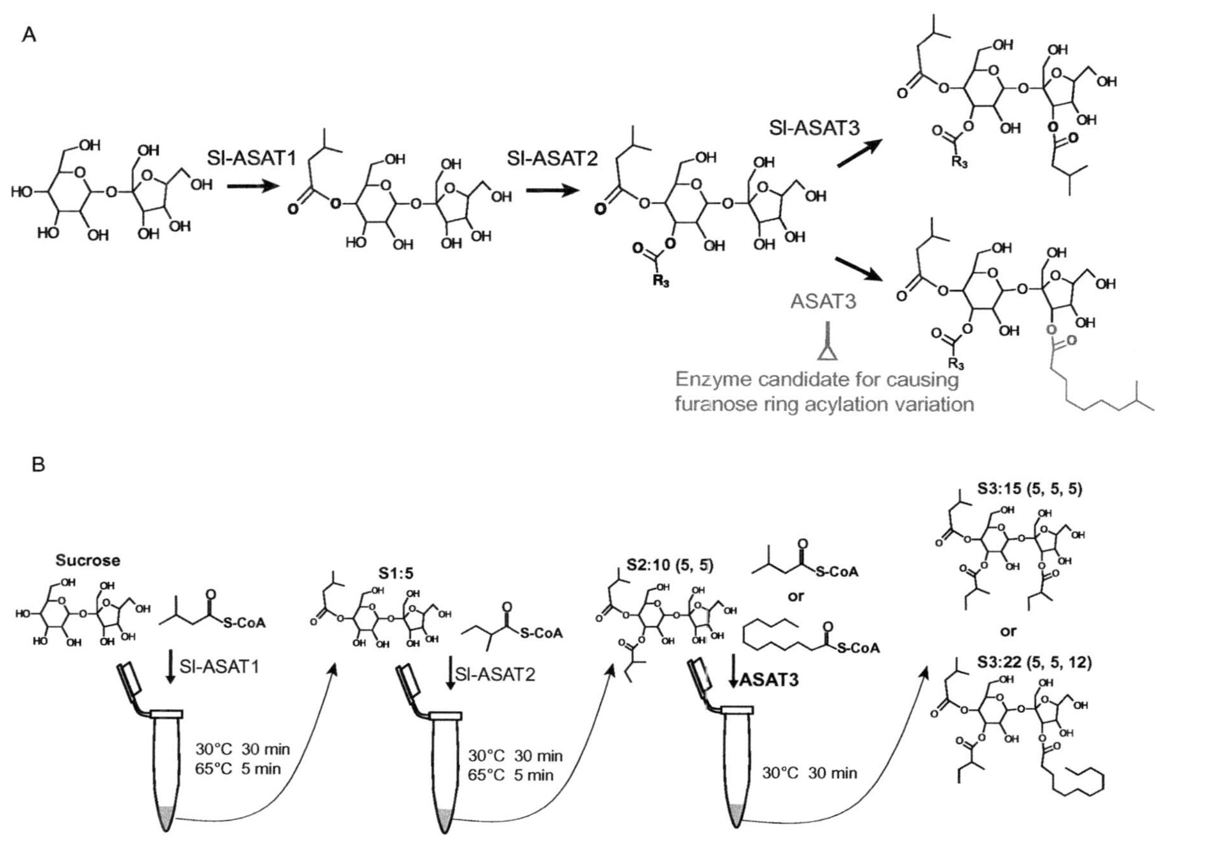

Fig. 4 Characterization of the basis for furanose ring acylation variation. (A) The function of SI-ASAT3 leads to the hypothesis that ASAT3 is the cause of furanose ring acylation diversity in *S. habrochaites*. (B) In vitro system used to test the activity of ASAT3 isoforms with acyl-CoAs of different chain lengths.

optical density at 600 nm reaches 0.6. Then, 0.05 m*M* isopropyl-*β*-D-1-thiogalactopyranoside (IPTG) is added, followed by 8–12 h growth at 16°C at 170 rpm. Ni-affinity chromatography is used to purify the soluble proteins (Schilmiller et al., 2012), and the purified enzymes are stored at −20°C. 30–50% glycerol is typically added to prevent freezing of the enzyme solution.

4. To test the in vitro activity of Sh-ASAT3, sequential enzyme reactions are set up using Sl-ASAT1 and Sl-ASAT2 to produce the acyl acceptor substrate for ASAT3 in 1.5 mL centrifuge tubes (Fig. 4B). Approximately 1 μg purified protein is used for each enzyme assay (estimated by Coomassie Brilliant Blue staining of samples on SDS-PAGE). Each incubation is at 30°C for 30 min. The first enzyme Sl-ASAT1 is incubated in 60 μL of 50 m*M* ammonium acetate (pH 6.0) with 1 m*M* sucrose and 50 μ*M* iC5-CoA followed by 5 min heat inactivation at 65°C. Sl-ASAT2 and 50 μ*M* aiC5-CoA are then added, followed by heat inactivation. After the first two steps, various ASAT3 enzyme preparations and 50 μ*M* aiC5-CoA or 50 μ*M* nC12-CoA are added. The reactions are quenched by addition of 120 μL of stop solution containing acetonitrile/isopropanol/formic acid (1:1:0.001, v/v/v) and 10 μ*M* propyl-4-hydroxybenzoate internal standard, followed by centrifugation at 12,000 × *g* for 10 min to remove precipitated proteins. Next, the supernatant solution is transferred to HPLC vials with 250 μL vial inserts and stored at −20°C.
5. Ten microliter of the enzyme reaction product is characterized by LC/MS using both negative- and positive-ion mode as described in Section 2.1.

3.3 Identifying a Key Polymorphism Driving Enzyme Function Divergence

Once novel enzyme activities are characterized, it is possible to identify amino acid differences that correlate with functional divergence. *S. habrochaites* and *S. lycopersicum* are separated by <7 million years of evolution, hence ASAT3 orthologs from the two species show large tracts of conserved residues interspersed with multiple instances of 1–2 amino acid differences. Specifically, there are only ~30 different amino acids (~7% of the protein lengths) in the LA1777 and LA1731 ASAT3 enzymes compared to their M82 ASAT3 ortholog. We analyzed the different amino acids using knowledge about the enzyme activities, homology modeling, site-directed mutagenesis, and enzyme kinetic properties to identify the mutations that result in ASAT3 functional diversification.

3.3.1 Discovery of an ASAT3 Polymorphism Associated with In Vitro Functional Variation

As shown in Fig. 5A, 19 amino acids in Sl-ASAT3 correlate with the observed enzyme activity difference between the short-chain Sl-ASAT3 and long-chain Sh-ASAT3. We performed homology modeling using the X-ray crystallographic structure for a trichothecene 3-*O*-acetyltransferase–acyl-CoA complex (Protein Data Bank ID: 3B2S) (Garvey et al., 2008) to identify amino acids Tyr^{41} and Phe^{35} as being strong candidates for amino acids that influence substrate specificity since they are close to the acyl-CoA binding site (Fig. 5B). Site-directed mutagenesis of Sl-ASAT3 changing from Tyr^{41} (in Sl-ASAT3) to Cys^{41} (in Sh-ASAT3) and analysis of in vitro activity showed that this substitution is sufficient to allow Sl-ASAT3 to perform long-chain furanose ring acylation (Fig. 5C). Enzyme kinetic analysis provided solid evidence that the Sl-ASAT3 mutant protein has a similar apparent K_m for nC12-CoA as Sh-ASAT3. This approach was shown to be generally applicable because it led to the identification of another key residue that causes ASAT2 discrimination between C5 acyl-CoAs that are subterminally branched (aiC5) and terminally branched (iC5) (Fan et al., 2016).

3.3.2 Protocol Used to Identify Key Polymorphisms Contributing to Variation in ASAT3 Acyl-CoA Substrate Specificity

1. Multiple sequence alignment of ASAT3 protein variants is done with MEGA version 5 (Tamura et al., 2011), using the default settings. Next, substrate specificities of the enzymes are aligned with the amino acid sequences (Fig. 5A). The amino acid residues that correlate with the ASAT3 enzyme activities are candidates for those that contribute to ASAT3 activity changes (see *green* typeface in Fig. 5A).
2. Structural homology modeling of Sl-ASAT3 was performed using a web-based protein homology/analogy recognition engine (http://www.sbg.bio.ic.ac.uk/phyre2) (Kelley, Mezulis, Yates, Wass, & Sternberg, 2015; Kelley & Sternberg, 2009). The predicted 3D structure was overlayed with the crystal structure of the BAHD acyltransferase trichothecene 3-*O*-acetyltransferase with its acyl donor ligands (Protein Data Bank ID: 3B2S) (Fig. 5B). The relative distance of the candidate residues to the putative acyl-CoA binding pocket was determined

A

	Residue positions																			Enzyme activity S2:10 (5, 5) as acyl acceptor	
	5	6	20	35	41	107	109	114	116	131	163	205	207	210	256	261	296	327	329	iC5-CoA	nC12-CoA
Sl-ASAT3	T	I	S	F	Y	L	H	V	D	T	L	A	R	Q	T	S	A	E	R	Yes	No
LA1777-ASAT3-F	K	M	P	L	C	I	N	A	G	N	V	T	H	K	A	K	T	K	M	Yes	Yes
LA1731-ASAT3-F	K	M	P	L	C	I	N	A	G	N	V	T	H	K	A	K	T	K	M	Yes	Yes

B

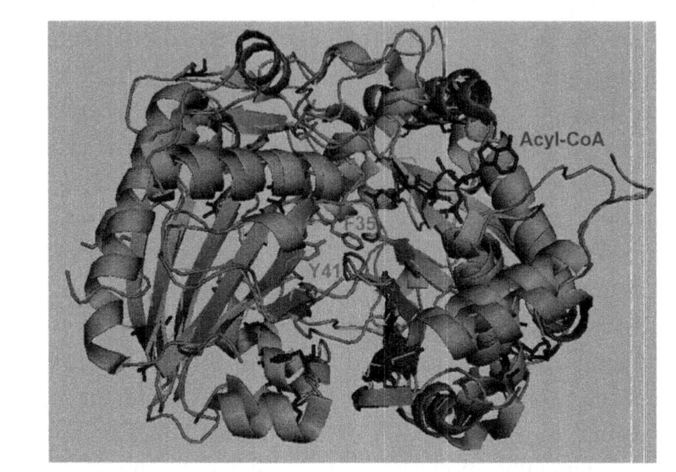

C

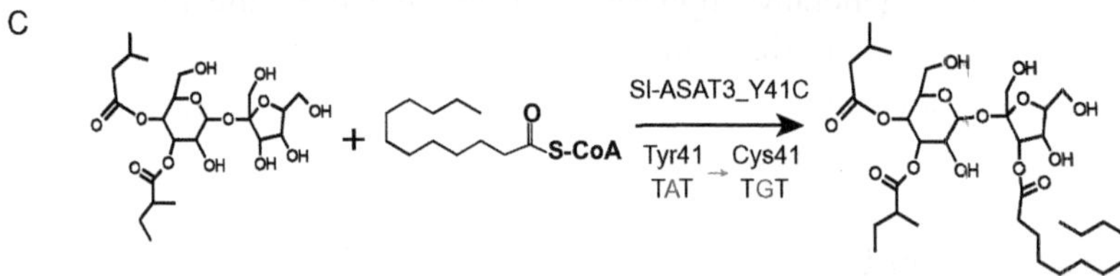

Fig. 5 Comparative sequence, in vitro function, and structure analysis lead to identification of a functionally important polymorphism in ASAT3. (A) The alignment of amino acid sequences to identify two positions (in *green*) that correlate with the enzyme function differences of three ASAT3 isoforms. (B) The homology-model-derived proposed structure of Sl-ASAT3 (*orange*) is superimposed on the crystallographic structure (*gray*) trichothecene 3-*O*-acetyltransferase–acyl-CoA complex (PDB ID: 3B2S) (Garvey, McCormick, & Rayment, 2008), which has the acyl-CoA (*red*) binding pocket resolved. The amino acid residues shown in (A) are labeled with *blue* and *green* in the Sl-ASAT3 modeled structure. The *green*-colored amino acids are those that correlate with activity and map close to the acyl-CoA binding pocket. (C) In vitro mutagenesis verified that a single nucleic acid substitution from A to G leads to the amino acid substitution from Tyr^{41} to Cys^{41} and enables ASAT3 to use the long-chain acyl-CoAs as substrates. (See the color plate.)

using PyMOL (version 1.7.4 Schrödinger, https://www.pymol.org). The working hypothesis is that residues close to the substrate-binding pocket are more likely to affect the enzyme activity and are nominated for experimental testing.

3. To test the importance of specific residues, amino acid substitutions are made using the PCR-based Q5-Site-Directed Mutagenesis Kit (NEB) (Fisher & Pei, 1997). The primers to introduce the desired mutation are designed using the web-based software NEBaseChanger (version 1.2.2, NEB, http://nebasechanger.neb.com). The presence of a mutation is verified by DNA sequencing. Protein expression of the Sl-ASAT3 mutants, ASAT3 enzyme assays, and LC/MS analysis are performed as described earlier (Sections 2.1.2 and 3.2.1).
4. Wild-type and mutagenized Sl-ASAT3 enzyme kinetics are determined for the various acyl-CoA substrates. To measure the apparent K_m and K_i of nC12-CoA, purified S2:10 (5,5) is used as the acyl acceptor substrate, and the concentration of nC12-CoA is varied from 0 to 200 μM. All enzyme reactions are done in triplicate at 30°C for 5 min and stopped using the enzyme stop solution described in Section 3.2.1. After LC/MS analysis, the enzyme assay product peak areas divided by the internal standard peak area (normalized peak response) are plotted for each concentration of nC12-CoA. Apparent K_m and K_i are calculated by the nonregression model in the GraphPad Prism 5 software (http://www.graphpad.com/scientific-software/prism/).

4. CONCLUSIONS

The comparative biochemical approach described in this chapter involves characterizing metabolite diversity, generating relevant hypotheses regarding the biochemical basis of this diversity using prior knowledge, selecting appropriate species—and finally—testing the hypothesis using appropriate approaches including site-directed mutagenesis and enzyme assays. This approach demonstrates the power of studying phenotypic diversity and its biochemical basis, which is a useful strategy to elucidate the causation of the evolutionary variation of specialized metabolic pathways. In addition, validating key residues shaping the biosynthetic enzyme activities is an important step in creating designer plants and for reconstruction of enzymatic pathways in bacteria or yeast. Given the rapid advancements in genomic tools in recent years, adoption of the approach that we have outlined here would be a powerful way to expand the repertoire of enzyme activities for synthetic biology.

ACKNOWLEDGMENTS

We thank the members of the *Solanum* Trichome Project, especially people in Dr. A. Daniel Jones' lab for their great work on resolving structures of the trichome metabolites, and Abigail M. Miller, Anthony L. Schilmiller for their contributions to cloning *ASAT* genes and characterizing the enzyme functions. We thank Xiaoxiao Liu for her helpful comments on description of the LC/MS and NMR methods. This work was funded by National Science Foundation Grant IOS-1025636.

REFERENCES

Alonso-Gutierrez, J., Chan, R., Batth, T. S., Adams, P. D., Keasling, J. D., Petzold, C. J., et al. (2013). Metabolic engineering of *Escherichia coli* for limonene and perillyl alcohol production. *Metabolic Engineering, 19*, 33–41.

Fan, P., Miller, A. M., Schilmiller, A. L., Liu, X., Ofner, I., Jones, A. D., et al. (2016). In vitro reconstruction and analysis of evolutionary variation of the tomato acylsucrose metabolic network. *Proceedings of the National Academy of Sciences of the United States of America, 113*, E239–E248.

Fisher, C. L., & Pei, G. K. (1997). Modification of a PCR-based site-directed mutagenesis method. *Biotechniques, 23*, 570–574.

Garvey, G. S., McCormick, S. P., & Rayment, I. (2008). Structural and functional characterization of the TRI101 trichothecene 3-*O*-acetyltransferase from *Fusarium sporotrichioides* and *Fusarium graminearum*: Kinetic insights to combating Fusarium head blight. *Journal of Biological Chemistry, 283*, 1660–1669.

Ghosh, B., Westbrook, T. C., & Jones, A. D. (2014). Comparative structural profiling of trichome specialized metabolites in tomato (*Solanum lycopersicum*) and *S. habrochaites*: Acylsugar profiles revealed by UHPLC/MS and NMR. *Metabolomics, 10*, 496–507.

Gu, L., Jones, D., & Last, R. L. (2010). Broad connections in the Arabidopsis seed metabolic network revealed by metabolite profiling of an amino acid catabolism mutant. *Plant Journal, 61*, 579–590.

Hawthorne, D. J., Shapiro, J. A., Tingey, W. M., & Mutschler, M. A. (1992). Trichome-borne and artificially applied acylsugars of wild tomato deter feeding and oviposition of the leafminer *Liriomyza trifolii*. *Entomologia Experimentalis et Applicata, 65*, 65–73.

Ignea, C., Ioannou, E., Georgantea, P., Loupassaki, S., Trikka, F. A., Kanellis, A. K., et al. (2015). Reconstructing the chemical diversity of labdane-type diterpene biosynthesis in yeast. *Metabolic Engineering, 28*, 91–103.

Kelley, L. A., Mezulis, S., Yates, C. M., Wass, M. N., & Sternberg, M. J. E. (2015). The Phyre2 web portal for protein modeling, prediction and analysis. *Nature Protocols, 10*, 845–858.

Kelley, L. A., & Sternberg, M. J. E. (2009). Protein structure prediction on the Web: A case study using the Phyre server. *Nature Protocols, 4*, 363–371.

Kim, J., Kang, K., Gonzales-Vigil, E., Shi, F., Jones, A. D., Barry, C. S., et al. (2012). Striking natural diversity in glandular trichome acylsugar composition is shaped by variation at the acyltransferase2 locus in the wild tomato *Solanum habrochaites*. *Plant Physiology, 160*, 1854–1870.

Lau, W., & Sattely, E. S. (2015). Six enzymes from mayapple that complete the biosynthetic pathway to the etoposide aglycone. *Science, 349*, 1224–1228.

Pichersky, E., & Gershenzon, J. (2002). The formation and function of plant volatiles: Perfumes for pollinator attraction and defense. *Current Opinion in Plant Biology, 5*, 237–243.
Schilmiller, A. L., Charbonneau, A. L., & Last, R. L. (2012). Identification of a BAHD acetyltransferase that produces protective acyl sugars in tomato trichomes. *Proceedings of the National Academy of Sciences of the United States of America, 109*, 16377–16382.
Schilmiller, A. L., Last, R. L., & Pichersky, E. (2008). Harnessing plant trichome biochemistry for the production of useful compounds. *Plant Journal, 54*, 702–711.
Schilmiller, A. L., Moghe, G. D., Fan, P., Ghosh, B., Ning, J., Jones, A. D., et al. (2015). Functionally divergent alleles and duplicated loci encoding an acyltransferase contribute to acylsugar metabolite diversity in *Solanum* trichomes. *Plant Cell, 27*, 1002–1017.
Schilmiller, A., Shi, F., Kim, J., Charbonneau, A. L., Holmes, D., Daniel Jones, A., et al. (2010). Mass spectrometry screening reveals widespread diversity in trichome specialized metabolites of tomato chromosomal substitution lines. *Plant Journal, 62*, 391–403.
Shepherd, R. W., Bass, W. T., Houtz, R. L., & Wagner, G. J. (2005). Phylloplanins of tobacco are defensive proteins deployed on aerial surfaces by short glandular trichomes. *Plant Cell, 17*, 1851–1861.
Simmons, A. T., & Gurr, G. M. (2005). Trichomes of *Lycopersicon* species and their hybrids: Effects on pests and natural enemies. *Agricultural and Forest Entomology*, 7, 265–276.
Tamura, K., Peterson, D., Peterson, N., Stecher, G., Nei, M., & Kumar, S. (2011). MEGA5: Molecular evolutionary genetics analysis using maximum likelihood, evolutionary distance, and maximum parsimony methods. *Molecular Biology and Evolution, 28*, 2731–2739.
Tissier, A. (2012). Glandular trichomes: What comes after expressed sequence tags? *Plant Journal, 70*, 51–68.
Wagner, G. J. (1991). Secreting glandular trichomes: More than just hairs. *Plant Physiology, 96*, 675–679.
Weinhold, A., & Baldwin, I. T. (2011). Trichome-derived O-acyl sugars are a first meal for caterpillars that tags them for predation. *Proceedings of the National Academy of Sciences of the United States of America, 108*, 7855–7859.

CHAPTER TWO

De Novo Deep Transcriptome Analysis of Medicinal Plants for Gene Discovery in Biosynthesis of Plant Natural Products

R. Han*, A. Rai*, M. Nakamura*, H. Suzuki†, H. Takahashi‡, M. Yamazaki*, K. Saito*,§,1

*Graduate School of Pharmaceutical Sciences, Chiba University, Chiba, Japan
†Kazusa DNA Research Institute, Kisarazu, Japan
‡Medical Mycology Research Center, Chiba University, Chiba, Japan
§RIKEN Center for Sustainable Resource Science, Yokohama, Japan
[1]Corresponding author: e-mail address: ksaito@faculty.chiba-u.jp

Contents

1. Methods for Transcriptomic Study Prior to High-Throughput Sequencing 20
2. Deep Transcriptome Outline 22
3. Preparation of the Plant Materials and RNA Extraction 24
4. cDNA Library Construction 27
5. High-Throughput Sequencing 30
6. Data Interpretation 31
   6.1 Trimming and Quality Validation of Raw Reads 31
   6.2 De Novo Assembly 32
   6.3 Expression Analysis of Transcripts 34
   6.4 Sequence Similarity Search 36
   6.5 Gene Annotation 37
   6.6 An Example of RNA-Seq 38
7. Application of RNA-Seq on Medicinal Plants and Perspectives 38
Acknowledgments 39
References 39

Abstract

Study on transcriptome, the entire pool of transcripts in an organism or single cells at certain physiological or pathological stage, is indispensable in unraveling the connection and regulation between DNA and protein. Before the advent of deep sequencing, microarray was the main approach to handle transcripts. Despite obvious shortcomings, including limited dynamic range and difficulties to compare the results from distinct experiments, microarray was widely applied. During the past decade, next-generation sequencing (NGS) has revolutionized our understanding of genomics in a fast,

Methods in Enzymology, Volume 576
ISSN 0076-6879
http://dx.doi.org/10.1016/bs.mie.2016.03.001

high-throughput, cost-effective, and tractable manner. By adopting NGS, efficiency and fruitful outcomes concerning the efforts to elucidate genes responsible for producing active compounds in medicinal plants were profoundly enhanced. The whole process involves steps, from the plant material sampling, to cDNA library preparation, to deep sequencing, and then bioinformatics takes over to assemble enormous—yet fragmentary—data from which to comb and extract information. The unprecedentedly rapid development of such technologies provides so many choices to facilitate the task, which can cause confusion when choosing the suitable methodology for specific purposes. Here, we review the general approaches for deep transcriptome analysis and then focus on their application in discovering biosynthetic pathways of medicinal plants that produce important secondary metabolites.

The unique feature of transcriptome, the entire RNA molecules that are transcribed from a genome and the key bond between information embodied in genotype and phenotype, is what determines its exclusive importance in studies of elucidating functional genes. The scope of transcriptomics covers: archiving all transcripts, including mRNAs, small RNAs, and microRNAs, in target organisms or even one single cell; tackling the structural composition of genes and their related characteristics, such as posttranscriptional modifications; measuring expression levels in different developmental stages, or to quantify impacts of environmental factors. To completely obtain the transcriptome in an individual plant is inherently challenging. Taking *Arabidopsis thaliana* as an example, the 5700 epidermal cells on the adaxial side of a petal and the average gene length of 2200 bp (Autran et al., 2002; Wortman et al., 2003), along with the estimation that the absolute number of RNA molecules per cell for each gene is between 50,000 and 300,000 (Marinov et al., 2014) enables us to peek into a tiny fraction of the full transcriptome that the mRNA bases in the confined petal area of about 1.9×10^{12} ($5700 \times 2200 \times 150{,}000$) by using 150,000 as the total RNA molecules in a petal cell. Moreover, mRNA only comprises between 1% and 3% of entire cellular RNA.

1. METHODS FOR TRANSCRIPTOMIC STUDY PRIOR TO HIGH-THROUGHPUT SEQUENCING

Prior to the introduction of microarray, studies on the transcript level had usually followed the routine protocol that involved sampling, homogenization, and RNA extraction. In order to assess the quantity of specific transcripts, radioactive DNA probes were used for hybridization and, subsequently, the intensity of radioactivity was monitored with the aid of electrophoresis as a mean of separation. Such an approach was painfully slow, labor intensive, and costly. The attempt to analyze transcriptome was first made possible by DNA microarray technology. In 1995, Mark Schena and coworkers from Stanford University adopted high-speed robotic printing of complementary DNAs on glass to quantify expressions of 45 genes from roots and leaves of *A. thaliana* in parallel (Schena, Shalon, Davis, & Brown, 1995). This ingenious approach inspired the bioindustry to produce convenient DNA chips, and the solid-phase DNA sequencing technology is also attributed to this work published in the periodical *Science*, which has been cited for more than 10,000 times thus far.

To perform a microarray experiment, a set of orderly arranged probes of known identity consisting of DNA, cDNA, or oligonucleotides are fixed on a solid substrate like silicon chips, or a glass surface where distinct clusters of spots come into being. Designing probes usually requires the knowledge of genome sequence or unraveled open reading frames to allow resultant probes complementary to the extracted RNA (or cDNA through reverse transcription). Those transcripts are then labeled with fluorescent dyes that are commonly Cy5 and Cy3 fluorescing at red and green wavelengths, respectively (Tang et al., 2007), and hybridized to the matching probes immobilized at certain spots. After the removal of nonspecific bonding sequences, the DNA microarray is ready for measurement of gene expression by capturing and calculating light intensity, because the strength of the signal relies on the quantity of dyed transcripts binding to the probes. With continuous development of this method, it exhibits versatile merits in gene discovery (Douglas & Ehlting, 2005), analyzing gene expression among different tissues in developmental stages (Girke et al., 2000; Wang, Yin, Wang, & To, 2009) as well as between distinct, yet related, species, (Moore, Payton, Wright, Tanksley, & Giovannoni, 2005), and for authentication of pharmaceutical plants (Zhu, Fushimi, & Komatsu, 2008).

Nevertheless, technical issues accompanying this approach are obvious. Existing knowledge of genome sequence is heavily relied on to design probes, which hinders the study of unknown transcripts—especially when majority of the genome information of nonmodel plants are not available. To achieve ideal microarray results, fluorescent signal and RNA concentration should be linearly dependent. But, when the dynamic range of microarrays or the amount of RNA that can bind to the exclusive array spot is exceeded, these two forms of saturation jeopardize the outcome (Scott, VanWye, McDonald, & Crawford, 2009). Regarding sensitivity, more than 1400 *Arabidopsis* transcription factors (TFs) were measured and compared by quantitative real-time PCR (qRT-PCR) and 22 K *Arabidopsis* Affymetrix array (Czechowski, Bari, Stitt, Scheible, & Udvardi, 2004). Around 87% of the TFs were detectable using qRT-PCR, while the microarray method reported less than 55% of TFs in the same samples, suggesting the unsatisfactory performance of microarray while dealing with a low abundance of transcripts that were often at the level of less than 100 copies per cell. Last but not least, to compare expression profiles among different experiments conducted in the same or different laboratories can be challenging, and sophisticated normalization approaches are needed, bringing about the issue of reproducibility.

As opposed to microarray approaches, over the years DNA sequencing utilizes various strategies to directly determine cDNAs. Serial analysis of gene expression (SAGE) facilitated the quantitative and simultaneous measurement of thousands of transcripts (Velculescu, Zhang, Vogelstein, & Kinzler, 1995). Extracted mRNA is used to synthesize cDNA, which is subsequently immobilized to streptavidin beads. Then, the cDNA is cleaved by an anchoring enzyme and divided into two halves to be ligated to one of the two linkers, which contain a type IIS restriction site. Applying tagging enzyme (type IIS restriction endonucleases) results in the linker with a cDNA fragment of up to 10 bp long, which provides enough information to identify initial intact transcripts. After ligation of the two parts of released tags and PCR amplification with specific primers, cleavage of the PCR product releases the sequences that can be ligated, cloned, and analyzed. Massively parallel signature sequencing (Brenner et al., 2000) features a similar concept as SAGE by using the sequence signature of ~20 bp to assess gene expression levels. Such tag sequencing methods did provide practical and relatively accurate measurements for transcripts; however, the expression of many transcripts that lack a restriction enzyme recognition site will not be detected, and the presence of only a very short fragment of the initial mRNA is useless to distinguish isoforms.

2. DEEP TRANSCRIPTOME OUTLINE

By summarizing previous technologies, in addition to the consideration of cost, efficiency, accuracy, and reproducibility, an ideal method should allow researchers to directly obtain the frequency of RNA molecules as well as sequence information of all transcripts with high or low expression levels presented in an organism or even a cell—not just a portion or the limited known parts. Deep transcriptome, also called RNA sequencing

(RNA-Seq), provides a new way to address the challenges. It employs deep sequencing technologies that sequence the arrayed amplified cDNA many times, and the coverage or depth indicating the times for which a nucleotide is detected can be up to several thousand. For example, in a sequencing process a cDNA fragment with the length of 1500 bp yielding 3000 reads with the length of 100 bp (on average), bears the coverage of 200 ($3000 \times 100/1500$). A typical sequencing workflow of RNA-Seq includes: preparation of cDNA from extracted total or fractionated RNA by means of, eg, poly(A) tail; ligation of adapters to fragmented cDNA with suitable length for sequencing; hybridization of ligated cDNA to solid substrate and amplify the target copies of interest to make clusters; capturing the signals while sequencing in real time with the common length of obtained short sequences ranging from 30 to 1000 bp.

Since the advent of RNA-Seq, it has been used to study the transcriptome of many model organisms, including novel regulatory RNAs from *Escherichia coli* (Raghavan, Groisman, & Ochman, 2011), *Saccharomyces cerevisiae* (Nagalakshmi et al., 2008), *A. thaliana* (Lister et al., 2008), *Lotus japonicus* (Deguchi et al., 2007), and *Medicago truncatula* (Lelandais-Briere et al., 2009). More recently, an international consortium that initiated the 1KP (1000 plants) project aimed at generating transcriptome profiles for more than one thousand plant species, covering all the major lineages across the clade of green plants (Matasci et al., 2014). Such an epic endeavor would not have been possible without the support of powerful deep-sequencing technologies.

According to current theories, within 4.6 billion years of the earth's history, multicellular eukaryotes started to evolve more than 650 million years ago, and they began adapting to terrestrial life around 450 million years ago (Raven, Evert, & Eichhorn, 2013). Nowadays, plants, especially vascular plants, flourish on our planet and provide oxygen, food, and numerous other applications covering almost every aspect of our lives. Profound characteristics of medicinal plants are the drugs they produce in nature. Just as Hippocrates stated, natural forces within us are the true healers of disease. Human beings have started to investigate and apply effective plants to cure diseases since ancient times. The oldest excavated written proof concerning application of medicinal plants was found on some Sumerian clay tablets that dated back approximately 6000 years ago. There were 300 medicinal herbs recorded on the tablets left by Assyrian king Ashurbanipal, which included myrrh and opium (Sumner, 2000). Then, the Chinese book *Shennong Bencao*, the Vedas (Indian holy books), Theophrastus's book that founded botanical science *Plant Etiology* and *Plant History* and many other classical works contributed to the better understanding of herbal medicines (Petrovska, 2012). Due to the gigantic pool of plant resources and complicated mechanisms, even when equipped with the latest cutting-edge technologies we are far from the goal of extracting all of the potent chemicals—many of which are plant secondary metabolites. A case in point is the estimation that angiosperms alone comprise over 250,000 species. The merits of interrogating biosynthetic pathways are: for the sake of plants themselves, because the products of use to patients are originally for the plants to survive and adapt to their environment; and for the better usage of the natural medicines since the concentration of many chemicals in the plants is very low and hard to synthesize using chemical approaches. The yield of taxol from the Pacific yew tree (*Taxus brevifolia*) was only 0.004% and from 12 kg of dry stem and bark, only 0.5 g was obtained, not to mention the removal of the bark will kill the tree (Wall & Wani, 1995). Taxol can now be synthesized, but the methods still need to be improved.

Depending on the purpose of the study, subjects of RNA-Seq can be a bulk cell population of different tissues or organs providing average evaluation of the transcripts, or single-cell types which help to reveal more detailed information on a subset basis. Here, we go through the whole process concerning deep transcriptome analysis of medicinal plants to discover promising genes in biosynthetic pathways of interest.

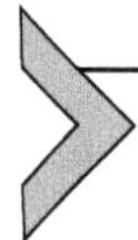

3. PREPARATION OF THE PLANT MATERIALS AND RNA EXTRACTION

Sampling tissues of medicinal plants requires fieldwork and, therefore, special treatment should be employed to prevent RNA from degradation or any change in its abundance in the tissues. As the foundation of the whole experiment on which all the subsequent steps rely, any degradation or damage in its inherent profile can compromise the results considerably (Ibberson, Benes, Muckenthaler, & Castoldi, 2009) because it is not clear whether transcript degradation of RNA samples occurs congruously (and thus can be corrected through data normalization), or if distinct transcripts undergo degradation at different rates (Parker, 2012; Romero, Pai, Tung, & Gilad, 2014). To secure high quality transcripts, materials should be processed immediately after sampling or stored temporarily in RNA stabilization reagents. In field conditions, cryopreservation is usually unavailable and applying this method to assure stability of transcripts is sometimes doubtful.

A kind of homemade nucleic acid preservation buffer consisting of EDTA disodium salt dihydrate, sodium citrate trisodium salt dihydrate, and ammonium sulfate was proposed for preserving RNA and DNA (Camacho-Sanchez, Burraco, Gomez-Mestre, & Leonard, 2013). What follows is the general protocol in our laboratory. We trim the plant tissues into pieces with the thickness of roughly 5 mm in every dimension right after sampling in the wild, and make sure all the subjects are submerged in RNA*later* (Life Technologies, USA) for future use. Even with this protective measure, prompt extraction of RNA is always a necessity. It is also highly recommended to store all samples on ice to prevent changes in the RNA profile of different tissues/samples. RNA*later* will be removed and the remaining solution attached to tissues will be cleaned with a lab wiper. Then, the samples are homogenized using Multi-beads Shocker (Yasui Kikai, Japan) in the presence of liquid nitrogen to prevent samples from thawing out. After a suitable quantity (50–100 mg) of powdered tissues is weighed, a mortar and pestle is used to obtain finer powders to enhance the efficiency of RNA isolation, which typically involves an acidic solution comprising guanidinium thiocyanate, sodium acetate, phenol, and chloroform, which separates RNA in the upper aqueous phase from proteins and DNA in either the interphase or organic phase after extraction (Bird, 2005; Chomczynski & Sacchi, 2006). This "single-step" method was first introduced in 1987 to enable a preparation of high-quality total RNA in less than 4 h and the published work has had more than 65,000 citations (Chomczynski & Sacchi, 1987). We use a combination of commercially available RNA preparation kits, namely, trizol reagent (Invitrogen, USA) and RNeasy Plant Mini Kit (Qiagen, Germany). Either a RLT buffer, which comprises guanidine isothiocyanate, or a RLC buffer containing guanidine hydrochloride from the Qiagen kit is used—based on the resultant RNA quality. Although the manufacturer suggests a RLT buffer to be the first priority, it is advisable to test both buffers to ascertain the best RNA extraction protocol for a specific tissue/sample. Prior to the routine isolation procedure, when we focus on high-molecular-weight RNAs, 20–30 mg polyethylene glycol 8000 (PEG8000) or PEG20000 will be added to the powdered sample and this helps improve the quality of extracted RNA because differential precipitation keeps those low-molecular-weight RNAs in supernatant. Trizol reagent is applied when the alternative method fails. In addition to proteins, fats, and polysaccharides, plants—especially medicinal plants—accumulate a large number of secondary metabolites, which need to be removed. For RNA purification, we use the following protocol: (1) Add 10% of the volume of 3 *M* sodium acetate (NaOAc) prepared in RNase-free water into RNA solution and put the mixture on ice for 1 h. (2) Centrifuge the mixture at 22,000 × *g* (15,000 rpm) for 15 min at 4°C. (3) Remove the supernatant into a new tube (RNase free) and add twice the volume of cold 100% ethanol. (4) Mix and centrifuge at 22,000 × *g* (15,000 rpm) for 15 min at 4°C. (5) Discard supernatant from the tube and then wash the pellet with 500 μL of 70% cold ethanol. (6) Centrifuge as described above. (7) Discard the wash and air dry for 5 min. (8) Resuspend the RNA pellet in RNase-free water for storage at −80°C for following applications.

The evaluation of RNA quality involves two concerns: purity and intactness. A 40 μg/mL solution of RNA has the optical density of 1.0 at 260 nm (OD_{260}) when a spectrophotometer with a 1-cm path length is used, and for pure RNA, the ratio of OD_{260}/OD_{280} can be up to 2.0 (Barbas, Burton, Scott, & Silverman, 2007). A secondary measurement is the OD_{260}/OD_{230} ratio, which is expected to be slightly higher than that of OD_{260}/OD_{280}, in the range of 2.0–2.2. Low readings suggest partially dissolved RNA samples, contamination of protein, phenol, or any other substance that has considerable absorbance at wavelength 280 or 230 nm. For measuring RNA integrity, we employ Agilent 2100 bioanalyzer, which was introduced in 1999 (initially for separating protein, DNA, and RNA). By selecting features like total RNA ratio, and information about 28 s region after measurement of RNA sizes, integrity categories termed as RIN (RNA integrity number), which are graded from 10 (intact RNA) to 1 (entirely decayed), of RNA samples are determined (Schroeder et al., 2006). For RNA-Seq, we set the threshold of RIN at 7.5 for our own samples. When the target is small RNAs, such as small interfering RNAs and microRNAs (miRNAs), which help regulate physiological and developmental process and gene expression in plants, PEG8000 can be used to retain small RNAs in supernatant for following ethanol precipitation (Peng et al., 2014) with the desired RIN value to be at least 7 (Ibberson et al., 2009).

The cell wall (comprising middle lamella, primary wall, and three-layered secondary wall) largely defines the appearance of plant cells, and the middle lamella holds adjacent cells together. For single-cell type RNA-Seq, suitable enzyme solution is prepared for digesting cell walls according to different cell types. The dissociated cells, often fluorescently marked, can then be singled out manually under microscopy. An experiment on *Arabidopsis* used digestion solution to dissociate the cells after the collection of root tips. Then, a fluorescence dissecting scope was applied to pick out the fluorescently marked cells, 24 quiescent center cells marked by *WUSCHEL-Related Homeobox 5* (*WOX5:GFP*) and 7 stele cells labeled with *WOODEN LEG* (*WOL*) *WOL:CRE-GR 35S:lox-CFP*, by using a glass mouth pipette (Efroni, Ip, Nawy, Mello, & Birnbaum, 2015). In addition to fluorescent-activated cell sorting (Zhang, Barthelson, Lambert, & Galbraith, 2008), as an alternative way to increase efficiency in collecting small tissue parts or single cells, laser-assisted microdissection not only excises but also transfers the cells to reduce human error in handling samples (Schutze & Lahr, 1998). When a single-cell type is harvested at large scale, for example, 60 root systems of *Glycine max* provide 0.8–1 g root hair cells, the total RNA can be isolated and purified using conventional approaches (Qiao & Libault, 2013). Nevertheless, in most cases the obtained cells with the number ranging from only one to dozens, to several thousands require another approach to handle RNA. Usually, cells are lysed by various agents, which can be grouped based on function i.e., carriers, detergent, enzymatic enhancers (Svec et al., 2013), or osmotic pressure, mechanic forces, and so on. To achieve a high yield of transcripts and compatibility with the following reverse transcription procedure, optimal handling should be determined experimentally. Because of the assumption that a very small quantity of debris, protein, or inorganic salts in the lysate do not affect reverse transcription, RNA purification can be omitted and this is also adopted by many researchers (Bengtsson, Hemberg, Rorsman, & Stahlberg, 2008; Tang, Lao, & Surani, 2011). When removal of cell debris and tRNA or rRNA is required, cellular mRNA can be collected using oligo-dT-coated magnetic beads (Klein et al., 2002).

4. cDNA LIBRARY CONSTRUCTION

Different types of RNAs, such as miRNA, antisense RNA, or noncoding RNA, demand distinct methods for analysis. We focus on mRNA that is responsible for the biosynthesis of critical chemicals accumulated in medicinal plants. Direct RNA sequencing was reported in 2009 (Ozsolak et al., 2009). However, commercial sequencing platforms tend to sequence DNA other than RNA molecules. To acquire a transcriptomic profile, cDNA is commonly synthesized from mRNA for sequencing. Typically, it involves the following steps: purifying mRNA molecules; producing fragmented RNA with desired length (alternatively, fragmentation is performed after cDNA is synthesized from intact RNA); synthesizing first-strand and then double-stranded (ds) cDNA; end repair and ligating oligonucleotide adapters to the ends of the ds cDNA; validation and quantification of the cDNA library for high-throughput sequencing (Fig. 1).

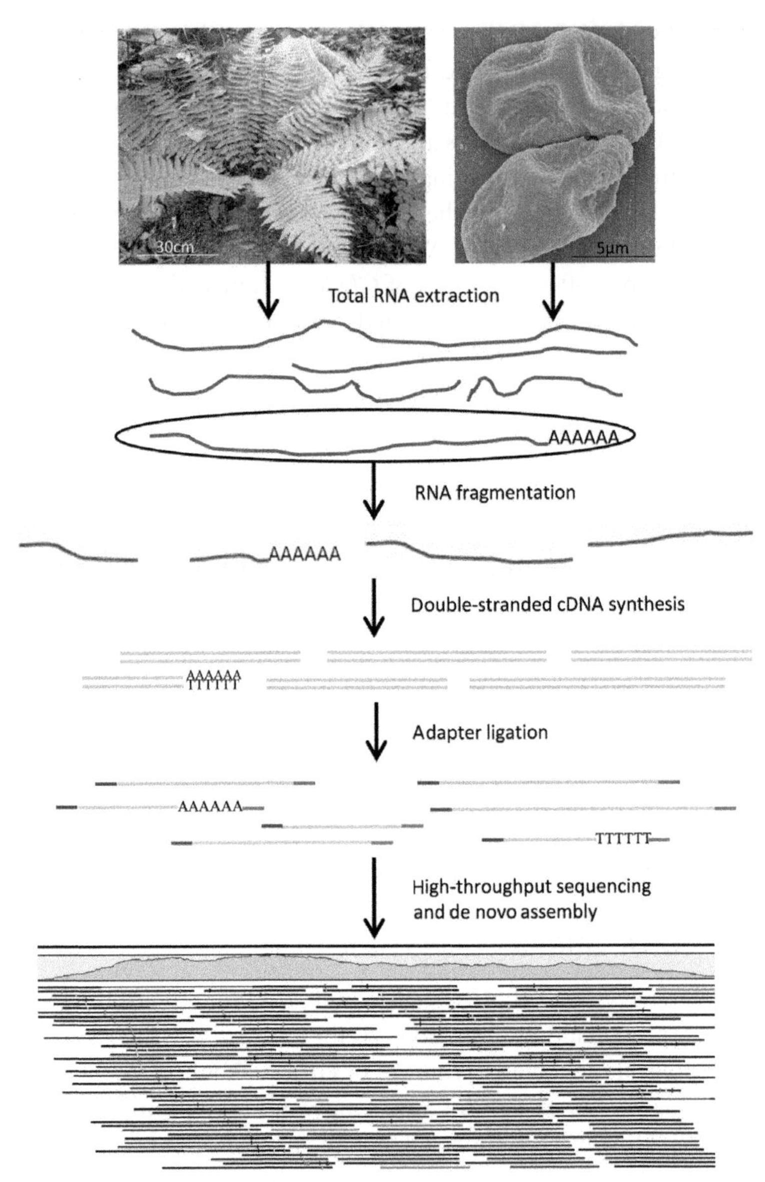

Fig. 1 A typical workflow for RNA-Seq. From plant tissues or single cells, total RNA is extracted. According to various purposes, specific types of RNA, eg, mRNA, are isolated and fragmented. After the synthesis of double-stranded cDNA, adapters are ligated for single or paired-end high-throughput sequencing. De novo assembly constructs transcripts from the obtained raw reads. *Plant photographs (*Dryopteris crassirhizoma *and its spores under scanning electron microscope) courtesy of Dr. Liang Xu, Liaoning University of TCM.* (See the color plate.)

Poly-A containing mRNAs are hybridized to oligo-dT-attached carriers, such as magnetic beads, and then the nonbound irrelevant material can be washed away using commercial mRNA purification kits. Finally, mRNA is eluted from the oligo-dT carrier. Carboxyl-coated magnetic nanoparticles were also used to capture mRNA, which was subsequently purified using wash buffer containing Tris–HCl, NaCl, $MgCl_2$, EDTA, and DTT (Sarkar & Irudayaraj, 2008). Different technologies/platforms have distinct sequencing abilities. Currently, the read length that can be sequenced by Illumina (San Diego, USA) and Ion PGM system (Life technologies, USA) is under 500 bases. GS FLX+ system (454 Life Sciences, USA) delivers up to 1000 bp in read length and PacBio RS II (Pacific Biosciences, USA) is capable of producing reads with length of 15 kb. The size of input RNA has to be decided before constructing a cDNA library. To fragment nucleotide sequences, physical methods like sonication, acoustic shearing, nebulization, and heat, as well as enzymatic ways including restriction endonucleases and nonspecific nuclease cocktails, can be used (Knierim, Lucke, Schwarz, Schuelke, & Seelow, 2011; Wang et al., 2011). A zinc-mediated RNA fragmentation method was reported to show better performance than RNase III on noncoding RNA analysis (Wery, Descrimes, Thermes, Gautheret, & Morillon, 2013). Subsequently, mRNA fragments are primed with random oligomers and reverse transcribed to produce first stand cDNA. Sheared RNA templates are replaced to generate ds cDNA. Before adapters can be ligated to cDNA for sequencing, overhangs caused by fragmentation have to be repaired. When a pair of adapters is ligated to both ends of a cDNA fragment, it therefore can be synthesized from both sides, resulting in "paired-end" reads. Finally, the size and purity of the cDNA library will be checked to ensure its reliability.

In contrast to the above-mentioned nonstrand-specific method, the strand-specific method provides directional information of initial RNA templates. Either strand of the double-stranded DNA may be the template strand for any given gene, and transcription may take place from both directions—which cannot solve whether the reads are from the original transcripts or their complementary forms. Strand orientation of RNA is useful for analyzing mRNA and indispensable for long noncoding RNA study (Hangauer, Vaughn, & McManus, 2013). To achieve this goal, deoxy-UTP is used instead of dTTP to synthesize the second strand of cDNA, which can be selectively destructed by uracil-*N*-glycosylase (UNG) after ligating to a Y-shaped adaptor (Parkhomchuk et al., 2009) or kept from amplification by specific PCR polymerases (Head et al., 2014). The intact first strand of cDNA that will be sequenced retains the desired message.

5. HIGH-THROUGHPUT SEQUENCING

DNA structure was discovered in 1953 by James Watson and Francis Crick. Twenty years later, the first report of a nucleotide sequence with the length of 24 bp appeared (Gilbert & Maxam, 1973). In 1977, both Frederick Sanger and Walter Gilbert published their methods of determining DNA sequence using enzymatic and chemical strategies (Maxam & Gilbert, 1977; Sanger, Nicklen, & Coulson, 1977), respectively, and they shared the 1980 Nobel Prize in chemistry. In 1998, another first-generation sequencing approach, pyrosequencing, was introduced. With the help of DNA polymerase, four types of nucleotides were sequentially added to extend the sequence, and, if reaction occurs, the released pyrophosphate (PPi) is captured by ATP sulfurylase and luciferase, resulting in light that can be detected to confirm the newly incorporated nucleotide in real time (Ronaghi, Uhlen, & Nyren, 1998). If the template DNA has successively identical nucleotides, the intensity of light emitted will not increase proportionally, weakening the popularity of this method.

As of 2005, next-generation sequencing (NGS), adopting different strategies, started to emerge, which features sequencing by synthesis. Researchers from 454 Life Sciences Corporation amplified DNA fragments in emulsion and then loaded the beads containing single-stranded DNA clones into microwells to obtain solid support for an optimized pyrosequencing procedure. Compared to the classical Sanger method, the achieved throughput reached up to 100 times, with the accuracy of 25 million bases being 99% (Margulies et al., 2005). The use of reversible terminator chemistry paved the way for the making of Illumina Inc., which has already taken up 70% of the sequencing market (Zimmerman, Regalado, LaMonica, Guan, & Simonite, 2014). Single DNA molecules are fixed to a flat solid substrate and "clusters" achieved in situ via bridge amplification serve as templates for sequencing, adopting fluorescently labeled terminator deoxyribonucleotides. Fluorescence emitted after laser excitation is detected to reveal the incorporated nucleotides at certain locations on the solid phase of a flow cell. The key point of this approach is the terminator and fluorescent dye that can be cleaved, allowing the following steps to continue (Bentley et al., 2008). The SOLiD sequencing technology relies on ligation. Following emulsion PCR and bead deposition onto a glass slide, the fluorescently labeled di-base probe, which has first and second bases complementary to the template, can be ligated to the sequencing primer to extend the synthesized sequence. After recording the fluorescence, the probe—consisting of eight bases—is cleaved from the fifth base, which eliminates fluorescent dyes attached to the last base. Now, the sequence is ready for a next ligation cycle. Five rounds of sequencing can puzzle out the whole template fragment by resetting the primers complementary to the adapter sequence (Valouev et al., 2008).

Deep sequencing (applying reversible terminators) requires removal of terminators every run after the light signal is captured and, hence, the sequencing procedure is punctuated. The single molecule, real-time (SMRT) sequencing promoted by Pacific Biosciences uses zero-mode waveguide (ZMW), which is a small cylindrical well of about 70 n*M* in width and assures the accuracy and sensitivity of signal detection, to observe the fluorescence when the phospho-linked nucleotide that can be incorporated into the extending primer strand by single DNA polymerases immobilized on the bottom surface of ZMW detectors enters the observation area (Metzker, 2010). Because the fluorescent dye is attached to the terminal phosphate instead of the base, the polymerase releases the dye during DNA replication, leaving the whole process uninterrupted.

The hydrogen ion released after incorporation of a base to a DNA strand with the function of enzymes causes the change of pH value in a well where the DNA template is held. The ion PGM system takes advantage of this, and turns the fluctuation of pH (ΔpH) to a change in potential (ΔV) of the field-effect transistor set underneath the well (Rothberg et al., 2011). Determining nucleotide sequences by semiconductor devices instead of optical systems is a significant feature that differs from other NGS methods.

6. DATA INTERPRETATION

6.1 Trimming and Quality Validation of Raw Reads

The common output file recording information obtained from deep sequencing is in FASTQ format. It was originally created at the Wellcome Trust Sanger Institute to combine nucleotide sequence and the quality score of every corresponding base (Cock, Fields, Goto, Heuer, & Rice, 2010). For each read (a fragment of sequence "read" by NGS instrumentation), FASTQ format gives four types of rows. The first row begins with the symbol @ followed by information as a record identifier. The second lists the sequence and, in case of ambiguous nucleotide(s), instead of space or tab, letter "N" may be used for indication. The third, starting with "+," separates the sequence which can be divided into multiple rows from the following quality scores that occupy the last type of rows and are often encoded in PHRED style.

The FASTQ output files may contain adapters and poor quality input data that are generated during sequencing process, and they have to be removed before any subsequent handling. To address this, both commercial (eg, CLC Genomics Workbench) and free softwares like SeqTrim (Falgueras et al., 2010), TagCleaner (Schmieder, Lim, Rohwer, & Edwards, 2010), and AdapterRemoval (Lindgreen, 2012) provide algorithms to detect and trim adapters, subsequences of Ns, nucleotides with low quality scores, yielding "clean" reads. Taking another popular trimming tool Trimmomatic, for example, the command line "`java -jar trimmomatic-0.35.jar PE 1_1.fastq 1_2.fastq F_P.fastq F_U.fastq R_P.fastq R_U.fastq ILLUMINACLIP:TruSeq3-PE.fa:2:25:9 LEADING:5 TRAILING:5 SLIDINGWINDOW:4:20 MINLEN:30`" has the following function. It processes forward reads in file 1_1.fastq and reverse 1_2.fastq generated by Illumina system for which TruSeq3 adapters are used. With up to two mismatches, adapters will be determined and dropped if a score of 25 is met (around 42 bases) in a palindromic manner, or about 15 bases are matched in a simple alignment manner. Leading and trailing Ns, or bases with quality scores under 5, are also trimmed. The program scans reads with a 4-base sliding window and removes the targets when the average quality per base is below 20. Following the above-mentioned steps, discard the remaining part when its length is shorter than 30. Finally, four resultant files consisting of paired and unpaired forward and reverse reads are reported, respectively (Bolger, Lohse, & Usadel, 2014). The quality-filtered reads can be uploaded to the Sequence Read Archive of international nucleotide sequence databases comprising GenBank, EMBL (European Molecular Biology Laboratory), and DDBJ (DNA DataBank of Japan) for exchanging and sharing information.

6.2 De Novo Assembly

For the study of model plants, clean reads can be directly mapped back to their draft genomes to measure the expression of genes, such as *Medicago truncatula* (Larrainzar et al., 2015). For nonmodel plants whose genome information is usually unavailable, the inconvenient fact concerning convenient NGS is that relatively short reads have to be assembled to restore sequence information sacrificed by fragmentation for the massive parallel process. Several approaches can be applied to do de novo assembly, which commonly utilizes *de Bruijn* graphs as the underlying algorithm (Compeau, Pevzner, & Tesler, 2011). The concept is to list all substrings with a certain length of *k* (*k-mer*) found in any single read is decisive because the uniqueness of every read, except for the rare overrepresented ones, make it impossible to directly assemble from the raw reads—even if a typical one may consist of only 100 bp in length. Suppose we have the following substrings: CTAAC, ATAGC, GCCTA, CCTAA, TAACG, AGCCT, TAGCC, and AACGA. We can search in this pool all the possible neighboring elements to extend the substring. For instance, CTAAC and TAACG differ only at the left and right border, respectively, from which CTAACG can be deduced. By this means, the original read sequence is ATAGCCTAACGA and using a *k-mer* of 5, we get 8 (12–5 + 1) substrings. The number of reads resulting from five tissues of *Pueraria lobata* can be 94 million, with average length of 99 bp, which will produce 7.5 billion substrings given a *k-mer* of 20 (Han, Takahashi, Nakamura, Yoshimoto, et al., 2015). To simultaneously assemble from the enormous input data can be challenging and requires considerable computer capacity. Over the years, several methods have been developed to achieve draft assemblies as reference sequences for the analysis of unexplored transcriptome. Short oligonucleotide analysis package 2 (SOAPdenovo2), which was designed to deal with paired-end data uses six modules which correct read error, establish *de Bruijn* graphs, assemble contigs, realign paired-end reads back to contigs, construct scaffolding and fill in intrascaffold gaps (Luo et al., 2012). This software requires a 64-bit Linux system with at least 5 G physical memory. For short read assembly, Velvet uses four stages: to hash the reads into predefined *k-mers*, to construct the *de Bruijn* graph, to correct errors and remove bubbles, and to resolve repeats (Zerbino & Birney, 2008). Transcriptomic assembly by short sequences (Trans-AbySS) employs its first package "transabyss" to assemble reads into contigs with various *k-mer* sizes, respectively. Then the second module "transabyss-merge" integrates multiple assemblies resulting from the same dataset using distinct *k-mer* sizes and keeps longer sequences for downstream analyses. With a reference genome on hand, structural variants and splice variants can be detected with the third package "transabyss-analyze" (Robertson et al., 2010). Although cluster work can be done, by default, all the processes of AbySS will be run on the local machine. Trinity has three independent modules, the first of which, Inchworm, uses the most abundant *k-mer* generated from all reads as a "seed" that incorporates the other single most abundant *k-mer* with *k-1* overlap to produce a single base extension; and this process goes recursively, resulting in a linear contig when finally no *k-mer* can be found to extend the existing one. By removing the used *k-mers* and adopting the same strategy, the rest contigs are made to ensure each *k-mer* appears only once in the assembled contigs. The second module, Chrysalis, regroups contigs that share at least one *k-1-mer* into individual components and create *de Bruijn* graphs for the linear inchworm contigs, followed by mapping reads to the matching components. The last one, Butterfly, reports validated contigs by selecting the paths that are best supported by the physical read sequences (Grabherr et al., 2011). Trinity basically requires 1 G of RAM for every one million reads.

Apart from some commercial workbenches like CLC running on Windows or Mac OS system, softwares supporting de novo assembly of massive input data often demand a high-performance Linux system. Researchers lack of sufficient equipment may use volunteer computing, like Berkeley Open Infrastructure for Network Computing (BOINC), or public computer resources from institutes such as Supercomputer Facilities of National Institute of Genetics in Japan, Pittsburgh Supercomputing Center (PSC), or Data Intensive Academic Grid (DIAG).

The output file of de novo assembly is commonly in FASTA format, which comprises two types of lines for every nucleotide sequence. The first line is a single-line description starting with a ">" symbol and the second recoding sequence data can be line wrapped. The contigs reported in a FASTA file may contain redundancy. CD-HIT clusters the data to remove duplicates by setting a threshold for reducing storage space, computational time, or noise interference (Fu, Niu, Zhu, Wu, & Li, 2012). Typical statistics for an assembled output file include the number and average length of assembled contigs, GC content, and N50, which is the length of the smallest contig in the set that contains the fewest (largest) contigs and whose combined length represents at least 50% of the assembly (Miller, Koren, & Sutton, 2010), as well as the number of total assembled bases.

6.3 Expression Analysis of Transcripts

With the assembled contigs at hand, reads can be mapped back to check the expression level of transcripts. Metrics include RPKM (reads per kilobase of transcript per million mapped reads), FPKM (substitute reads with fragments), and TPM (transcripts per million) (Wagner, Kin, & Lynch,

2012). Sequencing depth is represented by the number of mapped reads and the length of genes is normalized by "per kilobase of transcript." The formula for calculating RPKM is as follows: (read count of a transcript $\times 10^9$)/(length of a transcript $\times$ number of total mapped reads). For single-end RNA-Seq data, RPKM and FPKM are the same because each fragment is sequenced only once. However, for paired-end data, given the fact that each fragment generates two reads from both ends and some of the reads will be removed during quality control procedures due to their poor quality, use of RPKM may well bias the expression interpretation. FPKM is introduced to address this concern by counting fragments rather than reads. TPM is modified from RPKM to make the expression data comparable across replicates or samples. The entire transcript count is used to normalize TPM, whose formula is as follows: (read count of a transcript $\times$ average read length $\times 10^6$)/(length of a transcript $\times$ number of total transcripts). Several softwares provide the option for quantification of transcripts including RSEM (RNA-Seq by Expectation-Maximization) (Li & Dewey, 2011), eXpress (Roberts & Pachter, 2013), Sailfish (Patro, Mount, & Kingsford, 2014), and kallisto, which utilizes the idea of pseudoalignment to rapidly quantify abundances of transcripts without actual alignment (http://pachterlab.github.io/kallisto/).

The expression of any transcript is quantified by the number of sequenced reads or fragments that are mapped back, and it is dependent on sequencing depth as well as the expression levels of other transcripts. For differential expression (DE) analysis, the raw reads are typically modeled using negative binomial or Poisson distribution, followed by various methods for normalization (Rapaport et al., 2013). Free softwares utilizing R statistical programming language for this task include DESeq (Anders & Huber, 2010), edgeR (Robinson, McCarthy, & Smyth, 2010), NOISeq (Tarazona, Garcia-Alcalde, Dopazo, Ferrer, & Conesa, 2011), and so forth. Cuffdiff (Trapnell et al., 2013) is a program incorporated in Cufflinks to pinpoint significant changes in gene expression using command line. To enhance the ability to detect DE genes, sequencing depth and biological replicates are the two practical issues for design, handling, and cost of the experiment. Studies suggest including more biological replication outweighs sequencing more reads in order to increase accuracy and power in DE RNA-Seq studies (Liu, Zhou, & White, 2014). Due to the algorithms required, analysis on samples with or without technical replicates, biological replicates should use appropriate software options. For example, NOISeq provides NOISeq-real for samples with technical replicates, NOISeq-sim handles samples with no any replicates, and NOISeqBIO processes samples with biological replicates. In our laboratory, a candidate gene preferably expressed in an alkaloid-producing bitter cultivar of *Lupinus angustifolius* compared to a sweet cultivar that does not accumulate alkaloids was targeted and identified as a lysine decarboxylase, catalyzing the initial step of quinolizidine alkaloid biosynthesis (Bunsupa et al., 2012).

Enzymes usually coexpress with their upstream or downstream counterparts in a biosynthetic pathway and this feature is useful for coexpression analysis and to find new genes responsible for producing chemicals of interest. The study on monoterpene-derived indole alkaloid biosynthetic pathways using *Catharanthus roseus* discovered three coexpression clusters correlating to distinct stages in the vinblastine/vincristine pathway, suggesting coordinated regulation of the whole process (Kellner et al., 2015). Coexpression data were also used to identify new genes involved in flavonoid biosynthetic pathways in *Arabidopsis,* with the aid of liquid chromatography-mass spectrometry (LC-MS) to identify related compounds between wild-type and mutant lines (Yonekura-Sakakibara et al., 2008).

6.4 Sequence Similarity Search

Similarity between experimental transcripts and the identified genes deposited in public databases provides hints for their possible function. BLAST (Basic Local Alignment Search Tool) and its newer version BLAST+ are algorithms for this purpose (Camacho et al., 2009). Different programs suit specific requirements. For example, BLASTx uses a translated nucleotide query to search against a protein database and BLASTn searches a nucleotide database with a nucleotide query. Depending on the number and complexity of the query sequences, a BLAST search may take several days. It is advisable to download the databases, such as NCBI nr or UniProt, and run BLAST on your local machine. A simplified command line using BLAST+ to do a similarity search against the NCBI nr database built locally on your own server could be:

```
blastx -query input.fasta -db /path/to/your/nr/database -evalue 1e-10 -num_threads 10 -out output.txt -num_alignments 20 -outfmt 6 -seg no
```

This will use the input.fasta file that consists of DNA sequences to run BLASTx with the computational capacity of 10 threads, resulting in the output.txt file in a tab separated format. Low-complexity filtering will be disabled and up to 20 target sequences that show E-value (expectation value) less than 1e-10 will be retrieved. Depending on the substitution matrix of proteins and gap penalty, a numerical value (score) will be assigned to show the similarity between the query and target sequence. A bit score is a log-scaled type of score and this normalized value allows you to compare against the results obtained from databases of different sizes. Associated with a corresponding score, E-value is the number of hits you expect to find in a database solely by chance. For starters, one may try 1e-5, but there is no golden standard to set a certain E-value for achieving best analysis result because biological significance involves more than pure mathematical numbers.

6.5 Gene Annotation

Gene annotation concerns several aspects to make specific definitions about function or structure. Gene Ontology (GO) describes gene products with three independent categories: biological process, cellular component, and molecular function (Ashburner et al., 2000), which may produce multiple GO terms assigned to one query sequence. To collect GO information, gene identifiers (GI) from NCBI or accession number (AC) can be extracted from BLAST outputs and used to retrieve GO through online tools (http://www.uniprot.org/mapping/). Then, the GO information of corresponding sequences can be displayed using web gene ontology annotation plot (WEGO) (Han, Takahashi, Nakamura, Bunsupa, et al., 2015; Ye et al., 2006). Adopting Fisher's exact test, researchers can investigate the overexpressed GO terms by incorporating expression levels of contigs with associated GO terms (Takahashi et al., 2011). Alternatively, commercial software blast2go (Conesa et al., 2005) provides an algorithm to annotate sequences against public databases, including GO, from which enrichment analysis can be conducted. KEGG (Kyoto Encyclopedia of Genes and Genomes) (Kanehisa et al., 2014), InterPro (Hunter et al., 2012), and CathaCyc (Van Moerkercke et al., 2013) are also popular for gene annotation. Among the 17 groups affiliated with KEGG, the pathway database is frequently used to study secondary metabolic profiles of medicinal plants. The identified entire, or partial, pathways were drawn with enzyme information connecting chemical intermediates and products. To do a pathway search, KEGG supports various ways, including KO (KEGG Orthology), and EC (Enzyme Commission) number. Known enzymes presented in metabolic procedures provide important clues for plants of interest due to similarity in amino acid sequences.

6.6 An Example of RNA-Seq

Due to multiple strategies in RNA-Seq applications, with each featuring distinct merit, there is no set standard. We propose one possible approach to get it done.

(1) Fresh tissues of the target plant collected in the field are cut into little pieces and submerged in RNA*later* for preservation. Additional samples from the same origin are also collected at the same time for quantifying their chemical profiles. Extract total RNA as soon as possible in lab and check OD value as well as RIN value which is greater than 7.5. (2) Adopting Illumina Hiseq 4000 system for paired-end sequencing using standard procedure provided by the manufacturer. (3) The resultant FASTQ format reads are trimmed using Trimmomatic, followed by Trinity de novo assembly. (4) Optional: use CD-HIT software to remove the redundancy from assembled contigs. (5) Use software RSEM to calculate FPKM for every contigs and then perform DE analysis with NOISeq-sim when there is no replicate for each plant sample. (6) Meanwhile, the contigs are subjected to BLASTx search against NCBI nr and UniProt KB databases and subsequently, the retrieved identifiers or ACs are used for annotation from which the EC numbers are selected to map against KEGG pathway. (7) Compare the concentration of interested chemicals in different samples with the corresponding candidate or putative transcripts, along with the information of known secondary metabolic pathways, to speculate the possible missing links. (8) Conduct coexpression analysis to exploit the relationship between identified genes and the ones hidden in the dark.

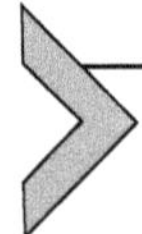

7. APPLICATION OF RNA-Seq ON MEDICINAL PLANTS AND PERSPECTIVES

The enumerated examples here represent only a tiny fraction of studies adopting RNA-Seq. To study the biosynthesis of artemisinin in *Artemisia annua*, glandular and filamentous trichomes were subjected to Illumina deep-sequencing instrumentation. The result demonstrated upregulated terpene biosynthetic pathways in glandular trichomes and suggested candidate genes for sesquiterpenoid biosynthesis (Soetaert et al., 2013). High-throughput sequencing on green and red forms of *Perilla frutescens var. crispa* provided insights into two varietal forms of anthocyanin accumulation (Fukushima, Nakamura, Suzuki, Saito, & Yamazaki, 2015). One methyl jasmonate-responsive transcription factor concerning tanshinone production was proposed by studying RNA-Seq data of *Salvia miltiorrhiza* using combined metabolomics and a transcriptomic approach (Gao et al., 2014). Deep transcriptome analysis with untargeted metabolic profiling in *Ophiorrhiza pumila* illustrated possible transcripts involved in the biosynthesis of the anticancer alkaloid camptothecin (Yamazaki et al., 2013). Searching candidate genes from opium poppy (*Papaver somniferum*), deep transcriptome data resulted in the discovery of a flavoprotein oxidase related to papaverine biosynthesis (Hagel et al., 2012).

For the past decade, NGS has revolutionized our view of how transcriptome can be sequenced and analyzed. In addition to the rapid increase in data output, accuracy, and the reduced time span for sequencing, the drastically lowered price for RNA-Seq really opened the door for ordinary labs with limited budgets to participate and contribute for better understanding of the sophisticated mechanisms regarding plant metabolism. Cutting-edge analytical methods including ultra-high resolution MS and LC-triple quadrupole MS reinforce the power of RNA-Seq by providing information on the substances catalyzed by the enzymes of interest (Rai & Saito, 2015; Saito, 2013). On the other hand, the gradual maturation of single-cell whole-transcriptome analysis allows researchers to tackle certain genes on a very tiny scale. RNA-Seq has exerted its massive impact and will unfold further potentials for elucidating plant genes responsible for producing effective natural products.

ACKNOWLEDGMENTS

This study was supported in part by Grants-in-Aid for Scientific Research from the Ministry of Education, Culture, Sports, Science, and Technology (MEXT) of Japan and by the Health Labor Sciences Research Grant from the Ministry of Health, Labor, and Welfare of Japan, as well as the Chiba University Strategic Priority Research Promotion Program.

Conflict of Interest: The authors declare no conflict of interest.

REFERENCES

Anders, S., & Huber, W. (2010). Differential expression analysis for sequence count data. *Genome Biology*, *11*(10), R106. http://dx.doi.org/10.1186/gb-2010-11-10-r106.

Ashburner, M., Ball, C. A., Blake, J. A., Botstein, D., Butler, H., Cherry, J. M., … Sherlock, G. (2000). Gene ontology: Tool for the unification of biology. The Gene Ontology Consortium. *Nature Genetics*, *25*(1), 25–29. http://dx.doi.org/10.1038/75556.

Autran, D., Jonak, C., Belcram, K., Beemster, G. T., Kronenberger, J., Grandjean, O., … Traas, J. (2002). Cell numbers and leaf development in Arabidopsis: Functional analysis of the STRUWWELPETER gene. *The EMBO Journal*, *21*(22), 6036–6049.

Barbas, C. F., 3rd, Burton, D. R., Scott, J. K., & Silverman, G. J. (2007). Quantitation of DNA and RNA. *CSH Protocols*, *2007*, pdb ip47. http://dx.doi.org/10.1101/pdb.ip47.

Bengtsson, M., Hemberg, M., Rorsman, P., & Stahlberg, A. (2008). Quantification of mRNA in single cells and modelling of RT-qPCR induced noise. *BMC Molecular Biology*, *9*, 63. http://dx.doi.org/10.1186/1471-2199-9-63.

Bentley, D. R., Balasubramanian, S., Swerdlow, H. P., Smith, G. P., Milton, J., Brown, C. G., … Smith, A. J. (2008). Accurate whole human genome sequencing using reversible terminator chemistry. *Nature*, *456*(7218), 53–59. http://dx.doi.org/10.1038/nature07517.

Bird, I. M. (2005). Extraction of RNA from cells and tissue. *Methods in Molecular Medicine*, *108*, 139–148.

Bolger, A. M., Lohse, M., & Usadel, B. (2014). Trimmomatic: A flexible trimmer for Illumina sequence data. *Bioinformatics*, *30*(15), 2114–2120. http://dx.doi.org/10.1093/bioinformatics/btu170.

Brenner, S., Johnson, M., Bridgham, J., Golda, G., Lloyd, D. H., Johnson, D., … Corcoran, K. (2000). Gene expression analysis by massively parallel signature sequencing (MPSS) on microbead arrays. *Nature Biotechnology*, *18*(6), 630–634. http://dx.doi.org/10.1038/76469.

Bunsupa, S., Katayama, K., Ikeura, E., Oikawa, A., Toyooka, K., Saito, K., & Yamazaki, M. (2012). Lysine decarboxylase catalyzes the first step of quinolizidine alkaloid biosynthesis and coevolved with alkaloid production in leguminosae. *Plant Cell*, *24*(3), 1202–1216. http://dx.doi.org/10.1105/tpc.112.095885.

Camacho, C., Coulouris, G., Avagyan, V., Ma, N., Papadopoulos, J., Bealer, K., & Madden, T. L. (2009). BLAST+: Architecture and applications. *BMC Bioinformatics*, *10*, 421. http://dx.doi.org/10.1186/1471-2105-10-421.

Camacho-Sanchez, M., Burraco, P., Gomez-Mestre, I., & Leonard, J. A. (2013). Preservation of RNA and DNA from mammal samples under field conditions. *Molecular Ecology Resources*, *13*(4), 663–673. http://dx.doi.org/10.1111/1755-0998.12108.

Chomczynski, P., & Sacchi, N. (1987). Single-step method of RNA isolation by acid guanidinium thiocyanate-phenol-chloroform extraction. *Analytical Biochemistry*, *162*, 156–159.

Chomczynski, P., & Sacchi, N. (2006). The single-step method of RNA isolation by acid guanidinium thiocyanate-phenol-chloroform extraction: Twenty-something years on. *Nature Protocols*, *1*(2), 581–585. http://dx.doi.org/10.1038/nprot.2006.83.

Cock, P. J. A., Fields, C. J., Goto, N., Heuer, M. L., & Rice, P. M. (2010). The Sanger FASTQ file format for sequences with quality scores, and the Solexa/Illumina FASTQ variants. *Nucleic Acids Research*, *38*(6), 1767–1771. http://dx.doi.org/10.1093/nar/gkp1137.

Compeau, P. E. C., Pevzner, P. A., & Tesler, G. (2011). How to apply de Bruijn graphs to genome assembly. *Nature Biotechnology*, *29*(11), 987–991.

Conesa, A., Gotz, S., Garcia-Gomez, J. M., Terol, J., Talon, M., & Robles, M. (2005). Blast2GO: A universal tool for annotation, visualization and analysis in functional genomics research. *Bioinformatics*, *21*(18), 3674–3676. http://dx.doi.org/10.1093/bioinformatics/bti610.

Czechowski, T., Bari, R. P., Stitt, M., Scheible, W. R., & Udvardi, M. K. (2004). Real-time RT-PCR profiling of over 1400 Arabidopsis transcription factors: Unprecedented sensitivity reveals novel root- and shoot-specific genes. *Plant Journal*, *38*(2), 366–379. http://dx.doi.org/10.1111/j.1365-313X.2004.02051.x.

Deguchi, Y., Banba, M., Shimoda, Y., Chechetka, S. A., Suzuri, R., Okusako, Y., … Hata, S. (2007). Transcriptome profiling of Lotus japonicus roots during arbuscular mycorrhiza development and comparison with that of nodulation. *DNA Research*, *14*(3), 117–133. http://dx.doi.org/10.1093/dnares/dsm014.

Douglas, C. J., & Ehlting, J. (2005). Arabidopsis thaliana full genome longmer microarrays: A powerful gene discovery tool for agriculture and forestry. *Transgenic Research*, *14*(5), 551–561. http://dx.doi.org/10.1007/s11248-005-8926-x.

Efroni, I., Ip, P. L., Nawy, T., Mello, A., & Birnbaum, K. D. (2015). Quantification of cell identity from single-cell gene expression profiles. *Genome Biology*, *16*, 9. http://dx.doi.org/10.1186/S13059-015-0580-X.

Falgueras, J., Lara, A. J., Fernandez-Pozo, N., Canton, F. R., Perez-Trabado, G., & Claros, M. G. (2010). SeqTrim: A high-throughput pipeline for pre-processing any type of sequence read. *BMC Bioinformatics*, *11*, 38. http://dx.doi.org/10.1186/1471-2105-11-38.

Fu, L., Niu, B., Zhu, Z., Wu, S., & Li, W. (2012). CD-HIT: Accelerated for clustering the next-generation sequencing data. *Bioinformatics*, *28*(23), 3150–3152. http://dx.doi.org/10.1093/bioinformatics/bts565.

Fukushima, A., Nakamura, M., Suzuki, H., Saito, K., & Yamazaki, M. (2015). High-throughput sequencing and de novo assembly of red and green forms of the Perilla frutescens var. crispa transcriptome. *Plos One*, *10*(6), e0129154. http://dx.doi.org/10.1371/journal.pone.0129154.

Gao, W., Sun, H. X., Xiao, H., Cui, G., Hillwig, M. L., Jackson, A., … Huang, L. (2014). Combining metabolomics and transcriptomics to characterize tanshinone biosynthesis in Salvia miltiorrhiza. *BMC Genomics*, *15*, 73. http://dx.doi.org/10.1186/1471-2164-15-73.

Gilbert, W., & Maxam, A. (1973). The nucleotide sequence of the lac operator. *Proceedings of the National Academy of Sciences of the United States of America*, *70*(12), 3581–3584.

Girke, T., Todd, J., Ruuska, S., White, J., Benning, C., & Ohlrogge, J. (2000). Microarray analysis of developing Arabidopsis seeds. *Plant Physiology*, *124*(4), 1570–1581.

Grabherr, M. G., Haas, B. J., Yassour, M., Levin, J. Z., Thompson, D. A., Amit, I., … Regev, A. (2011). Full-length transcriptome assembly from RNA-Seq data without a reference genome. *Nature Biotechnology*, *29*(7), 644–652. http://dx.doi.org/10.1038/nbt.1883. U130.

Hagel, J. M., Beaudoin, G. A. W., Fossati, E., Ekins, A., Martin, V. J. J., & Facchini, P. J. (2012). Characterization of a flavoprotein oxidase from opium poppy catalyzing the final steps in sanguinarine and papaverine biosynthesis. *Journal of Biological Chemistry*, *287*(51), 42972–42983. http://dx.doi.org/10.1074/jbc.M112.420414.

Han, R., Takahashi, H., Nakamura, M., Bunsupa, S., Yoshimoto, N., Yamamoto, H., … Saito, K. (2015a). Transcriptome analysis of nine tissues to discover genes involved in the biosynthesis of active ingredients in Sophora flavescens. *Biological & Pharmaceutical Bulletin*, *38*(6), 876–883. http://dx.doi.org/10.1248/bpb.b14-00834.

Han, R., Takahashi, H., Nakamura, M., Yoshimoto, N., Suzuki, H., Shibata, D., … Saito, K. (2015b). Transcriptomic landscape of Pueraria lobata demonstrates potential for phytochemical study. *Frontiers in Plant Science*, *6*, 426. http://dx.doi.org/10.3389/fpls.2015.00426.

Hangauer, M. J., Vaughn, I. W., & McManus, M. T. (2013). Pervasive transcription of the human genome produces thousands of previously unidentified long intergenic noncoding RNAs. *PLos Genetics*, *6*, e1003569. http://dx.doi.org/10.1371/journal.pgen.1003569.

Head, S. R., Komori, H. K., LaMere, S. A., Whisenant, T., Van Nieuwerburgh, F., Salomon, D. R., & Ordoukhanian, P. (2014). Library construction for next-generation sequencing: Overviews and challenges. *Biotechniques*, *56*(2), 61–77. http://dx.doi.org/10.2144/000114133.

Hunter, S., Jones, P., Mitchell, A., Apweiler, R., Attwood, T. K., Bateman, A., … Yong, S. Y. (2012). InterPro in 2011: New developments in the family and domain prediction database. *Nucleic Acids Research*, *40*(Database issue), D306–D312. http://dx.doi.org/10.1093/nar/gkr948.

Ibberson, D., Benes, V., Muckenthaler, M. U., & Castoldi, M. (2009). RNA degradation compromises the reliability of microRNA expression profiling. *BMC Biotechnology*, *9*, 102. http://dx.doi.org/10.1186/1472-6750-9-102.

Kanehisa, M., Goto, S., Sato, Y., Kawashima, M., Furumichi, M., & Tanabe, M. (2014). Data, information, knowledge and principle: Back to metabolism in KEGG. *Nucleic Acids Research, 42*(Database issue), D199–D205. http://dx.doi.org/10.1093/nar/gkt1076.

Kellner, F., Kim, J., Clavijo, B. J., Hamilton, J. P., Childs, K. L., Vaillancourt, B., ... O'Connor, S. E. (2015). Genome-guided investigation of plant natural product biosynthesis. *Plant Journal, 82*(4), 680–692. http://dx.doi.org/10.1111/tpj.12827.

Klein, C. A., Seidl, S., Petat-Dutter, K., Offner, S., Geigl, J. B., Schmidt-Kittler, O., ... Riethmuller, G. (2002). Combined transcriptome and genome analysis of single micrometastatic cells. *Nature Biotechnology, 20*(4), 387–392. http://dx.doi.org/10.1038/Nbt0402-387.

Knierim, E., Lucke, B., Schwarz, J. M., Schuelke, M., & Seelow, D. (2011). Systematic comparison of three methods for fragmentation of long-range PCR products for next generation sequencing. *Plos One, 6*(11), e28240. http://dx.doi.org/10.1371/journal.pone.0028240.

Larrainzar, E., Riely, B. K., Kim, S. C., Carrasquilla-Garcia, N., Yu, H. J., Hwang, H. J., ... Cook, D. R. (2015). Deep sequencing of the Medicago truncatula root transcriptome reveals a massive and early interaction between nodulation factor and ethylene signals. *Plant Physiology, 169*(1), 233–265. http://dx.doi.org/10.1104/pp. 15.00350.

Lelandais-Briere, C., Naya, L., Sallet, E., Calenge, F., Frugier, F., Hartmann, C., ... Crespi, M. (2009). Genome-wide Medicago truncatula small RNA analysis revealed novel microRNAs and isoforms differentially regulated in roots and nodules. *Plant Cell, 21*(9), 2780–2796. http://dx.doi.org/10.1105/tpc.109.068130.

Li, B., & Dewey, C. N. (2011). RSEM: Accurate transcript quantification from RNA-Seq data with or without a reference genome. *BMC Bioinformatics, 12*, 323. http://dx.doi.org/10.1186/1471-2105-12-323.

Lindgreen, S. (2012). AdapterRemoval: Easy cleaning of next-generation sequencing reads. *BMC Research Notes, 5*, 337. http://dx.doi.org/10.1186/1756-0500-5-337.

Lister, R., O'Malley, R. C., Tonti-Filippini, J., Gregory, B. D., Berry, C. C., Millar, A. H., & Ecker, J. R. (2008). Highly integrated single-base resolution maps of the epigenome in Arabidopsis. *Cell, 133*(3), 523–536. http://dx.doi.org/10.1016/j.cell.2008.03.029.

Liu, Y., Zhou, J., & White, K. P. (2014). RNA-Seq differential expression studies: More sequence or more replication? *Bioinformatics, 30*(3), 301–304. http://dx.doi.org/10.1093/bioinformatics/btt688.

Luo, R., Liu, B., Xie, Y., Li, Z., Huang, W., Yuan, J., ... Lam, T. W. (2012). SOAPdenovo2: An empirically improved memory-efficient short-read de novo assembler. *Gigascience, 1*(1), 18. http://dx.doi.org/10.1186/2047-217X-1-18.

Margulies, M., Egholm, M., Altman, W. E., Attiya, S., Bader, J. S., Bemben, L. A., ... Rothberg, J. M. (2005). Genome sequencing in microfabricated high-density picolitre reactors. *Nature, 437*(7057), 376–380. http://dx.doi.org/10.1038/nature03959.

Marinov, G. K., Williams, B. A., McCue, K., Schroth, G. P., Gertz, J., Myers, R. M., & Wold, B. J. (2014). From single-cell to cell-pool transcriptomes: Stochasticity in gene expression and RNA splicing. *Genome Research, 24*(3), 496–510. http://dx.doi.org/10.1101/gr.161034.113.

Matasci, N., Hung, L. H., Yan, Z., Carpenter, E. J., Wickett, N. J., Mirarab, S., ... Wong, G. K. (2014). Data access for the 1,000 plants (1KP) project. *Gigascience, 3*, 17. http://dx.doi.org/10.1186/2047-217X-3-17.

Maxam, A. M., & Gilbert, W. (1977). A new method for sequencing DNA. *Proceedings of the National Academy of Sciences of the United States of America, 74*(2), 560–564.

Metzker, M. L. (2010). Applications of next-generation sequencing sequencing technologies—The next generation. *Nature Reviews Genetics, 11*(1), 31–46. http://dx.doi.org/10.1038/nrg2626.

Miller, J. R., Koren, S., & Sutton, G. (2010). Assembly algorithms for next-generation sequencing data. *Genomics*, *95*(6), 315–327. http://dx.doi.org/10.1016/j.ygeno.2010.03.001.

Moore, S., Payton, P., Wright, M., Tanksley, S., & Giovannoni, J. (2005). Utilization of tomato microarrays for comparative gene expression analysis in the Solanaceae. *Journal of Experimental Botany*, *56*(421), 2885–2895. http://dx.doi.org/10.1093/jxb/eri283.

Nagalakshmi, U., Wang, Z., Waern, K., Shou, C., Raha, D., Gerstein, M., & Snyder, M. (2008). The transcriptional landscape of the yeast genome defined by RNA sequencing. *Science*, *320*(5881), 1344–1349. http://dx.doi.org/10.1126/science.1158441.

Ozsolak, F., Platt, A. R., Jones, D. R., Reifenberger, J. G., Sass, L. E., McInerney, P., ... Milos, P. M. (2009). Direct RNA sequencing. *Nature*, *461*(7265), 814–818. http://dx.doi.org/10.1038/nature08390. U873.

Parker, R. (2012). RNA degradation in Saccharomyces cerevisiae. *Genetics*, *191*(3), 671–702. http://dx.doi.org/10.1534/genetics.111.137265.

Parkhomchuk, D., Borodina, T., Amstislavskiy, V., Banaru, M., Hallen, L., Krobitsch, S., ... Soldatov, A. (2009). Transcriptome analysis by strand-specific sequencing of complementary DNA. *Nucleic Acids Research*, *37*(18), e123. http://dx.doi.org/10.1093/nar/gkp596.

Patro, R., Mount, S. M., & Kingsford, C. (2014). Sailfish enables alignment-free isoform quantification from RNA-Seq reads using lightweight algorithms. *Nature Biotechnology*, *32*(5), 462–464. http://dx.doi.org/10.1038/nbt.2862.

Peng, J., Xia, Z. H., Chen, L., Shi, M. J., Pu, J. J., Guo, J. R., & Fan, Z. F. (2014). Rapid and efficient isolation of high-quality small RNAs from recalcitrant plant species rich in polyphenols and polysaccharides. *Plos One*, *9*(5), e95687. http://dx.doi.org/10.1371/journal.pone.0095687.

Petrovska, B. B. (2012). Historical review of medicinal plants' usage. *Pharmacognosy Reviews*, *6*(11), 1–5. http://dx.doi.org/10.4103/0973-7847.95849.

Qiao, Z. Z., & Libault, M. (2013). Unleashing the potential of the root hair cell as a single plant cell type model in root systems biology. *Frontiers in Plant Science*, *4*, 484. http://dx.doi.org/10.3389/Fpls.2013.00484.

Raghavan, R., Groisman, E. A., & Ochman, H. (2011). Genome-wide detection of novel regulatory RNAs in E. coli. *Genome Research*, *21*(9), 1487–1497. http://dx.doi.org/10.1101/gr.119370.110.

Rai, A., & Saito, K. (2015). Omics data input for metabolic modeling. *Current Opinion in Biotechnology*, *37*, 127–134. http://dx.doi.org/10.1016/j.copbio.2015.10.010.

Rapaport, F., Khanin, R., Liang, Y., Pirun, M., Krek, A., Zumbo, P., ... Betel, D. (2013). Comprehensive evaluation of differential gene expression analysis methods for RNA-Seq data. *Genome Biology*, *14*(9), R95. http://dx.doi.org/10.1186/gb-2013-14-9-r95.

Raven, P. H., Evert, R. F., & Eichhorn, S. E. (2013). *Biology of plants* (8th ed.). New York: W.H. Freeman and Company Publishers.

Roberts, A., & Pachter, L. (2013). Streaming fragment assignment for real-time analysis of sequencing experiments. *Nature Methods*, *10*(1), 71–73. http://dx.doi.org/10.1038/nmeth.2251.

Robertson, G., Schein, J., Chiu, R., Corbett, R., Field, M., Jackman, S. D., ... Birol, I. (2010). De novo assembly and analysis of RNA-Seq data. *Nature Methods*, 7(11), 909–912. http://dx.doi.org/10.1038/nmeth.1517.

Robinson, M. D., McCarthy, D. J., & Smyth, G. K. (2010). edgeR: A bioconductor package for differential expression analysis of digital gene expression data. *Bioinformatics*, *26*(1), 139–140. http://dx.doi.org/10.1093/bioinformatics/btp616.

Romero, I. G., Pai, A. A., Tung, J., & Gilad, Y. (2014). RNA-Seq: Impact of RNA degradation on transcript quantification. *BMC Biology*, *12*, 42. http://dx.doi.org/10.1186/1741-7007-12-42.

Ronaghi, M., Uhlen, M., & Nyren, P. (1998). A sequencing method based on real-time pyrophosphate. *Science*, *281*(5375). 363, 365.

Rothberg, J. M., Hinz, W., Rearick, T. M., Schultz, J., Mileski, W., Davey, M., ... Bustillo, J. (2011). An integrated semiconductor device enabling non-optical genome sequencing. *Nature*, *475*(7356), 348–352. http://dx.doi.org/10.1038/nature10242.

Saito, K. (2013). Phytochemical genomics—A new trend. *Current Opinion in Plant Biology*, *16*(3), 373–380. http://dx.doi.org/10.1016/j.pbi.2013.04.001.

Sanger, F., Nicklen, S., & Coulson, A. R. (1977). DNA sequencing with chain-terminating inhibitors. *Proceedings of the National Academy of Sciences of the United States of America*, *74*(12), 5463–5467.

Sarkar, T. R., & Irudayaraj, J. (2008). Carboxyl-coated magnetic nanoparticles for mRNA isolation and extraction of supercoiled plasmid DNA. *Analytical Biochemistry*, *379*(1), 130–132. http://dx.doi.org/10.1016/j.ab.2008.04.016.

Schena, M., Shalon, D., Davis, R. W., & Brown, P. O. (1995). Quantitative monitoring of gene-expression patterns with a complementary-DNA microarray. *Science*, *270*(5235), 467–470. http://dx.doi.org/10.1126/science.270.5235.467.

Schmieder, R., Lim, Y. W., Rohwer, F., & Edwards, R. (2010). TagCleaner: Identification and removal of tag sequences from genomic and metagenomic datasets. *BMC Bioinformatics*, *11*, 341. http://dx.doi.org/10.1186/1471-2105-11-341.

Schroeder, A., Mueller, O., Stocker, S., Salowsky, R., Leiber, M., Gassmann, M., ... Ragg, T. (2006). The RIN: An RNA integrity number for assigning integrity values to RNA measurements. *BMC Molecular Biology*, *7*, 3. http://dx.doi.org/10.1186/1471-2199-7-3.

Schutze, K., & Lahr, G. (1998). Identification of expressed genes by laser-mediated manipulation of single cells. *Nature Biotechnology*, *16*(8), 737–742. http://dx.doi.org/10.1038/nbt0898-737.

Scott, C. P., VanWye, J., McDonald, M. D., & Crawford, D. L. (2009). Technical analysis of cDNA microarrays. *Plos One*, *4*(2), e4486. http://dx.doi.org/10.1371/journal.pone.0004486.

Soetaert, S. S., Van Neste, C. M., Vandewoestyne, M. L., Head, S. R., Goossens, A., Van Nieuwerburgh, F. C., & Deforce, D. L. (2013). Differential transcriptome analysis of glandular and filamentous trichomes in Artemisia annua. *BMC Plant Biology*, *13*, 220. http://dx.doi.org/10.1186/1471-2229-13-220.

Sumner, J. (2000). *The natural history of medicinal plants*. Portland, OR: Timber Press.

Svec, D., Andersson, D., Pekny, M., Sjoback, R., Kubista, M., & Stahlberg, A. (2013). Direct cell lysis for single-cell gene expression profiling. *Frontiers in Oncology*, *3*, 274. http://dx.doi.org/10.3389/fonc.2013.00274.

Takahashi, H., Morioka, R., Ito, R., Oshima, T., Altaf-Ul-Amin, M., Ogasawara, N., & Kanaya, S. (2011). Dynamics of time-lagged gene-to-metabolite networks of Escherichia coli elucidated by integrative omics approach. *OMICS*, *15*(1–2), 15–23. http://dx.doi.org/10.1089/omi.2010.0074.

Tang, T., Francois, N., Glatigny, A., Agier, N., Mucchielli, M. H., Aggerbeck, L., & Delacroix, H. (2007). Expression ratio evaluation in two-colour microarray experiments is significantly improved by correcting image misalignment. *Bioinformatics*, *23*(20), 2686–2691. http://dx.doi.org/10.1093/bioinformatics/btm399.

Tang, F. C., Lao, K. Q., & Surani, M. A. (2011). Development and applications of single-cell transcriptome analysis. *Nature Methods*, *8*(4), S6–S11. http://dx.doi.org/10.1038/Nmeth.1557.

Tarazona, S., Garcia-Alcalde, F., Dopazo, J., Ferrer, A., & Conesa, A. (2011). Differential expression in RNA-Seq: A matter of depth. *Genome Research*, *21*(12), 2213–2223. http://dx.doi.org/10.1101/gr.124321.111.

Trapnell, C., Hendrickson, D. G., Sauvageau, M., Goff, L., Rinn, J. L., & Pachter, L. (2013). Differential analysis of gene regulation at transcript resolution with RNA-Seq. *Nature Biotechnology*, *31*(1), 46–53. http://dx.doi.org/10.1038/nbt.2450.

Valouev, A., Ichikawa, J., Tonthat, T., Stuart, J., Ranade, S., Peckham, H., ... Johnson, S. M. (2008). A high-resolution, nucleosome position map of C. elegans reveals a lack of universal sequence-dictated positioning. *Genome Research, 18*(7), 1051–1063. http://dx.doi.org/10.1101/gr.076463.108.

Van Moerkercke, A., Fabris, M., Pollier, J., Baart, G. J., Rombauts, S., Hasnain, G., ... Goossens, A. (2013). CathaCyc, a metabolic pathway database built from Catharanthus roseus RNA-Seq data. *Plant & Cell Physiology, 54*(5), 673–685. http://dx.doi.org/10.1093/pcp/pct039.

Velculescu, V. E., Zhang, L., Vogelstein, B., & Kinzler, K. W. (1995). Serial analysis of gene-expression. *Science, 270*(5235), 484–487. http://dx.doi.org/10.1126/science.270.5235.484.

Wagner, G. P., Kin, K., & Lynch, V. J. (2012). Measurement of mRNA abundance using RNA-Seq data: RPKM measure is inconsistent among samples. *Theory in Biosciences, 131*(4), 281–285. http://dx.doi.org/10.1007/s12064-012-0162-3.

Wall, M. E., & Wani, M. C. (1995). Camptothecin and taxol—Discovery to clinic—13th Bruce-F-Cain-Memorial-Award-Lecture. *Cancer Research, 55*(4), 753–760.

Wang, L., Si, Y. Q., Dedow, L. K., Shao, Y., Liu, P., & Brutnell, T. P. (2011). A low-cost library construction protocol and data analysis pipeline for illumina-based strand-specific multiplex RNA-Seq. *Plos One, 6*(10), e26426. http://dx.doi.org/10.1371/journal.pone.0026426.

Wang, H. M., Yin, W. C., Wang, C. K., & To, K. Y. (2009). Isolation of functional RNA from different tissues of tomato suitable for developmental profiling by microarray analysis. *Botanical Studies, 50*(2), 115–125.

Wery, M., Descrimes, M., Thermes, C., Gautheret, D., & Morillon, A. (2013). Zinc-mediated RNA fragmentation allows robust transcript reassembly upon whole transcriptome RNA-Seq. *Methods, 63*(1), 25–31. http://dx.doi.org/10.1016/j.ymeth.2013.03.009.

Wortman, J. R., Haas, B. J., Hannick, L. I., Smith, R. K., Jr., Maiti, R., Ronning, C. M., ... Town, C. D. (2003). Annotation of the Arabidopsis genome. *Plant Physiology, 132*(2), 461–468. http://dx.doi.org/10.1104/pp. 103.022251.

Yamazaki, M., Mochida, K., Asano, T., Nakabayashi, R., Chiba, M., Udomson, N., ... Saito, K. (2013). Coupling deep transcriptome analysis with untargeted metabolic profiling in Ophiorrhiza pumila to further the understanding of the biosynthesis of the anti-cancer alkaloid camptothecin and anthraquinones. *Plant and Cell Physiology, 54*(5), 686–696. http://dx.doi.org/10.1093/pcp/pct040.

Ye, J., Fang, L., Zheng, H., Zhang, Y., Chen, J., Zhang, Z., ... Wang, J. (2006). WEGO: A web tool for plotting GO annotations. *Nucleic Acids Research, 34*(Web Server issue), W293–W297. http://dx.doi.org/10.1093/nar/gkl031.

Yonekura-Sakakibara, K., Tohge, T., Matsuda, F., Nakabayashi, R., Takayama, H., Niida, R., ... Saito, K. (2008). Comprehensive flavonol profiling and transcriptome coexpression analysis leading to decoding gene-metabolite correlations in Arabidopsis. *Plant Cell, 20*(8), 2160–2176. http://dx.doi.org/10.1105/tpc.108.058040.

Zerbino, D. R., & Birney, E. (2008). Velvet: Algorithms for de novo short read assembly using de Bruijn graphs. *Genome Research, 18*(5), 821–829. http://dx.doi.org/10.1101/gr.074492.107.

Zhang, C. Q., Barthelson, R. A., Lambert, G. M., & Galbraith, D. W. (2008). Global characterization of cell-specific gene expression through fluorescence-activated sorting of nuclei. *Plant Physiology, 147*(1), 30–40. http://dx.doi.org/10.1104/pp. 107.115246.

Zhu, S., Fushimi, H., & Komatsu, K. (2008). Development of a DNA microarray for authentication of ginseng drugs based on 18S rRNA gene sequence. *Journal of Agricultural and Food Chemistry, 56*(11), 3953–3959. http://dx.doi.org/10.1021/jf0732814.

Zimmerman, E., Regalado, A., LaMonica, M., Guan, X. Y., & Simonite, T. (2014). The 50 Smartest Companies 2014. *Technology Review, 117*(2), 26–46.

CHAPTER THREE

Genomics-Based Discovery of Plant Genes for Synthetic Biology of Terpenoid Fragrances: A Case Study in Sandalwood oil Biosynthesis

J.M. Celedon, J. Bohlmann[1]
Michael Smith Laboratories, University of British Columbia, Vancouver, BC, Canada
[1]Corresponding author: e-mail address: bohlmann@msl.ubc.ca

Contents

1. Introduction 48
2. Prior Knowledge of Sandalwood TPSs and P450s and Development of a Hypothesis 50
3. Replication, Sampling, and Statistical Design 51
4. Defining Temporal and Spatial Variables for Tissue Sampling 52
5. Tissue Sampling 53
6. Metabolite Profiling 53
7. Isolation of High-Quality RNA from Recalcitrant Tissues 55
8. Transcriptome Sequencing and De Novo Assembly 56
9. Transcriptome Mining and Annotation 57
10. Expression Analysis and Candidate Gene Selection 58
11. Functional Characterization of Candidate Genes 59
   11.1 Yeast In Vivo Assays 60
   11.2 Microsomes In Vitro Assays 62
12. Product Identification 63
   12.1 MS–MS 64
   12.2 Nuclear Magnetic Resonance 64
Acknowledgments 65
References 65

Abstract

Terpenoid fragrances are powerful mediators of ecological interactions in nature and have a long history of traditional and modern industrial applications. Plants produce a great diversity of fragrant terpenoid metabolites, which make them a superb source of biosynthetic genes and enzymes. Advances in fragrance gene discovery have enabled new approaches in synthetic biology of high-value speciality molecules toward

Methods in Enzymology, Volume 576
ISSN 0076-6879
http://dx.doi.org/10.1016/bs.mie.2016.03.008

applications in the fragrance and flavor, food and beverage, cosmetics, and other industries. Rapid developments in transcriptome and genome sequencing of nonmodel plant species have accelerated the discovery of fragrance biosynthetic pathways. In parallel, advances in metabolic engineering of microbial and plant systems have established platforms for synthetic biology applications of some of the thousands of plant genes that underlie fragrance diversity. While many fragrance molecules (eg, simple monoterpenes) are abundant in readily renewable plant materials, some highly valuable fragrant terpenoids (eg, santalols, ambroxides) are rare in nature and interesting targets for synthetic biology. As a representative example for genomics/transcriptomics enabled gene and enzyme discovery, we describe a strategy used successfully for elucidation of a complete fragrance biosynthetic pathway in sandalwood (*Santalum album*) and its reconstruction in yeast (*Saccharomyces cerevisiae*). We address questions related to the discovery of specific genes within large gene families and recovery of rare gene transcripts that are selectively expressed in recalcitrant tissues. To substantiate the validity of the approaches, we describe the combination of methods used in the gene and enzyme discovery of a cytochrome P450 in the fragrant heartwood of tropical sandalwood, responsible for the fragrance defining, final step in the biosynthesis of (*Z*)-santalols.

1. INTRODUCTION

Terpenoids constitute one of the largest and most diverse classes of plant specialized metabolites, with a wide range of chemoecological functions (Gershenzon & Dudareva, 2007) and industrial applications (Arendt, Pollier, Callewaert, & Goossens, 2016; Bohlmann & Keeling, 2008). Industrial applications of specialized plant terpenoids include medicinal compounds such as the anticancer drug taxol (Croteau, Ketchum, Long, Kaspera, & Wildung, 2006) and the antimalarial agent artemisin (Paddon, Westfall, Pitera, et al., 2013), as well as fragrance compounds such as ambroxides (Zerbe & Bohlmann, 2015a) or santalols (Jones et al., 2011). Terpene fragrances are produced in many different plant tissues and in all of the major plant organs. For example, terpenes represent the largest class of floral volatiles (Knudsen, Eriksson, Gershenzon, & Ståhl, 2006; Muhlemann, Klempien, & Dudareva, 2014). Flowers of species, such as snapdragon, petunia, or roses, have been explored for the discovery of fragrance genes (Dudareva et al., 2003; Magnard, Roccia, Caissard, et al., 2015; Qualley, Widhalm, Adebesin, Kish, & Dudareva, 2012), while organs such as roots, rhizomes, or mature heartwood represent another less explored and valuable source of such genes. Biosynthesis of terpenoid fragrances may occur in highly specialized cell types, such as floral glandular trichomes in lavender (Demissie et al., 2012; Demissie, Erland, Rheault, & Mahmoud,

2013) or clary sage (Caniard et al., 2012), or may be present in epidermal or parenchymatic tissues. The sequestration and release of terpenoid fragrances are thought to involve both active and passive processes (Widhalm, Jaini, Morgan, & Dudareva, 2015).

The chemical diversity of terpenoid fragrance compounds is the results of a modular pathway system that is present with variations in all plant species (Chen, Tholl, Bohlmann, & Pichersky, 2011). Terpenoid precursors of 10, 15, or 20 carbon atoms, namely, geranyl diphosphate (GPP, C_{10}), neryl diphosphate (NPP, C_{10}), farnesyl diphosphate (FPP, C_{15}), geranylgeranyl diphosphate (GGPP, C_{20}), and nerylneryl diphosphate (NNPP, C_{20}), are assembled by prenyltransferases from the two isomeric 5-carbon building blocks isopentenyl diphosphate (IPP) and dimethylallyl diphosphate (DMAPP). IPP and DMAPP are produced in the mevalonic acid (MEV) pathway, which is located across the cytosol, peroxisome, and endoplasmic reticulum, or in the plastidial methylerythritol phosphate (MEP) pathway. DMAPP, GPP, NPP, FPP, GGPP, and NNPP are the substrates of terpene synthases (TPSs) that produce acyclic and cyclic terpene olefins and alcohols, which may undergo additional regio- and stereo-specific oxidations, peroxidations, acylations, methylations, or glycosylations. The many different variations of producing terpenoid scaffolds by TPSs, combined with subsequent modification by other enzymes, most commonly involving cytochrome P450-dependent reactions, result in the many different terpenoid fragrance profiles of different plant species. In most plant species, TPS and P450 enzymes are encoded by large gene families, and the many possible combinations of these enzymes represent a nearly unlimited resource for the natural biosynthesis, and synthetic biology, of known and novel terpenoid molecules as demonstrated, for example, for the diterpenoids (Andersen-Ranberg, Kongstad, Nielsen, et al., 2016; Zerbe & Bohlmann, 2015b; Zerbe, Hamberger, Yuen, et al., 2013).

Here, we describe general approaches for the discovery and characterization of terpenoid fragrance genes and enzymes in nonmodel plants, which are defined here as plant species that have traditionally not been easily accessible with molecular and genetic tools, and therefore have been difficult to explore for gene discovery. Most of the world's plant species that are of interest for fragrances are nonmodel species, and the same is true for most medicinal plants. Some of these species, such as sandalwood (*Santalum* ssp.), may have been overexploited for extraction of terpenoids. High-throughput DNA sequencing, most commonly applied for transcriptome sequencing, has made nonmodel plant species accessible to gene discovery, and by extension has enabled new opportunities for synthetic biology of specialized plant metabolites, including fragrances (Arendt et al., 2016; Facchini, Bohlmann, Covello, et al., 2012; Zerbe & Bohlmann, 2015a). In general, for nonmodel plant systems, gene discovery by transcriptome sequencing has substantial advantages over genome sequencing. Specifically, transcriptomes are generally less costly to produce and easier to assemble than genome sequences, sequence coverage of transcriptomes is focused on expressed genes and not distributed across excessive nonexpressed or nonprotein-coding genomic regions, and protein-coding open reading frame sequences can be predicted from transcriptome assemblies without consideration of intron/exon structures. To illustrate general concepts and methods with specific examples, we describe the discovery of a novel P450 gene and the encoded enzyme that is critical in the biosynthesis of the fragrance defining sesquiterpenoid alcohol components of sandalwood oil, namely, (*Z*)-α-santalol, (*Z*)-β-santalol, (*Z*)-α-*exo*-bergamotol, and (*Z*)-*epi*-β-santalol (Celedon et al., 2016). The principles described here are broadly applicable to the discovery of specific genes in large gene families, such as TPSs and P450s.

2. PRIOR KNOWLEDGE OF SANDALWOOD TPSs AND P450S AND DEVELOPMENT OF A HYPOTHESIS

The first report of sandalwood (*Santalum album*) TPS gene discovery (Jones et al., 2008) used a PCR cloning strategy with degenerate primers targeting conserved TPS motifs followed by RACE to recover full-length cDNAs. This approach resulted in the cDNA cloning and functional characterization of the *SaMonoTPS1* and *SaSesquiTPS1* genes that produced, respectively, (+)-α-terpineol and (−)-limonene, and germacrene D-4-ol and helminthogermacrene as the main products. In a subsequent and different approach, cDNAs of TPS genes encoding santalene synthases (SSy) were discovered in three different *Santalum* species, *S. album* (*SaSSy*), *S. spicatum* (*SpiSSy*), and *S. austrocaledonicum* (*SauSSy*) (Jones et al., 2011). The breakthrough discovery of SaSSy was based on candidate TPS identification in Sanger-sequenced EST libraries from *S. album* wood cores followed by functional characterization, which identified SaSSy as a multiproduct sesquiterpene synthase that produces α-santalene, β-santalene, *epi*-β-santalene, and α-*exo*-bergamotene. These sesquiterpene olefins are the specific precursors for the major sesquiterpene alcohols of sandalwood oil, (*Z*)-α-santalol,

(*Z*)-β-santalol, (*Z*)-*epi*-β-santalol, and (*Z*)-α-*exo*-bergamotol (Jones et al., 2011).

The presence of a hydroxyl group at the C12 position of α-santalol, β-santalol, *epi*-β-santalol, and α-*exo*-bergamotol was indicative of P450-dependent oxidation of SaSSy products. Mining the same Sanger EST library and additional 454 sequences for P450 candidates of the CYP71 clan, Diaz-Chavez, Moniodis, Madilao, et al. (2013) identified nine different members of the sandalwood CYP76F subfamily, which were expressed in baker's yeast (*Saccharomyces cerevisiae*) and shown to produce predominantly the (*E*) stereoisomers of α-santalol, β-santalol, *epi*-β-santalol, and α-*exo*-bergamotol (Diaz-Chavez et al., 2013). Since sandalwood oil is mostly composed of the opposite (*Z*) stereoisomers of these sesquiterpene alcohols, it was hypothesized that the ability of P450s to hydroxylate α-santalene, β-santalene, *epi*-β-santalene, and α-*exo*-bergamotene evolved in different subfamilies of the P450 gene family in *S. album* and that different P450s preferentially produce either the (*Z*)- or the (*E*)-stereoisomers of α-santalol, β-santalol, *epi*-β-santalol, and α-*exo*-bergamotol. It was further hypothesized that transcripts of P450s that produce the (*Z*) isomers would be coexpressed (ie, similar spatial and temporal patterns of gene expression) with the early santalol pathway genes, including SaSSy, FPP synthase, and the MEV pathway.

With the goal to discover the elusive P450 enzyme involved in (*Z*) selective sesquiterpene oxidation, these hypotheses were tested by tissue-specific RNA-seq analysis across the gradient of developing *S. album* wood tissues, including the outer sapwood (SW), the intermediate transition zone (TZ), and the inner oil-accumulating heartwood (HW). To rule out possible temporal or developmental separation of the TPS and P450 steps in the biosynthesis of sandalwood oil, targeted sesquiterpene metabolite analysis was performed to track pathway intermediates and final products in each tissue (Celedon et al., 2016).

3. REPLICATION, SAMPLING, AND STATISTICAL DESIGN

Replication and experimental conditions are critical for statistical analysis of variation for RNA-seq-based gene discovery if comparison of multiple conditions and treatments is key to the analysis. The selection of an appropriate number of biological replicates and the control of sources of variability (eg, intraspecific genetic variation, tissue type and developmental variation, environmental and life-history variation) are important aspects when analyzing transcriptomes of plants, as many different factors may influence gene expression as well as metabolite biosynthesis and accumulation. This is of particular importance when working with nonmodel plant species collected from natural populations or grown under field conditions. A common observation, although not always the case, is that a high abundance of a metabolite of interest (eg, santalols) in a given tissue is correlated with, or preceded in time by, a high abundance of the corresponding biosynthetic transcripts. Therefore, comparison of transcriptomes of tissues that are contrasted for low and high levels of metabolite accumulation may lead to successful gene discovery via identification of differentially expressed transcripts. The general objective of the statistical design of tissue sampling is to minimize and control sources of variability (noise) and maximize the chances of capturing the natural and developmental variation of gene expression in fragrance biosynthesis.

The sandalwood trees in our case study were all of the same age of 15 years grown in a field plantation in the Kununurra area in Northern Australia. Six trees grown under the same conditions and exhibiting a similar visual phenotype (ie, foliage, trunk diameter, no signs of disease, or major herbivore damage) were selected for tissue sampling, and all samples were kept separate throughout the analyses to maintain proper biological replication and allow for statistical assessment of variations between samples.

4. DEFINING TEMPORAL AND SPATIAL VARIABLES FOR TISSUE SAMPLING

Given their roles in plant ecological interactions with other organisms, the biosynthesis (and transcript expression) of fragrance molecules may be restricted in space to specialized cell types, specific tissues, or organs. Biosynthesis may also be restricted in time to specific developmental stages, time of day, or response to external stimuli. Knowledge of the biological system of interest therefore can be critical in defining the relevant samples for transcriptome-based discovery of fragrance biosynthetic genes. At the minimum, metabolite profiling of different tissues and developmental stages is recommended to identify temporal and spatial patterns, as well as conditions, under which biosynthetic pathway genes are mostly likely to be expressed.

In sandalwood trees, the fragrant oil accumulates in the inner HW tissue of stems and roots, typically at a tissue age of 15 years and older. Contributions of the outer and younger SW, as well as the TZ (the tissue between SW and HW) to oil formation were previously not well characterized. For these reasons, we sampled all three tissues to assess their sesquiterpene content and composition and to gain insight into the potential differences with regard to gene expression of sesquiterpene biosynthesis.

5. TISSUE SAMPLING

The logistics of tissue sampling for RNA isolation range from straight forward, if plants can be grown in growth chambers or greenhouses in a laboratory or research facility; or they may require major planning and large efforts, if plants have to be accessed in remote field sites. Commercial reagents are available that may allow for less stringent conditions of sample handling. However, to preserve RNA quality for transcriptome sequencing we consider it essential that all instruments and reagents that come in contact with the plant tissue are treated to be RNase free and that tissues are immediately flash-frozen in liquid nitrogen, followed by cryogenic transportation to the laboratory. Storage of samples at −80°C is recommended especially if samples are not immediately extracted. Dry ice or nitrogen dry shippers are frequently used for temporary storage and transportation of field-collected samples. Special care must be taken when sampling involves procedures that damage cell integrity, as this increases exposure to RNA degrading enzymes.

Sampling of sandalwood tissues was performed in a remote tropical plantation, requiring prior shipment of liquid nitrogen cooled dry shippers over 3000 km by airplane and return shipment of samples in dry shippers back to the laboratory. Wood tissues were sampled from the stem of living sandalwood trees. The high value of these trees required nondestructive and minimally invasive sampling, which was performed with a manual drill early in the morning to avoid the high midday temperatures of Northern Australia. Trees were sampled at 50 cm height from the base of the trunk. Samples from the outer SW, intermediate TZ, and inner HW were visually separated by color (SW: white/yellow; TZ: pink/red; HW: red/dark-brown) as they were collected and immediately frozen in liquid nitrogen.

6. METABOLITE PROFILING

Metabolite profiles are often used with one or both of two main goals in a gene discovery study, both upstream and downstream of transcriptome sequencing: First, to select the appropriate tissue and conditions for sampling and second, to identify candidate genes based on correlations between gene expression patterns and patterns of metabolite distributions across tissues and conditions. Volatile fragrance compounds are generally profiled by gas chromatography–mass spectrometry (GC–MS) using liquid extractions from isolated tissue samples or headspace analysis from intact living plants or dissected plant samples. Solvent extraction of tissues allows quantitative analysis of small amounts of isolated tissues, providing a desired resolution of tissue- or even cell-type-specific metabolite profiles. Solvent extraction typically results in complex metabolite profiles, requiring additional data analysis efforts for identification of target compounds in the recorded GC–MS data. Cell-type-specific analysis can be achieved by isolation of surface cells such as those of glandular or nonglandular trichomes and epidermis (Lange, 2016), or by laser microdissection of inner tissues that accumulate terpenes (Abbott, Hall, Hamberger, & Bohlmann, 2010; Hamberger, Ohnishi, Hamberger, Séguin, & Bohlmann, 2011). Headspace analysis of the volatile emissions allows monitoring of fragrances emitted from whole organs (ie, flowers, leaves) or intact plants, making possible, for example, to measure temporal patterns of volatile emission profiles or changes in response to external stimuli. Metabolites represented in GC–MS profiles are identified by comparison of their mass spectra with reference mass spectral libraries, comparison of retention index with reference data, and whenever possible by comparison of mass spectra and retention indices with those of authentic standards.

In our case study, sandalwood metabolites were extracted from four biological replicates for each tissue type (SW, TZ, HW) in technical duplicates using ~50 mg of ground tissue with 1 mL of pentane spiked with isobutyl benzene (0.1 μg mL^{-1}) as internal standard. Extractions were performed in glass vials by end-over-end mixing for 24 h at room temperature (RT). Samples were centrifuged at 1000 × g for 15 min, and the pentane phase was transferred to a new GC vial. GC–MS analysis of sesquiterpenoids was performed on an Agilent 7890A/5975C GC–MS system operating in electron ionization selected ion monitoring (SIM) scan mode and equipped with a DB-Wax fused silica column (30 m, 250 μm ID, 0.25-μm film thickness). The conditions for a typical analysis of sandalwood sesquiterpenoids were: Injector was operated in pulsed splitless mode with the injector temperature kept at 250°C. Helium was used as the carrier gas with a flow rate of 0.8 mL min^{-1} and pulsed pressure set at 25 psi for 0.5 min. Scan range: m/z 40–500; SIM: m/z 93, 94, 105, 107, 119, 122, and 202 (dwell time 50 ms). The oven program was: 40°C for 3 min; ramp of 10°C min^{-1} to 130°C, ramp of 2°C min^{-1} to 200°C, ramp of 50°C min^{-1} to 250°C; hold at 250°C for 15 min. ChemStation software was used for data acquisition and processing. Metabolite identification was done by comparison of mass spectra with the NIST/EPA/NIH mass spectral library v2.0. Relative quantities of metabolites were calculated by manual integration of peak areas followed by normalization to the internal standard and dry weight of the tissue used in each extraction.

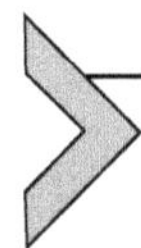

7. ISOLATION OF HIGH-QUALITY RNA FROM RECALCITRANT TISSUES

Challenges associated with isolation of high-quality RNA are highly dependent on the nature of the plant material. Comparison of different RNA isolation protocols, including available commercial kits, is highly recommended to achieve optimal results. Plant tissues containing high amounts of phenolic metabolites are particularly challenging for isolation of RNA with standard methods. A CTAB-based method (Kolosova et al., 2004) was specifically developed to isolate RNA from plant tissue with high content of polysaccharides, phenolics, and terpenoids. This protocol has been tested in recalcitrant tissues yielding high-quality RNA. PureLink® Plant RNA Reagent (Invitrogen, USA) is a fast, simple, and robust method to isolate RNA from most plant tissues yielding large amounts of high-quality RNA from small amounts of tissue (100 mg). Depending on the tissue of interest, this protocol can be modified to find an optimal ratio of reagent to tissue. The quality of isolated RNA should be assessed with quantitative methods such as the RIN number (Schroeder, Mueller, Stocker, et al., 2006) obtained with an Agilent 2100 Bioanalyzer or the RQI index obtained with an Experion™ Bio-Rad. Regardless of the method used, RNA samples should be analyzed at the temperature at which cDNA synthesis will be performed to represent the quality of the input RNA used for library construction.

The SW, TZ, and HW samples of sandalwood represent extremely difficult tissues for RNA isolation, since much of the material consists of highly lignified dead cells making up the wood core of the trunk of the tree, plus phenolic parenchyma cells. Frozen sandalwood tissue was ground to a fine powder for RNA extraction using PureLink® Plant RNA Reagent. Glycogen was added in the final precipitation step at a final concentration of 3.3 ng μL^{-1} to increase stability of the RNA pellet. Given that only very small amounts of RNA could be extracted from HW tissue, the addition of glycogen was also important as it made the final RNA pellet apparent by eye, which reduced the risk of accidentally discarding the RNA.

RNA quality and concentration were assessed on a 2100 Bioanalyzer after incubation of the isolated RNA at 70°C for 2 min as required in the Illumina cDNA library preparation protocol.

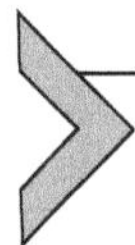

8. TRANSCRIPTOME SEQUENCING AND DE NOVO ASSEMBLY

The reduced costs for DNA sequencing and reduced amounts of RNA required for library construction have made transcriptome sequencing a common approach for gene discovery of secondary metabolism in non-model plant species (Facchini et al., 2012). The use of microfluidics in the preparation of cDNA libraries has reduced the amounts of nucleic acid required to as low as 25 ng of total RNA (NeoPrep™ System, Illumina), which makes it possible to obtain transcriptomes from small samples, such as those obtained by laser microdissection of individual cell types, or isolated trichomes and secretory glands. Removal of ribosomal RNA prior to library preparation improves the quality of the downstream transcriptome sequence assembly and increases sequence coverage of low-abundance transcripts. Presently, one of the most commonly used sequencing platforms used for transcriptome sequencing is the Illumina platform due to competitive per-base cost, sequence quality, and coverage.

To sequence the transcriptomes of different wood samples of sandalwood, we aimed for deep coverage that would allow us not only to identify expressed biosynthetic genes but also transcription factors and other low-abundance transcripts with potential regulatory functions in the developmental progression from SW to HW. cDNA libraries were constructed after removal of ribosomal RNA and ligation of strand-specific adapters. Paired-end (PE) reads, 100 bp long with an average insert size of 300 bases, were chosen for sequencing. Libraries were multiplexed, three libraries per sequencing lane, with the goal to achieve a sequencing depth of ~66 million reads per library. For reference, an RNA-seq study in *Arabidopsis* showed that 50 million reads represented a near-saturation coverage of expressed genes (Van Verk, Hickman, Pieterse, & Wees, 2013). Read quality of sandalwood RNA-seq data was inspected with FASTQC. Bases with a quality lower than 3 and adapter sequences were removed with Trimmomatic (Lohse et al., 2012). Overlapping PE-reads were merged with BBmerge (BBmap software; http://sourceforge.net/projects/bbmap/) improving the quality of the assembly and reducing time and computational resources required. Merged and unmerged reads from 12 libraries (four independent biological replicates each for SW, TZ, and HW) were combined and assembled with Trinity (Haas, Papanicolaou, Yassour, et al., 2013). Other de novo assemblers including Trans-ABySS (Robertson, Schein, Chiu, et al., 2010) and Velvet-Oases (Schulz, Zerbino, Vingron, & Birney, 2012) should also be considered. Detailed comparisons of assemblers can be found in several excellent review articles (Martin & Wang, 2011; Zhao et al., 2011). Prediction of protein-coding transcripts was done with Trans-Decoder (http://transdecoder.github.io/). To assess the quality of a de novo transcriptome assembly, consideration should be given to a number of parameters including: transcript median length, number of predicted peptides and peptide average length, and the representation of the Core Eukaryotic Genes Mapping Approach (CEGMA) with *Arabidopsis* sequences (Parra, Bradnam, & Korf, 2007). Specifically for the quality assessment of sandalwood transcriptome assemblies as a resource for discovery of terpenoid pathway genes, we also assessed the representation of the core terpenoid biosynthetic genes of the MEP and MEV pathways.

9. TRANSCRIPTOME MINING AND ANNOTATION

Using the catalogue of predicted protein-coding sequences in the transcriptome assemblies, transcripts are initially annotated based on their similarity to other annotated genes in databases such as those of the NCBI and EMBL. In addition, for discovery of genes of biosynthetic pathways, expert annotation of extracted gene sets has proven extremely valuable. This latter approach requires biochemical knowledge of metabolic pathways, enzymes, and gene families. For example, Zerbe et al. (2013) developed expert reference databases used for the annotation of TPS and P450 genes of terpenoid biosynthesis in transcriptomes of nonmodel plant species allowing for the focused and rapid discovery of members of these two large gene families in newly established transcriptomes and genomes. This approach can be extended to other gene families commonly involved in terpenoid and other secondary metabolism, such as glycosyl transferases, acyltransferases, reductases, or transporters. Additional whole-transcriptome annotations can be performed using tools such as Blast2go (Conesa et al., 2005) or the Trinotate software suite (http://trinotate.sourceforge.net/) of the Trinity assembler. When appropriate, annotated genes of gene families of interest can be further classified into subfamilies or classes by phylogenetic analysis or presence of subfamily-specific motifs (Zerbe et al., 2013). A priori prediction of specific gene functions based on sequence homology alone, ie, prediction of specific substrates or products of enzymes, is not currently possible for TPS and P450 gene family members. This is mostly due to the fact that as a feature of divergent evolution in terpenoid secondary metabolism, minor variations in sequence of duplicated genes can have major effects on enzyme functions (Chen et al., 2011; Nelson & Werck-Reichhart, 2011).

The combined sandalwood SW, TZ, and HW transcriptomes were annotated for candidate P450s and TPS using expert curated databases (Zerbe et al., 2013). Sandalwood P450s longer than 400 aa were tentatively assigned to CYP families and subfamilies by phylogenetic analysis and bidirectional best BLAST hit. All genes in the MEP and MEV pathway were annotated using *Arabidopsis* reference sequences.

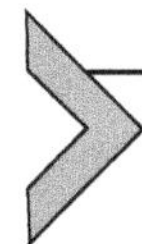

10. EXPRESSION ANALYSIS AND CANDIDATE GENE SELECTION

Differential expression (ie, transcript abundance) across tissues, developmental stages, or under different conditions can provide critical information to identify candidate genes associated with the biosynthesis of fragrance compounds. Methods to quantify gene expression based on kmer counting algorithms provide a rapid and resource-efficient approach for RNA-seq data (Patro, Mount, & Kingsford, 2014). This alignment-free method relies on creating an index of kmers in the assembled transcripts followed by counting of kmers in the raw reads. To detect statistically significant differences in gene expression, several R packages, including EdgeR and DESeq, have been developed and specifically adapted to analyze RNA-seq data (Anders & Huber, 2010; Robinson, McCarthy, & Smyth, 2010). Annotated transcripts and their expression can be used to create pathway expression maps that provide a broader picture of transcriptome dynamics across general and secondary metabolism. Similarly, all transcripts annotated in a gene family of interest can be used to generate family-specific expression maps. Unbiased analysis of gene expression data, such us hierarchical clustering, can be used to identify groups of genes and tissues that have similar overall expression patterns and include new gene families for further analysis and characterization. Candidate gene selection is accomplished by overlaying expression data with annotations and metabolite profiles. Top candidate genes should ideally be supported by relevant tissue-specific expression, metabolite profiles, and ontology.

Expression analysis of sandalwood transcripts identified a large number of them being differentially expressed between SW, TZ, and HW. These included transcripts annotated with core terpenoid pathways (eg, MEP and MEV pathways genes, FPP synthase), TPSs and P450s. Examination of HW-specific P450 transcripts identified members of the CYP736, CYP74, and CYP98 subfamilies as the most highly expressed P450 genes in sandalwood HW. Differential expression of MEP and MEV pathway genes across SW, TZ, and HW was captured in MEP and MEV pathway expression diagrams, revealing an informative developmental transcriptome profile with the MEP pathway (involved in mono- and diterpene biosynthesis) being preferentially expressed in SW, and the MEV pathway (involved in sesquiterpenoid biosynthesis) being preferentially expressed in HW (Celedon et al., 2016). In addition, expression analysis of TPS transcripts showed SaSSy as the overall most highly HW expressed TPS gene with a strong HW specificity of the expression. The combined and integrated analysis of metabolite profiles and gene expression identified the HW transcriptome (but not SW and TZ) as the target sequence source for candidate gene selection of P450s catalyzing the (*Z*)-hydroxylation of santalols.

11. FUNCTIONAL CHARACTERIZATION OF CANDIDATE GENES

Functional characterization of candidate fragrance genes of nonmodel systems typically requires cloning into suitable expression vectors and selection of a heterologous expression host system. Candidate genes can be cloned from cDNA or obtained by gene synthesis. Reduced costs of gene synthesis make this a time-efficient option with additional advantages of simultaneous codon and sequence optimization (ie, modification of targeting sequences) for expression in the host of choice. Different yeast strains have been engineered for the expression and characterization of genes of terpenoid fragrances and have been successfully employed in synthetic biology studies (Diaz-Chavez et al., 2013; Hansen, Møller, Kock, et al., 2009; Ignea, Pontini, Maffei, Makris, & Kampranis, 2014; Ignea et al., 2012). *Nicotiana benthamiana* and *Physcomitrella patens* are alternative plant expression platforms also successfully used for characterization of fragrance biosynthetic genes (Cankar et al., 2015; Zhan, Zhang, Chen, & Simonsen, 2014). Key advantages of using a plant host include utilization of plant tRNAs for translation, suitable ER for proper P450 and CPR insertion, and appropriate protein folding and trafficking machineries. Independent of the expression host, two complementary approaches exist to functionally characterize candidate genes: In vivo assays where whole pathways are reconstructed in a given host organism and in vitro assays where expressed proteins, such as TPSs or P450s, are first isolated and then tested with target substrates in enzyme assays under controlled conditions.

11.1 Yeast In Vivo Assays

Initial screening of sandalwood candidate P450s was done using in vivo assays in the yeast strain AM94 (Ignea et al., 2012) coexpressing sandalwood FPP Synthase (SaFPPS), SaSSy, and cytochrome P450 reductase (SaCPR2) (Celedon et al., 2016; Diaz-Chavez et al., 2013). The host yeast strain was engineered for improved production of sesquiterpenes and resulted in intracellular accumulation of SaSSy products, facilitating the functional screening of P450 candidates. Synthetic, codon-optimized P450 sequences were subcloned into the expression vector pYEDP60, which has been successfully used for yeast expression of many plant P450s (Duan & Schuler, 2006; Hamann & Møller, 2007). SaFPPS, SaSSy, and SaCPR2 were expressed in standard pESC vectors. All vectors were transformed into yeast following the procedure listed below:

Yeast transformation

1. Grow 5 mL overnight (ON) cultures from 3–5 yeast colonies in YPD media at 28°C.
2. Dilute cultures to $OD_{600} \sim 0.2$ in a total volume of 50 mL YPD and adenine. Grow yeast at 28°C to an OD_{600} 0.6–0.8.
3. Pellet the cells by centrifugation at 2000 × *g* for 2 min and discard supernatant.
4. Resuspend pellet in 25 mL sterile water.
5. Pellet the cells by centrifugation as before and discard supernatant.
6. Resuspend in 1 mL 100 m*M* LiAc and transfer to an Eppendorf tube.
7. Pellet the cells by centrifugation at 1000 × *g* for 1 min and discard supernatant.
8. Resuspend cells in 400 μL 100 m*M* LiAc.
9. Aliquot 50 μL of resuspended cells into Eppendorf tubes.
10. Pellet cells as before and remove supernatant.
11. In the following order add:
    240 μL PEG (MW 3350, 50% w/v)
    35 μL 1.0 *M* LiAc
    25 μL denatured Salmon sperm DNA (2.0 mg mL^{-1}, average size 5–7 kb)
    50 μL water with 0.5 μg plasmid DNA
12. Vortex cells vigorously until in suspension or for up to 1 min.

13. Shake tubes slowly at 28°C for 30 min.
14. Heat shock cells at 42°C, 15–25 min (optimal time may vary with strain).
15. Pellet cells by centrifugation at 1000 × *g* for 1 min and discard supernatant carefully.
16. Gently resuspend in 500 μL sterile water.
17. Apply 50 μL or more to selective medium (SD—selective aa) and incubate plates at 30°C for 3–4 days in order to see transformed colonies.

Depending on the number and the length of the genes to be transformed, options are to cotransform all of them at the same time or to perform sequential transformations. In the case of long genes and large plasmids, sequential transformation may be advised. Once all pathway genes are transformed into AM94 cells, expression cultures are prepared, and extraction of metabolites, as read out of the candidate P450 activities, is performed according to the following two consecutive protocols:

Expression cultures for in vivo assays

1. Start 5 mL ON culture from a single colony in selection media at 28°C, 220 RPM
2. Inoculate 50 mL culture (in 250-mL baffled flask) to a starting OD_{600} of 0.2
3. Incubate at 28°C, 220 RPM to an OD_{600} of 0.6–0.8
4. Centrifuge cultures in 50-mL Falcon tubes at 1000 × *g* for 5 min at RT
5. Resuspend cell pellet in 50 mL SG selective media to induce expression and incubate ON (12–16 h) at 28°C, 220 RPM
6. Centrifuge the cultures as before and transfer the supernatant to a new falcon tube for analysis of products potentially released to the media
7. Resuspend the pellet in 5 mL sterile water and transfer to ice
8. Transfer resuspended cells to preweighed a glass test tube and keep cells on ice
9. Centrifuge resuspended cells at 1000 × *g* for 5 min at 4°C
10. Remove and discard the supernatant and weigh pellets for normalization of metabolite amounts

Extraction of terpenoid metabolites

11. Start extraction of metabolites from weighted cell pellets by adding 250 μL of small acid-washed glass beads followed by 2 mL of organic solvent
12. Vortex for 1 min and decant the solvent to a new glass tube
13. Repeat the extraction a second time and pool the solvent fractions in the same tube

14. Add 250 μL sodium sulfate (anhydrous) to the tube containing the pooled solvents to remove residual water in the sample
15. Concentrate to ~1 mL under gentle flow of nitrogen gas and transfer to a GC vial
16. Store concentrated extracts at −80°C or proceed to GC–MS analysis

Extracts from in vivo assays are analyzed by GC–MS as described in Section 6.

11.2 Microsomes In Vitro Assays

Following the identification of P450 candidates which yield the target terpenoid metabolite (eg, santalols) in the yeast in vivo screening, in vitro assays with isolated microsomes containing the candidate P450 and CPR proteins allow for a more detailed biochemical characterization of P450 proteins including determination of kinetic constants, substrate specificity, optimum pH and temperature, activity with different CPR and cytB5 partners, etc. Microsomal membranes represent a mixture of native yeast protein and heterologously expressed proteins and, therefore, a negative control with WT microsomes should be included. Microsomes are isolated using the following procedures:

Microsome expression cultures

1. Start 5 mL cultures from a 3–5 colonies in SD selection media and incubate at 28°C, 220 RPM for 24 h.
2. Transfer cultures to 50 mL SD media (in 250-mL baffled flask) and grow ON at 28°C, 220 RPM.
3. Inoculate 200 mL YPD-E media with 50 mL culture and incubate for another 24 h at 28°C, 220 RPM.
4. Centrifuge cultures at 1000 × *g* for 10 min at RT.
5. Resuspend cell pellets in 200 mL YPG to induce expression and incubate ON (12–16 h) at 28°C, 220 RPM.

Microsome isolation

1. Centrifuge cells for 10 min at 4°C, 3000 × *g*, discard the supernatant, and keep cells on ice.
2. Resuspend the pellet with 5 mL of cold TEK buffer (50 m*M* Tris–HCl, pH 7.5, 1 m*M* EDTA, 100 m*M* KCl) and transfer to a Falcon tube.
3. Centrifuge cells again for another 10 min at 4°C, 3000 × *g* and discard the supernatant.
4. Resuspend pellet in 5 mL cold TES2 buffer (50 m*M* Tris–HCl, pH 7.5, 1 m*M* EDTA, 600 m*M* sorbitol, 5 m*M* DTT, 0.25 m*M* PMSF).
5. Add cold glass beads to the resuspended cells to ~5 mm from surface.

6. In a cold room, break cells by hand shaking vigorously in four cycles of 30 s shaking and 1 min rest on ice.
7. Transfer broken cells to a 50-mL Oakridge tube trying to avoid glass beads.
8. Wash beads twice with TES2 and pool with the supernatant from step 7. Aim for a final pooled volume of 25 mL.
9. Centrifuge for 15 min at 4°C, 10,000 × *g* to remove unbroken cells and glass beads.
10. Gently collect the supernatant through a nylon cloth and into an ultracentrifuge tube just until the surface of the pellet starts to loosen. Weigh tubes and balance with TES2 buffer.
11. Ultracentrifuge for 1 h at 4°C, 100,000 × *g* to pellet microsome membranes.
12. Discard the supernatant and wash the microsome pellet twice with TES buffer (50 m*M* Tris–HCl, pH 7.5, 1 m*M* EDTA, 600 m*M* sorbitol), and once in TEG (50 m*M* Tris–HCl, pH 7.5, 1 m*M* EDTA, 30% (v/v) glycerol) taking care to not loosen the pellet.
13. Resuspend pellet in 1 to 2 mL TEG buffer with a Potter-Elmer.
14. Microsomes can be directly used in assays or flash-frozen in liquid nitrogen and stored at −80°C.

The concentration of active P450 proteins in the microsome membranes can be estimated spectroscopically by the difference CO spectra method according to Guengerich, Martin, Sohl, and Cheng (2009).

Microsome assays

1. Microsome assays contain 50 m*M* potassium phosphate (pH 7.5), 1 m*M* NADPH, and 100 μ*M* sesquiterpene olefins (eg, SaSSy products) as substrates. Combine assay components in a GC vial and start reactions by adding 20 μL of purified microsomes (10 μg protein μL^{-1}) in a final volume of 400 μL.
2. Incubate reactions at 30°C for 20 min with gentle shacking at 30 RPM.
3. Terminate reactions by adding 0.5 mL of hexane spiked with isobutyl benzene as internal standard and vortexing immediately.
4. Transfer the hexane layer to a new GC vial for GC–MS analysis as described earlier.

12. PRODUCT IDENTIFICATION

Establishing the chemical identity of the products formed in an enzymatic reaction either in vitro or in vivo is critical for correct assignment of gene function and for our understanding of secondary metabolic pathways and networks. GC–MS profiles of all candidate gene assays should be compared with negative controls to rule out endogenous yeast activities that may lead to false-positive peaks or that may alter the profile generated by the target enzyme. These controls are critical both for in vivo assays and for microsome in vitro assays, both of which contain proteins from the host organism. The two most commonly used methods for product identification of fragrance compounds are described later.

12.1 MS–MS

In the case of previously characterized fragrance compounds, identification may rely on comparison of MS–MS spectra with reference mass spectral libraries, comparison of retention index with reference data, and whenever possible comparison of mass spectra and retention indices with those of authentic standards. We generally confirm metabolite identification by GC–MS analysis by performing GC injections at different temperatures to exclude artifacts of metabolites that are unstable at high temperature.

12.2 Nuclear Magnetic Resonance

Definitive identification of a reaction product is performed by nuclear magnetic resonance (NMR) analysis. This may be critical when the reaction product does not have a perfect match in reference libraries, or when the compound is a novel structure. In these cases, NMR analysis of purified reaction products can provide accurate determination of the chemical structure of the fragrance molecule. Depending on the yield and purity of the reactions, it may be necessary to perform large-scale cultures or enzyme assays to produce sufficient material for NMR analysis. However, this may not be a trivial task, since enzyme assays of TPS or terpenoid modifying P450s do not always yield sufficient amounts or purity of product for NMR analysis.

In our case study, the products of a newly discovered sandalwood P450, SaCYP736A167, were identified by GC–MS as (*Z*)-α-santalol, (*Z*)-β-santalol, (*Z*)-α-*exo*-bergamotol, and (*Z*)-*epi*-β-santalol (Celedon et al., 2016), successfully concluding the genomics enabled discovery of the final step in sandalwood fragrance biosynthesis. Yeast cells coexpressing SaFPPS, SaSSy, SaCPR, and SaCYP736A167 produced (*Z*)-α-santalol, (*Z*)-β-santalol, (*Z*)-α-*exo*-bergamotol, and (*Z*)-*epi*-β-santalol in proportions that resemble their relative abundance in authentic sandalwood oil.

ACKNOWLEDGMENTS

Research on sandalwood in the laboratory of J.B. was supported with funds from Evolva Inc., Allylix Inc., and the Natural Sciences and Engineering Research Council of Canada and was in collaboration with researchers from the University of Western Australia, including Elizabeth L. Barbour, Patrick M. Finnegan, Christopher Jones, and Julie Plummer. Their contributions are recognized with coauthorship on the original research papers on sandalwood that formed some of the foundation of this chapter. We acknowledge the expert technical assistance of Angela Chiang, Lina Madilao, and Macaire Yuen with the development or optimization of methods in J.B.'s laboratory described in this chapter, and Karen Reid for general project management support.

REFERENCES

Abbott, E., Hall, D., Hamberger, B., & Bohlmann, J. (2010). Laser microdissection of conifer stem tissues: Isolation and analysis of high quality RNA, terpene synthase enzyme activity and terpenoid metabolites from resin ducts and cambial zone tissue of white spruce (*Picea glauca*). *BMC Plant Biology*, *10*, 1–16.

Anders, S., & Huber, W. (2010). Differential expression analysis for sequence count data. *Genome Biology*, *11*, R106.

Andersen-Ranberg, J., Kongstad, K. T., Nielsen, M. T., et al. (2016). Expanding the landscape of diterpene structural diversity through stereochemically controlled combinatorial biosynthesis. *Angewandte Chemie International Edition in English*, *55*, 1–6.

Arendt, P., Pollier, J., Callewaert, N., & Goossens, A. (2016). Synthetic biology for production of natural and new-to-nature terpenoids in photosynthetic organisms. *The Plant Journal*, 1–22.

Bohlmann, J., & Keeling, C. I. (2008). Terpenoid biomaterials. *The Plant Journal*, *54*, 656–669.

Caniard, A., Zerbe, P., Legrand, S., Cohade, A., Valot, N., Magnard, J.-L., et al. (2012). Discovery and functional characterization of two diterpene synthases for sclareol biosynthesis in *Salvia sclarea* (L.) and their relevance for perfume manufacture. *BMC Plant Biology*, *12*, 119.

Cankar, K., Jongedijk, E., Klompmaker, M., Majdic, T., Mumm, R., Bouwmeester, H., et al. (2015). (+)-Valencene production in *Nicotiana benthamiana* is increased by down-regulation of competing pathways. *Biotechnology Journal*, *10*, 180–189.

Celedon, J. M., Chiang, A., Yuen, M. M. S., Diaz-Chavez, M. L., Madilao, L. L., Finnegan, P. M., et al. (2016). Heartwood specific transcriptome and metabolite signatures of tropical sandalwood (*Santalum album*) reveal the final step of (*Z*)-santalol fragrance biosynthesis. *The Plant Journal*. http://dx.doi.org/10.1111/tpj.13162. Published online.

Chen, F., Tholl, D., Bohlmann, J., & Pichersky, E. (2011). The family of terpene synthases in plants: A mid-size family of genes for specialized metabolism that is highly diversified throughout the kingdom. *The Plant Journal*, *66*, 212–229.

Conesa, A., Götz, S., García-Gómez, J. M., Terol, J., Talón, M., & Robles, M. (2005). Blast2GO: A universal tool for annotation, visualization and analysis in functional genomics research. *Bioinformatics*, *21*, 3674–3676.

Croteau, R., Ketchum, R. E. B., Long, R. M., Kaspera, R., & Wildung, M. R. (2006). Taxol biosynthesis and molecular genetics. *Phytochemistry Reviews*, *5*, 75–97.

Demissie, Z. A., Cella, M. A., Sarker, L. S., Thompson, T. J., Rheault, M. R., & Mahmoud, S. S. (2012). Cloning, functional characterization and genomic organization of 1,8-cineole synthases from *Lavandula*. *Plant Molecular Biology*, *79*, 393–411.

Demissie, Z. A., Erland, L. A. E., Rheault, M. R., & Mahmoud, S. S. (2013). The biosynthetic origin of irregular monoterpenes in *Lavandula*. *The Journal of Biological Chemistry*, *288*, 6333–6341.

Diaz-Chavez, M. L., Moniodis, J., Madilao, L. L., et al. (2013). Biosynthesis of sandalwood oil: *Santalum album* CYP76F cytochromes P450 produce santalols and bergamotol. *PloS One*, *8*, e75053.

Duan, H., & Schuler, M. A. (2006). Heterologous expression and strategies for encapsulation of membrane-localized plant P450s. *Phytochemistry Reviews*, *5*, 507–523.

Dudareva, N., Martin, D., Kish, C. M., Kolosova, N., Gorenstein, N., Fäldt, J., et al. (2003). (E)-β-ocimene and myrcene synthase genes of floral scent biosynthesis in snapdragon: Function and expression of three terpene synthase genes of a new terpene synthase subfamily. *Plant Cell*, *15*, 1227–1241.

Facchini, P. J., Bohlmann, J., Covello, P. S., et al. (2012). Synthetic biosystems for the production of high-value plant metabolites. *Trends in Biotechnology*, *30*, 127–131.

Gershenzon, J., & Dudareva, N. (2007). The function of terpene natural products in the natural world. *Nature Chemical Biology*, *3*, 408–414.

Guengerich, F. P., Martin, M. V., Sohl, C. D., & Cheng, Q. (2009). Measurement of cytochrome P450 and NADPH-cytochrome P450 reductase. *Nature Protocols*, *4*, 1245–1251.

Haas, B. J., Papanicolaou, A., Yassour, M., et al. (2013). *De novo* transcript sequence reconstruction from RNA-seq using the Trinity platform for reference generation and analysis. *Nature Protocols*, *8*, 1494–1512.

Hamann, T., & Møller, B. L. (2007). Improved cloning and expression of cytochrome P450s and cytochrome P450 reductase in yeast. *Protein Expression and Purification*, *56*, 121–127.

Hamberger, B., Ohnishi, T., Hamberger, B., Séguin, A., & Bohlmann, J. (2011). Evolution of diterpene metabolism: Sitka spruce CYP720B4 catalyses multiple oxidations in resin acid biosynthesis of conifer defense against insects. *Plant Physiology*, *157*, 1677–1695.

Hansen, E. H., Møller, B. L., Kock, G. R., et al. (2009). De novo biosynthesis of vanillin in fission yeast (*Schizosaccharomyces pombe*) and baker's yeast (*Saccharomyces cerevisiae*). *Applied and Environmental Microbiology*, *75*, 2765–2774.

Ignea, C., Pontini, M., Maffei, M. E., Makris, A. M., & Kampranis, S. C. (2014). Engineering monoterpene production in yeast using a synthetic dominant negative geranyl diphosphate synthase. *ACS Synthetic Biology*, *3*, 298–306.

Ignea, C., Trikka, F. A., Kourtzelis, I., Argiriou, A., Kanellis, A. K., Kampranis, S. C., et al. (2012). Positive genetic interactors of HMG2 identify a new set of genetic perturbations for improving sesquiterpene production in Saccharomyces cerevisiae. *Microbial Cell Factories*, *11*, 162.

Jones, C. G., Keeling, C. I., Ghisalberti, E. L., Barbour, E. L., Plummer, J. A., & Bohlmann, J. (2008). Isolation of cDNAs and functional characterisation of two multi-product terpene synthase enzymes from sandalwood, *Santalum album* L. *Archives of Biochemistry and Biophysics*, *477*, 121–130.

Jones, C. G., Moniodis, J., Zulak, K. G., Scaffidi, A., Plummer, J. A., Ghisalberti, E. L., et al. (2011). Sandalwood fragrance biosynthesis involves sesquiterpene synthases of both the terpene synthase (TPS)-a and TPS-b subfamilies, including santalene synthases. *The Journal of Biological Chemistry*, *286*, 17445–17454.

Knudsen, J. T., Eriksson, R., Gershenzon, J., & Ståhl, B. (2006). Diversity and distribution of floral scent. *The Botanical Review*, *72*, 1–120.

Kolosova, N., Miller, B., Ralph, S., Ellis, B. E., Douglas, C., Ritland, K., et al. (2004). Isolation of high-quality RNA from gymnosperm and angiosperm trees. *Biotechniques*, *36*, 821–824.

Lange, B. M. (2016). Online resources for gene discovery and biochemical research with aromatic and medicinal plants. *Phytochemistry Reviews*, 1–22.

Lohse, M., Bolger, A. M., Nagel, A., Fernie, A. R., Lunn, J. E., Stitt, M., et al. (2012). RobiNA: A user-friendly, integrated software solution for RNA-Seq-based transcriptomics. *Nucleic Acids Research*, *40*, W622–W627.

Magnard, J.-L., Roccia, A., Caissard, J.-C., et al. (2015). Biosynthesis of monoterpene scent compounds in roses. *Science*, *349*, 81–83.

Martin, J. A., & Wang, Z. (2011). Next-generation transcriptome assembly. *Nature Reviews. Genetics*, *12*, 671–682.

Muhlemann, J. K., Klempien, A., & Dudareva, N. (2014). Floral volatiles: From biosynthesis to function. *Plant, Cell & Environment*, *37*, 1936–1949.

Nelson, D., & Werck-Reichhart, D. (2011). A P450-centric view of plant evolution. *The Plant Journal*, *66*, 194–211.

Paddon, C. J., Westfall, P. J., Pitera, D. J., et al. (2013). High-level semi-synthetic production of the potent antimalarial artemisinin. *Nature*, *496*, 528–532.

Parra, G., Bradnam, K., & Korf, I. (2007). CEGMA: A pipeline to accurately annotate core genes in eukaryotic genomes. *Bioinformatics*, *23*, 1061–1067.

Patro, R., Mount, S. M., & Kingsford, C. (2014). Sailfish enables alignment-free isoform quantification from RNA-seq reads using lightweight algorithms. *Nature Biotechnology*, *32*, 462–464.

Qualley, A. V., Widhalm, J. R., Adebesin, F., Kish, C. M., & Dudareva, N. (2012). Completion of the core β-oxidative pathway of benzoic acid biosynthesis in plants. *Proceedings of the National Academy of Sciences of the United States of America*, *109*, 16383–16388.

Robertson, G., Schein, J., Chiu, R., et al. (2010). De novo assembly and analysis of RNA-seq data. *Nature Methods*, *7*, 909–912.

Robinson, M. D., McCarthy, D. J., & Smyth, G. K. (2010). edgeR: A Bioconductor package for differential expression analysis of digital gene expression data. *Bioinformatics*, *26*, 139–140.

Schroeder, A., Mueller, O., Stocker, S., et al. (2006). The RIN: An RNA integrity number for assigning integrity values to RNA measurements. *BMC Molecular Biology*, *7*, 3.

Schulz, M. H., Zerbino, D. R., Vingron, M., & Birney, E. (2012). Oases: Robust de novo RNA-seq assembly across the dynamic range of expression levels. *Bioinformatics*, *28*, 1086–1092.

Van Verk, M. C., Hickman, R., Pieterse, C. M. J., & Wees, S. C. M. Van. (2013). RNA-Seq: Revelation of the messengers. *Trends in Plant Science*, *18*, 175–179.

Widhalm, J. R., Jaini, R., Morgan, J. A., & Dudareva, N. (2015). Rethinking how volatiles are released from plant cells. *Trends in Plant Science*, *20*, 545–550.

Zerbe, P., & Bohlmann, J. (2015a). Enzymes for synthetic biology of ambroxide-related diterpenoid fragrance compounds. In J. Schrader & J. Bohlmann (Eds.), *Advances in biochemical engineering/biotechnology: Vol. 148* (pp. 427–447).

Zerbe, P., & Bohlmann, J. (2015b). Plant diterpene synthases: Exploring modularity and metabolic diversity for bioengineering. *Trends in Biotechnology*, *33*, 419–428.

Zerbe, P., Hamberger, B., Yuen, M. M. S., et al. (2013). Gene discovery of modular diterpene metabolism in nonmodel systems. *Plant Physiology*, *162*, 1073–1091.

Zhan, X., Zhang, Y.-H., Chen, D.-F., & Simonsen, H. T. (2014). Metabolic engineering of the moss *Physcomitrella patens* to produce the sesquiterpenoids patchoulol and α/β-santalene. *Frontiers in Plant Science*, *5*, 636.

Zhao, Q.-Y., Wang, Y., Kong, Y.-M., Luo, D., Li, X., & Hao, P. (2011). Optimizing de novo transcriptome assembly from short-read RNA-Seq data: A comparative study. *BMC Bioinformatics*, *12*, S2.

CHAPTER FOUR

A Workflow for Studying Specialized Metabolism in Nonmodel Eukaryotic Organisms

M.P. Torrens-Spence[*,1], **T.R. Fallon**[*,†,1], **J.K. Weng**[*,†,2]

[*]Whitehead Institute for Biomedical Research, Cambridge, MA, United States
[†]Massachusetts Institute of Technology, Cambridge, MA, United States
[2]Corresponding author: e-mail address: wengj@wi.mit.edu

Contents

1. Introduction 70
2. "Omics"-Based Novel Specialized Metabolic Pathway Discovery 72
   2.1 De Novo Transcriptome Sequencing and Assembly 73
   2.2 Making Use of the De Novo Transcriptome Data 74
   2.3 De Novo Genome Sequencing 76
   2.4 Metabolomics Approaches for Defining Intermediates of an Unknown Metabolic Pathway 78
3. Structure–Function Analysis of Specialized Metabolic Enzymes 80
   3.1 Evolutionary Approaches for Candidate Enzyme Identification 80
   3.2 Recombinant Expression and Purification of Candidate Enzymes 83
   3.3 Comparative Biochemical Analysis of Enzyme Functions 84
   3.4 Structural Basis for Specialized Metabolic Enzyme Evolution 86
4. Reconstitution of Specialized Metabolic Pathways in Heterologous Systems 87
   4.1 *Escherichia coli* 88
   4.2 *Saccharomyces cerevisiae* 89
   4.3 *Nicotiana benthamiana* 90
5. Summary 90
Acknowledgments 91
References 91

Abstract

Eukaryotes contain a diverse tapestry of specialized metabolites, many of which are of significant pharmaceutical and industrial importance to humans. Nevertheless, exploration of specialized metabolic pathways underlying specific chemical traits in nonmodel eukaryotic organisms has been technically challenging and historically lagged behind that of the bacterial systems. Recent advances in genomics, metabolomics, phylogenomics, and synthetic biology now enable a new workflow for interrogating

[1] These authors contributed equally to this manuscript.

Methods in Enzymology, Volume 576
ISSN 0076-6879
http://dx.doi.org/10.1016/bs.mie.2016.03.015

unknown specialized metabolic systems in nonmodel eukaryotic hosts with greater efficiency and mechanistic depth. This chapter delineates such workflow by providing a collection of state-of-the-art approaches and tools, ranging from multiomics-guided candidate gene identification to in vitro and in vivo functional and structural characterization of specialized metabolic enzymes. As already demonstrated by several recent studies, this new workflow opens up a gateway into the largely untapped world of natural product biochemistry in eukaryotes.

1. INTRODUCTION

As a universal means of adaptation, all life forms on earth evolved biosynthetic capacity to produce a broad variety of lineage-specific chemical compounds to mitigate stresses from biotic and abiotic environments (Luckner, 1984). These so-called specialized metabolites differ from those better-studied primary metabolites, because they are not absolutely vital for survival, but rather contribute to the fitness of the host organisms under diverse and ever-changing environmental niches (Pichersky & Lewinsohn, 2011; Weng, Philippe, & Noel, 2012). Within terrestrial plants, for example, the diverse array of specialized metabolites are produced to cope with unique challenges of life on land, such as UV radiation and drought, as well as to mediate a plethora of interspecies interactions, ranging from seduction of pollinators and seed dispersers to defense against pathogens and herbivores (Atkinson & Urwin, 2012; Jenkins, 2009; Pichersky & Gershenzon, 2002; Rasmann et al., 2005; Rudd & Hopwood, 1980).

The physicochemical features of specialized metabolites enabling interspecies communication also engender unique pharmacological properties that directly impact human health (Beutler, 2009). As such, organisms harboring therapeutic natural products have played important roles as human medications for thousands of years. In many cultures, through millennia of trial and error, plants with beneficiary medicinal properties were identified, systemized, and extracted to serve as traditional herbal medicines. This is exemplified in a 2600-year-old Sanskrit text on medicine (*Sushruta Samhita*) listing over 700 medicinal plants (Datta, Mitra, & Patwardhan, 2011; Dev, 1999). Many active compounds underlying the efficacy of medicinal plants have been identified through modern scientific approaches and play critical roles in modern medicine, including the important medications paclitaxel, morphine, and artemisinin (Rudd & Hopwood, 1980; Schiff & Horwitz, 1980; Ye et al., 1998). According to recent statistics, about 75% of antibacterial products and about 50% of anticancer compounds currently approved by the United States Food and Drug Administration (FDA) are either natural products or derived therefrom (Newman & Cragg, 2012). Natural products are disproportionally overrepresented in human drugs, considering that the number of currently known natural products (~170,000) (Seyedsayamdost & Clardy, 2014) account for less than 1% of the total chemicals on the Chemical Abstracts Service (CAS) Registry (~104,000,000). Despite being greatly outnumbered by synthetic chemicals, natural products represent a diverse array of unique structural scaffolds not represented in synthetic chemical libraries (Hert, Irwin, Laggner, Keiser, & Shoichet, 2009). These so-called privileged structures contained in natural product chemical libraries often yield higher discovery rate in drug screens compared to synthetic chemical libraries (Beutler, 2009; Newman & Cragg, 2009). Despite the significant therapeutic potential of natural products, their structural complexity makes their total synthesis through synthetic chemistry prohibitively expensive or impossible (Chemler & Koffas, 2008). Consequently, the majority of natural product-derived therapeutics are still harvested directly from their plant sources or more recently through semisynthesis. The inability to cultivate many natural product-containing species has also resulted in the overharvesting of these species from the wild, exemplified by the pacific yew (*Taxus brevifolia*), firmoss (*Huperzia serra*), and golden root (*Rhodiola rosea*) (Busing, Halpern, & Spies, 1995; Lan et al., 2013; Ma, Tan, Zhu, & Gang, 2006). Such myopic practices damage the biodiversity from which we attempt to benefit and necessitate critical advances in understanding natural product biosynthesis as well as metabolic engineering endeavors to produce high-value bioactive natural products to meet current medical and industrial needs.

Natural product research underwent an initial boom in the 1980s, when several new technologies led to a series of important discoveries in microbe-derived therapeutics (Datta et al., 2011; Newman & Cragg, 2007; Schiff & Horwitz, 1980; Weibel, Hadvary, Hochuli, Kupfer, & Lengsfeld, 1987). The identification of individual genes involved in bacterial natural product biosynthesis often led to full pathway elucidation thanks to the prevalence of operon structures in bacterial genomes (Chater, 1992; Rudd & Hopwood, 1980; Rutledge & Challis, 2015). Unfortunately, due to smaller effective population sizes and therefore less powerful selective pressures, biosynthetic genes of a given metabolic pathway in eukaryotic organisms often scatter throughout the genome, making metabolic pathway gene discovery process much more challenging than that of the bacterial systems (O'connor, 2009). Early success in defining eukaryotic specialized metabolic pathways benefited tremendously from studying genetically tractable model organisms through a combination of genetic and biochemical methods (Caelles, Ferrer, Balcells, Hegardt, & Boronat, 1989; Dittrich & Kutchan, 1991; Mandel, Feldmann, Herrera-Estrella, Rocha-Sosa, & Leon, 1996). However, interesting and valuable specialized metabolic traits of eukaryotic origin are widely spread in nonmodel species (O'connor, 2009). Major technological bottlenecks, including lack of genetic tools, tissue inaccessibility, and long life cycles, hamper efficient discovery of new natural products and their biosynthetic pathways from nonmodel eukaryotic hosts. On the other hand, recent advances in combinatorial chemistry have propelled the adoption of high-throughput drug screens using synthetic chemical libraries in the pharmaceutical industry instead of further expanding the current natural product libraries (Hajduk & Greer, 2007). Despite the notable success of such operations, which have yielded many synthetic drugs in the past two decades, expectations have been tempered due to a decline in high-throughput screen productivity (Perola, 2010).

With several recent technological advances in biology, elucidation of natural product biosynthesis in nonmodel eukaryotic organisms is becoming increasingly approachable. Integrated transcriptomics, metabolomics, and phylogenomics have greatly expedited the process of candidate gene identification. Subsequent in vitro and in vivo biochemical characterization of these new metabolic pathways, empowered by burgeoning structural and synthetic biology tools, can rapidly contribute to the dearth of biosynthetic understanding of eukaryotic specialized metabolic systems, paving the way toward metabolic engineering of high-value natural products of eukaryotic origin. This chapter provides a practical and generally applicable workflow summarized from recent literature that aids researchers to interrogate unknown specialized metabolic enzymes and pathways in nonmodel eukaryotic hosts possessing interesting chemical traits.

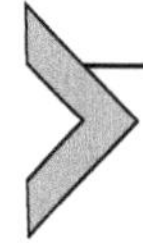

2. "OMICS"-BASED NOVEL SPECIALIZED METABOLIC PATHWAY DISCOVERY

The domain Eukaryota harbors a dazzling diversity of multicellular organisms. Although eukaryotic genomes do not contain metabolic operons as in bacterial genomes, specialized metabolites and the genes involved in their biosynthesis are often present in certain tissue types, and are regulated developmentally or elicited by various environmental factors. This feature can be exploited to identify and prioritize candidate genes responsible of a metabolic trait of interest by correlating the expression patterns of genes encoding possible enzymes with the spatial and temporal accumulation of the natural products of interest. This approach is greatly propelled by the rapid advance in several "omics"-type technologies. In contrast to the classical methods of enzymatic gene discovery, which were driven by initial biochemical studies in cell-free lysates followed by laborious purification, direct protein sequencing, and genomic cloning, modern DNA sequencing methods allow for almost all candidate enzyme-encoding genes in a given nonmodel organism to be discovered in a single experiment. Combined with powerful metabolomics approaches enabled by advances in mass spectrometry (MS) and nuclear magnetic resonance (NMR), the genetic basis of specialized metabolic traits can now be studied in an unprecedented diversity of eukaryotes.

2.1 De Novo Transcriptome Sequencing and Assembly

As eukaryotic genomes typically consist of considerable repetitive and nonfunctional sequence as well as splicing complexity that makes gene prediction in the absence of transcriptome data a nontrivial task, de novo transcriptome sequencing followed by transcriptome assembly and expression analysis represents a highly cost-effective method for generating hypothetical biosynthetic models and mining candidate enzyme-encoding genes. Currently, the most commonly used de novo transcriptome sequencing platform is Illumina (San Diego, CA) RNA-Seq. A typical RNA-Seq approach for de novo transcriptome sequencing and assembly requires tens of millions of paired end reads with 100×100 bp or greater length per sequencing library, which are prepared from either poly-T selected or rRNA depleted total RNA samples, ideally by a strand-specific sequencing library preparation method. As the typical unit of Illumina sequencing, the HiSeq flowcell "lane" produces hundreds of millions of reads. Multiple sequencing libraries are often pooled into a single sequencing lane by the addition of "barcode" sequences to each sample during the library prep so they can be distinguished and sorted during data analysis.

Assembly of the de novo transcriptome from short reads can then be performed by strand-specific de-Bruijn graph-based assemblers such as Trinity (Grabherr et al., 2011). Trinity does require access to a >100 GB memory server and significant computational time, though our assemblies typically complete within 1–3 days when running on a single high-performance server. The most cost-effective sequencing depths for eukaryotic de novo transcriptomes appear to be around 50 million reads, with the most critical aspect for transcriptome completeness being sequencing from multiple tissues (Francis et al., 2013). The assembly of a de novo transcriptome typically produces hundreds of thousands of transcript contigs, of which only a subset represents biologically interesting transcripts. An analysis of conserved gene content in the de novo transcriptome by the BUSCO tool provides a biologically interpretable measure of transcriptome quality (Simao, Waterhouse, Ioannidis, Kriventseva, & Zdobnov, 2015). In our experience, 80% or more single-copy conserved genes as reported by BUSCO represent a high-quality transcriptome, whereas 50% or lower is quite workable but could be improved with more sequencing.

As an alternative approach to Illumina-based de novo transcriptome sequencing and assembly, the IsoSeq RNA sequencing method, developed upon the Pacific Bioscience (PacBio, Menlo Park, CA) single-molecule real-time (SMRT) sequencing platform, produces long reads representing single cDNA molecules (Rhoads & Au, 2015). In cases where transcriptomic complexity, such as closely related paralogs or splicing complexity, might confuse short-read assemblers, long-read sequencing of cDNAs can represent an "assembly-free" method for generating de novo transcriptome. In species with large genome sizes and high levels of ploidy, such single-molecule transcriptome sequencing methods will likely become the preferred method for producing "canonical transcriptomes" in the near future.

2.2 Making Use of the De Novo Transcriptome Data

If multiple tissue types or the same tissue type treated with various elicitation conditions was included in one RNA-Seq experiment, expression analysis can be performed to identify genes that are differentially expressed between samples, which often relate to specific biological functions. RNA-Seq-based expression analysis is achieved by mapping the original reads to the assembled transcriptome and counting mapped reads through existing RNA-Seq pipelines. In our laboratory, we use a pipeline consisting of Trinity for de novo transcriptome assembly (Grabherr et al., 2011), bowtie/RSEM for transcript expression analysis (Li & Dewey, 2011), and EBSeq for statistical analysis of differential expression (Leng et al., 2013). By sampling from various tissues of a given nonmodel organism, it is possible to simultaneously assemble a high-quality transcriptome from the pooled reads from all tissues and perform expression analysis on individual tissues. Furthermore, by performing metabolomic profiling on the same tissue samples used for transcriptomic sequencing, the quantitative measurement of the target metabolites can be correlated with a list of enzyme-encoding genes to help identify and prioritize candidate enzymes for downstream biochemical characterization (Prosser, Larrouy-Maumus, & de Carvalho, 2014). Such approach has gained increasing popularity in the field, elegantly exemplified by two recent studies reporting the identification of iridoid synthase in monoterpene indole alkaloid biosynthesis from *Catharanthus roseus* (Geu-Flores et al., 2012) and the olivetolic acid cyclase in cannabinoid biosynthesis from *Cannabis sativa* (Gagne et al., 2012), respectively.

The TransDecoder in silico translation software is a useful tool to produce translated open reading frame (ORF) sequences from the assembled transcriptome (Haas et al., 2013). TransDecoder produces full-length protein sequences of likely complete ORFs and also annotates those ORF sequences that are incomplete. The TransDecoder-predicted proteome is a useful resource for querying protein sequences through BLAST searches as well as activity-based enzyme identification from the host organism by untargeted proteomics. In addition, we have found that the free and open-source BLAST server SequenceServer provides an user-friendly interface for routine interacting with both the assembled transcriptomes and the TransDecoder-predicted proteomes (Priyam et al., 2015). Simple annotation can be performed by reciprocal BLASTP of the translated transcriptome and reference protein sequences from a well-annotated model species such as *Arabidopsis thaliana* for plants, while the approaches for more sophisticated annotation can be achieved using the Trinotate annotation pipeline (Haas et al., 2013).

It is worth mentioning some of the caveats of interpreting de novo transcriptome data. With nonstrand-specific library preparation, we have observed common chimeric transcripts of two full-length transcripts containing high-quality ORFs. Follow-up genomic sequencing has revealed that the genomic loci giving rise to these chimeric transcripts are often tandemly arrayed (Peng et al., 2015). While such assembly artifacts neither change the sequence content nor affect PCR cloning attempts from cDNA, a fused transcript will not have an accurate expression value after read mapping. Additionally, while the prediction of an in-frame stop codon of an ORF by TransDecoder is straightforward, prediction of the correct start codon to begin the ORF is less clear. Therefore, an ORF reported as "5prime_partial" by TransDecoder should not be taken as a fragmented gene at face value. Similarly, some ORFs reported as "complete" may be 5′ truncations. Multiple sequence alignment (MSA) with related proteins can be helpful in determining the true coding sequence in preparation for cloning candidate genes. Given the above caveats, it is important to verify the coding sequence of candidate genes with significant interest by PCR amplification using mixed cDNA template prepared from the tissue of interest before proceeding into downstream functional analyses.

2.3 De Novo Genome Sequencing

Although transcriptome sequencing produces an immediately workable dataset containing coding sequences for candidate enzymes and their respective expression levels across samples, additional genetic information that aids specialized metabolic pathway discovery could be further gathered from de novo genomic sequencing. For example, it is difficult to call the absence of a given enzyme-encoding gene from transcriptome datasets, as the gene may be expressed at a low level or improperly assembled. Transcriptomes also provide no information on the genomic organization of genes belonging to a specialized metabolic pathway. Such knowledge is often useful for tracing the evolutionary history of a newly emerged metabolic trait. For example, an evolutionarily new metabolic enzyme-encoding gene could be derived from a tandem duplication of a nearby ancestral gene (Fraser et al., 2007), or represents an intronless retrotransposition (Matsuno et al., 2009). In some cases, specialized metabolic enzymes have been reported to form operon-like gene clusters (Osbourn, Papadopoulou, Qi, Field, & Wegel, 2012), indicating that discovery efforts for elucidating novel specialized metabolic pathways could be enhanced by mining additional candidate genes in the vicinity of genomic loci containing known genes from the pathway. That being said, prediction of gene structure from genomic sequence data is an involved process, typically using transcriptomic data combined with de novo gene prediction algorithms such as those embodied in the MAKER pipeline (Campbell, Holt, Moore, & Yandell, 2014).

In contrast to transcriptome sequencing, where deep short-read sequencing can produce a relatively complete assembled transcriptomic dataset, the repetitive sequence content of eukaryotic genomes ensures that an entirely short-read approach will produce a highly fragmented genomic dataset composed of a large number of genomic contigs in the order of tens of kilobases. Although this is useful for many purposes, such as analyses of intron structures, gene copy number, and local gene organization

(Kellner et al., 2015), the quality of these genomic assemblies is far from that of the polished genome sequences of model organisms. The diversity of genome sequencing and assembly approaches is quite broad. Not all of these approaches will be covered here. In our own experience, the continued improvement of PacBio SMRT sequencing platform presents a promising solution to the current limitations of the short-read-based de novo genome sequencing. The long reads generated from the PacBio platform using the current P6-C4 chemistry have a median read length of 10 kb and a maximum read length up to approximately 80 kb (Rhoads & Au, 2015), which allow for highly contiguous genome assemblies. In contrast to Illumina library preparation, where PCR biases can lead to greatly varying or no coverage of certain genomic regions, PacBio library prep is amplification-free and sequencing coverage follows a Poisson distribution. Despite a high per-base error rate, sequencing errors occur stochastically in PacBio reads, which can be compensated by repeated coverage of a given nucleotide to reach very high consensus accuracy (Ross et al., 2013). Combined with long-range scaffolding approaches, such as BioNano optical mapping (Levy-Sakin & Ebenstein, 2013), long-read sequencing technology is producing "platinum draft" genomes with high quality approaching that of the first publications of model organisms (VanBuren et al., 2015). PacBio sequencing is not without its downsides. The per-base error rate of a given read is currently around 15%, generally consisting of indels, while the per-base cost is roughly 1000 times greater than that of Illumina sequencing. Input DNA requirements for PacBio genomic sequencing are also quite stringent, requiring tens of micrograms of high-purity genomic DNA greater than 20 kb in size. In order to produce the 20 kb+ libraries most useful for genome assembly, high-molecular-weight genomic DNA has to be carefully prepared to prevent shearing from otherwise harmless laboratory manipulations such as pipetting or passing through spin columns. In our experience, careful and nonexcessive grinding of tissue under liquid N_2 followed by Qiagen's (Hilden, Germany) Genomic-tip anion-exchange DNA purification kit with a final alcohol precipitation step using glycogen as a carrier has produced high-quality genomic DNA suitable for PacBio SMRT sequencing.

Genomic sequencing represents a comparatively difficult assembly problem when compared to transcriptomic sequencing. It is worth noting that many genome assembly programs are developed in the context of 3Gbp or smaller inbred diploid or haploid genomes, whereas large genomes, outbred populations, and polyploid genomes represent the most difficult genomics projects and remain an ongoing research topic. As genome size can vary drastically even within a single taxonomic family and heavily influences sequencing approaches and cost, the first step of any genome project should ideally be measurement of the genome size, commonly done by fluorescence measurement of nuclei DNA content by calibrated propidium iodide flow cytometry (Hare & Johnston, 2011). In situations where there is sufficient PacBio coverage for a purely PacBio assembly approach (recommended 60–100 × coverage), overlap-layout-consensus-based genome assemblers, such as PacBio's experimental Falcon assembler and the Celera Assembler PBcR pipeline, represent the method to produce the best assembly quality (Koren & Phillippy, 2015).

2.4 Metabolomics Approaches for Defining Intermediates of an Unknown Metabolic Pathway

As described earlier, modern DNA sequencing technologies can now produce large lists of candidate genes potentially involved in a given specialized metabolic pathway. However, many specialized metabolic pathways are often incompletely described, and as such the potential substrates or products of promising candidate enzymes may be unknown. Recent advances in metabolomics, namely, high-resolution accurate-mass (HRAM) MS coupled with liquid chromatography (LC) or gas chromatography (GC), provide powerful tools for resolving potential biosynthetic intermediates of a novel specialized metabolic pathway. Linking these putative intermediates to respective enzymes remains a task of directed enzymology. Current HRAM instruments, such as the Thermo Scientific (Waltham, MA) Q-Exactive Orbitrap mass spectrometer, have high mass-to-charge (m/z) accuracy, typically below 5 ppm or around 0.001 Da for a singly charged ion. High mass accuracy allows higher confidence in detecting compounds of known chemical formula through the usage of smaller m/z ranges for extracted ion chromatograms, as well as in the predicting chemical formula for unknown features (Pluskal, Uehara, & Yanagida, 2012). Furthermore, high mass-to-charge resolution allows isotopic variants carrying the same nominal mass (eg, ^{15}N, ^{13}C, ^{2}H), to be distinguished, enabling fine analysis of isotopic spectra as well as unambiguous signals in stable isotope tracing experiments. Coupled with suitable chromatographic separation methods, the high capacity of the state-of-the-art HRAM instruments also allow thousands of mass features to be quantitatively measured in one run.

Our laboratory has routinely used two metabolomics approaches to help identify unknown biosynthetic intermediates of an unresolved pathway. In the first approach, untargeted metabolomics is used to profile tissue sample groups representing biosynthetically active and inactive states. For example, defense-related specialized metabolic pathway in plants is commonly elicited by exposure to pathogen or treatment with plant defense hormones (Klein & Sattely, 2015). For untargeted metabolomics data processing, our laboratory uses and contributes to the development of the open-source cross-platform metabolomics analysis tool MZmine2 for raw data processing, feature calling, quantification, and multisample alignment (Pluskal, Castillo, Villar-Briones, & Oresic, 2010). The resulted feature list is then exported to the online MetaboAnalyst tool for statistical analysis (Xia, Sinelnikov, Han, & Wishart, 2015). The subset of features that show significant differences between the biosynthetically active and inactive conditions may contain candidate biosynthetic intermediates of interest. In the second approach, we combine stable isotope tracing with untargeted metabolomics by feeding the organism of interest with stable isotope-labeled biosynthetic precursors and then searching for those features that gain appropriate labeling patterns. Ideally, tracing with multiple biosynthetic precursors as well as precursors that carry different labeled atoms can be performed. The features that become labeled in all of these conditions represent candidate biosynthetic intermediates. If possible, ^{2}H- or ^{15}N-labeled compounds are preferred, as these labels provide unambiguous tracing signals on an HRAM mass spectrometer. For plants, labeled tracers can be added to the growth media, which often show uptake through the roots (Weng, Li, Mo, & Chapple, 2012). However, in other cases, injection of biosynthetic precursors into specific organs of the target organism is necessary (Kutrzeba et al., 2007; Sheriha & Rapoport, 1976). Once optimal labeling condition is found, in which the end target metabolite gets labeled, computational analysis can be performed to search for other features that have also become labeled (eg, light ions accompanied by a corresponding heavy ion of expected mass). This heavy-light ion linking can be performed in either the OpenMS suite using their FeatureLinkerLabeled module (Sturm et al., 2008) or the MZmine2 suite using the adduct search module (Pluskal et al., 2010). Stable isotope tracing coupled with untargeted metabolomics can also be performed concurrently with the so-called TopN data-dependent fragmentation (ddMS^{2}) methods, which produces untargeted fragmentation data on the most abundant ions in the sample. While it is still an unsolved computational problem to produce candidate molecular structures from collision-induced dissociation/higher-energy collisional dissociation fragmentation spectra and vice versa, molecules of related structure often produce characteristic fragmentation ions. This information can be used to narrow down the list of putative intermediates to those with chemical structures related to previously characterized biosynthetic intermediates.

Putative structural assignment of candidate mass features representing metabolic pathway intermediates is performed manually in the context of a model biosynthetic pathway, wherein mass differences between the features can be potentially linked to biotransformations such as cyclization, hydroxylation, methylation, acylation, dehydrogenation, hydrogenation, glycosylation, and sulfonation. Confirmation of the putative structures remains a labor-intensive process, which typically involves natural product isolation from the host organism followed by NMR-based structural elucidation. Chemical total synthesis or in vivo synthesis of the compound through pathway reconstitution using candidate enzymes is two alternative approaches. Since biosynthetic intermediates of specialized metabolism often represent chemicals not available commercially, compounds purified or synthesized for the purpose of structural elucidation can later be used as substrates for examining candidate enzymatic functions.

3. STRUCTURE–FUNCTION ANALYSIS OF SPECIALIZED METABOLIC ENZYMES

Diverse specialized metabolic systems have evolved continuously along the radiating lineages of eukaryotes. An enhanced understanding of the evolutionary mechanisms underlying the eukaryotic metabolic evolution will contribute to our knowledge of how Darwinian evolution shapes complex traits in life (Weng, 2014). Although detailed biochemical and structural interrogation of enzymes remain one of the most time-consuming and technically challenging steps in the current workflow for elucidating new eukaryotic specialized metabolic pathways, these experiments have greatly benefited from the rapid expansion of genomic resources available for nonmodel eukaryotes as well as the steady advances of recombinant protein technologies and protein X-ray crystallography. Comprehensive analyses, such as comparative biochemical and structural characterizations of homologous enzymes within one species or across related species, are revealing the evolutionary trajectories of diverging specialized metabolic systems at unprecedented resolution and mechanistic depth.

3.1 Evolutionary Approaches for Candidate Enzyme Identification

Consistent with the fundamental principles of Darwinian evolution, the expansion of specialized metabolism did not occur through the invention of new protein folds, but rather through gradual structural modifications of the existing protein folds (Weng, Philippe, et al., 2012). Evolutionary exploitation of the sequence space within a fold alters an ancestral enzyme's stability, substrate specificity, promiscuity, catalytic speed, and catalytic mechanisms. Mutations that resulted in novel metabolic traits conferring selective advantage subsequently became fixed in the population. Additional mutations could then follow to further improve the nascent enzymatic activity. This process could occur independently in orthologous enzymes of different species, resulting in divergent evolution of enzymatic functions in orthologous enzymes. In specialized metabolic evolution, however, it is much more common to find neofunctionalized enzymes in paralogous enzymes derived from ancestral gene duplication events. The ancestral enzyme function is retained by at least one copy of the paralogs, allowing the other copies to freely explore new functional space. Understanding the common evolutionary mechanisms of specialized metabolic enzymes therefore enhances gene discovery efforts in studying new specialized metabolic pathways in nonmodel eukaryotes.

Under a biosynthetic model for a target metabolite, one can predict the involvement of specific enzymes in the proposed pathway. For example, the hydroxyl groups present in target natural products are likely installed by cytochromes P450 (P450s) or 2-oxoglutarate-dependent oxygenases, whereas the biosynthesis of ester and amide compounds probably involves consecutive catalysts consisting of an acyl-CoA ligase and an acyltransferase. With transcriptome dataset of the target species in hand, one can easily retrieve candidate-enzyme-encoding genes of a particular enzyme family by BLAST search using a query enzyme known to catalyze similar reactions. For comprehensive sequence analysis, it is important to also collect homologous sequences from other species that span a wide evolutionary distance. Besides sequences available from GenBank, it is worth noting that extensive transcriptomic resources for nonmodel eukaryotes have been generated and made available to the public by several consortiums as well as individual research groups in recent years (Matasci et al., 2014; Misof et al., 2014). These resources should be mined as well.

After sequence collection, large-scale MSA can be built using structure-based alignment algorithms, such as the PROMALS3D server (Pei, Kim, & Grishin, 2008). Structure-guided sequence analysis of catalytic and substrate-binding residues within the MSA can readily highlight unusual variations as signs of neofunctionalization. For instance, a single amino acid mutation in the enzyme active site was recently demonstrated to interconvert the catalytic mechanisms of phylogenetically related plant aromatic amino acid decarboxylases (AAADs) and aromatic acetaldehyde synthases (AASs) (Torrens-Spence et al., 2013). Substitutions at this residue have occurred multiple times in AAAD paralogs in various flowering plant lineages, resulting in the parallel evolution of AAS chemistry responsible for the production of volatile aromatic aldehydes (Torrens-Spence et al., 2013). In another example, the catalytic histidine residue, highly conserved in plant BAHD acetyltransferase family, is substituted to serine in Enhanced Pseudomonas Susceptibility 1 (EPS1), a BAHD acetyltransferase paralog unique to *Arabidopsis* with roles in salicylic acid metabolism (Zheng, Qualley, Fan, Dudareva, & Chen, 2009). Although the precise biochemical function of EPS1 is yet to be identified, this unusual substitution suggests its catalytic mechanism almost for certain has deviated from being a canonical acetyltransferase.

To gain insight into the evolutionary history of the candidate enzyme-encoding genes, it is necessary to perform comprehensive phylogenetic analyses based on an MSA. We routinely use the neighbor-joining and maximum-likelihood algorithms implemented in the Molecular Evolutionary Genetics Analysis (MEGA) package for phylogenetic inference (Tamura, Stecher, Peterson, Filipski, & Kumar, 2013). Analysis of various features in the resulted phylogenic tree, including lineage-specific gene duplication events, clustering pattern, and branch length, can be highly informative to illuminate plausible candidate enzymes. For example, relatively longer branch length in a tree suggests rapid accumulation of genetic differences between homologous sequences often associated with functional divergence. This phylogenetic pattern has been an important clue that has helped identify many plant specialized metabolic enzymes underlying taxonomically distributed chemical traits, including AAADs, terpene synthases, P450s, acyltransferases, and type III polyketide synthases (PKSs) (Berger, Meinhard, & Petersen, 2006; Cook et al., 2010; Facchini, Huber-Allanach, & Tari, 2000; Hamberger, Ohnishi, Hamberger, Seguin, & Bohlmann, 2011; Matsuba et al., 2013).

The phylogenomics approach for candidate enzyme identification is the most powerful when combined with information obtained from other complementary analyses, such as taxonomical distribution pattern of the target natural product, the tissue-specific accumulation of the relevant biosynthetic intermediates in the host organism, and the spatial and temporal transcript expression pattern deduced from transcriptome sequencing. Candidate genes that fulfill most of these criteria should be flagged with high priority for downstream biochemical tests.

3.2 Recombinant Expression and Purification of Candidate Enzymes

The extensive development of recombinant DNA and protein technology in the past decades has made it straightforward to attempt the expression of candidate enzymes as recombinant proteins in a wide selection of heterologous hosts. Candidate enzyme-encoding genes can be amplified directly from the cDNA or synthesized as codon-optimized ORFs, and cloned into expression vectors suitable for the desired expression hosts. Since enzyme families involved in specialized metabolism are very diverse and individual enzymes of the same enzyme family also behave differently when expressed recombinantly, it is still a low-throughput process for the researcher to identify and optimize a suitable protocol for producing enough active recombinant enzyme suitable for activity and structural analyses.

Escherichia coli remains the preferred host for producing recombinant candidate enzymes. Standard methods for recombinant expression and purification of polyhistidine-tagged soluble protein in *E. coli* have been reviewed previously (Weng & Noel, 2012b). If *E. coli* expression of the candidate enzyme mostly yields insoluble aggregations in inclusion bodies, N- or C-terminal fusion with maltose-binding protein or glutathione *S*-transferase can be attempted to increase protein solubility. A number of engineered *E. coli* expression strains are also available to overcome various technical difficulties. These include the Rosetta strain that enhances the expression of eukaryotic proteins with rare *E. coli* codons (Novy, Drott, Yaeger, & Mierendorf, 2001), the ArcticExpress strain that circumvents protein misfolding and insolubility by allowing bacterial growth and expression at 12°C (Gopal & Kumar, 2013), and the LOBSTR strain for reducing common *E. coli* protein contaminants from the His-tag purification process (Andersen, Leksa, & Schwartz, 2013).

Several enzyme families frequently involved in eukaryotic specialized metabolism are membrane-associated or require specific eukaryotic-type posttranslational modifications, creating challenges for expression in *E. coli*. For example, P450s, a ubiquitous family of heme-thiolate enzymes widely distributed in eukaryotic natural product biosynthesis, are anchored into the membrane of endoplasmic reticulum (ER) (Werck-Reichhart & Feyereisen, 2000). To enable expression of eukaryotic P450s in *E. coli*, the native N-terminal ER-membrane anchoring sequence can be replaced with a bacterial lipid bilayer-targeting sequence (Pritchard et al., 1998; Richardson et al., 1993). Alternatively, *Saccharomyces cerevisiae* has become the preferred system for recombinant eukaryotic P450 expression. To express functionally active eukaryotic P450s in *S. cerevisiae*, an NADPH-dependent cytochrome P450 reductase (CPR), preferably cloned from the same species as of the target P450, has to be coexpressed in the yeast cells (Chefson & Auclair, 2006; Pritchard et al., 1998). Microsomal fraction of the yeast cells containing active recombinant P450 enzyme can be prepared according to a previously described protocol (Urban et al., 1994) and used in biochemical assays. In addition to facilitating P450 expression, *S. cerevisiae* has also been widely used as the heterologous expression host for many other membrane-bound enzymes or enzymes that express poorly in *E. coli*, eg, plant membrane-bound prenyltransferases (Li et al., 2015), UDP-glucosyltransferases (Poppenberger et al., 2006), and serine carboxypeptidase-like proteins (Shirley & Chapple, 2003).

Although used less frequently, other heterologous expression systems, such as the yeast *Pichia pastoris* (Krainer et al., 2014) and insect cells (Qiu et al., 2012), have also been successfully adopted for recombinant expression of eukaryotic specialized metabolic enzymes. Furthermore, Agrobacterium-mediated transient protein production in *Nicotiana benthamiana* has emerged in recent years as a popular method not only for rapid examination of metabolic enzyme functions in vivo (detailed later) but also for industrial-scale production of recalcitrant eukaryotic proteins (Giritch et al., 2006).

3.3 Comparative Biochemical Analysis of Enzyme Functions

Although structure-guided phylogenomics analysis often uncovers sequence features suggestive of neofunctionalized enzymes, the breadth and complexity of the structure–function relationship in enzyme evolution make accurate prediction of the biochemical function of a candidate enzyme nearly impossible. Indeed, the high sequence identity between members of the same enzyme family has historically resulted in erroneous functional annotations of uncharacterized enzyme-encoding genes. Consequently, careful biochemical analysis of candidate enzymes to determine their in vivo substrates and catalytic mechanism as well as the characterization of full kinetic parameters are of significant importance toward ultimate elucidation of new specialized metabolic pathways. Comparative biochemical analysis of a set of phylogenetically related enzymes from the same species or related species is a particularly informative approach to shed light on unique catalytic functions among tested candidate enzymes.

Under a proposed biosynthetic model, knowledge about the previously characterized enzymes homologous to the candidate enzymes may suggest a limited number of putative substrates and relatively defined catalytic mechanisms. Ideally, the putative substrates are commercially available or are feasible to be synthesized chemically or isolated from the host species. For example, despite difficulty in predicting substrate selectivity, plant AAADs, as their name implies, implicitly catalyze the decarboxylation of a selection of aromatic amino acid substrates (Torrens-Spence, Lazear, von Guggenberg, Ding, & Li, 2014). As such, systematic biochemical assays can be carried out to measure product formation using an array of aromatic amino acid substrates. Nowadays, the formation of products from an enzyme assay are typically separated and quantitated by GC–MS or LC–MS. The experimental design and execution of kinetic characterization of specific reactions should follow fundamental principles for enzyme kinetics (Cornish-Bowden, 2013). An agreement between kinetic properties of a candidate enzyme and the biosynthetic model strengthens the model, whereas a disagreement often results in modification of the model or testing additional candidate enzymes. Comparative enzyme assays demonstrating unique biochemical activities in one or a few candidate enzymes among a larger selection of homologous enzymes is a preferred approach to show convincing evidence supporting the role of a particular candidate enzyme in a specialized metabolic pathway (Weng, Li, Stout, & Chapple, 2008).

It is important to verify the chemical identity of the products formed in enzyme assays, as many specialized metabolic enzymes occasionally exhibit novel catalytic functions unexpected for the enzyme family to which they belong (Weng & Noel, 2012a). In the case for plant AAADs, early enzymology studies monitoring the release of CO_2 instead of the formation of monoamine products as an indication for AAAD activity therefore likely misidentified some neofunctionalized AASs as AAADs (Facchini & De Luca, 1995; Lovenberg, Weissbach, & Udenfriend, 1962). This is because AASs also release CO_2 through decarboxylation but further catalyze oxidative deamination to produce aldehyde, ammonia, and hydrogen peroxide, which were missed in early enzyme assays (Kaminaga et al., 2006). In a recent study of cyanogenic glucoside biosynthesis in *Arabidopsis*, detailed in vitro biochemical assays uncovered a CYP71 family enzyme that catalyzes the conversion of indole aldoxime to its corresponding cyanohydrin, an unprecedented catalytic activity known for P450s (Rajniak, Barco, Clay, & Sattely, 2015).

In certain situations, the exact substrates for candidate enzymes cannot be predicted, as the hypothetical biosynthetic model could be incomplete. Facilitated by untargeted metabolomics, these ambiguous enzymes can be screened directly against crude extract of metabolites prepared from their parent organism (Prosser et al., 2014). Necessary cofactors for the target enzyme should be supplemented into the assays, and a wide range of dilution of the crude extract should also be tested. Enzyme-dependent conversion of metabolite features likely provides valuable insight into the chemical identify of in vivo substrates and products of the candidate enzyme. This approach has recently been applied to identify a highly specific *N*-methyltransferase involved in indole alkaloid metabolism in *C. roseus* (Liscombe, Usera, & O'Connor, 2010).

3.4 Structural Basis for Specialized Metabolic Enzyme Evolution

Natural enzymes are remarkable molecular machines that catalyze diverse chemical reactions as explicitly dictated by their three-dimensional structures and dynamics. Yet natural enzymes are evolvable through mutational trajectories, as their substrate selectivity and reaction mechanism can be significantly altered by only a small number of mutations (Austin, Bowman, Ferrer, Schroder, & Noel, 2004; O'Maille et al., 2008). Understanding the structure–function relationship underlying the natural evolution of enzymes is one of the most pressing quests of modern biochemistry, which also has significant practical implication in guiding enzyme engineering for industrial or pharmaceutical purposes. Structural biology followed by structure-guided mutagenesis studies offers a powerful approach to probe this question. In recent years, the continuing development of protein X-ray crystallography has made it increasingly accessible for researchers in the field of specialized metabolism who are not structural biology aficionados (Shi, 2014).

Using a microfluidic liquid handling robot (eg, Formulatrix Formulator, Bedford, MA), initial crystallization screens can be carried out on 96-well crystallization plates using commercially available crystallization screens. Currently, about 15 μL of ~10 mg/mL purified protein solution is necessary per 96-well plate. Crystal plates are then stored in a crystal storage hotel with scheduled imaging capacity (eg, Formulatrix Rock Imager). Once initial crystallization conditions are identified, they are optimized manually to obtain single crystals suitable for X-ray diffraction through a combination of fine grid of buffer conditions, additive screens, and seeding. In our laboratory, we routinely ship frozen crystals to a synchrotron beamline and collect X-ray diffraction data via its remote data collection pipeline. The recent implementation of the state-of-the-art silicon pixel detectors, eg, the Dectris Pilatus 6MF pixel array detector available at the Argonne National Laboratory, now allows shutter-less data collection, noiseless readout, and significantly improved data quality compared to normal CCD detectors. X-ray diffraction data indexing, integration, scaling, phasing with molecular replacement (MR), model building, refinement, and validation can be streamlined using several X-ray crystallography software packages, including HKL-2000 (Otwinowski & Minor, 1997), CCP4 (Winn et al., 2011), and Phenix (Adams et al., 2010). If phasing cannot be achieved by MR, crystals prepared from selenomethionine-substituted recombinant protein are used for experimental phasing by single-wavelength anomalous diffraction (Rice, Earnest, & Brunger, 2000). To gain a comprehensive view of enzyme catalytic mechanism, crystal structures of the target enzyme in complex with appropriate substrates, cofactors, products, and analogs of catalytic intermediates should also be obtained if possible.

Comparative structural analyses of divergent paralogous enzymes from diverse specialized metabolic enzyme families continue to uncover structural features responsible for alterations in activity at atomic resolution (Austin et al., 2004; Kries et al., 2016; O'Maille et al., 2008; Westfall et al., 2012). As an elegant example, the crystal structure of the recently discovered iridoid cyclase from *C. roseus* illuminated key structural features and active site dynamics required for its unusual reductive terpene cyclase activity likely neofunctionalized from an ancestral progesterone 5β-reductase progenitor (Kries et al., 2016). Guided by in-depth structural comparison, site-directed mutagenesis followed by enzyme assays is a routine approach to probe the respective contributions of important residues in substrate binding and catalysis. Moreover, systematic substitutions of active-site residues with variable residues naturally occurring at the same positions within an enzyme superfamily can be another efficient way to deduce possible mutational trajectories underlying evolutionary divergence (Goldsmith & Tawfik, 2013).

The structural and mechanistic understanding of molecular roles of individual amino acids also facilitates rational enzyme modification and design. For example, structural comparison of paralogous enzymes has enabled the production of mutant enzymes with enhanced or altered catalytic activities for desired purposes (Bornscheuer & Kazlauskas, 2004; Kazlauskas, 2005). Similar approaches have also enabled the sculpting of an enzyme's active site to accommodate novel substrates or a broader substrate profile (Cho et al., 2008; Lauble et al., 2002).

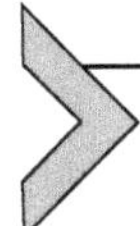

4. RECONSTITUTION OF SPECIALIZED METABOLIC PATHWAYS IN HETEROLOGOUS SYSTEMS

Biochemical and structural analyses of enzymes are important tools toward ultimately determining their in vivo functions. However, the ability to measure a particular activity in vitro does not necessitate that this activity is physiologically relevant. For example, promiscuous in vitro activity has historically confounded the principle biochemical pathway for the important plant hormone auxin, which was resolved only recently (Zhao, 2012). This potential pitfall is particularly pertinent to natural product biosynthesis, where there is a growing understanding that specialized metabolic enzymes are inherently promiscuous (Weng & Noel, 2012a). RNA interference and virus-induced gene silencing are two possible ways to test the in vivo biochemical function of a particular candidate enzyme (Liscombe & O'Connor, 2011; Stansbury & Moczek, 2014); however, these methods may not be feasible for many nonmodel eukaryotic species. Alternatively, partial or total reconstitution of a specialized metabolic pathway in a heterologous host can serve as the ultimate proof of the elucidation of the pathway. Moreover, such process is also a metabolic engineering excise, which generates prototype strains that can be further optimized to produce high-value natural products.

4.1 *Escherichia coli*

Recent advance in synthetic biology and metabolic engineering tools has transformed many organisms into production vehicles for producing desirable chemicals from simple sugars. *E. coli* has emerged as an attractive host due to its simple culture requirements, rapid growth rate, and the availability of a plethora of genetic tools. For example, *E. coli* is a convenient in vivo system to test the function of plant acyl-CoA ligases pairing with a downstream type III PKS or BAHD acyltransferase. A wide selection of acyl donor molecules and the acyl acceptor molecules (in the case of BAHD acyltransferase) can be directly supplemented to the medium, while the formation of potential polyketide or ester products in the bacterial cell extract can be assayed by LC–MS. More targeted systems have also engineered the upstream precursor pathways necessary for the synthesis of specific downstream polyketide or ester (Cha, Kim, Kim, & Ahn, 2014; Lim, Fowler, Hueller, Schaffer, & Koffas, 2011; Wu, Zhou, Du, Zhou, & Chen, 2014). Furthermore, an *E. coli* platform has been recently engineered to produce the high-value plant benzylisoquinoline alkaloid (BIA), (*S*)-reticuline, through expressing four consecutive BIA biosynthetic enzymes from *Coptis japonica* in a synthetically engineered background strain that supplies 3,4-dihydroxyphenylacetaldehyde and dopamine precursors (Nakagawa et al., 2011). *E. coli* has also been used as a heterologous host for the production of diterpene taxadiene, an key intermediate in the biosynthesis of anticancer natural product taxol from *T. brevifolia*, with a notable yield at approximately 1 g/L culture medium (Ajikumar et al., 2010). The multivariate-modular approach as described in this study not only significantly increases the titers of taxadiene in *E. coli* but also helps to unlock *E. coli's* potential to synthesize other terpene products derived from the methylerythritol phosphate pathway in the future. Indeed, engineering *E. coli* to accumulate common primary metabolic precursors is a useful strategy, which benefits many similar metabolic engineering efforts that draw from the same precursors for downstream natural product synthesis. An additional example of this forward-thinking metabolic engineering approach involves the overproduction of aromatic amino acids in *E. coli*, which facilitates the subsequent engineering of many aromatic natural products (Koma, Yamanaka, Moriyoshi, Ohmoto, & Sakai, 2012; Rodriguez et al., 2014).

4.2 *Saccharomyces cerevisiae*

As a eukaryotic host, *S. cerevisiae* possesses a number of advantages compared to *E. coli*, such as the presence of eukaryotic biomembranes and organelles, machineries for posttranslation modifications, as well as resistance to certain toxic antibacterial metabolites (Guo et al., 2015). *S. cerevisiae* is also very well suited for industrial-scale fermentation, exemplified by its predominant use in the production of bioethanol as well as several pharmaceutically and nutraceutically important commodity chemicals consumed directly by humans (Hong & Nielsen, 2012). In a recent study, Paddon et al. reported the metabolic engineering effort for producing artemisinic acid, a key biosynthetic intermediate of the antimalarial drug artemisinin, in *S. cerevisiae* (Paddon et al., 2013). This effort led to the discovery of two new artemisinin biosynthetic genes from *Artemisia annua*, encoding an artemisinic dehydrogenase and a cytochrome b_5, respectively. Reconstitution of the complete artemisinic acid biosynthetic pathway in *S. cerevisiae* provides an efficient biosynthetic route to artemisinic acid from simple sugars with impressive fermentation titers of approximately 25 g/L culture medium.

To reconstitute eukaryotic specialized metabolic pathways in *S. cerevisiae*, our laboratory uses a highly characterized and modular yeast toolkit developed by the Dueber lab, which contains characterized promoters, terminators, peptide tags, copy number machinery, and genome-editing modules that can be assembled at will with the Golden Gate cloning technique (Lee, DeLoache, Cervantes, & Dueber, 2015). This system allows tremendous flexibility to integrate multiple heterologous enzyme-encoding genes into the yeast genome in combination with custom-targeted mutagenesis of any yeast endogenous gene by CRISPR/Cas9 (Lee et al., 2015).

4.3 *Nicotiana benthamiana*

As mentioned previously, the use of plant systems, such as *N. benthamiana*, for transient production of recombinant enzymes and reconstitution of multistep metabolic pathways has emerged as an attractive approach to study in vivo biochemical functions of uncharacterized specialized metabolic enzymes, especially for those that do not express properly in other heterologous expression hosts. Currently, the pEAQ series of vectors developed by the Lomonossoff lab are among the most commonly used vector systems for transient expression of exogenous proteins in *N. benthamiana* (Peyret & Lomonossoff, 2013). The pEAQ vectors feature a cowpea mosaic virus hypertranslational (CPMV-HT) expression system, which results in extremely high translational efficiency for recombinant protein expression without the need for viral replication.

The *N. benthamiana* transient expression system also allows combinatorial expression of multiple candidate enzymes by coinfiltrating an *N. benthamiana* leaf with multiple *Agrobacterium* strains, each harboring a different expression construct. Putative substrates can also be coinfiltrated into the leaf tissue, if they are not available endogenously in *N. benthamiana*, allowing rapid evaluation of hypothesized biosynthetic models and candidate enzymes. This system has been recently employed to help elucidate the podophyllotoxin biosynthetic pathway in mayapple (*Podophyllum hexandrum*) (Lau & Sattely, 2015).

5. SUMMARY

Eukaryotic organisms constitute a substantial portion of the remarkable biodiversity existing on the planet Earth, and a rich source for discovering new natural products and the specialized metabolic systems that support their biosynthesis. Due to multiple layers of technical challenges, research in exploring novel specialized metabolic pathways in nonmodel eukaryotic species has been very limited. Not until very recently, significant advances in multiple frontiers in biological sciences have allowed renewed enthusiasm and capacity to study specialized metabolic systems in those otherwise difficult organisms. In this paper, we summarize a general workflow starting from omics-guided candidate pathway identification to in-depth comparative biochemical and structural characterization of rapidly evolving specialized metabolic enzymes. Although most of the examples we raised here are based on studies in the plant metabolic systems, the most exploited eukaryotic systems in terms of natural product research, the work should be generally applicable to other less studied eukaryotic groups such as fungi and animals.

ACKNOWLEDGMENTS

This chapter is based in part upon work supported by the Pew Scholars Program in the Biomedical Sciences and the Searle Scholars Program.

REFERENCES

Adams, P. D., Afonine, P. V., Bunkoczi, G., Chen, V. B., Davis, I. W., Echols, N., et al. (2010). PHENIX: A comprehensive Python-based system for macromolecular structure solution. *Acta Crystallographica. Section D, Biological Crystallography*, *66*, 213–221.

Ajikumar, P. K., Xiao, W. H., Tyo, K. E., Wang, Y., Simeon, F., Leonard, E., et al. (2010). Isoprenoid pathway optimization for Taxol precursor overproduction in Escherichia coli. *Science*, *330*, 70–74.

Andersen, K. R., Leksa, N. C., & Schwartz, T. U. (2013). Optimized E. coli expression strain LOBSTR eliminates common contaminants from His-tag purification. *Proteins*, *81*, 1857–1861.

Atkinson, N. J., & Urwin, P. E. (2012). The interaction of plant biotic and abiotic stresses: From genes to the field. *Journal of Experimental Botany*, *63*, 3523–3543.

Austin, M. B., Bowman, M. E., Ferrer, J. L., Schroder, J., & Noel, J. P. (2004). An aldol switch discovered in stilbene synthases mediates cyclization specificity of type III polyketide synthases. *Chemistry & Biology*, *11*, 1179–1194.

Berger, A., Meinhard, J., & Petersen, M. (2006). Rosmarinic acid synthase is a new member of the superfamily of BAHD acyltransferases. *Planta*, *224*, 1503–1510.

Beutler, J. A. (2009). Natural products as a foundation for drug discovery. *Current Protocols in Pharmacology/Editorial Board, S.J. Enna*, *46*, 9.11.1–9.11.21.

Bornscheuer, U. T., & Kazlauskas, R. J. (2004). Catalytic promiscuity in biocatalysis: Using old enzymes to form new bonds and follow new pathways. *Angewandte Chemie*, *43*, 6032–6040.

Busing, R. T., Halpern, C. B., & Spies, T. A. (1995). Ecology of Pacific Yew (Taxus-Brevifolia) in Western Oregon and Washington. *Conservation Biology*, *9*, 1199–1207.

Caelles, C., Ferrer, A., Balcells, L., Hegardt, F. G., & Boronat, A. (1989). Isolation and structural characterization of a cDNA encoding Arabidopsis thaliana 3-hydroxy-3-methylglutaryl coenzyme A reductase. *Plant Molecular Biology*, *13*, 627–638.

Campbell, M. S., Holt, C., Moore, B., & Yandell, M. (2014). Genome Annotation and Curation using MAKER and MAKER-P. *Current Protocols in Bioinformatics/Editoral Board, Andreas D. Baxevanis ... [et al.]*, *48*, 4.11.1–4.11.39.

Cha, M. N., Kim, H. J., Kim, B. G., & Ahn, J. H. (2014). Synthesis of chlorogenic acid and p-coumaroyl shikimates from glucose using engineered Escherichia coli. *Journal of Microbiology and Biotechnology*, *24*, 1109–1117.

Chater, K. F. (1992). Genetic regulation of secondary metabolic pathways in Streptomyces. *Ciba Foundation Symposium*, *171*, 144–156. discussion 156–162.

Chefson, A., & Auclair, K. (2006). Progress towards the easier use of P450 enzymes. *Molecular Biosystems*, *2*, 462–469.

Chemler, J. A., & Koffas, M. A. G. (2008). Metabolic engineering for plant natural product biosynthesis in microbes. *Current Opinion in Biotechnology, 19*, 597–605.

Cho, B. K., Park, H. Y., Seo, J. H., Kim, J., Kang, T. J., Lee, B. S., et al. (2008). Redesigning the substrate specificity of omega-aminotransferase for the kinetic resolution of aliphatic chiral amines. *Biotechnology and Bioengineering, 99*, 275–284.

Cook, D., Rimando, A. M., Clemente, T. E., Schroder, J., Dayan, F. E., Nanayakkara, N. P., et al. (2010). Alkylresorcinol synthases expressed in Sorghum bicolor root hairs play an essential role in the biosynthesis of the allelopathic benzoquinone sorgoleone. *Plant Cell, 22*, 867–887.

Cornish-Bowden, A. (2013). *Fundamentals of enzyme kinetics*. Hoboken, NJ: Wiley.

Datta, H. S., Mitra, S. K., & Patwardhan, B. (2011). Wound healing activity of topical application forms based on ayurveda. *Evidence-Based Complementary and Alternative Medicine: eCAM, 2011*, 134378.

Dev, S. (1999). Ancient-modern concordance in ayurvedic plants: Some examples. *Environmental Health Perspectives, 107*, 783–789.

Dittrich, H., & Kutchan, T. M. (1991). Molecular cloning, expression, and induction of berberine bridge enzyme, an enzyme essential to the formation of benzophenanthridine alkaloids in the response of plants to pathogenic attack. *Proceedings of the National Academy of Sciences of the United States of America, 88*, 9969–9973.

Facchini, P. J., & De Luca, V. (1995). Expression in Escherichia coli and partial characterization of two tyrosine/dopa decarboxylases from opium poppy. *Phytochemistry, 38*, 1119–1126.

Facchini, P. J., Huber-Allanach, K. L., & Tari, L. W. (2000). Plant aromatic L-amino acid decarboxylases: Evolution, biochemistry, regulation, and metabolic engineering applications. *Phytochemistry, 54*, 121–138.

Francis, W. R., Christianson, L. M., Kiko, R., Powers, M. L., Shaner, N. C., & Haddock, S. H. (2013). A comparison across non-model animals suggests an optimal sequencing depth for de novo transcriptome assembly. *BMC Genomics, 14*, 167.

Fraser, C. M., Thompson, M. G., Shirley, A. M., Ralph, J., Schoenherr, J. A., Sinlapadech, T., et al. (2007). Related Arabidopsis serine carboxypeptidase-like sinapoylglucose acyltransferases display distinct but overlapping substrate specificities. *Plant Physiology, 144*, 1986–1999.

Gagne, S. J., Stout, J. M., Liu, E., Boubakir, Z., Clark, S. M., & Page, J. E. (2012). Identification of olivetolic acid cyclase from Cannabis sativa reveals a unique catalytic route to plant polyketides. *Proceedings of the National Academy of Sciences of the United States of America, 109*, 12811–12816.

Geu-Flores, F., Sherden, N. H., Courdavault, V., Burlat, V., Glenn, W. S., Wu, C., et al. (2012). An alternative route to cyclic terpenes by reductive cyclization in iridoid biosynthesis. *Nature, 492*, 138–142.

Giritch, A., Marillonnet, S., Engler, C., van Eldik, G., Botterman, J., Klimyuk, V., et al. (2006). Rapid high-yield expression of full-size IgG antibodies in plants coinfected with noncompeting viral vectors. *Proceedings of the National Academy of Sciences of the United States of America, 103*, 14701–14706.

Goldsmith, M., & Tawfik, D. S. (2013). Enzyme engineering by targeted libraries. *Methods in Enzymology, 523*, 257–283.

Gopal, G. J., & Kumar, A. (2013). Strategies for the production of recombinant protein in Escherichia coli. *The Protein Journal, 32*, 419–425.

Grabherr, M. G., Haas, B. J., Yassour, M., Levin, J. Z., Thompson, D. A., Amit, I., et al. (2011). Full-length transcriptome assembly from RNA-Seq data without a reference genome. *Nature Biotechnology, 29*, 644–652.

Guo, Y., Dong, J., Zhou, T., Auxillos, J., Li, T., Zhang, W., et al. (2015). YeastFab: The design and construction of standard biological parts for metabolic engineering in Saccharomyces cerevisiae. *Nucleic Acids Research, 43*, e88.

Haas, B. J., Papanicolaou, A., Yassour, M., Grabherr, M., Blood, P. D., Bowden, J., et al. (2013). De novo transcript sequence reconstruction from RNA-seq using the Trinity platform for reference generation and analysis. *Nature Protocols*, *8*, 1494–1512.

Hajduk, P. J., & Greer, J. (2007). A decade of fragment-based drug design: Strategic advances and lessons learned. *Nature Reviews. Drug Discovery*, *6*, 211–219.

Hamberger, B., Ohnishi, T., Hamberger, B., Seguin, A., & Bohlmann, J. (2011). Evolution of diterpene metabolism: Sitka spruce CYP720B4 catalyzes multiple oxidations in resin acid biosynthesis of conifer defense against insects. *Plant Physiology*, *157*, 1677–1695.

Hare, E. E., & Johnston, J. S. (2011). Genome size determination using flow cytometry of propidium iodide-stained nuclei. *Molecular Methods for Evolutionary Genetics*, *772*, 3–12.

Hert, J., Irwin, J. J., Laggner, C., Keiser, M. J., & Shoichet, B. K. (2009). Quantifying biogenic bias in screening libraries. *Nature Chemical Biology*, *5*, 479–483.

Hong, K. K., & Nielsen, J. (2012). Metabolic engineering of Saccharomyces cerevisiae: A key cell factory platform for future biorefineries. *Cellular and Molecular Life Sciences: CMLS*, *69*, 2671–2690.

Jenkins, G. I. (2009). Signal transduction in responses to UV-B radiation. *Annual Review of Plant Biology*, *60*, 407–431.

Kaminaga, Y., Schnepp, J., Peel, G., Kish, C. M., Ben-Nissan, G., Weiss, D., et al. (2006). Plant phenylacetaldehyde synthase is a bifunctional homotetrameric enzyme that catalyzes phenylalanine decarboxylation and oxidation. *The Journal of Biological Chemistry*, *281*, 23357–23366.

Kazlauskas, R. J. (2005). Enhancing catalytic promiscuity for biocatalysis. *Current Opinion in Chemical Biology*, *9*, 195–201.

Kellner, F., Kim, J., Clavijo, B. J., Hamilton, J. P., Childs, K. L., Vaillancourt, B., et al. (2015). Genome-guided investigation of plant natural product biosynthesis. *The Plant Journal*, *82*, 680–692.

Klein, A. P., & Sattely, E. S. (2015). Two cytochromes P450 catalyze S-heterocyclizations in cabbage phytoalexin biosynthesis. *Nature Chemical Biology*, *11*, 837–839.

Koma, D., Yamanaka, H., Moriyoshi, K., Ohmoto, T., & Sakai, K. (2012). Production of aromatic compounds by metabolically engineered Escherichia coli with an expanded shikimate pathway. *Applied and Environmental Microbiology*, *78*, 6203–6216.

Koren, S., & Phillippy, A. M. (2015). One chromosome, one contig: Complete microbial genomes from long-read sequencing and assembly. *Current Opinion in Microbiology*, *23*, 110–120.

Krainer, F. W., Pletzenauer, R., Rossetti, L., Herwig, C., Glieder, A., & Spadiut, O. (2014). Purification and basic biochemical characterization of 19 recombinant plant peroxidase isoenzymes produced in Pichia pastoris. *Protein Expression and Purification*, *95*, 104–112.

Kries, H., Caputi, L., Stevenson, C. E., Kamileen, M. O., Sherden, N. H., Geu-Flores, F., et al. (2016). Structural determinants of reductive terpene cyclization in iridoid biosynthesis. *Nature Chemical Biology*, *12*, 6–8.

Kutrzeba, L., Dayan, F. E., Howell, J., Feng, J., Giner, J. L., & Zjawlony, J. K. (2007). Biosynthesis of salvinorin A proceeds via the deoxyxylulose phosphate pathway. *Phytochemistry*, *68*, 1872–1881.

Lan, X., Chang, K., Zeng, L., Liu, X., Qiu, F., Zheng, W., et al. (2013). Engineering salidroside biosynthetic pathway in hairy root cultures of Rhodiola crenulata based on metabolic characterization of tyrosine decarboxylase. *PloS One*, *8*, e75459.

Lau, W., & Sattely, E. S. (2015). Six enzymes from mayapple that complete the biosynthetic pathway to the etoposide aglycone. *Science*, *349*, 1224–1228.

Lauble, H., Miehlich, B., Forster, S., Kobler, C., Wajant, H., & Effenberger, F. (2002). Structure determinants of substrate specificity of hydroxynitrile lyase from Manihot esculenta. *Protein Science: A Publication of the Protein Society*, *11*, 65–71.

Lee, M. E., DeLoache, W. C., Cervantes, B., & Dueber, J. E. (2015). A highly characterized Yeast Toolkit for modular, multipart assembly. *ACS Synthetic Biology*, *4*, 975–986.

Leng, N., Dawson, J. A., Thomson, J. A., Ruotti, V., Rissman, A. I., Smits, B. M., et al. (2013). EBSeq: An empirical Bayes hierarchical model for inference in RNA-seq experiments. *Bioinformatics*, *29*, 1035–1043.

Levy-Sakin, M., & Ebenstein, Y. (2013). Beyond sequencing: Optical mapping of DNA in the age of nanotechnology and nanoscopy. *Current Opinion in Biotechnology*, *24*, 690–698.

Li, H., Ban, Z., Qin, H., Ma, L., King, A. J., & Wang, G. (2015). A heteromeric membrane-bound prenyltransferase complex from hop catalyzes three sequential aromatic prenylations in the bitter acid pathway. *Plant Physiology*, *167*, 650–659.

Li, B., & Dewey, C. N. (2011). RSEM: Accurate transcript quantification from RNA-Seq data with or without a reference genome. *BMC Bioinformatics*, *12*, 323.

Lim, C. G., Fowler, Z. L., Hueller, T., Schaffer, S., & Koffas, M. A. (2011). High-yield resveratrol production in engineered Escherichia coli. *Applied and Environmental Microbiology*, *77*, 3451–3460.

Liscombe, D. K., & O'Connor, S. E. (2011). A virus-induced gene silencing approach to understanding alkaloid metabolism in Catharanthus roseus. *Phytochemistry*, *72*, 1969–1977.

Liscombe, D. K., Usera, A. R., & O'Connor, S. E. (2010). Homolog of tocopherol C methyltransferases catalyzes N methylation in anticancer alkaloid biosynthesis. *Proceedings of the National Academy of Sciences of the United States of America*, *107*, 18793–18798.

Lovenberg, W., Weissbach, H., & Udenfriend, S. (1962). Aromatic L-amino acid decarboxylase. *The Journal of Biological Chemistry*, *237*, 89–93.

Luckner, M. (1984). *Secondary metabolism in microorganisms, plants, and animals*. Berlin, Germany: Springer-Verlag.

Ma, X., Tan, C., Zhu, D., & Gang, D. R. (2006). A survey of potential huperzine A natural resources in China: The Huperziaceae. *Journal of Ethnopharmacology*, *104*, 54–67.

Mandel, M. A., Feldmann, K. A., Herrera-Estrella, L., Rocha-Sosa, M., & Leon, P. (1996). CLA1, a novel gene required for chloroplast development, is highly conserved in evolution. *The Plant Journal: For Cell and Molecular Biology*, *9*, 649–658.

Matasci, N., Hung, L. H., Yan, Z., Carpenter, E. J., Wickett, N. J., Mirarab, S., et al. (2014). Data access for the 1,000 Plants (1KP) project. *GigaScience*, *3*, 17.

Matsuba, Y., Nguyen, T. T., Wiegert, K., Falara, V., Gonzales-Vigil, E., Leong, B., et al. (2013). Evolution of a complex locus for terpene biosynthesis in solanum. *The Plant Cell*, *25*, 2022–2036.

Matsuno, M., Compagnon, V., Schoch, G. A., Schmitt, M., Debayle, D., Bassard, J. E., et al. (2009). Evolution of a novel phenolic pathway for pollen development. *Science*, *325*, 1688–1692.

Misof, B., Liu, S., Meusemann, K., Peters, R. S., Donath, A., Mayer, C., et al. (2014). Phylogenomics resolves the timing and pattern of insect evolution. *Science*, *346*, 763–767.

Nakagawa, A., Minami, H., Kim, J. S., Koyanagi, T., Katayama, T., Sato, F., et al. (2011). A bacterial platform for fermentative production of plant alkaloids. *Nature Communications*, *2*, 326.

Newman, D. J., & Cragg, G. M. (2007). Natural products as sources of new drugs over the last 25 years. *Journal of Natural Products*, *70*, 461–477.

Newman, D. J., & Cragg, G. M. (2009). Natural product scaffolds as leads-to drugs. *Future Medicinal Chemistry*, *1*, 1415–1427.

Newman, D. J., & Cragg, G. M. (2012). Natural products as sources of new drugs over the 30 years from 1981 to 2010. *Journal of Natural Products*, *75*, 311–335.

Novy, R., Drott, D., Yaeger, K., & Mierendorf, R. (2001). Overcoming the codon bias of E. coli for enhanced protein expression. *Innovations*, *12*, 1–3.

O'connor, S. (2009). Methods for molecular identification of biosynthetic enzymes in plants. In A. E. Osbourn & V. Lanzotti (Eds.), *Plant-derived natural products: Synthesis, function, and application* (pp. 165–179). New York, NY: Springer.

O'Maille, P. E., Malone, A., Dellas, N., Andes Hess, B., Jr., Smentek, L., Sheehan, I., et al. (2008). Quantitative exploration of the catalytic landscape separating divergent plant sesquiterpene synthases. *Nature Chemical Biology, 4*, 617–623.

Osbourn, A., Papadopoulou, K. K., Qi, X., Field, B., & Wegel, E. (2012). Finding and analyzing plant metabolic gene clusters. *Methods in Enzymology, 517*, 113–138.

Otwinowski, Z., & Minor, W. (1997). Processing of X-ray diffraction data collected in oscillation mode. In C. W. Carter & R. M. Sweet (Eds.), *Macromolecular crystallography, part A: Vol. 276. Methods in enzymology* (pp. 307–326). New York: Academic Press.

Paddon, C. J., Westfall, P. J., Pitera, D. J., Benjamin, K., Fisher, K., McPhee, D., et al. (2013). High-level semi-synthetic production of the potent antimalarial artemisinin. *Nature, 496*, 528–532.

Pei, J., Kim, B. H., & Grishin, N. V. (2008). PROMALS3D: A tool for multiple protein sequence and structure alignments. *Nucleic Acids Research, 36*, 2295–2300.

Peng, Z. Y., Yuan, C. F., Zellmer, L., Liu, S. Q., Xu, N. Z., & Liao, D. J. (2015). Hypothesis: Artifacts, including spurious chimeric RNAs with a short homologous sequence, caused by consecutive reverse transcriptions and endogenous random primers. *Journal of Cancer, 6*, 555–567.

Perola, E. (2010). An analysis of the binding efficiencies of drugs and their leads in successful drug discovery programs. *Journal of Medicinal Chemistry, 53*, 2986–2997.

Peyret, H., & Lomonossoff, G. P. (2013). The pEAQ vector series: The easy and quick way to produce recombinant proteins in plants. *Plant Molecular Biology, 83*, 51–58.

Pichersky, E., & Gershenzon, J. (2002). The formation and function of plant volatiles: Perfumes for pollinator attraction and defense. *Current Opinion in Plant Biology, 5*, 237–243.

Pichersky, E., & Lewinsohn, E. (2011). Convergent evolution in plant specialized metabolism. *Annual Review of Plant Biology, 62*, 549–566.

Pluskal, T., Castillo, S., Villar-Briones, A., & Oresic, M. (2010). MZmine 2: Modular framework for processing, visualizing, and analyzing mass spectrometry-based molecular profile data. *BMC Bioinformatics, 11*, 395.

Pluskal, T., Uehara, T., & Yanagida, M. (2012). Highly accurate chemical formula prediction tool utilizing high-resolution mass spectra, MS/MS fragmentation, heuristic rules, and isotope pattern matching. *Analytical Chemistry, 84*, 4396–4403.

Poppenberger, B., Berthiller, F., Bachmann, H., Lucyshyn, D., Peterbauer, C., Mitterbauer, R., et al. (2006). Heterologous expression of Arabidopsis UDP-glucosyltransferases in Saccharomyces cerevisiae for production of zearalenone-4-O-glucoside. *Applied and Environmental Microbiology*, 72, 4404–4410.

Pritchard, M. P., Glancey, M. J., Blake, J. A., Gilham, D. E., Burchell, B., Wolf, C. R., et al. (1998). Functional co-expression of CYP2D6 and human NADPH-cytochrome P450 reductase in Escherichia coli. *Pharmacogenetics, 8*, 33–42.

Priyam, A., Woodcroft, B. J., Rai, V., Munagala, A., Moghul, I., Ter, F., et al. (2015). Sequenceserver: A modern graphical user interface for custom BLAST databases. *bioRxiv*. http://biorxiv.org/content/early/2015/11/27/033142.

Prosser, G. A., Larrouy-Maumus, G., & de Carvalho, L. P. (2014). Metabolomic strategies for the identification of new enzyme functions and metabolic pathways. *EMBO Reports, 15*, 657–669.

Qiu, Y., Tittiger, C., Wicker-Thomas, C., Le Goff, G., Young, S., Wajnberg, E., et al. (2012). An insect-specific P450 oxidative decarbonylase for cuticular hydrocarbon biosynthesis. *Proceedings of the National Academy of Sciences of the United States of America, 109*, 14858–14863.

Rajniak, J., Barco, B., Clay, N. K., & Sattely, E. S. (2015). A new cyanogenic metabolite in Arabidopsis required for inducible pathogen defence. *Nature, 525*, 376–379.

Rasmann, S., Kollner, T. G., Degenhardt, J., Hiltpold, I., Toepfer, S., Kuhlmann, U., et al. (2005). Recruitment of entomopathogenic nematodes by insect-damaged maize roots. *Nature, 434*, 732–737.

Rhoads, A., & Au, K. F. (2015). PacBio sequencing and its applications. *Genomics, Proteomics & Bioinformatics, 13*, 278–289.

Rice, L. M., Earnest, T. N., & Brunger, A. T. (2000). Single-wavelength anomalous diffraction phasing revisited. *Acta Cystallographica. Section D, Biological crystallography, 56*, 1413–1420.

Richardson, T. H., Hsu, M. H., Kronbach, T., Barnes, H. J., Chan, G., Waterman, M. R., et al. (1993). Purification and characterization of recombinant-expressed cytochrome P450 2C3 from Escherichia coli: 2C3 encodes the 6 beta-hydroxylase deficient form of P450 3b. *Archives of Biochemistry and Biophysics, 300*, 510–516.

Rodriguez, A., Martinez, J. A., Flores, N., Escalante, A., Gosset, G., & Bolivar, F. (2014). Engineering Escherichia coli to overproduce aromatic amino acids and derived compounds. *Microbial Cell Factories, 13*, 126.

Ross, M. G., Russ, C., Costello, M., Hollinger, A., Lennon, N. J., Hegarty, R., et al. (2013). Characterizing and measuring bias in sequence data. *Genome Biology, 14*, R51.

Rudd, B. A., & Hopwood, D. A. (1980). A pigmented mycelial antibiotic in Streptomyces coelicolor: Control by a chromosomal gene cluster. *Journal of General Microbiology, 119*, 333–340.

Rutledge, P. J., & Challis, G. L. (2015). Discovery of microbial natural products by activation of silent biosynthetic gene clusters. *Nature Reviews. Microbiology, 13*, 509–523.

Schiff, P. B., & Horwitz, S. B. (1980). Taxol stabilizes microtubules in mouse fibroblast cells. *Proceedings of the National Academy of Sciences of the United States of America*, 77, 1561–1565.

Seyedsayamdost, M. R., & Clardy, J. (2014). Natural products and synthetic biology. *ACS Synthetic Biology, 3*, 745–747.

Sheriha, G. M., & Rapoport, H. (1976). Biosynthesis of Camptotheca-acuminata alkaloids. *Phytochemistry, 15*, 505–508.

Shi, Y. G. (2014). A glimpse of structural biology through X-ray crystallography. *Cell, 159*, 995–1014.

Shirley, A. M., & Chapple, C. (2003). Biochemical characterization of sinapoylglucose:choline sinapoyltransferase, a serine carboxypeptidase-like protein that functions as an acyltransferase in plant secondary metabolism. *The Journal of Biological Chemistry, 278*, 19870–19877.

Simao, F. A., Waterhouse, R. M., Ioannidis, P., Kriventseva, E. V., & Zdobnov, E. M. (2015). BUSCO: Assessing genome assembly and annotation completeness with single-copy orthologs. *Bioinformatics, 31*, 3210–3212.

Stansbury, M. S., & Moczek, A. P. (2014). The function of Hox and appendage-patterning genes in the development of an evolutionary novelty, the Photuris firefly lantern. *Proceedings of the Royal Society B: Biological Sciences, 281*, 1782.

Sturm, M., Bertsch, A., Gropl, C., Hildebrandt, A., Hussong, R., Lange, E., et al. (2008). OpenMS—An open-source software framework for mass spectrometry. *BMC Bioinformatics, 9*, 163.

Tamura, K., Stecher, G., Peterson, D., Filipski, A., & Kumar, S. (2013). MEGA6: Molecular evolutionary genetics analysis version 6.0. *Molecular Biology and Evolution, 30*, 2725–2729.

Torrens-Spence, M. P., Lazear, M., von Guggenberg, R., Ding, H., & Li, J. (2014). Investigation of a substrate-specifying residue within Papaver somniferum and Catharanthus roseus aromatic amino acid decarboxylases. *Phytochemistry, 106*, 37–43.

Torrens-Spence, M. P., Liu, P., Ding, H., Harich, K., Gillaspy, G., & Li, J. (2013). Biochemical evaluation of the decarboxylation and decarboxylation-deamination activities of plant aromatic amino acid decarboxylases. *The Journal of Biological Chemistry, 288*, 2376–2387.

Urban, P., Werck-Reichhart, D., Teutsch, H. G., Durst, F., Regnier, S., Kazmaier, M., et al. (1994). Characterization of recombinant plant cinnamate 4-hydroxylase produced in yeast. Kinetic and spectral properties of the major plant P450 of the phenylpropanoid pathway. *European Journal of Biochemistry, 222*, 843–850.

VanBuren, R., Bryant, D., Edger, P. P., Tang, H., Burgess, D., Challabathula, D., et al. (2015). Single-molecule sequencing of the desiccation-tolerant grass Oropetium thomaeum. *Nature, 527*, 508–511.

Weibel, E. K., Hadvary, P., Hochuli, E., Kupfer, E., & Lengsfeld, H. (1987). Lipstatin, an inhibitor of pancreatic lipase, produced by Streptomyces toxytricini. I. Producing organism, fermentation, isolation and biological activity. *The Journal of Antibiotics, 40*, 1081–1085.

Weng, J. K. (2014). The evolutionary paths towards complexity: A metabolic perspective. *The New Phytologist, 201*, 1141–1149.

Weng, J. K., Li, Y., Mo, H., & Chapple, C. (2012). Assembly of an evolutionarily new pathway for α-pyrone biosynthesis in Arabidopsis. *Science, 337*, 960–964.

Weng, J. K., Li, X., Stout, J., & Chapple, C. (2008). Independent origins of syringyl lignin in vascular plants. *Proceedings of the National Academy of Sciences of the United States of America, 105*, 7887–7892.

Weng, J. K., & Noel, J. P. (2012a). The remarkable pliability and promiscuity of specialized metabolism. *Cold Spring Harbor Symposia on Quantitative Biology, 77*, 309–320.

Weng, J. K., & Noel, J. P. (2012b). Structure-function analyses of plant type III polyketide synthases. *Methods in Enzymology, 515*, 317–335.

Weng, J. K., Philippe, R. N., & Noel, J. P. (2012). The rise of chemodiversity in plants. *Science, 336*, 1667–1670.

Werck-Reichhart, D., & Feyereisen, R. (2000). Cytochromes P450: A success story. *Genome Biology, 1*. REVIEWS3003. http://genomebiology.biomedcentral.com/articles/10.1186/gb-2000-1-6-reviews3003.

Westfall, C. S., Zubieta, C., Herrmann, J., Kapp, U., Nanao, M. H., & Jez, J. M. (2012). Structural basis for prereceptor modulation of plant hormones by GH3 proteins. *Science, 336*, 1708–1711.

Winn, M. D., Ballard, C. C., Cowtan, K. D., Dodson, E. J., Emsley, P., Evans, P. R., et al. (2011). Overview of the CCP4 suite and current developments. *Acta Crystallographica. Section D, Biological Crystallography, 67*, 235–242.

Wu, J., Zhou, T., Du, G., Zhou, J., & Chen, J. (2014). Modular optimization of heterologous pathways for de novo synthesis of (2S)-naringenin in Escherichia coli. *PloS One, 9*, e101492.

Xia, J., Sinelnikov, I. V., Han, B., & Wishart, D. S. (2015). MetaboAnalyst 3.0—Making metabolomics more meaningful. *Nucleic Acids Research, 43*, W251–W257.

Ye, K. Q., Ke, Y., Keshava, N., Shanks, J., Kapp, J. A., Tekmal, R. R., et al. (1998). Opium alkaloid noscapine is an antitumor agent that arrests metaphase and induces apoptosis in dividing cells. *Proceedings of the National Academy of Sciences of the United States of America, 95*, 1601–1606.

Zhao, Y. (2012). Auxin biosynthesis: A simple two-step pathway converts tryptophan to indole-3-acetic acid in plants. *Molecular Plant, 5*, 334–338.

Zheng, Z., Qualley, A., Fan, B., Dudareva, N., & Chen, Z. (2009). An important role of a BAHD acyl transferase-like protein in plant innate immunity. *The Plant Journal: For Cell and Molecular Biology, 57*, 1040–1053.

CHAPTER FIVE

Gene Discovery for Synthetic Biology: Exploring the Novel Natural Product Biosynthetic Capacity of Eukaryotic Microalgae

E.C. O'Neill*[,1], G. Saalbach[†], R.A. Field[†,1]
*University of Oxford, Oxford, United Kingdom
[†]John Innes Centre, Norwich, United Kingdom
[1]Corresponding authors: e-mail address: ellis.oneill@plants.ox.ac.uk; rob.field@jic.ac.uk

Contents

1. Introduction 100
2. Natural Product Synthases 102
   2.1 Polyketide Synthases 103
   2.2 Nonribosomal Peptide Synthetases 105
3. Genome Mining for the Identification of Natural Products 106
   3.1 Identification of PKSs 106
   3.2 Identification of NRPSs 110
   3.3 Hybrid NRPS/PKSs 111
   3.4 Identification of Other Components of Microalgal Natural Product Biosynthetic Pathways 112
4. Natural Product Discovery 113
5. Conclusions 115
Acknowledgments 116
References 116

Abstract

Eukaryotic microalgae are an incredibly diverse group of organisms whose sole unifying feature is their ability to photosynthesize. They are known for producing a range of potent toxins, which can build up during harmful algal blooms causing damage to ecosystems and fisheries. Genome sequencing is lagging behind in these organisms because of their genetic complexity, but transcriptome sequencing is beginning to make up for this deficit. As more sequence data becomes available, it is apparent that eukaryotic microalgae possess a range of complex natural product biosynthesis capabilities. Some of the genes concerned are responsible for the biosynthesis of known toxins, but there are many more for which we do not know the products. Bioinformatic

Methods in Enzymology, Volume 576
ISSN 0076-6879
http://dx.doi.org/10.1016/bs.mie.2016.03.005

and analytical techniques have been developed for natural product discovery in bacteria and these approaches can be used to extract information about the products synthesized by algae. Recent analyses suggest that eukaryotic microalgae produce many complex natural products that remain to be discovered.

1. INTRODUCTION

Eukaryotic algae represent an extremely diverse group of photosynthetic organisms, which produce a wide range of low molecular weight organic compounds. Currently some of these organisms are used for the industrial production of lipids and dyes, and others are under investigation for the manufacture of next generation biofuels. Algae have long been known as producers of highly bioactive metabolites, due to the lethal effects of harmful algal blooms. These occur when environmental conditions, such as nutrient supply or loss of predators, favor the rapid growth of one particular species which then produces toxins that accumulate throughout the food chain. These toxins include isoprenoids, such as domoic acid produced by diatoms (Ramsey, Douglas, Walter, & Wright, 1998), large polyethers such as the ciguatoxin produced by dinoflagellates (Murata, Legrand, Ishibashi, Fukui, & Yasumoto, 1990), and alkaloids, such as euglenophycin produced by a *Euglena* (Zimba, Moeller, Beauchesne, Lane, & Triemer, 2010). Harmful algal blooms are caused by a range of species, including dinoflagellates, diatoms, euglenozoa, and haptophytes (see Fig. 1). The structural diversity of toxic compounds and the wide range of producing organisms illustrate the complexity of the largely unexplored natural product biosynthetic capacity that is evident among the eukaryotic algae.

There are few genomic sequences currently available for the eukaryotic microalgae, and those that are available are focused on the green alga, as relatives of the plants. The microalgal genomes can be extremely complex, ranging in size from tens of megabases to hundreds of gigabases, with highly repetitive regions, accompanied by a propensity for gene splicing as well as unusual DNA modifications. This has limited the application of established genome sequencing approaches to this group of organisms. Researchers have recently turned their attention to transcriptomics, in which the mRNA is sequenced instead of the DNA. While this strategy only uncovers genes expressed under the defined experimental conditions, much of the complexity is removed and the amount of data to be analyzed is reduced by orders of magnitude.

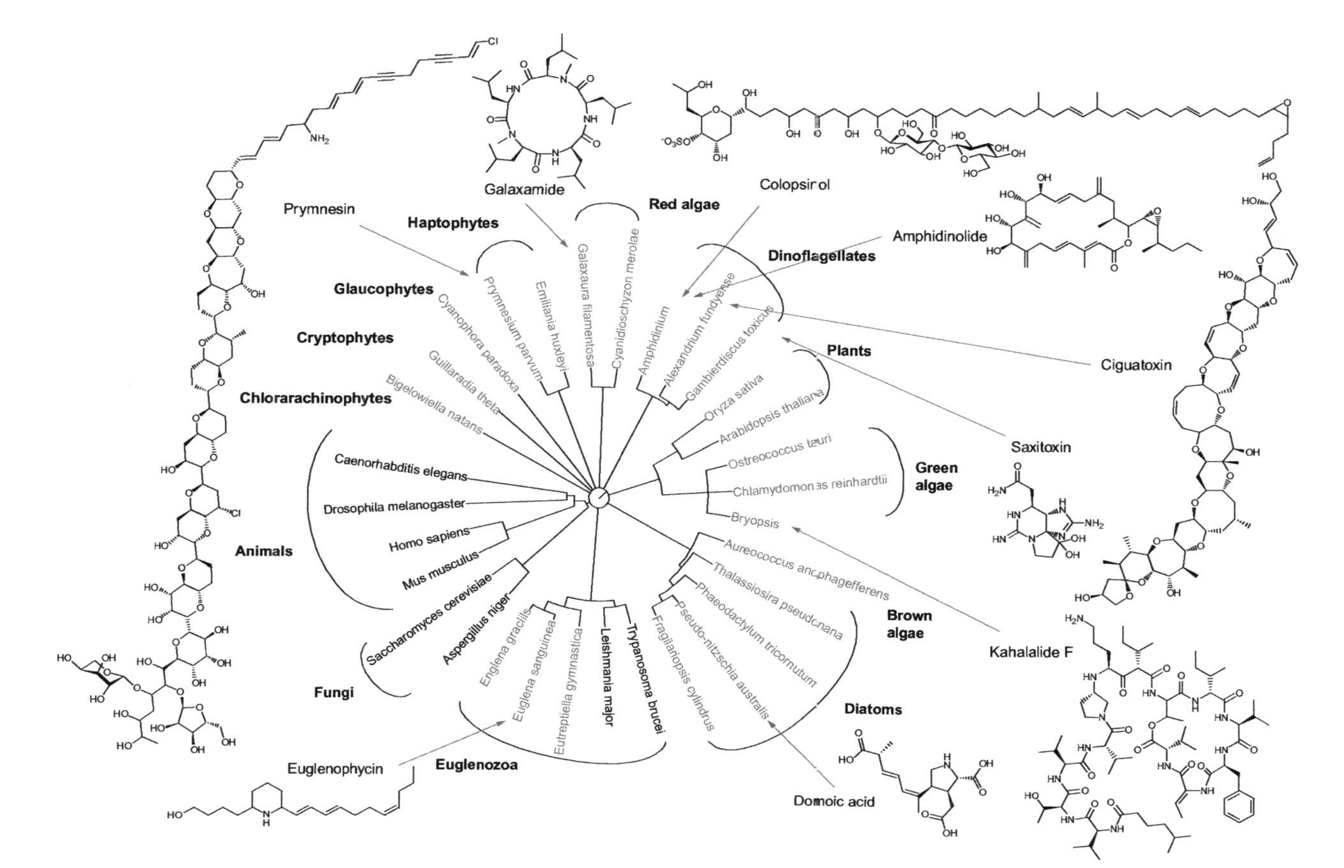

Fig. 1 Toxins produced by eukaryotic algae. Phylogeny showing relationship between algae and model organisms (Letunic & Bork, 2011) and a selection of the toxins produced by alga. Species shown in *green* are photosynthetic. (See the color plate.)

The few natural products known to be produced by microalgae are those that affect humans or animals, including the potent toxins saxitoxin and brevetoxin. The saxitoxin gene cluster has been characterized in cyanobacteria and the major protein involved, a polyketide synthase (PKS), was also identified in a eukaryotic diatom algae (Stüken et al., 2011). Bioactive metabolites produced by microalgae are being explored for possible applications in the life sciences, including as antiinflammatories, antimicrobials, and antioxidants (de Morais, Vaz, de Morais, & Costa, 2015). Other metabolites produced by alga, such as the putative siderophore produced by *Euglena gracilis* (O'Neill et al., 2015), have not received the same attention. The diversity and ability to adapt to environmental conditions make Eukaryotic microalga an important untapped source of new bioactives, potentially including antibiotics (Senhorinho, Ross, & Scott, 2015). In the limited full genome sequences available, there is clear capacity for complex natural product biosynthesis, but hardly any metabolites have been identified, structurally characterized, or assessed for their biological activity. This great capacity for complex natural product biosynthesis provides huge potential for the exploration and exploitation of these algae.

2. NATURAL PRODUCT SYNTHASES

There are many different classes of natural products and their biosynthesis can be complex. Most research on the biosynthesis of secondary metabolites has focused on bacteria, as the most genetically tractable organisms. Fungi synthesize a similar range of products to the bacteria, but produce more terpenoids (Quin, Flynn, & Schmidt-Dannert, 2014), while plants do not have "megasynthases"—multimodular enzyme assemblies—to produce nonribosomal peptides (NRPs) and polyketides (PKs) but instead have a greatly expanded repertoire of alkaloid-based natural products (Ziegler & Facchini, 2008). The biosynthetic principles elucidated in bacteria seem to apply to similar pathways in eukaryotes, although there are additional complexities in the latter. Of the many classes of natural products, the biosynthesis of PKs and NRPs is particularly interesting, as the enzymes involved have a unique structure and the products cannot be mistaken for primary metabolites. In these cases, biosynthesis is carried out by large multimodular megasynthases which can be over 1000 kDa in size. Typically the biosynthesis proceeds in a linear, iterative manner, with each module carrying out its reaction before transferring the growing product onto the next module in an N- to C-terminal direction, though there are exceptions such as the stalling and domain skipping seen in thalassospiramide biosynthesis (Ross et al., 2013).

2.1 Polyketide Synthases

PKs comprise a huge range of compounds, including antibiotics, such as tetracycline (Pickens & Tang, 2009), and toxins, including the algal ciguatoxin (Murata et al., 1990), formed by repeated decarboxylative condensation of carboxylic acids. Broadly speaking, PKSs can be large multidomain proteins (type I) or composed of discrete proteins with individual functions (type II); they can also be iterative, repeating the use of one module (Shen, 2003). Type III PKS, such as chalcone synthase, are single proteins that catalyze the repeated condensation of malonyl-CoA to form aromatic natural products and are found mainly in land plants (Abe & Morita, 2010). The PKS reaction proceeds in one module by selection and attachment of the extender unit onto the acyl carrier protein (ACP) by the acyltransferase (AT), a decarboxylative condensation to elongate the PK by the ketosynthase (KS), reduction of the ketone to an alcohol by the ketoreductase (KR), dehydration to form a double bond by the dehydratase (DH), and reduction of the double bond by the enoylreductase (ER) (see Fig. 2A and B) (Keatinge-Clay, 2012). A module can be missing specific domains, resulting in alterations in the product formed and there can be different accessory domains, such as methyl transferases, embedded within the megasynthase sequence and adding functionality (Hertweck, 2009). The final reaction, catalyzing the release of the PK from the PKS, is catalyzed by a thioesterase (TE) domain, which can also modify the PK by, for example, reduction or cyclization. A specific subset of PKSs, the trans-AT PKSs, do not contain the AT domain, which is instead encoded as a separate protein and delivers the activated carboxylic acid to all of the modules. The extender carboxylic acid is typically malonyl-CoA, extending the chain by two carbons at a time (as a result of accompanying decarboxylation), in the same manner as for fatty acid biosynthesis (Keatinge-Clay, 2012). However, other dicarboxylic acids can also be used, such as methylmalonyl-CoA or more exotic units, and these need to be synthesized in the cell (Chan, Podevels, Kevany, & Thomas, 2009; Ray & Moore, 2016). The extender unit can, to a certain extent, be predicted in bacterial species, based on the sequence of the AT domain, using programs such as SBSPKS (Anand et al., 2010). The products are often highly elaborated by further enzymes,

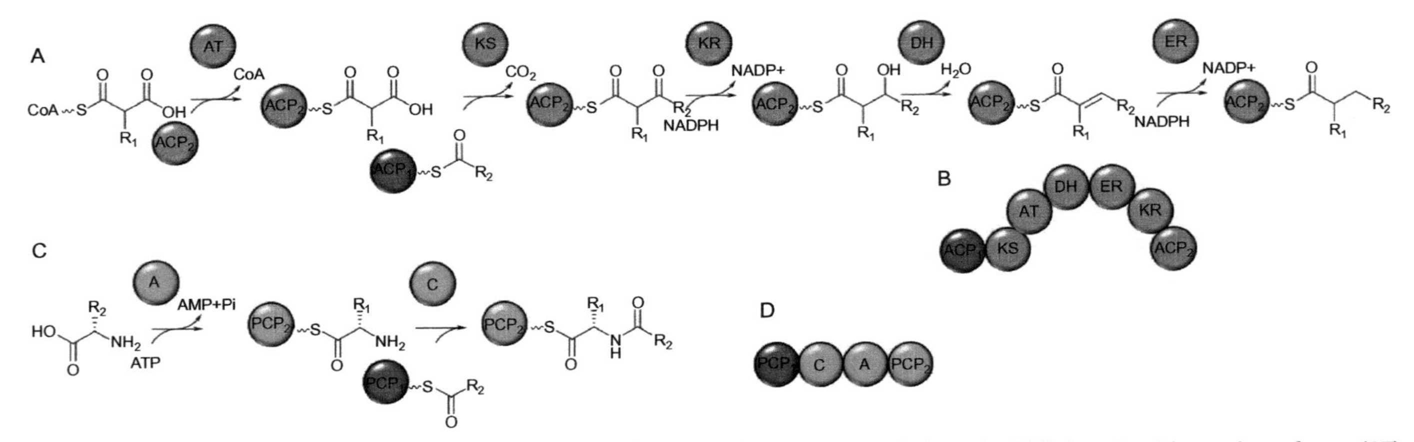

Fig. 2 The reaction catalyzed by natural product synthase domains. (A) Reactions carried out by PKS domains. The acyltransferase (AT) domain selects an acyl CoA and transfers the dicarboxylic acid onto the acyl carrier protein (ACP). R_1 is typically H for malonyl-CoA but a range of other extender units can be used, selected by the AT. The ketosynthase (KS) catalyzes the decarboxylative condensation between the acyl group and the acyl held on the previous ACP. The ketoreductase (KR) reduces the ketone to a hydroxyl, the dehydratase (DH) domain removes water, leaving a double bond, and the enoyl reductase (ER) reduces this double bond (Keatinge-Clay, 2012). (B) The order of domains in a typical PKS—note this is not the order in which the reactions are carried out. Any of these domains may be missing or inactive. (C) Reactions carried out by NRPS domains. The adenylation (A) domain activates an amino acid and attaches it to the peptidyl carrier protein (PCP). The condensation (C) domain catalyzes the condensation between the new amino acid and the acid held on the previous PCP, forming a new amide bond. (D) The order of domains in a typical NRPS. (See the color plate.)

adding extra functionality through oxidation and introduction of heteroatoms or pendant sugars, as is found in the toxic prymnesins produced by the haptophyte alga *Prymnesium parvum* (Manning & La Claire, 2010) (Fig. 1). Many of the larger compounds that have been isolated from microalgae are proposed to be synthesized in this manner, such as the colopsinols (Kobayashi & Kubota, 2007), and feeding of isotopically labeled precursors has confirmed this in some cases, such as the amphidinolides (Kobayashi & Tsuda, 2004) (Fig. 1).

2.2 Nonribosomal Peptide Synthetases

NRPs are a range of small molecules with a diversity of functions that include siderophores such as enterobactin (Walsh, Liu, Rusnak, & Sakaitani, 1990), antibiotics such as actinomycin (Schauwecker, Pfennig, Schröder, & Keller, 1998), and toxins such as microcystin (Noguchi et al., 2009). NRPs are formed by the sequential condensation of amino acids, including both proteinogenic and nonproteinogenic examples. They are synthesized by nonribosomal peptide synthetases (NRPSs), multimodular megasynthases: each module comprises an adenylation domain (A), which selects the incoming amino acid, activates it by adenylation, and attaches it to a peptidyl carrier protein (PCP) via a thiol-containing phosphopantetheine arm, and a condensation domain (C) which catalyzes the formation of an amide linkage in the growing peptide (see Fig. 2) (Walsh, 2016). As for PKSs, there is a TE domain which catalyzes the off-loading of the completed peptide and there can be accessory domains within the megasynthases, encodings *N*-methyltransferases or amino acid epimerases, for instance. The amino acid specificity of the A domain can be predicted with reasonable certainty based on the amino acid residues that line the binding pocket of this domain (Stachelhaus, Mootz, & Marahiel, 1999). NRPs can be modified by, for example, initiation with a lipid chain or formation of an ester between side chains and the C-terminus to cyclize the product, as is seen in arthrofactin, a member of the large class of cyclic lipopetide antibiotics (Roongsawang et al., 2003). There are very few examples of bioactive peptides that have been purified from eukaryotic algae: those that have been, such as the galaxamide from the red algal *Galaxaura* (Xu et al., 2008) and kahalalide F from the green alga *Bryopsis* (Hamann & Scheuer, 1993), were purified from macroalgae, collected from the wild, and so a bacterial origin for these compounds cannot be ruled out. There is no data on the biosynthesis of these compounds, thought their structure, including, *N*-methylation in the former and D-amino acids in the latter, and cyclization suggests an NRPS origin for these compounds.

As both PKSs and NRPSs have modular and linear biosynthetic logic, hybrid arrangements can occur in which an amino acid can be extended with a ketide or an amino acid incorporated into a PK. For example, in the former case the proteasome inhibitor epoxomicin has a peptide backbone extended by a ketide to form the epoxide warhead, which forms a covalent adduct with its' target enzyme (Schorn et al., 2013). Curacin is an example of the latter, in which a cysteine residue is added into a growing PK and then decarboxylated to continue the chain extension (Chang et al., 2004).

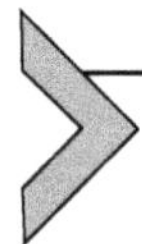

3. GENOME MINING FOR THE IDENTIFICATION OF NATURAL PRODUCTS

Having established the validity of eukaryotic microalgae as a source of natural products, it remains to be seen how these compounds are biosynthesized and to link the structures with their cognate biosynthetic machinery. Most of the eukaryotic genome sequencing efforts have to date been focused on a few Kingdoms (namely the animals, plants, and fungi), with very few algal genomes available (<20 to date). Among the few available, there are nevertheless examples of the complex megasynthases, including PKSs and NRPSs (Sasso, Pohnert, Lohr, Mittag, & Hertweck, 2012). There are however several hundred transcriptomes available, sequencing just the expressed mRNA. These datasets can be queried to uncover the presence and genetic architecture for these megasynthases which can lead to predictions of the type of molecule produced. There are programs which are designed for the identification of gene clusters encoding secondary metabolites, such as antiSMASH (Weber et al., 2015), and these can be used for the analysis of algal transcriptomes. The evolutionary distance of the eukaryotic microalgae from the organisms these programs are designed for (ie, leaping from bacteria to algae), combined with their unusual gene architecture and only having expressed sequences available, means that these programs cannot be relied upon, and manual curation of sequences is therefore necessary.

3.1 Identification of PKSs

In order to identify PKSs, the predicted proteome can be searched for the key KS catalytic domain using BLASTP (Johnson et al., 2008). The proteins identified as containing these domains can then be analyzed further by identifying other domains, searching for homologues and predicting activity. This strategy can recover fatty acid synthases, to which PKSs are related, but these can be readily identified at this stage by homology to known fatty acid biosynthesis components. DELTA-BLAST can be used to identify and compare the domains to the domains in known proteins, even if the primary sequences are quite divergent. For bacterial sequences, programs such as SBSPKS (Anand et al., 2010) and PKS/NRPS analysis website (Bachmann & Ravel, 2009) can be used, but in our experience they are not reliable for analysis of sequences from eukaryotic microalgae as they frequently miss domains (particularly, DH domains) which can be identified in the NCBI Conserved Domain Database. The monomer building blocks selected by the AT domain also cannot be predicted reliably, probably because of the evolutionary distance from the bacterial species that this software was designed to analyze (Yadav, Gokhale, & Mohanty, 2003). Once the domains have been identified, the biosynthetic logic of the PKS can then be used to predict the likely modification carried out by each module. For example, in the transcriptome of *E. gracilis* six multidomain PKSs could be identified (O'Neill et al., 2015). The largest (lm.8157) contains, in addition to a fully reducing PKS module, two putative enoyl hydratases and an HMGCoA synthase. These proteins, characterized in bacterial gene clusters as single domain proteins, are projected to add a β-methyl branch to PKs (Butcher et al., 2007) and this domain architecture suggests the formation of a methyl branched alkane.

It has been noted that the KS domains in PKSs within a given gene cluster tend to be more closely related than those from other gene clusters, and this relationship can be used to predict PKSs associated with a particular pathway. They also tend to group within pathway-specific clades according to the class of product formed; phylogenetic analysis performed using the NaPDoS webserver can aid pathway elucidation (Ziemert et al., 2012). Because the sequences from Eukaryotic microalga are so divergent and also distant from more studied organisms, their investigation is complicated, but our analysis indicates that KS domains do cluster according to the protein of origin. For example, in the genome of the polar microalga *Coccomyxa subellipsoidea* there are 10 KS domain-containing proteins (Blanc et al., 2012). It has previously been noted that four of these are adjacent to each other on the genome, indicative of a gene cluster (Sasso et al., 2012). The KS domains from these four proteins do indeed form a phylogenetic cluster, slightly more distantly related to three other PKSs in this alga

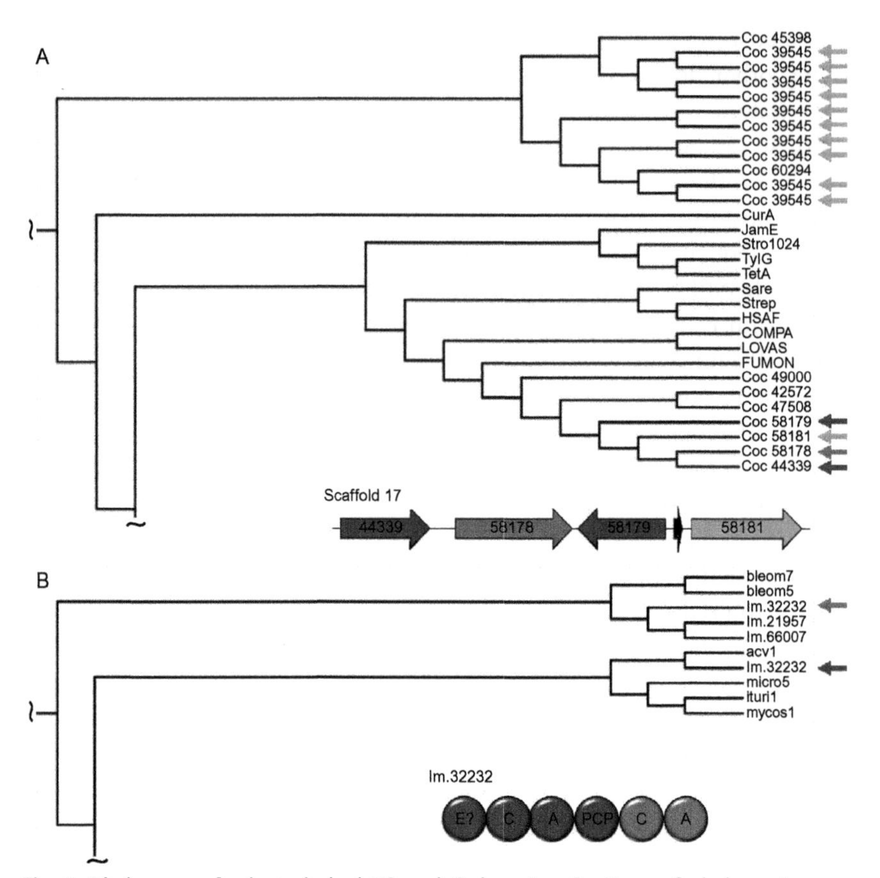

Fig. 3 Phylogeny of selected algal KS and C domains. Sections of phylogenies constructed using NaPDoS (Ziemert et al., 2012). (A) KSs from *Coccomyxa* PKSs cluster in two clades among the standard KSs, with all the KSs from the 10 domain protein (Coc 39545, *gray arrows*) in one and all the KSs from the four PKS gene cluster (shown below as a 120-kb region of scaffold 17) in the other (*colored arrows*). (B) C domains from *E. gracilis* multidomain NRPSs are scattered among the standard C domains, with the two C domains from one protein (lm.32232, shown below, *colored arrows*) separated. (See the color plate.)

(see Fig. 3). *Coccomyxa* also encodes a large PKS with 10 KS domains, which forms a separate clade and this has the remaining two KS domains embedded in it. It is unclear if these two are involved in the biosynthesis of the same compound as the large PKS, but it is suggestive of this situation. None of these PKSs contain an AT domain, which is presumably encoded in trans and may deliver the same extender unit to all modules. This analysis supports the phylogenetic relationship between KS domains within the same PKS and in gene clusters within Eukaryotic microalga, and supports the use of this type of analysis in evaluating biosynthetic pathways.

The PKSs in microalgae have some unusual features, indicating additional complexity unavailable to simpler organisms. For instance, the largest PKS encoded in the transcriptome of the Euglenid *Eutreptiella gymnastica* (Keeling et al., 2014), which is lacking some residues at the N-terminus, contains two hydroxymethylglutaryl-CoA synthase domains and a partial enoyl-CoA hydratase at the N-terminus, and a C-terminal decarboxylative aminotransferase, which is presumably involved in unloading (see Fig. 4). The first of these catalyzes the condensation of a carboxylic acid to a ketone, followed by dehydration, and the latter adds an amino acid side chain onto small molecules (such as in biotin biosynthesis) (Webster et al., 1999).

Type II PKSs encode each domain as a separate, individual protein and are thus extremely difficult to analyze, particularly in the absence of gene clustering or information about such clusters (viz transcriptome vs genome). However, it has been noted that *Karenia brevis*, a toxic dinoflagellate, encodes novel individual domains which are phylogenetically more similar to type I PKSs (Monroe & Van Dolah, 2008). The transcriptome sequences of two species of *Gambierdiscus* that produce maitotoxins were found to encode 192 KS domain transcripts, which formed five unique phylogenetic

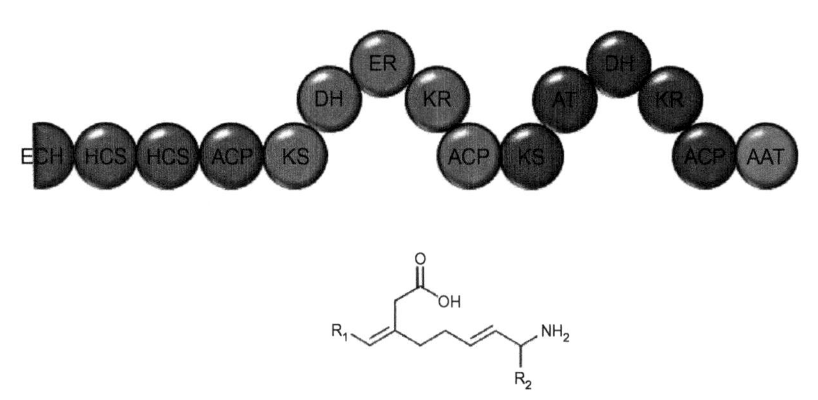

Fig. 4 *Eutreptiella* PKS. The largest PKS in the *Eutreptiella* transcriptome does not appear to be complete (Keeling et al., 2014). At the N-terminus is a partial enoyl-CoA-hydratase (ECH) and two hydroxymethylglutaryl-CoA synthase domains (HCS). It is unclear what the compound on which these domains act (R_1). This is followed by two PKS domains, one fully reducing without an AT domain and one leaving a double bond. This is followed by a decarboxylative amino acid transferase (AAT) which can take an amino acid and add it to a small molecule, decarboxylating it and adding the side chain (R_2), and the terminal nitrogen, to give the possible product shown. (See the color plate.)

clusters (Kohli et al., 2015). This implies that maitotoxin, which contains 164 carbons, is not biosynthesized by an iterative PKS, as has been suggested, but is constructed on a module-by-module basis and that this may be the case for other polycyclic polyether toxins, such as ciguatoxin and brevetoxin.

3.2 Identification of NRPSs

Putative NRPSs can be recognized based on the presence of the amino acid condensation (C) domain. This search strategy reveals many sequences which are most closely related to enzymes involved in fatty acid biosynthesis. Only those proteins which have more than one domain, including at least one A domain and one C domain, may be considered as candidate NRPSs. NRPS genes seem to be considerably rarer among the eukaryotic microalgae than PKSs, though this may be due to the search strategy here being suboptimal. The amino acid selected by the A domain can be predicted in bacteria and fungi, for example, using NRPSpredictor (Röttig et al., 2011) or PKS/NRPS analysis website (Bachmann & Ravel, 2009), but these are not successful in predicting the microalgal specificity, probably due to the evolutionary distance from the species these programs are designed to deal with. Those that are identified seem to be smaller and less complex than are found among bacteria and fungi, in contrast to the PKSs. For example, *Bigelowiella natans* has one of the largest microalgal NRPSs, consisting of 10 modules (C-A-PCP-C-A-PCP-C-A-PCP-TE) but this would only make a tripeptide, an extremely small example of an NRP (Curtis et al., 2012).

NaPDoS can also be used to map the phylogeny of the C domains from NRPSs in order to link enzymes associated with the same biosynthetic pathway (Ziemert et al., 2012). Due to the limited number of identified NRPSs in the eukaryotic algae, we have been unable to validate this strategy in these organisms. In *E. gracilis*, which encodes five multidomain NRPSs (O'Neill et al., 2015), the relationship between C domains within a single megasynthase (lm.32232) does not hold (see Fig. 3). However, this may be due to the additional N-terminal domain being in reality an epimerase, meaning the two C domains in this protein actually act to join D–L and L–L amino acids, respectively (Rausch, Hoof, Weber, Wohlleben, & Huson, 2007). Other domains can be found within the algal NRPS megasynthases. For example, *E. gracilis* transcript lm.9669 encodes an epimerase domain upstream of the first condensation domain, as defined by antiSMASH, though this is identified as an additional C domain by CDD homology.

3.3 Hybrid NRPS/PKSs

Hybrid NRPS/PKS gene clusters are well known among the bacteria, but are quite rare among the eukaryotic microalga, with currently only a few examples of single NRPS modules within large PKSs (Sasso et al., 2012). By screening a *K. brevis* fosmid library using PCR, a 16-kb gene cluster containing three NRPSs, one PKS, and a TE, with each gene encoding a single module, was successfully cloned (López-Legentil, Song, DeTure, & Baden, 2010). The genes are arranged in a bacteria-like operon and this is consistent with the identification of these genes in DNA purified from the plastid. This is hard to reconcile with knowledge of genome evolution, but it is possible this gene arrangement has been recently transferred in (or that it is actually derived from a bacterial contaminant). As this cluster seems to be bacterial in origin, it is possible to predict a structure for the produced compound (Fig. 5). We can predict that the structure contains glutamine, proline, and isoleucine in that order, assuming normal colinearity is employed, followed by a two carbon extension and reduction of the ketone (Weber et al., 2015). It is unclear if the TE merely releases the product by hydrolysis or if there is some cyclization involved and it is not possible to identify tailoring enzymes.

Recently three trans-AT PKS genes, one with an NRPS module, were identified in the genome of the haptophyte *Chrysochromulina tobin* in the same orientation, with only a stop codon between them (Hovde et al., 2015). This is reminiscent of a bacterial operon and is somewhat surprising in a eukaryotic cell. It is probable that there is a sequencing error in this

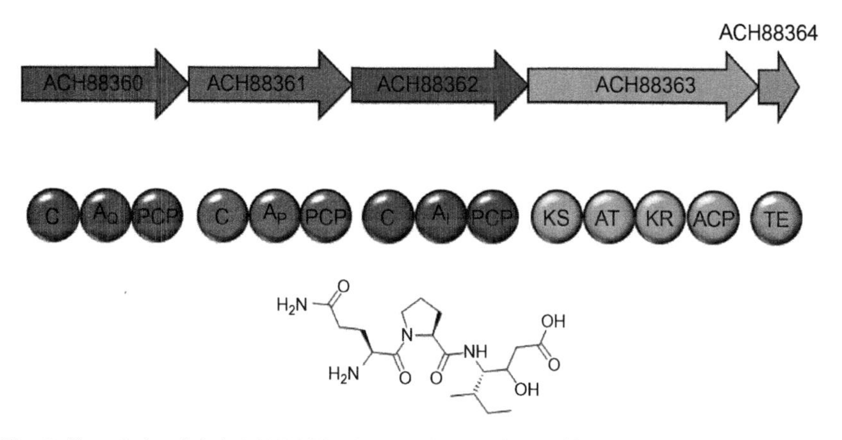

Fig. 5 *Karenia brevis* hybrid NRPS/PKS gene cluster. The 16-kb gene cluster from *K. brevis* contains five genes, which encode three NRPS modules, one PKS module and a thioesterase (López-Legentil et al., 2010). This is tentatively predicted to produce the compound shown. (See the color plate.)

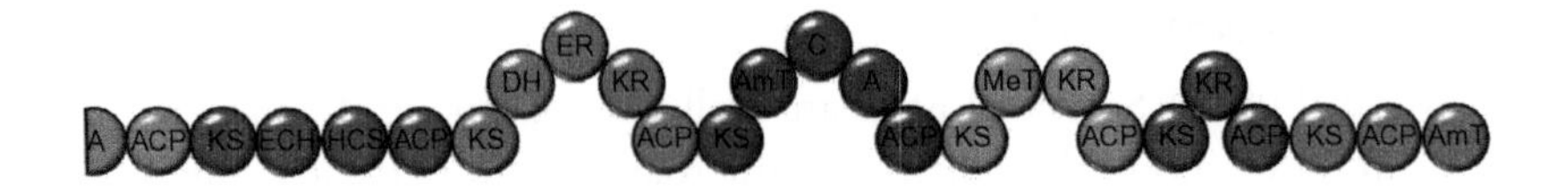

Fig. 6 *Chrysochromulina* PKS. There is a hybrid PKS/NRPS encoded in the *Chrysochromulina* genome (Hovde et al., 2015), although it is apparently interrupted by two stop codons, within domains. Assuming these are sequencing errors and this is in reality one gene, the protein has the domain architecture shown with six ketosynthases, one NRPS domain and is initiated by an A domain activating an unknown substrate (R_1). There is also an enoyl-CoA-hydratase (ECH) and a hydroxymethylglutaryl-CoA synthase domain (HCS), two amino transferases (AmT), which likely replace a ketone with a nitrogen from glutamate, and a C-methyltransferase (MeT). Assuming each domain acts once and the product is released by hydrolysis (by no means certain) the product may resemble the structure shown, with R_2 representing the amino acid side chain which cannot be predicted. (See the color plate.)

contig, and it is in fact a single hybrid PKS/NRPS. There is an initiating adenylation domain, which activates a starter molecule, followed by six KSs and one NRPS (see Fig. 6). There is also a C-methyltransferase which catalyzes the addition of a methyl group to the ketide α-position. In order to leave this position unoccupied, this module probably uses malonate as the extender unit, delivered by a trans-AT which probably supplies the other modules of this PKS. There is a hydroxymethylglutaryl-CoA synthase and an enoyl-CoA hydratase domain, which add a carboxylate branch, and if the latter acts twice form a β-methyl group, and also two amino transferase domains. Although the substrates cannot be predicted, it is possible to predict some features of the structure, as outlined in Fig. 6.

3.4 Identification of Other Components of Microalgal Natural Product Biosynthetic Pathways

Genes encoding bacterial and fungal natural product biosynthetic machinery are typically clustered, with genes involved in the biosynthesis of a single molecule adjacent within the genome. It is less clear to what extent this also occurs in other eukaryotes, although in plants some of the genes for the synthesis of a single product may be located near to each other (Osbourn, Papadopoulou, Qi, Field, & Wegel, 2012). Whether clusters are reliable enough to be useful in the elucidation of biosynthetic pathways in the eukaryotic microalgae remains to be seen, but there is some evidence, at least in *Coccomyxa*, where a cluster of four PKS genes and a desaturase were found adjacent to each other in the genome (Sasso et al., 2012).

If a transcriptome sequencing approach is used then physical proximity of genes on the genome cannot be evaluated. However, transcriptomics can be used to identify genes with an expression pattern mirroring the production of the metabolite, as has been used successfully in plants (Góngora-Castillo et al., 2012). As these are single-celled organisms, tissue-specific variation cannot be considered, but either temporal or environment-dependent variation in toxin production in a single strain, or comparison between strains provides valuable insights. The toxicity and gene expression profiles of the haptophyte *Prymnesium pavum* were measured under different nutrient stresses and, by correlating the two, some hints as to the genes involved in toxicity can be explored (Beszteri et al., 2012). Comparison of the transcriptome of a toxic wild type and a nontoxic mutant strain of the dinoflagellate *Alexandrium catenella* confirmed alterations in genes known to be involved in the biosynthesis of saxitoxin (Zhang, Zhang, Lin, & Wang, 2014). The transcriptome of a nontoxic strain of *Heterocapsa circularisquama* revealed that it contained more natural product synthases than the toxic strains, to the authors' surprise (Salcedo, Upadhyay, Nagasaki, & Bhattacharya, 2012). This illustrates the diversity of the natural products that can be made by algae, which has been greatly ignored due to preoccupation with toxin production.

4. NATURAL PRODUCT DISCOVERY

The majority of algal natural products identified so far are toxins and their purification is typically achieved by bioactivity-guided fractionation (Potterat & Hamburger, 2013). For compounds for which there is no bioassay, other attributes, such as spectrophotometric profile, can be used. Projecting ahead, high-throughput metabolomic analyses and bioinformatic techniques developed for bacterial natural products will enable the rapid discovery of new compounds in microalgae based on mass spectrometry profiling (Duncan et al., 2015). By utilizing natural products libraries, known compounds can be identified and data shared through the community to facilitate discovery projects and prevent rediscovery of known compounds (Wang et al., 2016). These data can be used to rapidly identify known compounds and to prioritize structure elucidation projects on novel compounds which can be purified, once production conditions have been optimized, through mass-guided fractionation. These strategies can be readily deployed for discovery projects in any culturable microorganism, including microalgae. Imaging mass spectrometry can also be used to identify microbial compounds from colonies grown on agar plates and shows potential for roll out beyond bacteria and fungi (Yang et al., 2012). For example, using a MALDI-ToF mass spectrometer to image colonies of *E. gracilis* grown on agar plates revealed a peak that was localized to the colony with a $m/z = 870.553$ (see Fig. 7) (E.C. O'Neill, G. Saalbach, & R.A. Field, unpublished observations); MS2 fragmentation could then be used to investigate this peak further, though the structure of the compound in question has not yet been elucidated.

It has been noted that the choice of the cultivation parameters is important in the range of secondary metabolites produced by microorganisms and by altering growth conditions the production of cryptic natural products can be triggered (Scherlach & Hertweck, 2009). By challenging microorganisms through varying culture conditions, nutrient availability, and stress, the

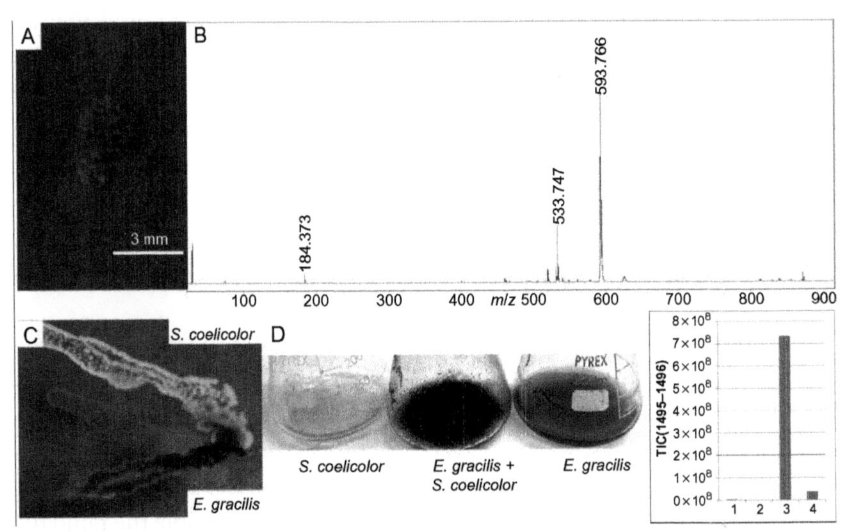

Fig. 7 *Euglena* and natural products biosynthesis. (A) Imaging mass spectrometry of *Euglena gracilis* colonies grown on EG:JM+Glc agar, imaged using MALDI-ToF. The pixels containing ions corresponding to $m/z = 870–871$ Da are highlighted. (B) MS2 of the corresponding ion at $m/z = 870.553$. (C) *E. gracilis* and *Streptomyces coelicolor* streaked on EG:JM+Glc agar. (D) Cultures of *E. gracilis* and *S. coelicolor* grown in EG: JM+Glc media. After 7 days the coculture is darkly pigmented. LCMS analysis of the methanol extract of these cultures shows that the peak corresponding to CDA ($m/z = 1495.52$) is not detectable in either the *Euglena* (1) or *Streptomyces* (2) only extracts but is easily detectable in the coculture (3), corresponding to the extract from *S. coelicolor* grown on its preferred media (LB, 4). (See the color plate.)

profile of secondary metabolites can be altered and this is critical in the production of many compounds (Bode, Bethe, Hofs, & Zeeck, 2002). Coculturing of organisms with other species can be used to influence metabolite profiles, possibly by the simulation of competition or predation triggering the production of defense molecules. For example, pestalone was only produced by a marine fungus when a marine bacterium was also present during fermentation (Cueto et al., 2001). These techniques could also be considered for inducing the production of natural products in algae; it is also possible to use algae to alter the expression of natural products in bacteria. We found that on coculturing *Streptomyces coelicolor* in the presence of *E. gracilis* a large increase in the production of CDA (calcium-dependent antibiotic) could be observed, while the other products were unchanged (see Fig. 7) (Hojati et al., 2002; E.C. O'Neill, G. Saalbach, & R.A. Field, unpublished observations). There is also a clear increase in the production of the red pigment prodigiosin when the two species are grown on agar in close proximity. It is unclear if this is due to a signaling mechanism, inducing production of these compounds as a response to the alga, or if there are other alterations in the local environment of the alga. These could include nutrient release, pH changes, or depletion of an environmental factor, such as take up of iron by the *Euglena*, as has previously been noted (O'Neill et al., 2015). These results suggest that the coculturing of eukaryotic algae with other species may be useful in facilitating the unlocking of cryptic biosynthetic pathways.

5. CONCLUSIONS

Eukaryotic microalgae have a complex evolutionary history and the genetic contributions from their endosymbiont-derived plastids give them highly complex metabolism (O'Neill, Trick, Henrissat, & Field, 2015). They have long been known to produce a range of toxins with intricate structures, including some of the largest nonbiopolymer compounds known. With new nucleic acid sequencing technologies, it is becoming apparent that there is a great capacity for natural product biosynthesis that has not been explored in these organisms. By utilizing the techniques developed for interrogating bacterial natural product biosynthesis, it is now possible to identify and characterize these genes. Although genetic manipulation techniques lag behind in these nonmodel organisms, there are still techniques that appear to be applicable to many algae, such as RNAi (Cerutti, Ma, Msanne, & Repas, 2011), that can be explored to link genes to products. With advances in technologies, there are now sophisticated high-throughput natural product identification methodologies and databases which can assist in the identification of novel compounds from algae. By moving away from model organisms there is, quite literally, an ocean of potential hidden away in the eukaryotic microalgae.

ACKNOWLEDGMENTS

The authors would like to thank L.H. for assistance with the mass spectrometry and M.R. for general advice on natural product chemistry. The authors gratefully acknowledge support from the BBSRC Institute Strategic Programme Grant on Understanding and Exploiting Metabolism (MET) (BB/J004561/1), the BBSRC Open Plant Synthetic Biology Centre (BB/L014130/1), and the John Innes Foundation.

REFERENCES

Abe, I., & Morita, H. (2010). Structure and function of the chalcone synthase superfamily of plant type III polyketide synthases. *Natural Product Reports*, *27*(6), 809–838. http://dx.doi.org/10.1039/b909988n.

Anand, S., Prasad, M. V. R., Yadav, G., Kumar, N., Shehara, J., Ansari, M. Z., & Mohanty, D. (2010). SBSPKS: Structure based sequence analysis of polyketide synthases. *Nucleic Acids Research*, *38*(Suppl. 2), W487–W496. http://dx.doi.org/10.1093/nar/gkq340.

Bachmann, B. O., & Ravel, J. (2009). Methods for in silico prediction of microbial polyketide and nonribosomal peptide biosynthetic pathways from DNA sequence data. In D. A. Hopwood (Ed.), *Complex enzymes in microbial natural product biosynthesis, part A: Overview articles and peptides: Vol. 458.* (pp. 181–217). Amsterdam: Elsevier.

Beszteri, S., Yang, I., Jaeckisch, N., Tillmann, U., Frickenhaus, S., Glöckner, G., … John, U. (2012). Transcriptomic response of the toxic prymnesiophyte *Prymnesium parvum* (N. Carter) to phosphorus and nitrogen starvation. *Harmful Algae*, *18*, 1–15. http://dx.doi.org/10.1016/j.hal.2012.03.003.

Blanc, G., Agarkova, I., Grimwood, J., Kuo, A., Brueggeman, A., Dunigan, D. D., … Van Etten, J. L. (2012). The genome of the polar eukaryotic microalga *Coccomyxa subellipsoidea* reveals traits of cold adaptation. *Genome Biology*, *13*(5), R39. http://dx.doi.org/10.1186/gb-2012-13-5-r39.

Bode, H. B., Bethe, B., Hofs, R., & Zeeck, A. (2002). Big effects from small changes: Possible ways to explore nature's chemical diversity. *ChemBioChem*, *3*(7), 619–627. http://dx.doi.org/10.1002/1439-7633(20020703)3:7<619::aid-cbic619>3.0.co;2-9.

Butcher, R. A., Schroeder, F. C., Fischbach, M. A., Straight, P. D., Kolter, R., Walsh, C. T., & Clardy, J. (2007). The identification of bacillaene, the product of the PksX megacomplex in *Bacillus subtilis*. *Proceedings of the National Academy of Sciences*, *104*(5), 1506–1509. http://dx.doi.org/10.1073/pnas.0610503104.

Cerutti, H., Ma, X., Msanne, J., & Repas, T. (2011). RNA-mediated silencing in algae: Biological roles and tools for analysis of gene function. *Eukaryotic Cell*, *10*(9), 1164–1172. http://dx.doi.org/10.1128/ec.05106-11.

Chan, Y. A., Podevels, A. M., Kevany, B. M., & Thomas, M. G. (2009). Biosynthesis of polyketide synthase extender units. *Natural Product Reports*, *26*(1), 90–114. http://dx.doi.org/10.1039/B801658P.

Chang, Z., Sitachitta, N., Rossi, J. V., Roberts, M. A., Flatt, P. M., Jia, J., … Gerwick, W. H. (2004). Biosynthetic pathway and gene cluster analysis of curacin A, an antitubulin natural product from the tropical marine cyanobacterium *Lyngbya majuscula*. *Journal of Natural Products*, *67*(8), 1356–1367. http://dx.doi.org/10.1021/np0499261.

Cueto, M., Jensen, P. R., Kauffman, C., Fenical, W., Lobkovsky, E., & Clardy, J. (2001). Pestalone, a new antibiotic produced by a marine fungus in response to bacterial challenge. *Journal of Natural Products*, *64*(11), 1444–1446. http://dx.doi.org/10.1021/np0102713.

Curtis, B. A., Tanifuji, G., Burki, F., Gruber, A., Irimia, M., Maruyama, S., … Archibald, J. M. (2012). Algal genomes reveal evolutionary mosaicism and the fate of nucleomorphs. *Nature*, *492*(7427), 59–65. http://dx.doi.org/10.1038/nature11681.

de Morais, M. G., Vaz, B. d. S., de Morais, E. G., & Costa, J. A. V. (2015). Biologically active metabolites synthesized by microalgae. *BioMed Research International*, *2015*, 15. http://dx.doi.org/10.1155/2015/835761.

Duncan, K. R., Crüsemann, M., Lechner, A., Sarkar, A., Li, J., Ziemert, N., … Jensen, P. R. (2015). Molecular networking and pattern-based genome mining improves discovery of biosynthetic gene clusters and their products from *Salinispora* species. *Chemistry & Biology*, *22*(4), 460–471. http://dx.doi.org/10.1016/j.chembiol.2015.03.010.

Góngora-Castillo, E., Childs, K. L., Fedewa, G., Hamilton, J. P., Liscombe, D. K., Magallanes-Lundback, M., … Buell, C. R. (2012). Development of transcriptomic resources for interrogating the biosynthesis of monoterpene indole alkaloids in medicinal plant species. *PLoS One*, 7(12), e52506. http://dx.doi.org/10.1371/journal.pone.0052506.

Hamann, M. T., & Scheuer, P. J. (1993). Kahalalide F: A bioactive depsipeptide from the sacoglossan mollusk *Elysia rufescens* and the green alga *Bryopsis* sp. *Journal of the American Chemical Society*, *115*(13), 5825–5826. http://dx.doi.org/10.1021/ja00066a061.

Hertweck, C. (2009). The biosynthetic logic of polyketide diversity. *Angewandte Chemie (International Ed. in English)*, *48*(26), 4688–4716. http://dx.doi.org/10.1002/anie.200806121.

Hojati, Z., Milne, C., Harvey, B., Gordon, L., Borg, M., Flett, F., … Micklefield, J. (2002). Structure, biosynthetic origin, and engineered biosynthesis of calcium-dependent antibiotics from *Streptomyces coelicolor*. *Chemistry & Biology*, *9*(11), 1175–1187. http://dx.doi.org/10.1016/S1074-5521(02)00252-1.

Hovde, B. T., Deodato, C. R., Hunsperger, H. M., Ryken, S. A., Yost, W., Jha, R. K., … Cattolico, R. A. (2015). Genome sequence and transcriptome analyses of *Chrysochromulina tobin*: Metabolic tools for enhanced algal fitness in the prominent order Prymnesiales (Haptophyceae). *PLoS Genetics*, *11*(9), e1005469. http://dx.doi.org/10.1371/journal.pgen.1005469.

Johnson, M., Zaretskaya, I., Raytselis, Y., Merezhuk, Y., McGinnis, S., & Madden, T. L. (2008). NCBIBLAST: A better web interface. *Nucleic Acids Research*, *36*, W5–W9. http://dx.doi.org/10.1093/nar/gkn201.

Keatinge-Clay, A. T. (2012). The structures of type I polyketide synthases. *Natural Product Reports*, *29*(10), 1050–1073. http://dx.doi.org/10.1039/C2NP20019H.

Keeling, P. J., Burki, F., Wilcox, H. M., Allam, B., Allen, E. E., Amaral-Zettler, L. A., … Worden, A. Z. (2014). The Marine Microbial Eukaryote Transcriptome Sequencing Project (MMETSP): Illuminating the functional diversity of eukaryotic life in the oceans through transcriptome sequencing. *PLoS Biology*, *12*(6), e1001889. http://dx.doi.org/10.1371/journal.pbio.1001889.

Kobayashi, J., & Kubota, T. (2007). Bioactive macrolides and polyketides from marine dinoflagellates of the genus *Amphidinium*. *Journal of Natural Products*, *70*(3), 451–460. http://dx.doi.org/10.1021/np0605844.

Kobayashi, J., & Tsuda, M. (2004). Amphidinolides, bioactive macrolides from symbiotic marine dinoflagellates. *Natural Product Reports*, *21*(1), 77–93. http://dx.doi.org/10.1039/b310427n.

Kohli, G. S., John, U., Figueroa, R. I., Rhodes, L. L., Harwood, D. T., Groth, M., ... Murray, S. A. (2015). Polyketide synthesis genes associated with toxin production in two species of *Gambierdiscus* (Dinophyceae). *BMC Genomics*, *16*(1), 410. http://dx.doi.org/10.1186/s12864-015-1625-y.

Letunic, I., & Bork, P. (2011). Interactive tree of life v2: Online annotation and display of phylogenetic trees made easy. *Nucleic Acids Research*, *39*(Suppl. 2), W475–W478. http://dx.doi.org/10.1093/nar/gkr201.

López-Legentil, S., Song, B., DeTure, M., & Baden, D. (2010). Characterization and localization of a hybrid non-ribosomal peptide synthetase and polyketide synthase gene from the toxic dinoflagellate *Karenia brevis*. *Marine Biotechnology*, *12*(1), 32–41. http://dx.doi.org/10.1007/s10126-009-9197-y.

Manning, S. R., & La Claire, J. W., II. (2010). Prymnesins: Toxic metabolites of the golden alga, Prymnesium parvum Carter (Haptophyta). *Marine Drugs*, *8*(3), 678–704. http://dx.doi.org/10.3390/md8030678.

Monroe, E. A., & Van Dolah, F. M. (2008). The toxic dinoflagellate *Karenia brevis* encodes novel type I-like polyketide synthases containing discrete catalytic domains. *Protist*, *159*(3), 471–482. http://dx.doi.org/10.1016/j.protis.2008.02.004.

Murata, M., Legrand, A. M., Ishibashi, Y., Fukui, M., & Yasumoto, T. (1990). Structures and configurations of ciguatoxin from the moray eel *Gymnothorax javanicus* and its likely precursor from the dinoflagellate *Gambierdiscus toxicus*. *Journal of the American Chemical Society*, *112*(11), 4380–4386. http://dx.doi.org/10.1021/ja00167a040.

Noguchi, T., Shinohara, A., Nishizawa, A., Asayama, M., Nakano, T., Hasegawa, M., ... Shirai, M. (2009). Genetic analysis of the microcystin biosynthesis gene cluster in *Microcystis* strains from four bodies of eutrophic water in Japan. *The Journal of General and Applied Microbiology*, *55*(2), 111–123. http://dx.doi.org/10.2323/jgam.55.111.

O'Neill, E. C., Trick, M., Henrissat, B., & Field, R. A. (2015). Euglena in time: Evolution, control of central metabolic processes and multi-domain proteins in carbohydrate and natural product biochemistry. *Perspectives in Science*, *6*, 84–93. http://dx.doi.org/10.1016/j.pisc.2015.07.002.

O'Neill, E. C., Trick, M., Hill, L., Rejzek, M., Dusi, R. G., Hamilton, C. J., ... Field, R. A. (2015). The transcriptome of *Euglena gracilis* reveals unexpected metabolic capabilities for carbohydrate and natural product biochemistry. *Molecular Biosystems*, *11*(10), 2808–2820. http://dx.doi.org/10.1039/c5mb00319a.

Osbourn, A., Papadopoulou, K. K., Qi, X., Field, B., & Wegel, E. (2012). Finding and analyzing plant metabolic gene clusters. *Methods in Enzymology*, *517*, 113–138. http://dx.doi.org/10.1016/b978-0-12-404634-4.00006-1.

Pickens, L. B., & Tang, Y. (2009). Decoding and engineering tetracycline biosynthesis. *Metabolic Engineering*, *11*(2), 69–75. http://dx.doi.org/10.1016/j.ymben.2008.10.001.

Potterat, O., & Hamburger, M. (2013). Concepts and technologies for tracking bioactive compounds in natural product extracts: Generation of libraries, and hyphenation of analytical processes with bioassays. *Natural Product Reports*, *30*(4), 546–564. http://dx.doi.org/10.1039/C3NP20094A.

Quin, M. B., Flynn, C. M., & Schmidt-Dannert, C. (2014). Traversing the fungal terpenome. *Natural Product Reports*, *31*(10), 1449–1473. http://dx.doi.org/10.1039/C4NP00075G.

Ramsey, U. P., Douglas, D. J., Walter, J. A., & Wright, J. L. (1998). Biosynthesis of domoic acid by the diatom *Pseudo-nitzschia* multiseries. *Natural Toxins*, *6*(3-4), 137–146. http://dx.doi.org/10.1002/(SICI)1522-7189(199805/08)6:3/4<137::AID-NT28>3.0.CO;2-L.

Rausch, C., Hoof, I., Weber, T., Wohlleben, W., & Huson, D. H. (2007). Phylogenetic analysis of condensation domains in NRPS sheds light on their functional evolution. *BMC Evolutionary Biology*, 7, 78. http://dx.doi.org/10.1186/1471-2148-7-78.

Ray, L., & Moore, B. S. (2016). Recent advances in the biosynthesis of unusual polyketide synthase substrates. *Natural Product Reports*, *33*, 150–161. http://dx.doi.org/10.1039/C5NP00112A.

Roongsawang, N., Hase, K.-I., Haruki, M., Imanaka, T., Morikawa, M., & Kanaya, S. (2003). Cloning and characterization of the gene cluster encoding arthrofactin synthetase from *Pseudomonas* sp. MIS38. *Chemistry & Biology*, *10*(9), 869–880. http://dx.doi.org/10.1016/j.chembiol.2003.09.004.

Ross, A. C., Xu, Y., Lu, L., Kersten, R. D., Shao, Z., Al-Suwailem, A. M., ... Moore, B. S. (2013). Biosynthetic multitasking facilitates thalassospiramide structural diversity in marine bacteria. *Journal of the American Chemical Society*, *135*(3), 1155–1162. http://dx.doi.org/10.1021/ja3119674.

Röttig, M., Medema, M. H., Blin, K., Weber, T., Rausch, C., & Kohlbacher, O. (2011). NRPSpredictor2—A web server for predicting NRPS adenylation domain specificity. *Nucleic Acids Research*, *39*(Suppl. 2), W362–W367. http://dx.doi.org/10.1093/nar/gkr323.

Salcedo, T., Upadhyay, R. J., Nagasaki, K., & Bhattacharya, D. (2012). Dozens of toxin-related genes are expressed in a nontoxic strain of the dinoflagellate *Heterocapsa circularisquama*. *Molecular Biology and Evolution*, *29*(6), 1503–1506. http://dx.doi.org/10.1093/molbev/mss007.

Sasso, S., Pohnert, G., Lohr, M., Mittag, M., & Hertweck, C. (2012). Microalgae in the postgenomic era: A blooming reservoir for new natural products. *FEMS Microbiology Reviews*, *36*(4), 761–785. http://dx.doi.org/10.1111/j.1574-6976.2011.00304.x.

Schauwecker, F., Pfennig, F., Schröder, W., & Keller, U. (1998). Molecular cloning of the actinomycin synthetase gene cluster from *Streptomyces chrysomallus* and functional heterologous expression of the gene encoding actinomycin synthetase II. *Journal of Bacteriology*, *180*(9), 2468–2474.

Scherlach, K., & Hertweck, C. (2009). Triggering cryptic natural product biosynthesis in microorganisms. *Organic & Biomolecular Chemistry*, 7(9), 1753–1760. http://dx.doi.org/10.1039/B821578B.

Schorn, M., Zettler, J., Noel, J. P., Dorrestein, P. C., Moore, B. S., & Kaysser, L. (2013). Genetic basis for the biosynthesis of the pharmaceutically important class of epoxyketone proteasome inhibitors. *ACS Chemical Biology*, *9*(1), 301–309. http://dx.doi.org/10.1021/cb400699p.

Senhorinho, G. N. A., Ross, G. M., & Scott, J. A. (2015). Cyanobacteria and eukaryotic microalgae as potential sources of antibiotics. *Phycologia*, *54*(3), 271–282. http://dx.doi.org/10.2216/14-092.1.

Shen, B. (2003). Polyketide biosynthesis beyond the type I, II and III polyketide synthase paradigms. *Current Opinion in Chemical Biology*, 7(2), 285–295. ISSN 1367-5931, http://dx.doi.org/10.1016/S1367-5931(03)00020-6.

Stachelhaus, T., Mootz, H. D., & Marahiel, M. A. (1999). The specificity-conferring code of adenylation domains in nonribosomal peptide synthetases. *Chemistry and Biology*, *6*(8), 493–505. http://dx.doi.org/10.1016/s1074-5521(99)80082-9.

Stüken, A., Orr, R. J. S., Kellmann, R., Murray, S. A., Neilan, B. A., & Jakobsen, K. S. (2011). Discovery of nuclear-encoded genes for the neurotoxin saxitoxin in dinoflagellates. *PLoS One*, *6*(5), e20096. http://dx.doi.org/10.1371/journal.pone.0020096.

Walsh, C. T. (2016). Insights into the chemical logic and enzymatic machinery of NRPS assembly lines. *Natural Product Reports*, *33*(2), 127–135. http://dx.doi.org/10.1039/C5NP00035A.

Walsh, C. T., Liu, J., Rusnak, F., & Sakaitani, M. (1990). Molecular studies on enzymes in chorismate metabolism and the enterobactin biosynthetic pathway. *Chemical Reviews*, *90*(7), 1105–1129. http://dx.doi.org/10.1021/cr00105a003.

Wang, M., Carver, J., Phelan, V., Sanchez, L., Garg, N., Peng, Y., ... Dorrestein, P. (2016). GNPS—Global Natural Products Social Molecular Networking. *Nature Biotechnology*, (accepted for publication).

Weber, T., Blin, K., Duddela, S., Krug, D., Kim, H. U., Bruccoleri, R., … Medema, M. H. (2015). antiSMASH 3.0—A comprehensive resource for the genome mining of biosynthetic gene clusters. *Nucleic Acids Research*, *43*(W1), W237–W243. http://dx.doi.org/10.1093/nar/gkv437.

Webster, S. P., Alexeev, D., Campopiano, D. J., Watt, R. M., Alexeeva, M., Sawyer, L., & Baxter, R. L. (1999). Mechanism of 8-amino-7-oxononanoate synthase: Spectroscopic, kinetic, and crystallographic studies. *Biochemistry*, *39*(3), 516–528. http://dx.doi.org/10.1021/bi991620j.

Xu, W.-J., Liao, X.-J., Xu, S.-H., Diao, J.-Z., Du, B., Zhou, X.-L., & Pan, S.-S. (2008). Isolation, structure determination, and synthesis of galaxamide, a rare cytotoxic cyclic pentapeptide from a marine algae *Galaxaura filamentosa*. *Organic Letters*, *10*(20), 4569–4572. http://dx.doi.org/10.1021/ol801799d.

Yadav, G., Gokhale, R. S., & Mohanty, D. (2003). Computational approach for prediction of domain organization and substrate specificity of modular polyketide synthases. *Journal of Molecular Biology*, *328*(2), 335–363. http://dx.doi.org/10.1016/S0022-2836(03)00232-8.

Yang, J. Y., Phelan, V. V., Simkovsky, R., Watrous, J. D., Trial, R. M., Fleming, T. C., … Dorrestein, P. C. (2012). Primer on agar-based microbial imaging mass spectrometry. *Journal of Bacteriology*, *194*(22), 6023–6028. http://dx.doi.org/10.1128/jb.00823-12.

Zhang, Y., Zhang, S.-F., Lin, L., & Wang, D.-Z. (2014). Comparative transcriptome analysis of a toxin-producing dinoflagellate *Alexandrium catenella* and its non-toxic mutant. *Marine Drugs*, *12*(11), 5698–5718. http://dx.doi.org/10.3390/md12115698.

Ziegler, J., & Facchini, P. J. (2008). Alkaloid biosynthesis: Metabolism and trafficking. *Annual Review of Plant Biology*, *59*(1), 735–769. http://dx.doi.org/10.1146/annurev.arplant.59.032607.092730.

Ziemert, N., Podell, S., Penn, K., Badger, J. H., Allen, E., & Jensen, P. R. (2012). The natural product domain seeker NaPDoS: A phylogeny based bioinformatic tool to classify secondary metabolite gene diversity. *PLoS One*, 7(3), e34064. http://dx.doi.org/10.1371/journal.pone.0034064.

Zimba, P. V., Moeller, P. D., Beauchesne, K., Lane, H. E., & Triemer, R. E. (2010). Identification of euglenophycin—A toxin found in certain euglenoids. *Toxicon*, *55*(1), 100–104. http://dx.doi.org/10.1016/j.toxicon.2009.07.004.

CHAPTER SIX

cis-Prenyltransferase and Polymer Analysis from a Natural Rubber Perspective

M. Kwon, E.-J.G. Kwon, D.K. Ro[1]
University of Calgary, Calgary, AB, Canada
[1]Corresponding author: e-mail address: daekyun.ro@ucalgary.ca

Contents

1. Introduction 122
2. Rationale: Observation of Revertants from *rer2* Mutant 125
3. Generation of *rer2* and *srt1* Double Knockout Yeast Strain 128
   3.1 Background on *rer2Δ* Strain 128
   3.2 Deletion of *SRT1* on *rer2Δ* Strain 129
4. Complementation of *rer2Δ srt1Δ* with *CPT* and *CBP* 131
   4.1 Construct Generation 131
   4.2 Yeast Transformation with Plasmids 132
   4.3 Selection in 5-FOA 133
   4.4 Results 134
5. CPT Biochemical Assay Using Yeast Microsomes 135
   5.1 In Vitro CPT Assays 135
   5.2 Yeast Microsome Preparation 136
   5.3 *cis*-Prenyltransferase Enzyme Assays Using Yeast Microsomes and ^{14}C-IPP 138
   5.4 Dephosphorylation 139
   5.5 Thin Layer Chromatography 140
   5.6 Results 140
6. General Discussion 142
Acknowledgments 143
References 144

Abstract

Dolichol and natural rubber are representative *cis*-polyisoprenoids in primary and secondary metabolism, respectively. Their biosynthesis is catalyzed by *cis*-prenyltransferase (CPT) by sequential condensations of isopentenyl diphosphates (IPPs) to a priming molecule. Although prokaryotic CPTs have been well characterized, the mechanism of eukaryotic CPTs in *cis*-polyisoprene biosynthesis was only recently revealed. It was shown that eukaryotes have evolved a unique protein complex, comprised of CPT and CPT-binding protein (CBP), to synthesize *cis*-polyisoprenoids. In the context of this

Methods in Enzymology, Volume 576
ISSN 0076-6879
http://dx.doi.org/10.1016/bs.mie.2016.02.026

new discovery, we found discrepancies in literature for CPT or CBP biochemical assays and in vivo CPT complementation using *rer2* (yeast *CPT*) yeast mutant. Our study here shows that *rer2* revertants occur at a frequency that cannot be disregarded and are likely accountable for the results that cannot be explained by the CPT/CBP heteroprotein complex model. To make a stable mutant, *SRT1* gene (secondary *CPT* expressed at a basal level in yeast) was additionally deleted in the *rer2Δ* mutant background. This stable *rer2Δ srt1Δ* strain was then used to individually or simultaneously express *Arabidopsis CPT1* (*AtCPT1*, *At2g17570*) and *CBP* (*AtLEW1*, *At1G11755*). We found that the simultaneous expression of *Arabidopsis CPT1* and *AtLEW1* effectively complements the *rer2Δ srt1Δ* strain, whereas the individual expression of *AtCPT1* alone or *AtLEW1* alone failed to rescue the yeast mutant. Microsomes from the dual expresser showed an efficient incorporation of IPPs into *cis*-polyisoprenoid (30% in 2 h). These results showed that the *CPT/CBP* heteroprotein complex model is valid in *Arabidopsis thaliana*. Experimental details of these results are described in this methodology paper.

1. INTRODUCTION

Isoprenoids (or terpenoids) are a diverse and large class of natural products with more than 70,000 known structures (Dictionary of Natural Products; http://dnp.chemnetbase.com). Some isoprenoids are indispensable in prokaryotic and eukaryotic organisms, playing essential physiological roles such as hormones (eg, testosterone, brassinolides, and gibberellin), membrane components (eg, cholesterols), and prenyl moieties of ubiquinone, plastoquinone, and prenylated proteins. Isoprenoid metabolism, therefore, forms an essential branch of primary metabolism that controls the growth and development of living organisms. On the other hand, an enormous structural diversity of isoprenoids has served as a treasured chemical reservoir, from which a number of pharmaceutical, nutraceutical, aroma, flavor, and industrial chemicals have been found or developed to benefit mankind (Gershenzon & Dudareva, 2007). Understanding the key principles governing the biosynthesis of isoprenoids has been a major research theme in the scientific community.

The two central precursors of all isoprenoids are five-carbon molecules, isopentenyl diphosphate (IPP) and its isomer dimethylallyl diphosphate (DMAPP) (Tholl, 2006). IPP and DMAPP are synthesized through either mevalonate or methylerythritol phosphate pathway, and one to three IPP molecules are sequentially condensed onto DMAPP to generate geranyl (GPP), farnesyl (FPP), and geranyl geranyl diphosphate (GGPP). Terpene synthase, a large and diverse enzyme family conserved across all the kingdoms, utilizes the prenyl diphosphates as substrates (ie, GPP, FPP, and GGPP) to produce structurally diverse terpenes by carbocation cascade reactions (Chappell, 2002). The GPP, FPP, and GGPP synthases condense the isoprene moiety of IPP onto the DMAPP priming molecule in *trans*-configuration, and these enzymes are named *trans*-prenyltransferase (TPT). A number of biochemical studies have been carried out on TPTs, leading to the consolidated model to explain the structure and reaction mechanism of TPTs (Koksal, Jin, Coates, Croteau, & Christianson, 2011; Lesburg, Zhai, Cane, & Christianson, 1997; Starks, Back, Chappell, & Noel, 1997).

In contrast to the *trans*-configured condensation by TPT, IPP can be condensed to allylic diphosphates in the *cis*-configuration, and such reaction is catalyzed by *cis*-prenyltransferase (CPT). The most studied *cis*-isoprenoids are undecaprenol in prokaryote and dolichol in eukaryote (Fig. 1). Undecaprenol is polyisoprene alcohol synthesized in bacteria and functions as a sugar carrier molecule for peptidoglycan cell wall biosynthesis (Teng & Liang, 2012); dolichol is another polyisoprene alcohol used as a sugar carrier molecule for posttranslational modification (ie, glycosylation) of proteins on the endoplasmic reticulum in eukaryotes (Spiro, 2002). Undecaprenol and dolichol are structurally analogous except that the α-position of dolichol is saturated (Fig. 1). In the biosynthesis of these *cis*-polyisoprenoids, CPT catalyzes the condensations of 8–20 IPPs onto FPP to synthesize *cis*-polyisoprenes built on the *trans*-, *trans*-FPP priming molecule. Although the reaction mechanism of TPT and CPT are similar, the isolation of the

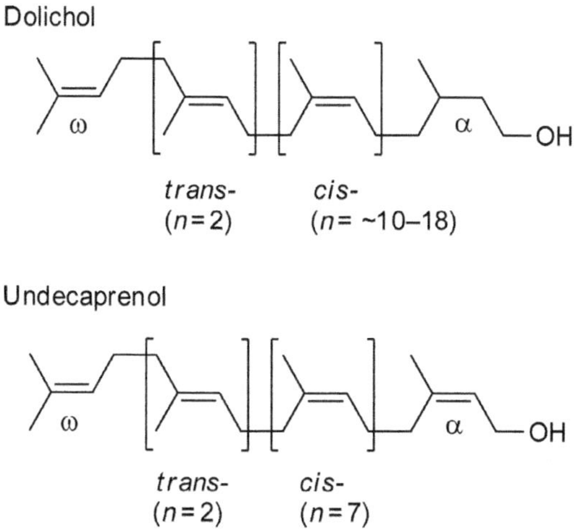

Dolichol

ω α OH

trans- ($n=2$) *cis*- ($n=$ ~10–18)

Undecaprenol

ω α OH

trans- ($n=2$) *cis*- ($n=7$)

Fig. 1 Structures of dolichol and undecaprenol.

first CPT from bacteria and its crystal structure showed that its primary and tertiary structures have no relationship to those of the TPT-family enzymes (Fujihashi et al., 2001), indicating that the TPT and CPT enzymes have evolved independently.

Another important *cis*-polyisoprenoid is natural rubber (NR), an important industrial biopolymer primarily produced from the Brazilian rubber tree (*Hevea brasiliensis*). NR possesses properties of superior resilience, heat dispersion, and abrasion resistance over synthetic rubber derived from petrochemicals, thereby being considered as an irreplaceable and strategically important industrial raw material (Cornish, 2001). Although more than 2500 plant species are known to produce NR, the Brazilian rubber tree is the only species that produces NRs in a commercially viable quantity and polymer length (van Beilen & Poirier, 2007). To be useful in industry, the average molecular weight of NR needs to be sufficiently large (>1 million Dalton). However, only a limited number of plant species, such as guayule, lettuce, and dandelion, are known to produce sufficiently lengthy NR biopolymers (Bushman et al., 2006; Mooibroek & Cornish, 2000). The simplest view of NR biosynthesis is that specialized CPTs emerged from undecaprenol and dolichol CPT catalyze the condensations of more than 10,000 IPPs without being terminated at the oligomeric length. Rubber elongation factor, small rubber particle protein, and CPT have been implicated in NR biosynthesis (Chakrabarty, Qu, & Ro, 2015; Collins-Silva et al., 2012; Hillebrand et al., 2012; Post et al., 2012), but reconstitution of NR biosynthesis in an in vitro condition or in surrogate hosts has not been achieved. The exact mechanism of NR biosynthesis remains still elusive at present.

Cumulative TPT studies have allowed consensual conclusions regarding the mechanism and metabolism of TPTs and their products, but the CPT-mediated biosynthesis of *cis*-isoprenoids including NR is not fully understood. Notably, recent four independent studies in human (Park et al., 2014), lettuce (Qu et al., 2015), dandelion (Epping et al., 2015), and tomato (Brasher et al., 2015) identified a new protein for CPT activity, conserved across the eukaryotic lineages. These works demonstrated that eukaryotes have developed a unique heteromeric CPT protein complex. Differently from the self-sufficient prokaryotic CPT, the heteroprotein complex is composed of a catalytic CPT and an unusual nonenzymatic protein, referred to as *CPTL* (CPT-like protein) in lettuce, *NgBR* (Nogo-B receptor) in human, *CPTBP* (CPT-binding protein) in tomato, and *RTA* (rubber transferase activator) in dandelion. These orthologous proteins do not possess the conserved motifs for CPT catalysis but directly interact with catalytic CPTs. Remarkably, silencing of the respective gene markedly reduced dolichol (tomato) or NR (lettuce and dandelion). In this chapter, we will refer this nonenzymatic protein as CBP (CTP-binding protein) for simplicity.

Collectively, the data from these reports concluded that two classes of CPTs are present; self-sufficient prokaryotic CPTs in bacteria and chloroplasts of plants, and self-insufficient CPTs which must form a heteroprotein complex with CBP to acquire the CPT enzymatic activity. These eukaryotic CPT/CBP complexes are conserved from yeast to human and are involved in dolichol and NR biosynthesis. The results from different organisms are complementary to each other and also validated data from different systems. Curiously, some previously reported results are not congruent to the model of eukaryotic CPT/CBP complexes that seems nonrebuttable based on the data from the four independent groups. Some examples are: (1) *Arabidopsis* CBP ortholog, *AtLEW1* showed in vitro CPT activity by itself when expressed in *Escherichia coli* (Zhang et al., 2008); (2) dandelion eukaryotic type *CPT* alone without *CBP* could functionally complement the mutant yeast strain that has a lesion in yeast *CPT* (*rer2* temperature sensitive strain, *rer2*^{ts}) (Schmidt et al., 2010); (3) rubber tree CPT purified from *E. coli* could synthesize NR of ~1 million Dalton without *CBP* coexpression with an addition of unknown latex components (Asawatreratanakul et al., 2003). Considering the substantial influences of these works in the community of plant biochemistry, it is necessary to discuss possible technical issues and incongruent results of such previous reports in the context of recent new findings.

In this chapter, we identified the key problem of using the *rer2* yeast strain for *CPT* complementation assays, and then developed a new yeast strain suitable for *CPT* complementation. Furthermore, in the new strain background, *Arabidopsis CPT1* (*At2g17570*) and *CBP* (*AtLEW1*) were expressed to examine their biochemical activities by in vivo complementation and in vitro enzyme assays using microsomes. The experimental rationales, detailed procedures, and *Arabidopsis CPT/CBP* (*AtCPT1/AtLEW1*) functional data are described here.

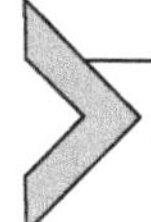

2. RATIONALE: OBSERVATION OF REVERTANTS FROM *rer2* MUTANT

In yeast (*Saccharomyces cerevisiae*), *RER2* encodes the dehydrodolichyl diphosphate synthase that catalyzes the CPT reaction in dolichol metabolic pathway. The null mutation of *RER2* in yeast (*rer2Δ*) is not entirely lethal, due to a weak expression of the second copy of *CPT* encoded in *SRT1*, and it grows extremely slowly (Sato, Fujisaki, Sato, Nishimura, & Nakano, 2001; Sato et al., 1999). However, such a slow growth pattern makes it cumbersome to use the *rer2Δ* for research. To overcome this, a temperature sensitive *rer2* mutant (*rer2*^{ts}) was identified that grows normally at a room temperature (RT) (23°C) but cannot grow at a nonpermissive temperature (37°C) (Sato et al., 1999). Since then, *rer2*^{ts} strain has been routinely used to evaluate in vivo activities of the *CPTs* (ie, dehydrodolichyl diphosphate synthases) from various plant species, including those of rubber tree and Russian dandelion.

We obtained the *rer2*^{ts} (SNH23-7D), which has *rer2* gene with a point mutation at the 209th residue (S209N), and assessed the feasibility to use this strain to measure CPT activities. To perform control experiments, the *rer2*^{ts} strain was first transformed with the empty yeast plasmid (p414) for a negative control or with the same vector expressing *RER2* under *GPD* promoter (p414-*RER2*) for a positive control. The two transformants were incubated on the selective solid medium at 23°C for 5–7 days. Both yeast transformants grew normally and formed a number of independent colonies on the plates at 23°C (Fig. 2A and B). These results confirmed that *rer2*^{ts} shows a normal growth phenotype at 23°C, as expected from other works.

In parallel, the same transformants were plated and incubated at a nonpermissive 37°C for 5–7 days. In these experiments, the *RER2*-expressing *rer2*^{ts} strain is expected to grow, but the empty plasmid-transformed *rer2*^{ts} strain is not expected to grow because the S209N point mutation makes RER2 unstable at 37°C. As predicted, the *RER2*-expressing *rer2*^{ts} showed a normal growth at 37°C (Fig. 2C). However, we consistently observed yeast colonies on the plates where the *rer2*^{ts} yeasts transformed with the empty plasmid were incubated at 37°C (Fig. 2D, arrow). The frequency of this apparent revertant was estimated to be 1 out of ~20,000 yeast cells, and this rate is not too low to ignore in complementation assays. To ensure that the revertant is not due to the problem in a specific batch of the *rer2*^{ts}, we received a new *rer2*^{ts} (SNH23-7D) strain again from Dr. Nakano laboratory (RIKEN, Japan) and repeated the experiments again with this freshly acquired strain. However, revertants still occurred in this independent experiment with a similar frequency.

To observe the growth rates of these transformants more accurately, colonies from each plate were selected and cultivated in liquid medium at a nonpermissive 37°C, and their growth rates were measured

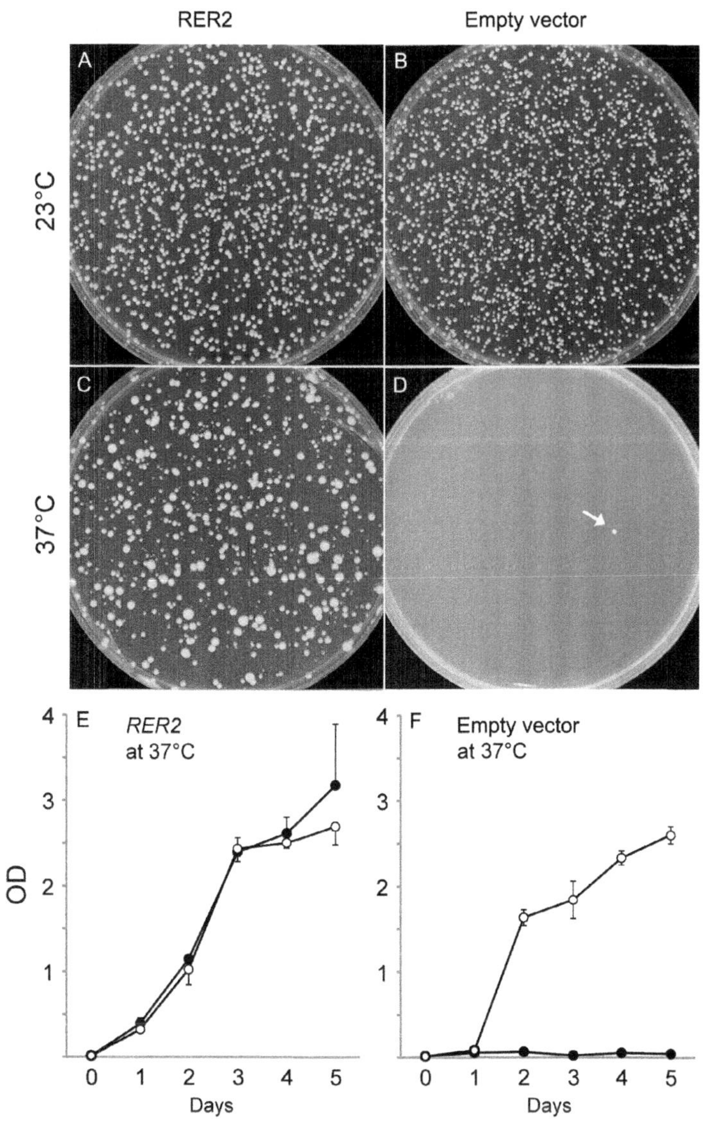

Fig. 2 Control experiments of *rer2*^{ts} mutant after transforming with the empty plasmid or expressing *RER2*. (A) and (B) Yeast *rer2*^{ts} mutants were grown at a permissive temperature (23°C) after transforming with the empty plasmid or *RER2*. (C) and (D) Yeast *rer2*^{ts} mutants were grown at a nonpermissive temperature (37°C) after transforming with the empty plasmid or *RER2*. The plate in D is one representative from multiple experiments, and the *arrow* in D indicates a colony from revertants. (E) and (F) Growth patterns of the yeasts at 37°C were plotted by measuring optical density (OD) at 600 nm. *Black circles* are from the yeasts preselected on solid medium at 23°C (A or B), and *white circles* are from the yeast preselected on solid medium at 37°C (C or D). Data are mean ± S.D. from three independent transformants.

spectroscopically at various time points. The *rer2^{ts}* yeasts expressing *RER2* showed similar growth rates at 37°C whether they were preincubated at 23 or 37°C (Fig. 2E). This result confirmed that the overexpression of *RER2* can competently complement the *rer2^{ts}* mutation at a nonpermissive temperature. On the other hand, the *rer2^{ts}* yeast harboring the empty vector, preselected on the plates at 23°C, showed no growth at 37°C (Fig. 2F, black circles), displaying a predicted temperature sensitive behavior. On the contrary, the empty vector harboring *rer2^{ts}* strain grown at 37°C (referred to as the revertant from here on) showed a growth rate almost identical to the *RER2*-expressing *rer2^{ts}* in liquid medium (Fig. 2F, white circles). These observations demonstrate that the revertant has an equally restored growth pattern as the *RER2*-expressing *rer2^{ts}*.

Although the observed revertant problem in complementation assays was not explicitly stated, an article reporting the occurrence of revertants from the null *rer2* mutant (*rer2Δ*) population can be found in literature (Schenk, Rush, Waechter, & Aebi, 2001). Intriguingly, in the lipid profile analysis of the revertants, dolichols produced from the *rer2Δ* revertants were longer (19–22 isoprene units) than those synthesized from wild-type yeast with an intact RER2 (14–18 isoprene units). Because it is known that SRT1 (the second dehydrodolichyl diphosphate synthase) synthesizes isoprene polymers with 19–22 isoprene units, authors proposed that the activation of *SRT1* is the direct cause of the revertants in *rer2Δ*. Based on this and our own observation from the *rer2^{ts}*, we concluded that the complementation assays using *rer2^{ts}* or *rer2Δ* strain must be performed with extreme cautions or is preferable to be abandoned. In addition, this implied that previous complementation results using *rer2^{ts}* strain need to be interpreted with cautions. Accordingly, we pursued to generate a double knockout yeast strain (*rer2Δ srt1Δ*) for CPT and CBP complementation and biochemical assays. Detailed procedures are given below.

3. GENERATION OF *rer2* AND *srt1* DOUBLE KNOCKOUT YEAST STRAIN

3.1 Background on *rer2Δ* Strain

To generate *rer2Δ* and *srt1Δ* double knockout strain, we deleted *SRT1* on the *rer2Δ* background. The *rer2* null mutant yeast strain (*MATα rer2Δ::HIS3 ade2 trp1 his3 leu2 ura3 lys2* pRS316-*RER2-URA3*) was obtained from Dr. Nakano (RIKEN, Japan). Since the *rer2Δ* strain grows extremely slowly, this strain is maintained by expressing *RER2* in pRS3-*URA3* plasmid, which can be removed on 5-fluoroorotic acid (5-FOA) plate where *URA3* gene product converts 5-FOA to a cytotoxic product.

3.2 Deletion of *SRT1* on *rer2Δ* Strain

3.2.1 Preparation of Homologous DNA

1. *SRT1* null mutation (*srt1Δ*) was generated in *rer2Δ* background by replacing *SRT1* open-reading frame (ORF) with a KanMX cassette. A linear piece of DNA for replacement was created by stitching together three pieces of DNA fragments—UP homologous fragment, DOWN homologous fragment, and KanMX fragment. Each of approximately 500 bp of UP and DOWN homologous regions has identical sequences at the S288C genome coordinates of 470205-469635 and 468681-468159, respectively, covering the *SRT1* ORF.
2. The three DNA fragments were PCR amplified using primer sets in Table 1 (primers are designed with T_m of ~60°C). Note that the UP reverse and DOWN forward primers contain overhangs (~60–62°C, T_m) complementary to 5′- and 3′-end of the KanMX fragment, respectively. Each PCR fragment must be gel purified for optimal stitching. For the stitching reaction, three purified fragments were combined in an equimolar ratio to a total DNA amount of 100–200 ng in a 20 μL volume containing 0.2 units of Q5 High-Fidelity DNA Polymerase (NEB), 1 × Q5 Reaction Buffer (NEB), and 0.2 m*M* dNTP. Five cycles of PCR were performed (98°C for 20 s, 60°C for 30 s, and 72°C for 30 s).
3. Once three fragments were stitched to a single fragment by the first five cycles of PCR, this stitched fragment was further PCR amplified by UP forward and DOWN reverse primers. The entire stitching reaction is combined with additional 0.3 units of Q5 High-Fidelity DNA Polymerase in a final volume of 50 μL in 1 × Q5 Reaction Buffer, 0.2 m*M* dNTP, 0.4 μ*M* UP forward primer, and 0.4 μ*M* DOWN reverse primer in a typical PCR program for Q5 DNA Polymerase (1 cycle of 98°C/30 s, 30 cycles of 98°C/10 s, 62°C/20 s, and 72°C/1 min, 1 cycle of 72°C/2 min). The final PCR product should be gel purified as there will be undesirable spurious products (as visible on gel) that are stitched or amplified during PCR. Finally, 2–3 μg of the correct size DNA was purified in the elution volume not exceeding 60 μL.

3.2.2 Yeast Transformation with Linear DNA

1. Transformation was carried out using a typical yeast transformation protocol with additional steps to select for antibiotic resistant yeast.

Table 1 Primers Used in This Study

Primer Names	Sequences
SRT1Up_F	GAGTCGAAGCTTCAACTCG
SRT1Up_R	GGACGAGGCAAGCTAAACAGATCTTTGGAGTAGTT ACCCTTAACAGG
SRT1Down_F	GATACTAACGCCGCCATCCATGCTATGGCAAGTAC ATGAAAATG
SRT1Down_R	GGGATGAAAATAGCATACCTGAG
KanMX_F	AGATCTGTTTAGCTTGCCTCG
KanMX_R	TGGATGGCGGCGTTAGTATC
SRT1KO_CF	GTGTTTAGAGCAGATATGCCC
AtCPT1_F	ATATACTAGTAACATGGCTGAACTTCCTGGTCAA
AtCPT1_R	CGTCATCCTTGTAATCCCCTAATTGTTTCTTCCTC TTTTCCAAGTATG
FLAG_R	ATATCTCGAGTCAGATCTTATCGTCGTCATCCTTG TAATCCCC
AtLEW1_F	ATATACTAGTAACATGGATTCGAATCAATCGATGC
AtLEW1_R	ATATCTCGAGTTAAGTTCCATAGTTTTGGT GGACTCC
RER2_F	ATATACTAGTAACATGGAAACGGATAGTGGTATAC
RER2_R	CGTCATCCTTGTAATCCCCATTCAACTTTTTTTCT TTCAAATCGAT

Note: Underlined sequences are complementary to KanMX sequence. *KO*, knockout. PCR amplification for FLAG-tagged genes (ie, AtCPT1-FLAG and RER2-FLAG) was performed twice, first with partial-FLAG reverse primer (primer pair: AtCPT1_F/AtCPT1_R, RER2_F/RER2_R) and second with complete-FLAG reverse primer (primer pair: AtCPT1_F/FLAG_R, RER2_F/FLAG_R). SRT1KO_CF (confirmation forward primer) was used with KanMX R primer for the PCR screening of yeast colonies to confirm *SRT1* deletion.

Mid-logarithmic phase yeast cells were harvested by centrifugation at $3000 \times g$, 3 min, RT. Yeast pellet was washed in sterile water once and centrifuged again as before. Approximately 10^7 yeast cells were transformed per DNA construct. Washed yeast cells were resuspended in a small volume of sterile water and divided up to the number of transformations being performed (ie, the number of constructs) in microfuge tubes. The yeast cells were spun down again at $10{,}000 \times g$, 30 s, RT, and the water is removed.

2. A mixture of sterile Polyethylene glycol [50% PEG (w/v), average Mn 3350], sterile 1 *M* lithium acetate, and sterile denatured salmon sperm

DNA (SSD, 10 mg mL^{-1} in water) was made in a ratio of 120:18:25 (v/v/v), respectively. The yeast cell pellet in the microfuge tube was combined with 2–3 μg (in <60 μL volume) of purified linear DNA and 163 μL of PEG/LiAc/SSD and mixed gently by pipetting. The yeast/DNA/PEG mixture was heat shocked in 42°C water bath for 30–40 min.

3. After incubation, yeast cells were centrifuged at 10,000 × *g*, 1 min, RT. The yeast pellet was resuspended in 1 mL of YPDA medium, transferred to a culture flask containing 10–20 mL of YPDA, and cultured for 18 h at 30°C. This recovery time in rich medium is necessary to allow for homologous integration and expression of the antibiotic resistance gene. If the recovery time is too short or too long, the transformation efficiency will suffer due to an inefficient integration or wild-type yeast override, respectively. Several 200 μL aliquots of the 18-h culture were plated on YPDA agar plates containing 300 μg mL^{-1} Geneticin (G418). If selection needs to be done on a minimal medium, the nitrogen source needs to be replaced with monosodium glutamate because ammonium sulfate will significantly reduce the effectiveness of the G418.
4. Yeast colonies were screened by colony PCR method using one primer within the selection marker (ie, KanMX reverse primer) and one primer corresponding to the region outside of the homologous regions (ie, forward sequence upstream of UP forward primer in the genome, Table 1). The positive yeast strain is selected on G418 medium once more prior to glycerol stock preparation. Among the positive yeast strains, a couple of strains were verified by sequencing the PCR product of *SRT1* locus amplified from the genomic DNA to ensure the deletion of *SRT1* ORF.
5. For subsequent manipulation of this yeast strain, such as yeast transformation for complementation assay and 5-FOA selection, minimal media without G418 was used. The original *rer2Δ*::*HIS3* [pRS316-*RER2*] strain displayed a slower growth than the wild-type yeast, despite of the presence of a plasmid copy of *RER2*. This slow growth was consistently present in our newly created strain, *rer2Δ*::*HIS3 srt1Δ*::*KanMX* [pRS316-*RER2*].

4. COMPLEMENTATION OF *rer2Δ srt1Δ* WITH *CPT* AND *CBP*

4.1 Construct Generation

Arabidopsis CPT numbering system suggested by Kera, Takahashi, Sutoh, Koyama, and Nakayama (2012) was used for this work (Kera et al., 2012). *Arabidopsis* encodes nine *cis*-prenyltransferase genes (*AtCPT1–9*)

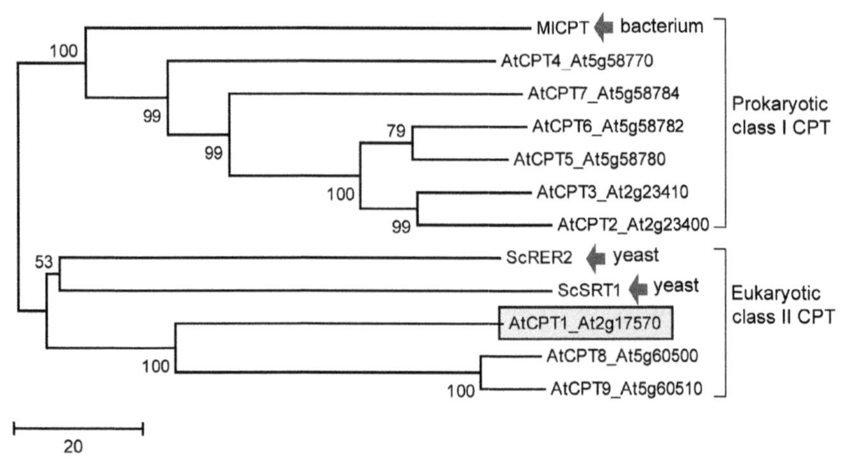

Fig. 3 A phylogenetic analysis of the nine *Arabidopsis* CPTs. The phylogenetic tree was generated by MEGA5.2 software using a neighbor-joining algorithm (1000 replicates). Bootstrap values are shown in each node in percentage. *Arabidopsis* gene names are given beside the *CPT* numbers assigned by Kera et al. (2012). ScRER2 and ScSRT1 are yeast dehydrodolichyl diphosphate synthases; MlCPT is *Micrococcus luteus* undecaprenyl diphosphate synthase (AB004319).

and one *CBP*, *AtLEW1*. A phylogenetic analysis of the nine AtCPTs showed that AtCPT2–7 are clustered with the bacterial *Micrococcus luteus* undecaprenyl diphosphate (MlCPT), while AtCPT1/8/9 are clustered with the yeast CPTs (RER2 and SRT1, Fig. 3). The phylogenetic separation of these two groups (indicated as class I and class II in Fig. 3) is strongly supported by the bootstrap value 100 by the Neighbor-Joining algorithm. In addition, AtCPT2–7 encode a putative chloroplast-targeting motif in their N-termini, while *AtCPT1/8/9* proteins do not have such a motif. These data indicated that AtCPT2–7 are likely to be self-sufficient prokaryotic type CPTs that localize to the chloroplast, whereas AtCPT1/8/9 are not self-sufficient, eukaryotic type CPTs that require CBP, according to the CPT/CBP protein complex model. We chose *AtCPT1* (At2g17570) for functional study in this work (gray box in Fig. 3). *AtCPT1* and *AtLEW1* were amplified from *Arabidopsis thaliana* cDNA using primers listed in Table 1. *AtCPT1* was cloned into p414-GPD and *AtLEW1* was cloned into p415-GPD.

4.2 Yeast Transformation with Plasmids

Yeast transformation was carried out as described in Section 3.2.2 with the exception of long recovery incubation. Instead, the recovery was done by

Table 2 Yeast Strains Before 5-FOA Selection (in *rer2Δ::HIS3 srt1Δ::KanMX* [pRS316-*RER2*] Background)

Strain Name	Trp Marker Plasmid	Leu Marker Plasmid	Ura Marker Plasmid
Control	p414-GPD	p415-GPD	pRS316 RER2
AtCPT1	p414-GPD AtCPT1-F	p415-GPD	pRS316 RER2
AtCPT1-LEW1	p414-GPD AtCPT1-F	p415-GPD AtLEW1	pRS316 RER2
AtLEW1	p414-GPD	p415-GPD AtLEW1	pRS316 RER2
ScRER2	p414-GPD RER2-F	p415-GPD	pRS316 RER2

Grown in SC-His-Ura-Trp-Leu medium.
Note: F indicates FLAG tag.

incubation in YPDA medium for 1 h at 30°C in culture tubes. Recovered yeast cells were transferred to microfuge tubes and spun down at 10,000 × *g*, 1 min, RT. The yeast pellet was washed twice each with 1 mL of sterile water or selection minimal medium by spinning down as above. The entire pellet was resuspended in 200 μL of water and plated on Synthetic Complete medium (SC-His-Ura-Trp-Leu, based on Hopkins recipe). The yeast strains generated are listed in Table 2.

4.3 Selection in 5-FOA

1. To make 500 mL of 5-FOA medium plates, 10 g glucose, 3.35 g yeast nitrogen base without amino acid, 10 mg L-methionine, 10 mg L-arginine, 15 mg L-lysine, 10 mg adenine, 10 mg uracil, and 500 mg 5-FOA were dissolved in 250 mL water by rotating between 50°C water bath and a stir plate until completely dissolved (~30 min). This warm mixture was filter-sterilized, combined with autoclaved agar (10 g in 250 mL water), and poured to plates. Plates were dried at RT for 1–2 days before use or storage at 4°C. Minimal amount of amino acids and uracil were used in 5-FOA medium (SC-His-Trp-Leu) to maximize 5-FOA effectiveness, while regular minimal medium (SC-His-Trp-Leu-Ura) contained higher amounts of amino acids and nucleic acids (Hopkins recipe).
2. To examine the complementation of the *rer2Δ srt1Δ* strain by *AtCPT1/AtLEW1*, the *RER2* gene in URA-selection marker needs to be removed by 5-FOA selection. To achieve this, yeast colonies after a transformation were streaked on a 5-FOA-containing agar plate as well as on a regular agar plate as a control. A single colony from the

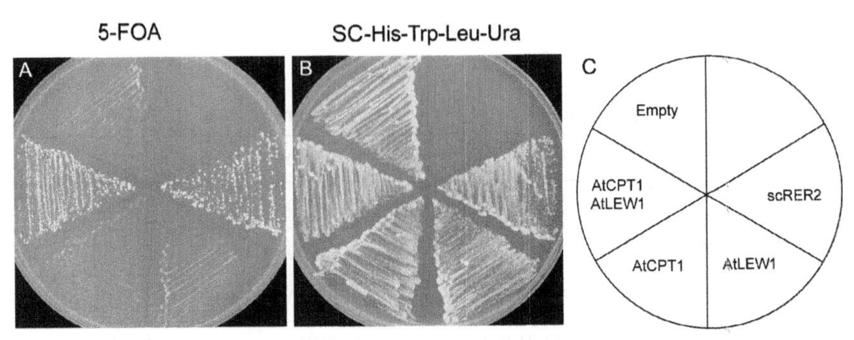

Fig. 4 Functional complementation assays by expressing *AtCPT1* (At2g17570) and/or *AtCBP* (*AtLEW1*) in the *rer2Δ srt1Δ* double knockout yeast. (A) Yeast cultivation on 5-FOA counter-selective medium (SC-His-Trp-Leu + 5-FOA). (B) Yeast cultivation under the nutritional selective condition to keep all the plasmids listed in Table 2. (C) Experimental design. The *rer2Δ srt1Δ* double knockout is lethal and thus maintained by expressing yeast *RER2* (*ScRER2*) in pRS316-URA. The *rer2Δ* and *srt1Δ* were made by HIS and G418 marker, respectively. *CPT1* and *CBP* were expressed by TRP and LEU marker, respectively.

transformation plate was picked using a sterile flat toothpick. Approximately 5% of the picked colony was dabbed on the regular plate first (later streaked using a new toothpick), and the rest was streaked on a 5-FOA plate (Fig. 4). We recommend streaking multiple sets as 5-FOA selection requires a lengthy incubation. Growth in 5-FOA will select for yeast cells that lose the plasmid copy of *RER2* in URA3 marker and are complemented by the CPT activity from the coexpression of *AtCPT1* and *AtLEW1*. 5-FOA-containing plates were grown at 30°C for 10 days. Colonies growing on 5-FOA plates were picked and streaked on SC-His-Trp-Leu agar plates to recover from 5-FOA selection. To avoid colony variations on gene expression, multiple 5-FOA-resistant colonies were picked for generating glycerol stocks or liquid culture for microsome preparation.

4.4 Results

As the *rer2Δ srt1Δ* double knockout strain is lethal, this mutant was maintained by expressing *RER2* in pRS316-URA plasmid. When the *RER2* was removed on the 5-FOA counter-selective medium, no colony was detected in *rer2Δ srt1Δ* yeast strains transformed with the empty vector, *AtLEW1* alone, or *AtCPT1* alone in multiple experiments (Fig. 4). In other words, we never observed occurrences of revertants from the *rer2Δ srt1Δ* double knockout strain, and individual expression of *AtLEW1* or *AtCPT1* cannot synthesize dolichol equivalent products to support the growth of *rer2Δ srt1Δ* yeast. These results suggested that the previously observed revertants were indeed caused by *SRT1* activation, and the *rer2Δ srt1Δ* double knockout strain is the ideal background to perform the *CPT* complementation assays. Yeast also encodes the *CBP* ortholog, *NUS1* (YDL193W), and NUS1 protein could interact with and activate eukaryotic CPTs, such as AtCPT1, to fully or partially rescue the lethal phenotype of *rer2Δ srt1Δ* strain. However, no sign of complementation was observed, when *AtCPT1* was expressed in *rer2Δ srt1Δ* strain (Fig. 4). This result indicates that yeast NUS1 cannot interact with and activate plant CPTs. The *rer2Δ srt1Δ* strain could be rescued only with the coexpression of *AtCPT1*/*AtLEW1* or with *RER2* on a different marker plasmid (positive control) (Fig. 4). The absence of complementation in the *rer2Δ srt1Δ* double knockout strain by *AtLEW1* alone or *AtCPT1* alone in this work is consistent with the recent reports of the CPT/CBP heteroprotein complex (Brasher et al., 2015; Epping et al., 2015; Qu et al., 2015) but is not congruent with the partial *rer2*^{ts} complementation by *AtLEW1* and in vitro CPT activity of the *E. coli*-produced AtLEW1 (Zhang et al., 2008).

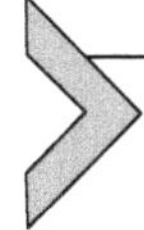

5. CPT BIOCHEMICAL ASSAY USING YEAST MICROSOMES

5.1 In Vitro CPT Assays

In order to measure eukaryotic CPT activity in vitro, a few facts from the recent studies need to be considered. First, since the *rer2* mutant (*rer2*^{ts} or *rer2Δ*) is not stable, and the *rer2* mutant population can be overridden by revertants at any time during liquid cultivation. Therefore, *rer2* is not a suitable system to measure CPT activity. On the other hand, wild-type yeast has endogenous RER2 and SRT1 activities, making it difficult to separate the recombinant CPT activity from the endogenous ones. Second, CPT alone appears to be unstable in vivo without forming a complex with CBP. Silencing *CBP* resulted in the reduction of CPT protein without perturbing *CPT* transcripts in lettuce (Qu et al., 2015). Third, no obvious membrane-bound domain is present in CPTs, but all CBPs have clear integrating membrane domains in their N-termini. Localization studies using green fluorescent protein showed that CBPs are localized to the endoplasmic reticulum (Brasher et al., 2015; Epping et al., 2015; Qu et al., 2015). Four, CPT and CBP directly interact with each other, organizing heteroprotein complexes on the endoplasmic reticulum. Five, if CPT and CBP are separately expressed and then mixed together in vitro, no CPT activity was observed, indicating these two must be simultaneously expressed in a single cell (Park et al., 2014). Considering all these factors together, the most appropriate approach for CPT in vitro assay is to perform assays using the microsomes isolated from the *CPT/CBP* coexpressing *rer2Δ srt1Δ* strain. Here we performed the in vitro assays of *AtCPT* and *AtLEW1* using the microsomes prepared from *RER2*-expressing (positive control) or *AtCPT1* and *AtLEW1* coexpressing *rer2Δ srt1Δ* strain.

5.2 Yeast Microsome Preparation

5.2.1 Yeast Culture

The yeast strains used to prepare microsomes are listed in Table 3. A starter culture of 1–2 mL SC-His-Trp-Leu was inoculated from freshly growing, multiple colonies on a plate and grown overnight at 30°C. A main culture of 50–100 mL SC-His-Trp-Leu was inoculated with 1 mL of the starter culture. The main culture was grown for approximately 15–20 h to a late logarithmic phase. Depending on the quality of the complementation, growth rate may be affected and may require longer culturing. The yeast culture was spun down at 3000 × *g*, 3 min, RT and washed once with 50 mL of water and once with 10 mL of TES-B buffer (0.6 *M* Sorbitol, 50 m*M* Tris–HCl, pH 7.4, 1 m*M* EDTA). The yeast pellet may be used immediately to prepare microsomes or stored at –80°C.

5.2.2 Bead Beating

1. The yeast pellet (from 50 mL culture) was resuspended in a final volume of 1 mL of cold TES-B and transferred to a prechilled 2-mL screw-cap vial. Prechilled glass beads (0.5 mm diameter, Biospec Products, Cat. 11079105) were added to the vial leaving approximately 15–20% void space in the vial. Bead beating was performed in a cold room by beating for 2 min at 3/4 of the intensity setting (Mini-Beadbeater-8, Biospec Products) and chilling in slurry of ice and water for 2 min. The cycle

Table 3 5-FOA-Selected Yeast Strains (in *rer2Δ::HIS3 srt1Δ::KanMX* Background)

Strain Name	Trp Marker Plasmid	Leu Marker Plasmid
AtCPT1-LEW1	p414-GPD AtCPT1-F	p415-GPD AtLEW1
RER2	p414-GPD RER2-F	p415-GPD

Grown in SC-His-Trp-Leu medium.
Note: F indicates FLAG tag to the C-terminus of CPT.

was repeated total of three times. The broken yeast cells were examined under microscope to ensure majority (~80%) was broken. Punctured yeast cells will appear ghostly (ie, less defined cell wall than live cells) and completely broken cells will not be visible as cells, but as numerous small vesicles floating around. It is better to have less than perfect breakage since excessive bead beating will result in increased heat and increased handling time, which will jeopardize the quality of microsomal proteins.

2. To separate broken yeast cells from glass beads, various approaches can be taken. One method is pipetting out from the top of the bead-beating vial by repeated washing with a few mL of TES-B and combining the washed lysate. A better method, which is more high throughput and technically consistent, is to centrifuge yeast out of the bead-beating vial. The bead-beating vial was placed on a 1.5-mL microfuge tube with its cap completely cut off, and this structure was nested inside a 15-mL Falcon™ tube to collect the flow through yeast lysate in the microfuge tube by centrifugation at 2000 × *g*, 3 min, 4°C. All tubes were prechilled. To introduce a hole at the bottom of the bead-beating vial containing the glass beads and the yeast, 25-gauge syringe needle was flamed and used to puncture a hole. The tubes were taken out using a pair of forceps. The bead-beating vials can be washed once if desired to collect remaining yeast lysate by adding 1 mL of TES-B to the vial and centrifuging again.

5.2.3 Microsome Preparation

1. In order to remove large cellular debris, collected yeast lysate from above was centrifuged twice at 10,000 × *g*, 10 min, 4°C, once in the microfuge tube (1 mL volume) and once in a larger volume tube (10 mL volume). The supernatant from the first 10,000 × *g* spin was transferred to a polycarbonate tube (the supernatant from the second wash combined) and combined with additional cold TES-B buffer to a total volume of ~10 mL, and the second 10,000 × *g*, 10 min, 4°C centrifugation was performed.
2. The supernatant is again transferred to a prechilled ultracentrifuge tube using serological pipet, ensuring not to disturb the pellet (to avoid contamination of microsomes with large cellular debris) and combined with additional cold TES-B buffer to a total volume of ~20 mL. The 100,000 × *g*, 1 h, 4°C centrifugation was performed in Beckman Coulter Ultracentrifuge with Type 70 Ti rotor. The reason for the increased

centrifugation volume especially in the ultracentrifugation step is to minimize precipitating soluble proteins with microsomes. In a highly dense solution of lysate, the soluble proteins may combine with microsomes and come down together during the ultracentrifugation. For the purpose of in vitro CPT assay, it is important to minimize the contamination by soluble proteins in the microsomal preparation, as it appears to contribute to some level of background activity.

3. The supernatant from the ultracentrifugation was discarded. The remaining microsomal pellet was washed twice gently each with 1 mL of cold TG buffer [50 m*M* Tris–HCl, pH 7.4, 20% (v/v) glycerol] to wash away soluble proteins in the tube. Note that the microsome pellet is sticky and will adhere to the pipet tip and microfuge tube. As much of the pellet as possible was picked up by a pipet tip and transferred to the microfuge tube. The remaining pellet was resuspended in cold 50 μL TG buffer in the ultracentrifuge tube and transferred to the same microfuge tube. Resuspend the microsome pellet gradually by soaking in TG buffer on ice for 10 min and pipetting up and down to resuspend. Repeat soaking/pipetting two to three times depending on how well the pellet resuspends. Ideally, no visible bits of the microsome pellet should be seen while holding the microfuge tube up against light and pipetting. Depending on the desired protein concentration, the microsome pellet can be resuspended in 50–200 μL cold TG buffer. Generally, we bring the total protein concentration to 10–15 μg μL^{-1}. Bradford assay was performed to determine the protein concentration and adequate aliquots were made for storage in –80°C and for in vitro enzyme assays.

5.3 *cis*-Prenyltransferase Enzyme Assays Using Yeast Microsomes and ^{14}C-IPP

All the reaction components except for microsomes were combined into a master mix—50 m*M* HEPES, pH 7.5, 5 m*M* $MgCl_2$, 2 m*M* NaF, 2 m*M* Na_3VO_4, 2 m*M* DTT, 20 μ*M* FPP, and 80 μ*M* total IPP (^{14}C-IPP, PerkinElmer NEC773050UC, 50–60 mCi $mmol^{-1}$). The reaction was performed in organic solvent-safe 1.7 mL microfuge tubes with 50 μg of yeast microsomes in 50 μL reaction volume by incubating at 30°C for 2 h. Each reaction was set up in triplicate and extracted separately. Reaction products were extracted by adding 400 μL of 0.9% NaCl and 1 mL of chloroform: methanol (2:1), vortexing for 20 s, and centrifuging at 10,000 × *g*, 1 min, RT. The upper methanol/water phase was discarded, and the lower chloroform phase was transferred to a new microfuge tube. The chloroform phase was washed to remove unincorporated IPPs by adding 500 μL of methanol:water:chloroform (48:47:3), vortexing and centrifuging as above. The lower chloroform phase was transferred to a new tube, and two more washing steps were performed. The chloroform phase after the final wash was transferred to a 2-mL screw-cap glass vial (gas chromatography vial). To count ^{14}C radioactivity, 60 μL (approximately 10% of the chloroform phase volume) was measured by scintillation counting with 4 mL scintillation cocktail. The remaining product was air-dried overnight in the fume hood.

5.4 Dephosphorylation

The dried product needs to be devoid of the phosphate group for size separation on reverse-phase thin layer chromatography (TLC). Dephosphorylation can be performed either enzymatically using potato acid phosphatase or chemically using acid hydrolysis.

5.4.1 Enzymatic Dephosphorylation

Two to five units of potato acid phosphatase in 500 μL of buffer (50 m*M* sodium acetate, 0.1% Triton X-100, pH 4.7, 60% methanol) were added to the dried product obtained from 50 μg of microsomes. The reaction was incubated at 37°C for approximately 18 h, followed by an equal-volume extraction with benzene. The potato acid phosphatase (Sigma, Cat. P1146) is crudely purified, lyophilized powder, and when dissolving in the buffer, significant impurities will remain as particulates. The impurities are centrifuged down and only the supernatant is used for dephosphorylation.

5.4.2 Chemical Dephosphorylation

Since enzymatic dephosphorylation is costly and time-consuming, we often use HCl hydrolysis. Alternative to using phosphatase, 300 μL of 1 *M* HCl was added to the dried product and incubated in 85°C water bath for 1 h. Once the reaction was cooled down to RT, the hydrolyzed product was extracted with benzene by adding 300 μL of benzene, vortexing for 30 s, and centrifuging at 4000 × *g*, 3 min, RT. Since the reactions are performed in a glass vial (ie, 2 mL screw-cap glass vials with PTFE closures) to avoid plastic contamination from benzene extraction, the glass vials are nested in 15 mL Falcon™ tubes for centrifugation. Two-third of the upper benzene phase was carefully transferred to a new glass vial without carrying over the bottom aqueous phase. To maximize recovery, additional 300 μL of benzene was added to the extraction vial, and 300 μL was transferred out to combine with the earlier transfer. The extracted product was either air-dried overnight or dried under N_2 in the fume hood until an adequate volume remained (~20 μL) for spotting on a TLC plate.

5.5 Thin Layer Chromatography

For separating *cis*-polyisoprenoids, C18 reverse-phase silica with preadsorbent area on glass support (Analtech Cat. P90011, 20 × 20 cm) was used. Spots of 3–5 mm diameter were placed on the preadsorbent area 1 cm below the start of the reverse-phase area using microcapillary pipettes. Multiple drops were placed on the same spot until all benzene-dissolved samples was used. Too much product on the spot can result in poor separation; therefore, the optimal amount of the products for TLC separation and ^{14}C detection needs to be determined by loading various amounts. Ensure that the benzene is completely dried off of the TLC plate before placing in the separation solvent. Acetone:water (39:1) solvent system was equilibrated in a TLC chamber with filter papers to help maximize the amount of solvent vapor. The TLC plate should be placed in the chamber ensuring the bottom edge of the plate goes into the solvent evenly to allow for horizontal solvent front. *RER2* is known to produce dolichols of 14–18 isoprene units and was used as a positive control of the enzyme assay. An isoprenoid standard of a known size can be also spotted (C45, Solanesol). The solvent level was 5–10 mm from the bottom of the TLC plate. The run was stopped when the solvent front reached 1 cm before the end of the silica area (mark this line before you start the run). Immediately lay the TLC plate down flat to prevent solvent from pulling down the separation and to help rapid drying of the solvent. Once the plate was completely dried of the solvent, it was placed against a BAS Storage Phosphor Screen (GE Healthcare) in a film cassette, and depending on the amount of the ^{14}C-labeled isoprenoid products, exposure was done for a few hours to a few days, followed by digitalization by scanning in a phosphorimager (Bio-Rad, Molecular Imager FX). Once adequate data are obtained, the plate can be stained in a chamber with iodine to visualize the solanesol standard. R_f values may be calculated, but in our case, comparison to the products of RER2 control was adequate.

5.6 Results

Using ^{14}C-IPP substrate and the microsomes from the *rer2Δ srt1Δ* yeast expressing *AtCPT1/AtLEW1* or *RER2*, we were able to measure IPP incorporation rates to *cis*-polyisoprenes as well as estimate the polymer profile on the reverse-phase TLC (Fig. 5). The microsomes containing AtCPT1

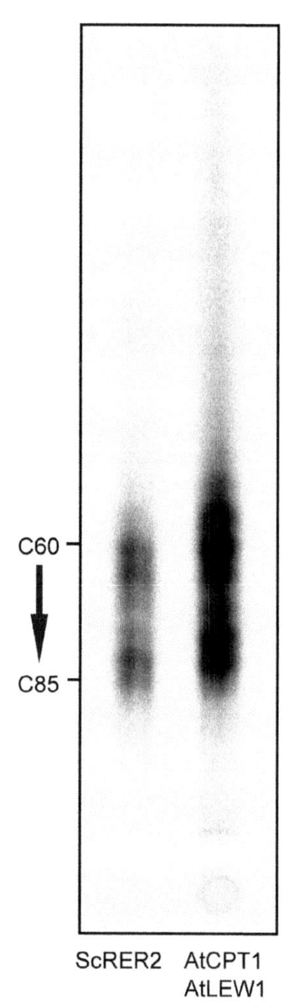

Fig. 5 Separation of *cis*-polyisoprenols by thin layer chromatography.

and AtLEW1 showed an IPP incorporation rate of 12.0 ± 0.3 pmol μg^{-1} h^{-1} ($n=3$) under the in vitro condition we used (20 μ*M* FPP, 80 μ*M* IPP, 30°C, 2 h). In 2-h reaction time, 29.9% of the total IPPs were incorporated into polyisoprene products, signifying a potent CPT activity using the microsomes having AtCPT1 and AtLEW1. In comparison, the microsomes from the yeast expressing *RER2* alone showed an IPP incorporation rate of 1.9 ± 0.1 pmol μg^{-1} h^{-1} ($n=3$) with 4.8% incorporation of total IPPs. This specific activity is approximately sixfold lower than that measured from the microsomes containing AtCPT1 and AtLEW1. The yeast CBP, NUS1, has been shown to be the partner for RER2 for the CPT activity (Park et al., 2014). Since *NUS1* was not overexpressed together with *RER2*, only the endogenous level of NUS1 is available and therefore is likely the limiting factor in CPT activity in the microsomes containing RER2. It appears that balancing the molar ratio of CPT and CBP is important for optimal CPT activity. When the radiolabeled *cis*-polyisoprene products were resolved on a reverse-phase TLC, no significant qualitative difference was observed for the polyisoprenes synthesized by RER2 or AtCPT1/AtLEW1 (Fig. 5). AtCPT1 is, therefore, a direct ortholog of yeast RER2 with respect to its biochemical property.

6. GENERAL DISCUSSION

Despite the long history of CPT research, it was only recently discovered that unique CPT/CBP complexes operate in the eukaryotic lineage to synthesize *cis*-polyisoprenes, including dolichol and NR. The fact that plants have two types of CPTs [the prokaryotic CPT in plastids (class I) and the eukaryotic CPT complexed with CBP on endoplasmic reticulum (class II); Fig. 3] together with the revertant occurrence in *rer2* mutants has complicated the finding of the CPT/CBP complex. However, it is now clear that the heteroprotein complexes comprised of class II CPT and CBP are required in eukaryotes to synthesize dolichol, NR, and possibly other *cis*-polyisoprenes (Brasher et al., 2015; Epping et al., 2015; Park et al., 2014; Qu et al., 2015). Since the unicellular eukaryote yeast has CPT/CBP complex, known as RER2/NUS1, the CPT/CBP complexes in higher eukaryotes are likely the direct orthologs of the RER2/NUS1, indicating the importance of such complexes in all eukaryotes. Of particular interest in plants is that lettuce has two sets of *CPT/CBP* genes—one for dehydrodolichyl diphosphate biosynthesis and the other for much longer NR biosynthesis. Thus, conservation and divergence of the CPT/CBP complexes for the biosynthesis of short to extremely long *cis*-polyisoprenes are the interesting subjects of further investigations.

Data prior to the discovery of CPT/CBP complexes for eukaryotic type CPTs (class II in Fig. 3) have puzzled us, and here we aimed to clarify the incongruent data found in literature by identifying the likely cause and providing a more reliable tool with which the activity of the eukaryotic type CPTs and CBPs can be assessed. Using the *rer2Δ srt1Δ* double knockout strain, we evaluated *Arabidopsis* CPT and CBP (AtCPT1 and AtLEW1, respectively). AtLEW1 was previously shown to partially complement *rer2*^{ts}

mutant, and *E. coli*-produced AtLEW1 enzyme showed dehydrodolichyl diphosphate synthase activity in vitro assays (Zhang et al., 2008). In our experiments, *rer2* revertants occur at a considerable frequency, leading to the conclusion that *rer2* mutants (*rer2*^{ts} or *rer2Δ*) cannot be used for in vivo complementation or to express heterologous *CPT*s for in vitro CPT assays. On the contrary, the *rer2Δ srt1Δ* double knockout strain generated from this work is stable and cannot revert to the wild-type growth pattern, making it suitable for both in vivo CPT complementation and the microsome preparation for in vitro assays. Although revertant occurrences from *rer2* have not been explicitly stated except for one early report (Schenk et al., 2001), recent studies has started employing *rer2Δ srt1Δ* double- or *rer2Δ srt1Δ nus1Δ* triple-deletion mutant to characterize eukaryotic type CPTs possibly due to the same revertant issue found in this work (Epping et al., 2015; Grabinska et al., 2010; Park et al., 2014). The possibility of AtCPT1 activity in *rer2Δ srt1Δ* double knockout yeast through interactions with yeast NUS1 can be ruled out as AtCPT1 alone could not complement the *rer2Δ srt1Δ* strain, and therefore we concluded that the *rer2Δ srt1Δ* yeast can be used to assess plant CPT and/or CBP functions.

The methods described here can be used to assess rubber-synthesizing enzyme activities. If the rubber polymers, significantly longer than undecaprenol and dolichol, are synthesized in vitro assays, these polymer products will be immobile on the reverse-phase TLC, thereby leaving intense radiolabeled signals on or near the sample loading spots. On the other hand, in yeast complementation assays, if only high-molecular-weight rubber is synthesized (in other words, no intermediate dolichol length polymers), the lethal phenotype of *rer2Δ srt1Δ* double knockout yeast may not be rescued. The methods described here are technically reliable and do not require high-end analytical instruments. Hence, these methods can be used as the first line of experiments to examine biochemical functions of CPT and CBP in vitro and in vivo for short to long *cis*-polyisoprenes (eg, dolichol and NR) in most laboratories.

ACKNOWLEDGMENTS

We thank Dr. Markus Aebi (Institute of Microbiology, ETH, Switzerland) for personal communication for the reported *Δrer2* revertants, Drs. Nakano and Hirata (RIKEN Advanced Science Institute, Japan) for providing us SNH23-7D and *Δrer2* strains, and Dr. Vanina Zaremberg (University of Calgary, Canada) for p414 and p415 yeast expression plasmids. We also thank Gillian MacNevin for critical reading of this manuscript. This work was supported by National Science and Engineering Research Council of Canada (NSERC) and Canada Research Chair program to D.K.R., and is also supported by the Next-Generation BioGreen 21 program (PJ00110552016) to M.K.

REFERENCES

Asawatreratanakul, K., Zhang, Y. W., Wititsuwannakul, D., Wititsuwannakul, R., Takahashi, S., Rattanapittayaporn, A., et al. (2003). Molecular cloning, expression and characterization of cDNA encoding *cis*-prenyltransferases from *Hevea brasiliensis*. *European Journal of Biochemistry, 270*, 4671–4680.

Brasher, M. I., Surmacz, L., Leong, B., Pitcher, J., Swiezewska, E., Pichersky, E., et al. (2015). A two-component enzyme complex is required for dolichol biosynthesis in tomato. *The Plant Journal, 82*, 903–914.

Bushman, B. S., Scholte, A. A., Cornish, K., Scott, D. J., Brichta, J. L., Vederas, J. C., et al. (2006). Identification and comparison of natural rubber from two *Lactuca* species. *Phytochemistry, 67*, 2590–2596.

Chakrabarty, R., Qu, Y., & Ro, D. K. (2015). Silencing the lettuce homologs of small rubber particle protein does not influence natural rubber biosynthesis in lettuce (*Lactuca sativa*). *Phytochemistry, 113*, 121–129.

Chappell, J. (2002). The genetics and molecular genetics of terpene and sterol origami. *Current Opinion in Plant Biology, 5*, 151–157.

Collins-Silva, J., Nural, A. T., Skaggs, A., Scott, D., Hathwaik, U., Woolsey, R., et al. (2012). Altered levels of the *Taraxacum kok-saghyz* (Russian dandelion) small rubber particle protein, TkSRPP3, result in qualitative and quantitative changes in rubber metabolism. *Phytochemistry, 79*, 46–56.

Cornish, K. (2001). Biochemistry of natural rubber, a vital raw material, emphasizing biosynthetic rate, molecular weight and compartmentalization, in evolutionarily divergent plant species. *Natural Product Reports, 18*, 182–189.

Epping, J., van Deenen, N., Niephaus, E., Stolze, A., Fricke, J., Huber, C., et al. (2015). A rubber transferase activator is necessary for natural rubber biosynthesis in dandelion. *Nature Plants, 1*, 15048.

Fujihashi, M., Zhang, Y. W., Higuchi, Y., Li, X. Y., Koyama, T., & Miki, K. (2001). Crystal structure of *cis*-prenyl chain elongating enzyme, undecaprenyl diphosphate synthase. *Proceedings of the National Academy of Sciences of the United States of America, 98*, 4337–4342.

Gershenzon, J., & Dudareva, N. (2007). The function of terpene natural products in the natural world. *Nature Chemical Biology, 3*, 408–414.

Grabinska, K. A., Cui, J., Chatterjee, A., Guan, Z., Raetz, C. R., Robbins, P. W., et al. (2010). Molecular characterization of the *cis*-prenyltransferase of *Giardia lamblia*. *Glycobiology, 20*, 824–832.

Hillebrand, A., Post, J. J., Wurbs, D., Wahler, D., Lenders, M., Krzyzanek, V., et al. (2012). Down-regulation of small rubber particle protein expression affects integrity of rubber particles and rubber content in *Taraxacum brevicorniculatum*. *PLoS One*, 7, e41874.

Kera, K., Takahashi, S., Sutoh, T., Koyama, T., & Nakayama, T. (2012). Identification and characterization of a *cis*, *trans*-mixed heptaprenyl diphosphate synthase from *Arabidopsis thaliana*. *FEBS Journal, 279*, 3813–3827.

Koksal, M., Jin, Y., Coates, R. M., Croteau, R., & Christianson, D. W. (2011). Taxadiene synthase structure and evolution of modular architecture in terpene biosynthesis. *Nature, 469*, 116–120.

Lesburg, C. A., Zhai, G., Cane, D. E., & Christianson, D. W. (1997). Crystal structure of pentalenene synthase: Mechanistic insights on terpenoid cyclization reactions in biology. *Science, 277*, 1820–1824.

Mooibroek, H., & Cornish, K. (2000). Alternative sources of natural rubber. *Applied Microbiology and Biotechnology, 53*, 355–365.

Park, E. J., Grabińska, K. A., Guan, Z., Stránecký, V., Hartmannová, H., Hodaňová, K., et al. (2014). Mutation of Nogo-B receptor, a subunit of *cis*-prenyltransferase, causes a congenital disorder of glycosylation. *Cell Metabolism, 20*, 448–457.

Post, J., van Deenen, N., Fricke, J., Kowalski, N., Wurbs, D., Schaller, H., et al. (2012). Laticifer-specific cis-prenyltransferase silencing affects the rubber, triterpene, and inulin content of *Taraxacum brevicorniculatum*. *Plant Physiology*, *158*, 1406–1417.

Qu, Y., Chakrabarty, R., Tran, H. T., Kwon, E.-J. G., Kwon, M., Nguyen, T.-D., et al. (2015). A lettuce (*Lactuca sativa*) homolog of human Nogo-B receptor interacts with *cis*-prenyltransferase and is necessary for natural rubber biosynthesis. *Journal of Biological Chemistry*, *290*, 1898–1914.

Sato, M., Fujisaki, S., Sato, K., Nishimura, Y., & Nakano, A. (2001). Yeast *Saccharomyces cerevisiae* has two *cis*-prenyltransferases with different properties and localizations. Implication for their distinct physiological roles in dolichol synthesis. *Genes to Cells*, *6*, 495–506.

Sato, M., Sato, K., Nishikawa, S., Hirata, A., Kato, J., & Nakano, A. (1999). The yeast *RER2* gene, identified by endoplasmic reticulum protein localization mutations, encodes *cis*-prenyltransferase, a key enzyme in dolichol synthesis. *Molecular and Cellular Biology*, *19*, 471–483.

Schenk, B., Rush, J. S., Waechter, C. J., & Aebi, M. (2001). An alternative *cis*-isoprenyltransferase activity in yeast that produces polyisoprenols with chain lengths similar to mammalian dolichols. *Glycobiology*, *11*, 89–98.

Schmidt, T., Lenders, M., Hillebrand, A., van Deenen, N., Munt, O., Reichelt, R., et al. (2010). Characterization of rubber particles and rubber chain elongation in *Taraxacum koksaghyz*. *BMC Biochemistry*, *11*, 11.

Spiro, R. G. (2002). Protein glycosylation: Nature, distribution, enzymatic formation, and disease implications of glycopeptide bonds. *Glycobiology*, *12*, 43R–56R.

Starks, C. M., Back, K., Chappell, J., & Noel, J. P. (1997). Structural basis for cyclic terpene biosynthesis by tobacco 5-epi-aristolochene synthase. *Science*, *277*, 1815–1820.

Teng, K.-H., & Liang, P.-H. (2012). Structures, mechanisms and inhibitors of undecaprenyl diphosphate synthase: A *cis*-prenyltransferase for bacterial peptidoglycan biosynthesis. *Bioorganic Chemistry*, *43*, 51–57.

Tholl, D. (2006). Terpene synthases and the regulation, diversity and biological roles of terpene metabolism. *Current Opinion in Plant Biology*, *9*, 297–304.

van Beilen, J. B., & Poirier, Y. (2007). Establishment of new crops for the production of natural rubber. *Trends in Biotechnology*, *25*, 522–529.

Zhang, H., Ohyama, K., Boudet, J., Chen, Z., Yang, J., Zhang, M., et al. (2008). Dolichol biosynthesis and its effects on the unfolded protein response and abiotic stress resistance in *Arabidopsis*. *The Plant Cell*, *20*, 1879–1898.

CHAPTER SEVEN

Generation and Functional Evaluation of Designer Monoterpene Synthases

N. Srividya, I. Lange, B.M. Lange[1]
Institute of Biological Chemistry, M. J. Murdock Metabolomics Laboratory, Washington State University, Pullman, WA, United States
[1]Corresponding author: e-mail address: lange-m@wsu.edu

Contents

1. Introduction 148
2. Equipment 150
3. Materials 150
   3.1 Molecular Biology 150
   3.2 Recombinant Enzyme Purification 151
   3.3 Protein Gel Electrophoresis 151
   3.4 Enzyme Assays 152
4. Step 1—Generation of Expression Constructs 153
   4.1 Quick Change PCR 154
   4.2 Overlap Extension Mutagenesis 155
5. Step 2—Production of Purified, Recombinant Target Enzyme 158
   5.1 *E. coli* Cultivation and Induction of Target Enzyme Expression 158
   5.2 Isolation of Target Enzyme and Assessment of Purity 158
6. Step 3—Functional Evaluation of Recombinant Monoterpene Synthases 161
   6.1 Enzymatic Reaction 161
   6.2 Product Quantitation 162
   6.3 Analysis of Kinetic Data 162
7. Conclusions 164
Acknowledgments 164
References 164

Abstract

Monoterpene synthases are highly versatile enzymes that catalyze the first committed step in the pathways toward terpenoids, the structurally most diverse class of plant natural products. Recent advancements in our understanding of the reaction mechanism have enabled engineering approaches to develop mutant monoterpene synthases that produce specific monoterpenes. In this chapter, we are describing protocols to introduce targeted mutations, express mutant enzyme catalysts in heterologous hosts,

Methods in Enzymology, Volume 576
ISSN 0076-6879
http://dx.doi.org/10.1016/bs.mie.2016.03.013

and assess their catalytic properties. Mutant monoterpene synthases have the potential to contribute significantly to synthetic biology efforts aimed at producing larger amounts of commercially attractive monoterpenes.

1. INTRODUCTION

Terpenoids are a structurally diverse class of natural products (also called secondary or specialized metabolites) with numerous applications in the flavor, fragrance, nutraceutical, and pharmaceutical industries (Lange & Ahkami, 2013). The first committed step in the biosynthesis of a specific terpenoid class is generally catalyzed by a terpene synthase with high affinity for a prenyl diphosphate substrate of a certain chain length. For example, those accepting a C10 precursor (geranyl diphosphate (GPP) or neryl diphosphate) are termed monoterpene synthases. Based on currently available experimental evidence plant genomes harbor large families of terpene synthase genes (Chen, Tholl, Bohlmann, & Pichersky, 2011). Terpene synthases are also notorious for producing multiple products, which further increases the capacity to generate terpenoid diversity. Synthetic biology efforts by various laboratories have resulted in the development of microbial platform strains for the assembly of different classes of terpenoids. Prominent examples are the production of artemisinic acid, a precursor to the antimalarial sesquiterpene artemisinin in yeast (Paddon et al., 2013), and of diverse pharmaceutically relevant diterpene skeletons in *Escherichia coli* (Morrone et al., 2010; Zerbe et al., 2013). In most cases, the formation of a particular high-value terpenoid is desired and enzyme engineering has the potential to provide enzymes with high catalytic rates and specificity. A critical challenge is presented by the scarcity of structure–function data that would enable the knowledge-based engineering of terpene synthases. In this chapter, we describe how to generate and functionally characterize monoterpene synthase mutants engineered to produce specific monoterpenoid end products.

We chose 4*S*-limonene synthase from spearmint (*Mentha spicata*) as a model monoterpene synthase (Fig. 1) because of the availability of substantial resources. A crystal structure of the pseudomature form of the enzyme (termed R58; with the plastidial targeting sequence removed) and a bound substrate analogue, obtained at 2.7 Å resolution (Hyatt et al., 2007), was used to determine the distance between amino acid residues and the

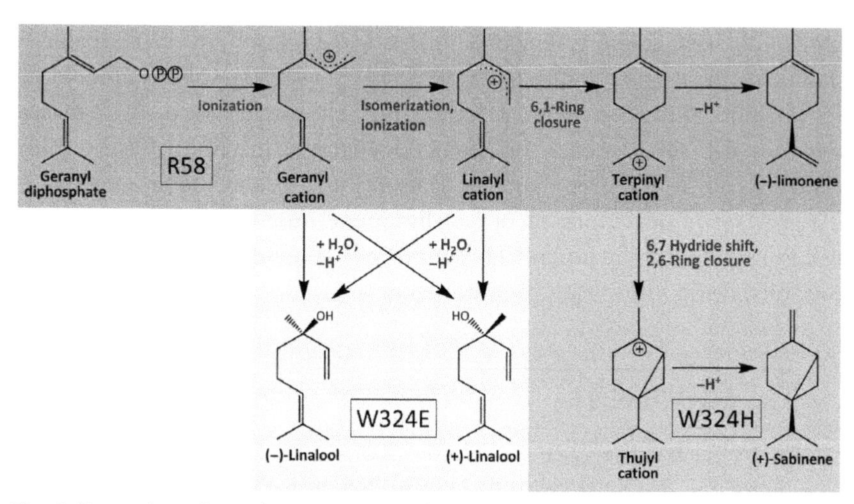

Fig. 1 Formation of novel monoterpenes by mutants of (4S)-limonene synthase. The primary reaction pathway toward (−)-limonene, as catalyzed by the pseudomature form of the wild-type enzyme (designated as R58), is highlighted in *dark gray*, while primary products formed by mutant enzymes W324H and W324E are highlighted with brighter shades of *gray*. (See the color plate.)

substrate. All amino acid residues that line the active site of the enzyme (and therefore have a high probability of being involved in catalysis) were exchanged with L-alanine and functionally characterized (Srividya, Davis, Croteau, & Lange, 2015). Mutants where a single amino acid exchange resulted in a dramatic loss of enzyme fidelity (measured as percentage of (−)-limonene among all monoterpenes in the enzyme reaction; as reference, the R58 enzyme generates 96% (−)-limonene) were considered for further mutagenesis. In particular, we were studying mutations that resulted in the release of a novel skeleton (acyclic (eg, (+)-linalool and/or (−)-linalool) or bicyclic (eg, (+)-sabinene) products) rather than the monocyclic (−)-limonene produced by R58 (Fig. 1).

To understand the implications of generating new products with R58 mutants, it is important to briefly mention the postulated catalytic mechanism of 4*S*-limonene synthase (Coates et al., 1997; Pyun, Coates, Wagschal, McGeady, & Croteau, 1993; Rajaonarivony, Gershenzon, & Croteau, 1992). The reaction involves the initial migration of the diphosphate moiety from GPP to form linalyl diphosphate (via a carbocation intermediate). Isomerization and reionization generate the linalyl cation, which then undergoes cyclization to the terpinyl cation (Fig. 1). Deprotonation yields (−)-limonene as the primary product of R58. Mutants that release acyclic alcohols (eg, (+)-linalool and/or (−)-linalool) are not capable of stabilizing the terpinyl cation and water capture instead results in a premature termination of the enzymatic reaction (Fig. 1). In contrast, bicyclic monoterpenes (eg, (+)-sabinene) are generated when the formation of the terpinyl cation is followed by a second cyclization prior to deprotonation (Fig. 1). The successful development of (−)-limonene mutants that catalyze novel reactions indicates the potential for enzyme engineering toward the production of specific high-value monoterpenes.

2. EQUIPMENT

Analytical Balance (XS105, Mettler Toledo, Columbus, OH)
Centrifuge (RC-5B, Sorvall-Thermo Scientific, Waltham, MA)
Electrophoresis System for DNA Gels (Thermo Scientific, Waltham, MA)
Electrophoresis System for Protein Gels (Bio-Rad, Hercules, CA)
Gas Chromatograph (7890, Agilent Technologies, Santa Clara, CA)
Gel Imager (MultiImage II, Alpha Innotech, San Leandro, CA)
Incubator Shaker (Series 25, New Brunswick, Edison, NJ)
Micro-Centrifuge (Z180M, Hermle, Wehingen, Germany)
Multi-Tube Rotator (Barnstead/Thermolyne, Dubuque, IA)
Orbital Shaker (Lab-Line)
pH Meter (430, Corning, Inc., Corning, NY)
SDS-PAGE Apparatus (Mini-PROTEAN, Bio-Rad, Hercules, CA)
Single Tube Shaker (Vortex Genie 2, IKA)
Microplate Reader (Synergy H1, BioTek, Winooski, VT)
Thermocycler (Mastercycler Pro, Eppendorf, Hauppauge, NY)
Thermomixer (5437, Eppendorf, Hauppauge, NY)
Ultrasonic Cell Disruptor (475, Virsonic, Mount Holly, NJ)
Water Bath (Lab-Line Instruments, Dubuque, IA)

3. MATERIALS

3.1 Molecular Biology

3.1.1 TE Stock Solution

Add 1214 mg tris(hydroxymethyl)aminomethane (Tris) and 372 mg ethylenediaminetetraacetic acid (EDTA) into a graduated cylinder. Fill to the 90 mL mark deionized water, adjust pH to 8.0, and add more deionized water to a final volume of 100 mL. Transfer the solution to a 250-mL glass bottle and sterilize in an autoclave. To prepare a working solution (1 × TE), add 9 mL of sterilized water to 1 mL of the TE stock solution.

3.1.2 DNA Gel Loading Dye

In a graduated cylinder, mix 25 mg bromophenol blue, 25 mg xylene cyanol, and 3 mL glycerol, and then add deionized water to a final volume of 10 mL. Prepare 1 mL aliquots and store these at 4°C until further use.

3.1.3 IPTG Solution

Dissolve 2.38 g isopropyl-β-D-1-thiogalactopyranoside (IPTG) in 10 mL sterilized water. Pass the solution through a 0.45 μm filter and store at −20°C until further use.

3.2 Recombinant Enzyme Purification

3.2.1 MOPSO Buffer

Dissolve 1126 mg 3-morpholino-2-hydroxypropanesulfonic acid (MOPSO) in 50 mL deionized water, adjust pH to 7.0, add 20 mL glycerol, and then add deionized water to a final volume of 100 mL. Add 200 μL of 0.5 *M* dithiothreitol (DTT) just before use.

3.2.2 Phosphate Buffer

Dissolve 136.09 g monobasic potassium phosphate (KH_2PO_4) and 174.2 g dibasic potassium phosphate (K_2HPO_4) in 900 mL deionized water. Adjust pH to 7.0 and add deionized water to a final volume of 1 L (1 *M* stock solution). To prepare a working solution (100 m*M*), add 9 mL deionized water to 1 mL phosphate buffer stock solution.

3.3 Protein Gel Electrophoresis

3.3.1 Protein Gel Loading Dye

Dissolve 1214 mg Tris in 9 mL deionized water, adjust pH to 6.8 with hydrochloric acid (HCl), and add deionized water to a final volume of 10 mL (10 × stock solution). Dissolve 10 mg bromophenol blue in 10 mL deionized water (12.5 × stock solution).

Mix 2.5 mL Tris stock solution, 0.8 mL bromophenol blue stock solution, 1.0 g sodium dodecyl sulfate (SDS), and 4 mL glycerol, and then add deionized water for a final volume of 10 mL. Prepare 1 mL aliquots and store these at −20°C until further use.

3.3.2 Protein Gel Running Buffer

In a graduated cylinder mix 30 g of Tris, 144 g glycine, and 10 g of SDS, and add deionized water to a final volume of 1 L (10× stock solution). To prepare a working solution, add 900 mL deionized water to 100 mL Tris–glycine stock solution.

3.3.3 Protein Staining Dye

Dissolve 3.2 mL phosphoric acid (H_3PO_4) and 16 g ammonium sulfate ($(NH_4)_2SO_4$) in 16 mL deionized water. Add 160 mg Coomassie Brilliant Blue G-250 and 40 mL methanol and mix well until the dye dissolves. Add deionized water to a final volume of 200 mL and allow to stand at 23°C for 12 h before use.

3.3.4 Protein Gel Destaining Solution

Mix 200 mL methanol and 600 mL deionized water, add 70 mL acetic acid, and then add deionized water to a final volume of 1 L.

Protein Gel (optional; no protocol provided here; for details see Brunelle & Green, 2014)

Acrylamide
1.5 *M* Tris Buffer, pH 8.8
0.5 *M* Tris Buffer, pH 6.8
10% (w/v) Sodium dodecyl sulfate (SDS)
10% (w/v) Ammonium persulfate (APS)
N,*N*,*N*′,*N*′-Tetramethylethane-1,2-diamine (TEMED)

3.4 Enzyme Assays

3.4.1 Geranyl Diphosphate Stock Solution

GPP was synthesized as described elsewhere (Davisson et al., 1986; Woodside, Huang, & Poulter, 1988). A GPP stock solution (10 m*M*) was prepared by dissolving 3.14 mg GPP in 1 mL deionized water. Aliquots (100 μL each) were stored at −80°C until further use.

3.4.2 Enzyme Assay Buffer

Dissolve 1.126 g MOPSO in 50 mL deionized water, adjust pH to 7.0, add 1.016 g $MgCl_2 \times 6H_2O$ and 20 mL glycerol, and add deionized water to a final volume of 100 mL. Filter-sterilize the solution before use.

4. STEP 1—GENERATION OF EXPRESSION CONSTRUCTS

The introduction of point mutations was achieved using one of two approaches. When possible, the fast and cost-effective quick change PCR method (based on Fisher & Pei, 1997 and modified from the Strategene protocol) was employed. However, for unknown reasons, this method sometimes failed and in those cases a traditional Overlap Extension Strategy was used successfully. For the convenience of the user, both protocols are listed here.

In this paragraph, we will comment briefly on primer design, which follows the same principles for both mutation approaches. The location of the desired exchange of a base-pair triplet should always be in the center of the primer sequence. As a general rule, leaving 15 unchanged nucleotides on both side of a mutation location always gave the desired results in our hands. The two primers cover the same stretch of cDNA (bold letters indicate the site of desired mutation):

(+)-Strand R58	---GTGGAATGCTACTTT**TGG**AATACTGGGATCATCG--->
Sense W324D mutation primer	GTGGAATGCTACTTT**GAT**AATACTGGGATCATCG--->
(−)-Strand R58	<---CACCTTACGATGAAA**ACC**TTATGACCCTAGTAGC---
(−)-Strand R58 (reversed)	---CGATGATCCCAGTATT**CCA**AAAGTAGCATTCCAC--->
Antisense W324D mutation primer	CGATGATCCCAGTATT**ATC**AAAGTAGCATTCCAC--->

If subcloning of the PCR amplicon by restriction digestion is desired, the user should ensure that suitable restriction sites are present in the amplified PCR product. Useful information about primer design can be found at the following URL: http://www.genomics.agilent.com/primerDesignProgram.jsp.

Note: The pSBET vector (Schenk, Baumann, Mattes, & Steinbiβ, 1995) was previously demonstrated to be an excellent choice for producing recombinant 4*S*-limonene synthase (R58) in *E. coli* (Williams, McGarvey, Katahira, & Croteau, 1998). Mutant versions of the gene were therefore obtained using a plasmid that contained R58 in pSBET.

4.1 Quick Change PCR

4.1.1 Amplification of Mutated cDNA Sequence

DNA template (plasmid containing target gene)	5–20 ng
5 × HF Phusion Polymerase Buffer (New England Biolabs)	10.0 μL
5 μ*M* gene-specific primer (forward)	2.5 μL
5 μ*M* gene-specific primer (reverse)	2.5 μL
Dimethyl sulfoxide (DMSO)	1.0 μL
10 m*M* deoxynucleotide mix (dNTPs)	1.0 μL
Phusion DNA polymerase (2.5 U) (New England Biolabs)	1.0 μL
Add deionized, sterile-filtered water to a final volume of 50 μL	

PCR conditions: initial denaturing at 98°C for 1.5 min; 20 cycles of 98°C for 15 s, 55°C for 30 s, 72°C for 5 min; final extension at 72°C for 10 min; chill to 4°C for 15 min.

The original plasmid (containing nonmutated target sequence), which was maintained in a bacterial host strain and is therefore partially methylated, is then digested by adding *Dpn*I (10 U) to the above PCR mixture and incubating the mixture at 37°C for 1 h. Only amplicons with a mutated target sequence are thus retained and further processed. An aliquot (10 μL) of the reaction is evaluated by agarose gel electrophoresis (Fig. 2). Note that the mutation PCR generates an amplicon of the size of the vector plus insert (not just the target cDNA).

4.1.2 Transformation of Vector Containing Mutated cDNA into Host Cells

An aliquot (3 μL) of the *Dpn*I-digested PCR solution was added to 50 μL ice-cold, highly competent XL-1 Blue cells (10^9/μg transformation efficiency or greater) and the mixture kept on ice for 30 min. To allow entry of the vector, the reaction tube was placed at 42°C for 45 s (heat shock) and returned to ice for 2 min. LB broth (350 μL) was added and the mixture incubated at 37°C for 1 h. An aliquot of the transformation reaction

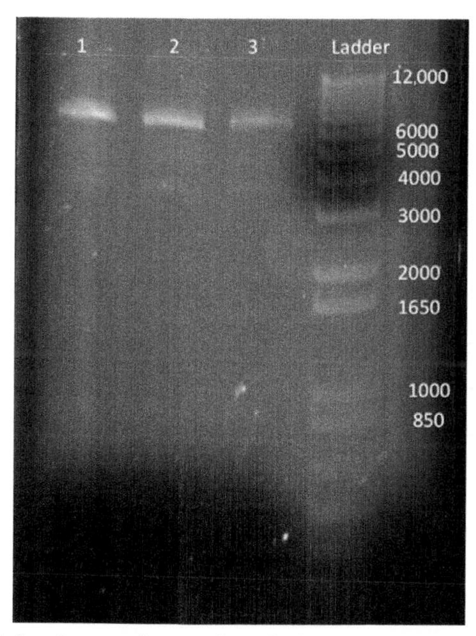

Fig. 2 Agarose gel showing amplicons of quick change PCRs using primers to amplify the wild-type enzyme (R58) (*lane 1*) and the mutations leading to the exchanges of W324H (*lane 2*) and W324E (*lane 3*). A DNA size standard ladder was loaded in *lane 4*. Note that PCR bands are often weak when using this method but, in our hands, that usually does not have negative effects on downstream manipulations.

(50–200 μL depending on experience values for transformation efficiency) was spread evenly onto LB-agar plates containing 50 μg/mL kanamycin and maintained at 37°C for 16 h. Single colonies were picked and grown overnight in culture tubes containing 5 mL liquid LB broth and 50 μg/mL kanamycin. Plasmids were isolated using a MiniPrep kit (these can be purchased from various suppliers) and the mutation verified by sequencing using commercial services.

Notes: Although the method above lists Phusion as a DNA polymerase with proofreading capacity that gives blunt ends, Pfu Turbo (Agilent Technologies) was also used successfully, but PCR conditions have to be adjusted. The *Dpn*I enzyme will perform satisfactorily in any Mg^{2+}-containing Polymerase Buffer.

4.2 Overlap Extension Mutagenesis

This method involves three rounds of PCR: in the first round, two partial cDNAs containing the desired mutation are generated; in the second round, the partial cDNA are self-annealed; and in the third round, a full-length, mutated cDNA is amplified that also contains restriction sites for convenient subcloning.

4.2.1 Introduction of Point Mutations

This first round of PCRs requires two consecutive PCRs, where one reaction (A) will amplify the nucleotides from the vector (sense primer) to the point mutation (antisense primer) and the second reaction (B) will amplify from the point mutation (sense primer) into the vector (antisense primer):

Template	5–20 ng
10 × Buffer	2.5 μL
dNTPs (10 m*M*)	0.2 μL
Primers (10 μ*M* each)	0.6 μL
Reaction A: vector sense primer and gene-specific antisense primer	
Reaction B: gene-specific sense primer and vector antisense primer	
Pfu Turbo Polymerase	0.4 μL
Add deionized, sterile-filtered water to a final volume of 25 μL	

Thermocycling was carried out with an initial denaturation at 95°C for 2 min, 40 cycles at 95°C for 30 s, 53°C for 40 s, 72°C for 2 min, and a final extension at 72°C for 10 min. The amplicons were gel-purified in a 1.3% (w/v) low melting point agarose gel, bands were excised, and DNA obtained using a gel extraction and purification kit (Qiagen) with 30 μL Elution Buffer.

4.2.2 Self-Annealing of Amplicons to Generate a Full-Length Mutated cDNA

The purified amplicons from Section 4.2.1 were mixed and self-annealed to create the full-length cDNA with vector anchors:

Template	5.0 μL each (reactions A and B)
10 × Buffer	2.5 μL
dNTPs (10 m*M*)	0.2 μL
Pfu Turbo Polymerase	0.5 μL
Add deionized, sterile-filtered water to a final volume of 25 μL	

Themocycling was performed with an initial denaturation at 95°C for 2 min, 25 cycles at 95°C for 30 s, 60°C for 5 min, 72°C for 5 min, and a final extension at 72°C for 20 min. The reaction mixture was diluted 1:20 with deionized, sterile-filtered water and then used as template in the last round of PCR:

Template (1:20 dilution)	1.0 μL
10 × Buffer	2.5 μL
dNTPs (10 m*M*)	0.2 μL
Vector Primer Sense (5 μ*M*)	1.5 μL
Vector Primer Antisense (5 μ*M*)	1.5 μL
Pfu Turbo Polymerase	0.4 μL
Add deionized, sterile-filtered water to a final volume of 25 μL	

Thermocycling conditions included and initial denaturation at 95°C for 2 min, 40 cycles at 95°C for 30 s, 53°C for 50 s, and 72°C for 2 min, and a final extension at 72°C for 10 min.

The resulting amplicon was gel-purified using a NucleoSpin Gel and PCR Clean-up kit (Macherey and Nagel), double-digested with *Nde*I (20 U; New England BioLabs) and *Bam*HI (20 U; New England BioLabs) at 37°C for 90 min and then gel-purified again as above. The purified insert was ligated into the pSBET vector (50–100 ng; predigested with *Nde*I and *Bam*HI as above) using T4 DNA ligase (20 U; New England BioLabs) at 20°C for 12 h. The ligated product was transformed into *E coli* XL-1 Blue competent cells, the cells were plated on LB-agar plates containing 50 μg/mL kanamycin, and the plates incubated at 37°C for 16 h. Single colonies were picked and cells grown overnight in 5 mL LB liquid broth containing 50 μg/mL kanamycin. Plasmids were isolated using a commercial MiniPrep kit according to the manufacturer's instructions and the mutation confirmed by sequencing using a commercial service provider.

Note: Plasmid yields in our hands were greatest when using XL-1 Blue competent cells, while lower yields were obtained with DH5α, Top10, or Omnimax competent cells.

5. STEP 2—PRODUCTION OF PURIFIED, RECOMBINANT TARGET ENZYME

5.1 *E. coli* Cultivation and Induction of Target Enzyme Expression

Plasmids containing cDNAs of R58 and mutated versions were transformed into chemically competent *E. coli* BL21 (DE3) cells, which were then plated on LB-agar plates (containing 50 μg/mL kanamycin for selection) and incubated for 16 h at 37°C. A single colony was picked from each plate, transferred into 150 mL liquid LB medium, and bacteria incubated at 37°C with shaking at 250 rpm in an Incubator Shaker (Series 25, New Brunswick). Upon reaching an optical density of 0.8 at 600 nm, the expression of the transgene was induced by the addition of 75 μL of 1 *M* IPTG solution, and the bacterial suspension was transferred to a different Incubator Shaker set to 16°C and 250 rpm for another 24 h.

Note: The induction of transgene expression under rapid growth conditions (37°C) results in the precipitation of recombinant enzyme in insoluble inclusion bodies. Incubation at reduced temperature (16°C) dramatically improves the solubility and yield of recombinant enzyme.

5.2 Isolation of Target Enzyme and Assessment of Purity

5.2.1 Cell Disruption

Bacterial cells were harvested by centrifugation (RC-5B, Sorvall) at 2500 × *g* for 10 min. The supernatant was discarded and the remainder resuspended in 500 μL MOPSO Buffer. The mixture was transferred to a 2-mL microcentrifuge tube and samples were kept on ice whenever possible for all subsequent steps. Cells were broken using an Ultrasonic Cell Disruptor (Model 475, Virsonic; 3 × 15 s bursts on ice with a 3.2-mm microprobe operated at 20% output power). The resulting homogenate was precipitated using a Micro-Centrifuge (Z180M, Hermle) at 15,000 × *g* for 30 min. The supernatant was used for the purification of recombinant enzyme as described under Section 5.2.2. An aliquot (15 μL) was mixed with 5 μL Protein Gel Loading Dye and the sample transferred to the slot of a protein gel. Sodium dodecyl sulfate-polyacrylaminde gel electrophoresis (SDS-PAGE) was performed as described under Section 5.2.3.

Note: Various methods for cell disruption were tested, including sonication, grinding with mortar and pestle, and the use of a glass homogenizer.

The most consistent results and highest recombinant enzyme yields were obtained with the sonication protocol outlined earlier.

5.2.2 Batch Purification

In a 2-mL microcentrifuge tube, 100 mg ceramic hydroxyapatite type II (CHT, Bio-Rad) was suspended in 500 μL MOPSO Buffer, the supernatant obtained under Section 5.2.1 added, and the suspension mixed gently on a Multi-Tube Rotator (Labquake, Barnstead/Thermolyne) for 1 h at 4°C. The tube was put on ice and the CHT support with bound recombinant enzyme allowed to settle for 10 min (do not centrifuge as this results in desorption of recombinant protein). The supernatant was removed carefully and 500 μL MOPSO Buffer were added. The contents were mixed gently on a Multi-Tube Rotator (Labquake, Barnstead/Thermolyne) for 10 min at 4°C (this wash step removes weakly adsorbed proteins). The tube was put on ice and the CHT support with bound recombinant enzyme allowed to settle for 10 min. The supernatant was removed carefully and 300 μL of Phosphate Buffer were added. Contents were mixed and CHT support subsequently allowed to settle as described earlier. Aliquots of the supernatant were used to quantify total protein content using the Bradford Assay (see Section 5.2.3 for details) and assess target enzyme purity by SDS-PAGE (see Section 5.2.4 for details).

Note: The majority of studies with recombinant enzymes are performed using polyhistidine or other tags for expedient purification. However, the expression of R58 using commercial vectors with a $(\text{histidine})_6$-tag led to the copurification of *E. coli* chaperones and a compromised target protein activity (Williams et al., 1998). Therefore, we obtained recombinant R58 and mutant enzymes by expressing the corresponding gene from the pSBET vector (Schenk et al., 1995) without tag and employing a purification protocol based on a conventional mixed-mode CHT support. The fastest and most efficient purification was achieved in batch mode rather than by gravity column chromatography.

5.2.3 Total Protein Quantitation

The Bradford Assay was employed to quantify total protein content by comparison to a calibration curve. A 1 mg/mL stock solution of bovine serum albumin (BSA) in deionized water was prepared and different volumes added to the Bradford Assay Reagent (Bio-Rad). A calibration curve was obtained by plotting the absorbance measured at 595 nm (Microplate

Table 1 Pipetting Scheme to Determine a Calibration Curve for Quantifying Total Protein Content Using the Bradford Assay

Standard (Three Replicates for Each Dilution)	BSA Stock Solution (μL)	Bradford Reagent (μL)	Deionized Water (μL)	Protein Concentration (mg/mL)
Dilution 1	2	40	158	0.01
Dilution 2	4	40	156	0.02
Dilution 3	8	40	152	0.04
Dilution 4	12	40	148	0.06
Dilution 5	16	40	144	0.08
Dilution 6	20	40	140	0.1

Reader Synergy H1, BioTek) against the concentration of BSA in the reagent solution (Table 1). The total protein content was calculated based on a linear equation ($y = m\,x + b$, where y is the concentration, m is the slope, x is the absorbance value, and b is the y-intercept).

5.2.4 Assessment of Recombinant Enzyme Purity

The purity of the recombinant enzyme was monitored using SDS-PAGE. The gels can be purchased in a precast form commercial manufacturers (which is a convenient but costly option) or can be prepared onsite (which is more cost-effective but requires extra effort and time). All materials needed to cast your own gels are listed under Section 3.3, but for experimental details the reader is referred to an excellent review article (Brunelle & Green, 2014). For assessing the purification of target enzymes, we generally use SDS-PAGE gels (102 mm × 74 mm × 0.75 mm) with 5% (w/v) stacking and 12% (w/v) resolving components.

Samples were prepared as follows: 5 μg of total protein (based on Bradford Assay outlined under Section 5.2.3) were mixed with 5 μL of the 4× Protein Gel Loading Dye and the solution was heated to 95°C for 5 min. Samples and the protein ladder control (Precision Plus Protein All Blue Prestained Standards, Bio-Rad) were loaded side by side into the wells of the stacking gel. The gel was run at 110 V in Protein Gel Running Buffer until the dye front reached the other end of the gel. The gel was then carefully transferred to a dish containing 50 mL Protein Staining Dye solution, and the dish rocked on a for 2 h. The stained gel was transferred to another dish containing 100 mL gel destaining solution and rocked until

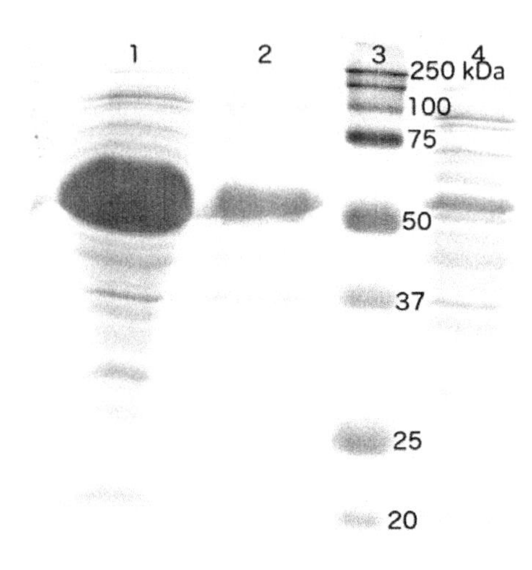

Fig. 3 Purification of recombinant R58 as indicated by SDS-PAGE. *Lane 1*, crude protein extract from *E. coli* cells expressing R58 (note the large band for the target protein); *lane 2*, purified protein fraction following hydroxyapatite fractionation (note that the target protein is >95% pure); *lane 3*, protein size standard ladder; and *lane 4*, crude protein extract from uninduced control *E. coli* cells (note that the target protein constitutes only one of many detectable bands).

the blue background color disappeared and individual protein bands became visible. The progress of purification was determined by comparing protein band intensities using a Gel Imager (MultiImage II, Alpha Innotech). Preparations with a purity of >90% target protein (band at ~50 kDa) (Fig. 3) were further evaluated in enzyme assays (details under Section 6).

6. STEP 3—FUNCTIONAL EVALUATION OF RECOMBINANT MONOTERPENE SYNTHASES

6.1 Enzymatic Reaction

In a 2-mL screw-cap glass vial, 200 μg of the purified enzyme (volume depends on enzyme concentration as determined under Section 5.2.3) and 20 μL GPP Stock Solution (final concentration of substrate in the assay is 0.5 m*M*) were mixed and Enzyme Assay Buffer was added to a final reaction volume of 400 μL. Each sample was overlaid with 100 μL *n*-hexanes (OmniSolv, EMD Millipore), the reaction allowed to proceed to completion at 30°C on a tube rotator for 16 h. The enzymatic reaction was stopped by vigorous mixing of the aqueous and organic phases on a Single Tube

Shaker (Vortex Genie 2, IKA) operated at the higher setting. To facilitate phase separation, reaction vials were kept at −20°C for 2 h. The supernatant containing the organic phase was transferred to 2 mL glass vial with conical insert and stored at −20°C until further use (maximum storage duration was less than 1 week).

6.2 Product Quantitation

The products formed during the enzymatic reaction described under Section 6.1 were quantified by Gas Chromatography (GC) (model 7890, Agilent Technologies) with a Flame Ionization Detection. Analytes were separated on a Cyclodex-B chiral column (J&W Scientific; 30 m × 0.25 mm, 0.25 μm film thickness). The following GC settings were used: injector temperature 250°C; injector split 1:20; sample volume 1 μL; flow rate 2 mL/min with He as carrier gas; oven heating program with an initial ramp from 40 to 120°C at 2°C/min, followed by a second ramp to 200°C at 50°C/min, and a final hold at 200°C for 2 min; detector temperature 250°C; detector gas flows of 30 mL/min for H_2, 400 mL/min for air, and 25 mL/min for He as makeup gas. Quantitation was achieved by comparing detector responses, using a linear equation, with those of a calibration curve, which was obtained by injecting varying quantities of authentic standards as described under Section 5.2.3 for protein quantitation. Representative chromatograms for enzyme assays are shown in Fig. 4.

Note: A chiral GC column was employed to enable the analysis of enantiomeric monoterpenes. The identities of monoterpenes formed in enzymatic reactions were verified by gas chromatography–mass spectrometry as described elsewhere (Adams, 2007).

6.3 Analysis of Kinetic Data

Kinetic assays were carried out to determine the maximum reaction rate (V_{max}), the turnover number (K_{cat}), the Michaelis constant K_m (inverse measure of the enzyme's affinity for the substrate), and the catalytic efficiency (calculated as k_{cat}/K_M) of recombinant monoterpene synthases. In a continuous assay, the enzyme concentration is kept constant and the substrate concentration is varied. Purified, recombinant monoterpene synthase (25 μ*M*) was reacted with 2.5–50 μ*M* substrate (GPP) by gentle mixing rotator at 30°C for 20 min (these conditions ensured that <20% of the available substrate was consumed). A Michaelis–Menten saturation curve was obtained by plotting substrate concentration against reaction rate (formation of product over time) (Fig. 5). The Michaelis constant K_m is defined

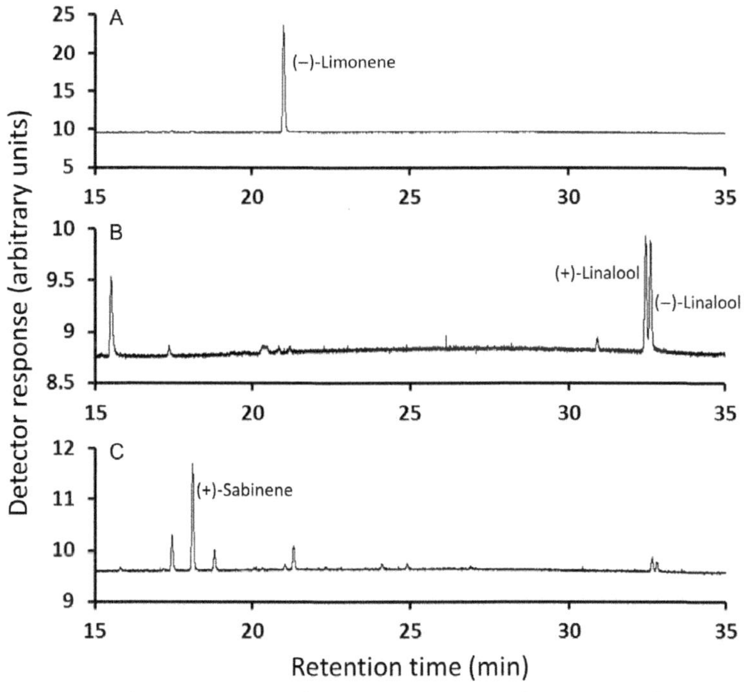

Fig. 4 GC–FID chromatograms of enzyme assay products. (A) Wild-type (R58), (B) W324E mutant, and (C) W324H mutant.

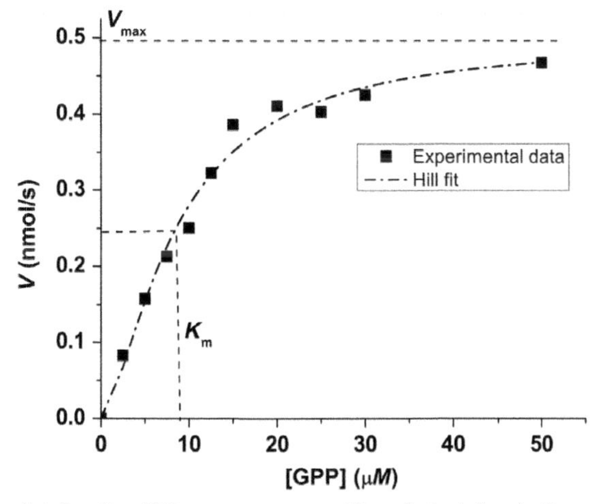

Fig. 5 Kinetic plot for the R58 enzyme assay. The *dotted line* indicates the curve fit obtained by nonlinear regression analysis.

as the substrate concentration at which the reaction rate is half of the maximal rate (V_{max}), but these constants were determined more accurately by nonlinear regression analysis using a Hill-type equation $y = V_{max} * \frac{x^n}{K_m^n + x^n}$ (Origin 8; OriginLab). The turnover number (K_{cat}) was then calculated by dividing V_{max} by the total enzyme concentration.

7. CONCLUSIONS

During recent years, various synthetic biology platforms have been developed for the production of commercially relevant terpenoids (from C5 to C40) by introducing, among others, plant-derived terpene synthase genes into microbial hosts. Starting with (4*S*)-limonene synthase from spearmint as a template, we generated a large number of mutant enzymes that produce specific monoterpenes, including acyclic, monocyclic, and bicyclic terpenoid hydrocarbons and alcohols. The methods described in this chapter encompass mutant generation by site-directed mutagenesis, the introduction of a mutant gene into a bacterial host cell, and the production, purification, and functional characterization of the recombinant mutant enzyme. Based on the strategies outlined here, readers can develop designer mutant enzymes optimized for the production of specific target monoterpenes.

ACKNOWLEDGMENTS

This work was supported by the Division of Chemical Sciences, Geosciences, and Biosciences, Office of Basic Energy Sciences, and US Department of Energy (Grant No. DE-SC0001553 to B.M.L.).

REFERENCES

Adams, R. P. (2007). *Identification of essential oil components by gas chromatography/mass spectrometry*. Carol Stream, IL: Allured Publishing Corp.

Brunelle, J. L., & Green, R. (2014). One-dimensional SDS-polyacrylamide gel electrophoresis (1D SDS-PAGE). *Methods in Enzymology, 541*, 151–159.

Chen, F., Tholl, D., Bohlmann, J., & Pichersky, E. (2011). The family of terpene synthases in plants: A mid-size family of genes for specialized metabolism that is highly diversified throughout the kingdom. *The Plant Journal, 66*, 212–229.

Coates, R. M., Elmore, C. C., Croteau, R. B., Williams, D. C., Morimoto, H., & Williams, P. G. (1997). Stereochemistry of the methyl→methylene elimination in the enzyme-catalyzed cyclization of geranyl diphosphate to (4S)-limonene. *Chemical Communications, 21*, 2079–2080.

Davisson, V. J., Woodside, A. B., Neal, T. R., Stremler, K. E., Muehlbacher, M., & Poulter, D. C. (1986). Phosphorylation of isoprenoid alcohols. *Journal of Organic Chemistry, 51*, 4768–4779.

Fisher, C. L., & Pei, G. K. (1997). Modification of a PCR-based site-directed mutagenesis method. *Biotechniques*, *23*, 570–574.

Hyatt, D. C., Youn, B., Zhao, Y., Santhamma, B., Coates, R. M., Croteau, R. B., et al. (2007). Structure of limonene synthase, a simple model for terpenoid cyclase catalysis. *Proceedings of the National Academy of Sciences of the United States of America*, 104, 5360–5365.

Lange, B. M., & Ahkami, A. (2013). Metabolic engineering of plant monoterpenes, sesquiterpenes and diterpenes—Current status and future opportunities. *Plant Biotechnology Journal*, *11*, 169–196.

Morrone, D., Lowry, L., Determan, M. K., Hershey, D. M., Xu, M., & Peters, R. J. (2010). Increasing diterpene yield with a modular metabolic engineering system in *E. coli*: Comparison of MEV and MEP isoprenoid precursor pathway engineering. *Applied Microbiology and Biotechnology*, *85*, 1893–1906.

Paddon, C. J., Westfall, P. J., Pitera, D. J., Benjamin, K., Fisher, K., McPhee, D., et al. (2013). High-level semi-synthetic production of the potent antimalarial artemisinin. *Nature*, *496*, 528–532.

Pyun, H. J., Coates, R. M., Wagschal, K. C., McGeady, P., & Croteau, R. B. (1993). Regiospecificity and isotope effects associated with the methyl → methylene eliminations in the enzyme-catalyzed biosynthesis of (R) and (S)-limonene. *Journal of Organic Chemistry*, *58*, 3998–4009.

Rajaonarivony, J. I., Gershenzon, J., & Croteau, R. (1992). Characterization and mechanism of (4*S*)-limonene synthase, a monoterpene cyclase from the glandular trichomes of peppermint (*Mentha* x *piperita*). *Archives of Biochemistry and Biophysics*, *296*, 49–57.

Schenk, P. M., Baumann, S., Mattes, R., & Steinbiβ, H. H. (1995). Improved high-level expression system for eukaryotic genes in *Escherichia coli* using T7 RNA polymerase and rare ^{Arg}tRNAs. *Biotechniques*, *19*, 196–200.

Srividya, N., Davis, E. M., Croteau, R. B., & Lange, B. M. (2015). Functional analysis of (4*S*)-limonene synthase mutants reveals determinants of catalytic outcome in a model monoterpene synthase. *Proceedings of the National Academy of Sciences in the United States of America*, *112*, 3332–3337.

Williams, D. C., McGarvey, D. J., Katahira, E. J., & Croteau, R. (1998). Truncation of limonene synthase preprotein provides a fully active 'pseudomature' form of this monoterpene cyclase and reveals the function of the amino-terminal arginine pair. *Biochemistry*, *37*, 12213–12220.

Woodside, A. B., Huang, Z., & Poulter, C. D. (1988). Trisammonium geranyl diphosphate [diphosphoric acid, mono(3,7-dimethyl-2,6-octadienyl) ester (*E*)-, trisammonium salt]. *Organic Synthesis*, *66*, 211.

Zerbe, P., Hamberger, B., Yuen, M. M. S., Chiang, A., Sandhu, H. K., Madilao, L. L., et al. (2013). Gene discovery of modular diterpene metabolism in nonmodel systems. *Plant Physiology*, *162*, 1073–1091.

CHAPTER EIGHT

Prequels to Synthetic Biology: From Candidate Gene Identification and Validation to Enzyme Subcellular Localization in Plant and Yeast Cells

E. Foureau*[,1], I. Carqueijeiro*[,1], T. Dugé de Bernonville*[,1], C. Melin*, F. Lafontaine*, S. Besseau*, A. Lanoue*, N. Papon†, A. Oudin*, G. Glévarec*, M. Clastre*, B. St-Pierre*, N. Giglioli-Guivarc'h*, V. Courdavault*[,2]

*Université François-Rabelais de Tours, EA2106 "Biomolécules et Biotechnologies Végétales", Tours, France
†Université d'Angers, Groupe d'Etude des Interactions Hôte-Pathogène, UPRES EA 3142, Angers, France
[2]Corresponding author: e-mail address: vincent.courdavault@univ-tours.fr

Contents

1. Introduction 168
2. Identification of Candidate Genes Through Transcriptomic Data Mining and Analysis 171
   2.1 Transcriptome Assembly, Annotation, and Transcript Abundance Estimation 172
   2.2 Transcriptome Postassembly Analysis 180
3. Validation of Candidate Gene Function by Biolistic-Mediated VIGS 187
   3.1 Plant Material and Growth Condition Pretransformation 188
   3.2 Silencing Constructs for VIGS 189
   3.3 Biolistic-Mediated Transformation of *C. roseus* 190
   3.4 Posttransformation Treatments and Analysis 191
4. Studying the Subcellular Localization of Biosynthetic Pathway Enzymes in Plant and Yeast Cells to Alleviate Bottlenecks in Bioengineering Approaches 191
   4.1 Protein Subcellular Localization in *C. roseus* Cells 192
   4.2 Protein Subcellular Localization in Yeast Cells 198
5. Concluding Remarks 202
Acknowledgments 203
References 203

[1] Equal contribution.

Methods in Enzymology, Volume 576
ISSN 0076-6879
http://dx.doi.org/10.1016/bs.mie.2016.02.013

Abstract

Natural compounds extracted from microorganisms or plants constitute an inexhaustible source of valuable molecules whose supply can be potentially challenged by limitations in biological sourcing. The recent progress in synthetic biology combined to the increasing access to extensive transcriptomics and genomics data now provide new alternatives to produce these molecules by transferring their whole biosynthetic pathway in heterologous production platforms such as yeasts or bacteria. While the generation of high titer producing strains remains per se an arduous field of investigation, elucidation of the biosynthetic pathways as well as characterization of their complex subcellular organization are essential prequels to the efficient development of such bioengineering approaches. Using examples from plants and yeasts as a framework, we describe potent methods to rationalize the study of partially characterized pathways, including the basics of computational applications to identify candidate genes in transcriptomics data and the validation of their function by an improved procedure of virus-induced gene silencing mediated by direct DNA transfer to get around possible resistance to Agrobacterium-delivery of viral vectors. To identify potential alterations of biosynthetic fluxes resulting from enzyme mislocalizations in reconstituted pathways, we also detail protocols aiming at characterizing subcellular localizations of protein in plant cells by expression of fluorescent protein fusions through biolistic-mediated transient transformation, and localization of transferred enzymes in yeast using similar fluorescence procedures. Albeit initially developed for the Madagascar periwinkle, these methods may be applied to other plant species or organisms in order to establish synthetic biology platform.

1. INTRODUCTION

For ages, humans have exploited natural compounds, notably those arising from plant specialized metabolisms, as dyes, herbicides, flavors, and scents, or as bioenergy sources, but above of all by taking advantage of their pharmacological properties (Hanson, 2003; Ragauskas et al., 2006). These broad biological activities resulted in the valorization of specialized metabolites in a variety of industries and pharmaceutical applications rendering these molecules inescapable to our life habits. Since the beginning of the 20th century and the progress of synthetic chemistry, chemists began to reproduce these natural products alike both to improve their supply and to get around the drawback of biological sourcing, and subsequently to include rational modifications to generate novel and more potent drugs. However, in spite of decades of efforts and the development of combinatorial chemistry or computer molecular modeling, this task remains challenging for numerous compounds and/or economically unviable given the high complexity of their structures. As a consequence, it is admitted that more than half of the approved drugs used over the past 30 years are still directly or indirectly extracted from natural sources (Newman & Cragg, 2012). This inextinguishable list of bioactive molecules includes for instance, the prominent quinine, extracted from *Cinchona* tree, which is still one of the major drug used against malaria in combination with artemisinin from *Artemisia annua* (Pollier, Moses, & Goossens, 2011); ajmaline and ajmalicine used, respectively, as antiarrhythmic and in the treatment of hypertension following extraction *Rauwolfia serpentina* (Drewes, George, & Khan, 2003; Pollier et al., 2011); camptothecin and its derivatives topotecan hydrochloride and irinotecan hydrochloride used in chemotherapy and produced from *Camptotheca acuminata* (Thomas, Rahier, & Hecht, 2004); sanguinarine (*Sanguinarea canadensis*)—as antibacterial; or cocaine (*Erythroxylum coca*) as anesthetic (Mano, 2006; Wink, 1999; Zhao & Dixon, 2009); conolidine from *Tabernaemontana* ssp., which was recently proved to have analgesic properties as powerful as the opiates (Tarselli et al., 2011); berberine from *Coptis* with antibiotic and antiinflammatory activities, and more recently shown to have anticancer and antidiabetic activities (Li et al., 2015; Stermitz, Lorenz, Tawara, Zenewicz, & Lewis, 2000); taxol isolated from *Taxus brevifolia* and widely used in chemotherapy cocktails, which was in difficult supply until alternative sources like suspension cell cultures were developed (Miller & Ojima, 2001); ephedrine from *Ephedra sinica* to treat asthma (Lee, 2011) and the antineoplastics vinblastine and vincristine from *Catharanthus roseus* (Madagascar periwinkle) that are still produced via extraction from the periwinkle leaves.

The complexity of the biosynthetic pathways of the specialized metabolites as well as their high degrees of organization in planta has prevented for years, the use of simplified models, such as dedifferentiated plant cell cultures in vitro, to produce these valuable compounds through biotechnological approaches. However, the recent progress in synthetic biology and the dramatic reduction in cost for development of genomic and transcriptomic datasets are now changing this hostile scenario and open new ways to produce these valuable compounds via metabolic engineering of microbial platforms. Over the last 10 years, numerous examples of this type of production in heterologous organisms have been detailed via partial and/or complete biosynthetic pathway reconstitutions through multiple plant gene transfers, starting with the synthesis of artemisinic acid, the artemisinin precursor, in both *Saccharomyces cerevisiae* and *Escherichia coli* (Chang, Eachus, Trieu, Ro, & Keasling, 2007; Paddon et al., 2013; Ro et al., 2006), the production of benzylisoquinoline alkaloids in yeast transformed with enzymes from three plant sources and humans (Cassels et al., 1995; Hawkins & Smolke, 2008) and the recent synthesis in yeast of the precursor of all monoterpene indole alkaloids (MIAs), strictosidine, achieved by the transfer of 14 MIA biosynthetic genes from *C. roseus* along with seven additional genes and three gene deletions (Brown, Clastre, Courdavault, & O'Connor, 2015).

The development of such strategies relies on a complete characterization of the biosynthetic pathways of specialized metabolites as a commitment for the effective transfer in heterologous organisms of the whole set of genes constituting these pathways. Except for a few number of major drugs, the knowledge of biosynthetic pathways still remains scarce as compared to their immense diversity and complexity. From this point of view, accurate exploitations of transcriptomic data now have the potential to rationalize the identification of candidate genes for the missing steps of a pathway, notably for enzyme coded by large multigene family. Albeit a consequent bench work is still required to validate candidate gene function, such type of approaches has permitted huge advances in deciphering plant specialized metabolisms as recently illustrated for MIAs (Dugé de Bernonville, Foureau, et al., 2015). Furthermore, the numerical complexity of pathways is frequently enriched by their complex organization in planta at both the cellular and subcellular levels. For instance, in *C. roseus*, no less than four distinct leaf cell types and more than five subcellular compartments host the almost 30–40 enzymatic steps required to synthesize MIAs (Courdavault et al., 2014). While the cellular organization of a pathway can be easily skirted for metabolic engineering approaches by coexpressing all genes in a single heterologous organism, much attention should be paid to subcellular organizations since the inherent metabolite translocations between organelles could impact metabolic fluxes and hampered the production of the desired compounds. This implies a rigorous characterization of the biosynthetic pathway subcellular architecture to foreshadow and remedy future potential bottlenecks in metabolite synthesis postgene transfers. In addition, several biases of localizations could arise when expressing plant proteins in yeast or bacteria necessitating the validation of each protein localization in the recipient organism.

With the intent to streamline/rationalize the production of valuable compounds through synthetic biology approaches, we describe hereafter a guideline of the main technical procedures that should be used to generate some prerequisites to the transfer of a biosynthetic pathway in heterologous organisms (Fig. 1). By using, as a framework, examples from the Madagascar periwinkle, we successively present (1) the basics of computational

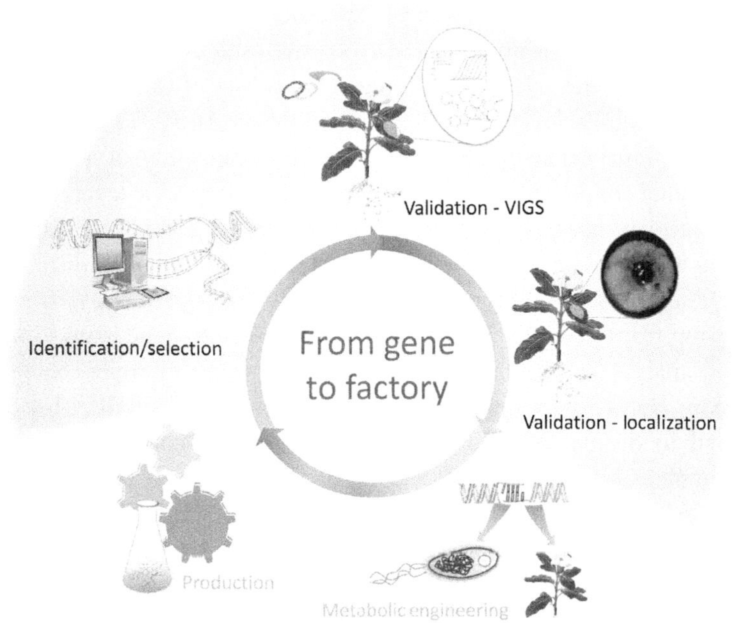

Fig. 1 Prequels to synthetic biology: from candidate gene identification and validation to their characterization and transfer into heterologous organisms. (See the color plate.)

applications to valorize transcriptomic data and to identify candidate genes, (2) a procedure of virus-induced gene silencing (VIGS) mediated by direct DNA transfer to get around resistance to *Agrobacterium*-delivery of viral vectors, (3) a procedure of protein subcellular localization study via particle bombardment and fluorescent protein (FP) imaging in planta, and (4) the validation of protein localization in heterologous organisms such as yeast.

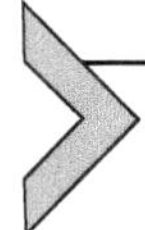

2. IDENTIFICATION OF CANDIDATE GENES THROUGH TRANSCRIPTOMIC DATA MINING AND ANALYSIS

Identification of all the genes of a specific biosynthetic pathway is the first prerequisite to its transfer in heterologous organisms. However, this characterization remains partial for many valuable compounds and implies prediction and validation of candidate genes. Basically, it is admitted that genes of a similar biosynthetic pathway are potentially subjected to coregulation, at least in distinct subparts of this pathway. Such coregulation can thus lead to the coexpression of all these genes in a wide set of conditions, which can be uncovered through analysis of transcriptomic data.

By consequence, the comparison of expression profiles between previously identified genes of the investigated pathway, used as baits, and the transcriptome allows the selection of a restricted subset of candidate genes displaying the highest level of expression profile similarity. The efficiency of this type of analysis strongly depends on the availability of a large set of experimental conditions, which display various and specific levels of biosynthetic pathway gene expression. Furthermore, other/previous RNA-seq runs can be included in these analyses via straightforward procedures of data reuse. The following sections will describe the basics of transcriptomic data processing from assembly and transcript abundance to gene expression correlation analysis and hierarchical clustering by using examples based on a consensus transcriptome built for *C. roseus* with previous RNA-seq data obtained in distinct plant organs and experimental conditions (Dugé de Bernonville, Clastre, et al., 2015).

2.1 Transcriptome Assembly, Annotation, and Transcript Abundance Estimation

2.1.1 Transcriptome Assembly

Short-read sequencing of transcriptomes generates raw FASTq files containing a sequence for each spot on the lane together with base quality indices. Careful examination of quality and trimming of raw reads constitutes essential prerequisites before assembling them into larger contigs. Several assembly methods and protocols allowing de novo assembly of transcripts from short read paired-end sequencing have already been reported (Góngora-Castillo et al., 2012; Haas et al., 2013). De novo assemblers are designed to resolve de Bruijn graphs built with raw reads splitted into shorter words of length *k*, named *k*-mers (Martin & Wang, 2011). Strategies to prepare a transcriptome include single assembly from combined reads, combination of assemblies prepared with different *k*-mer length using a same set of reads (Velvet/Oases, Schulz, Zerbino, Vingron, & Birney, 2012), or combination of single assemblies obtained from individual samples. In the last two strategies, redundancy may be reduced by using assemblers such as TGICL (Pertea et al., 2003) or CD-HIT-EST (Huang, Niu, Gao, Fu, & Li, 2010). Assessment of the quality of transcript reconstruction has to be performed to ensure that parameters used in the assembly process are appropriate. This can be checked by (i) analyzing transcript lengths, (ii) annotating transcripts, and (iii) comparing transcripts to known EST or full-length mRNA belonging to the same studied species or closely related ones.

Genome-guided assembly is also possible, albeit plant genomic sequences are still scarce. In addition, de novo assembly may be preferred in the case of low quality or incomplete genomic sequences and to ease identification of alternatively spliced transcripts. Genome-guided assembly starts by mapping reads to the genomic sequences. De Bruijn graphs are then applied to resolve transcript structures. Several programs are designed for that purpose, eg, Cufflinks/TopHat (Trapnell et al., 2012), Trinity (Haas et al., 2013), and StringTie (Pertea et al., 2015).

2.1.2 Transcriptome Annotation

Annotation of the transcriptome assembly is a fundamental step which should (i) improve transcriptome quality (see above) and (ii) help for post-assembly analyses. The most common procedures include the research of homologies against a database by BLAST (either blastx on mRNA sequence or blastp on predicted peptide sequences) and detection of functional domains with HMMER (hmmerscan). The free software Trinotate (https://trinotate.github.io/) offers an easy interface to incorporate different annotation layers in a transcriptome. We also recommend the use of perl script HpcRunningGrid collection written by B. Haas to parallelize BLAST and HMMERscan and significantly improve annotation speed (http://hpcgridrunner.github.io/). The following protocol describes an annotation process performed with Trinotate.

1. Download Uniprot database for local homology search and PfamA: follow instructions found on the Trinotate webpage in section "2. Sequence Databases Required"

2. Prepare a script for transcriptome annotation: transdecoder, blastp, blastx and hmmerscan

Don't forget to configure the hpc_running_grid file to speed up analysis.

```
#!/bin/bash

#SBATCH -J transdec1
#SBATCH -o transdec1.%j
#SBATCH -e transdec1.%j
#SBATCH -n 1
#SBATCH -t 24:00:00
#SBATCH -p defq
```

```
module purge
module load shared
module load gcc/4.9.0
module load slurm/14.03.0

export PATH=$BLAST+DIR/bin:$PATH

#predict ORFs
$HOME/TransDecoder-2.0/TransDecoder.LongOrfs -t CDF97.fa

#run blastp
mkdir /scratch/blastp_CDF97
cp CDF97.fa.transdecoder_dir/longest_orfs.pep /scratch/blastp_CDF97
cp uniprot* /scratch/blastp_CDF97
cd /scratch/blastp_CDF97

#parallelize search (configure SLURM.test.conf accordingly)
perl $HOME/HpcGridRunner-1.0.0/BioIfx/hpc_FASTA_GridRunner.pl
--cmd_template "blastp -query __QUERY_FILE__ -db uniprot_sprot.
trinotate_v2.0.pep -max_target_seqs 1 -outfmt 6 -evalue 1e-5
-num_threads 2" --query_fasta longest_orfs.pep -G $HOME/SLURM.test.
conf -N 1000 -O CDF97_blastp

#group results and move them to $HOME directory
find CDF97_blastp/ -name "*.fa.OUT" -exec cat {} \; > CDF97.blastp.out
mv CDF97.blastp.out $HOME
cd $HOME
rm -fr /scratch/blastp_rau_CDF97_v2

#run blastx
mkdir /scratch/blastx_CDF97
cp CDF97.fa /scratch/blastx_CDF97
cp uniprot* /scratch/blastx_CDF97
cd /scratch/blastx_CDF97

#parallelize search (configure SLURM.test.conf accordingly)
perl $HOME/HpcGridRunner-1.0.0/BioIfx/hpc_FASTA_GridRunner.pl
--cmd_template "blastx -query __QUERY_FILE__ -db uniprot_sprot.
trinotate_v2.0.pep -max_target_seqs 1 -outfmt 6 -evalue 1e-20
-num_threads 2" --query_fasta longest_orfs.pep -G $HOME/SLURM.test.
conf -N 1000 -O CDF97_blastx
```

```
#group results and move them to $HOME directory
find CDF97_blastx/ -name "*.fa.OUT" -exec cat {} \; > CDF97.blastx.out
mv CDF97.blastx.out $HOME
cd $HOME
rm -fr /scratch/blastx_rau_CDF97_v2

#run hmmerscan
mkdir /scratch/CDF97
cp $HOME/CDF97.fa.transdecoder_dir/longest_orfs.pep /scratch/
hmmer_CDF97
cp Pfam-A.* /scratch/hmmer_CDF97
cd /scratch/hmmer_CDF97

perl $HOME/HpcGridRunner-1.0.0/BioIfx/hpc_FASTA_GridRunner.pl
--cmd_template "hmmscan --cpu 2 --domtblout __QUERY_FILE__.domtblout
Pfam-A.hmm __QUERY_FILE__" --query_fasta longest_orfs.pep -G $HOME/
SLURM.test.conf -N 1000 -O hmmer_CDF97

find hmmer_CDF97/ -name "*.fa.domtblout" -exec cat {} \; > CDF97.
hmmer.out

mv CDF97.hmmer.out $HOME
cd $HOME
rm -fr /scratch/hmmer_CDF97

#end
```

3. run trinotate and output to xls file: initialize MySQL Database and load annotation results in it by following Trinotate basics instructions.

2.1.3 Transcript Abundance Estimation

The measurement of gene expression evaluated by transcript abundance estimation is essential for coexpression analysis in order to identify new genes associated within a same biosynthetic pathway. In addition, measuring transcript abundance is also an important complementary step of the assembly process. Indeed, setting appropriate abundance thresholds may significantly improve transcriptome quality by removing nonexpressed chimeric transcripts. Several algorithms are available to estimate transcript abundance. Cufflinks (Trapnell et al., 2010), eXpress (Roberts, Feng, & Pachter, 2013), and RSEM (Li & Dewey, 2011) are among the most commonly used.

The starting point of abundance estimation is the mapping of raw reads to the reference assembly by Burrows-Wheeler schemes like Bowtie (Langmead, Trapnell, Pop, & Salzberg, 2009), Bowtie2 (Langmead & Salzberg, 2012), or BWA algorithms (Li & Durbin, 2009). Expression levels are next estimated after determining the most probable contig origin in case of multiple read mapping, in particular by applying expectation-minimization algorithms.

In the next sections, we will describe how to use RSEM to estimate transcript abundance in a transcriptome (RSEM v1.2.15 procedure for CDF97 with paired-end reads according to Dugé de Bernonville, Clastre, et al., 2015). Readers are invited to refer to original articles for more explanations about the different algorithms presented here.

The following procedures are based on a transcriptome resulting from a clustering with CD-HIT-EST of single Trinity assemblies obtained for each RNA-seq run available on NCBI Sequence Read Archive (SRA) for *C. roseus*. To take into account polymorphisms linked to the diversity of samples, CD-HIT-EST representative cluster sequences were used as "gene" sequences and the remaining contigs in the clusters as "transcripts" corresponding to these genes. This allowed attributing correct expression levels to "genes" given the more exact mapping of reads on each "transcript." In the following example, publicly available data were reused. SRA SRR accessions were downloaded by ftp (access with wget, for example) and the resulting .sra files convert to fastq with the SRA Toolkit (./fastq-dump -I –split-files SRRxxxxx.sra).

TIP: Transcriptome assembly and read assignment to transcripts intensively consume computational resources that cannot be efficiently managed on a single computer. The optimal situation is to use a computing grid with several nodes containing several CPUs with memory. For the following parts, computations were performed on the CCSC Computing Grid of Orléans, France. Nodes are composed of 20 CPUs (Intel Xeon processors) with 64 Go of RAM. Jobs are scheduled with SLURM under Scientific Linux 6. In addition, computations were performed on a high-speed reading partition named "scratch" when possible.

2.1.3.1 Prepare Reference for Abundance Estimation on CDF97

The input file shall be a transcriptome containing all transcript (not only gene) sequences. For the CDF97 example, we used a multifasta file (CDF97_allcontigs.fasta) containing all contig sequences (not only representative sequences).

```
#!/bin/bash
### This is a SLURM submission file.
```

```
#SBATCH -J prep_ref
#SBATCH -o out_prep_ref.%j
#SBATCH -e err_prep_ref.%j
#SBATCH --ntasks=1
#SBATCH --cpus-per-task=5
#SBATCH --tasks-per-node=1
#SBATCH -t 96:00:00
#SBATCH -p kernel3

export PATH=$BOWTIE_DIR/bowtie2-2.2.4:$PATH
$RSEM_DIR/rsem-prepare-reference --bowtie2 --transcript-to-gene-map CDF97_reference_map CDF97_allcontigs.fasta CDF97_allcontigs

#copy reference to the «scratch» partition
mkdir /scratch/ref/
cp CDF97_allcontigs* /scratch/ref/

#end
```

In this script, the "--transcript-to-gene-map" corresponds to a two-column text file. The second column contains each «transcript» represented once, and the first one includes «genes» for which are found the corresponding «transcript». In the case of CDF97 example, «genes» correspond to CD-HIT-EST representative sequences and «transcripts» to other sequences found in sequence clusters. For other transcriptomes, current assemblers such as Trinity provide both genes and transcripts, as well as the corresponding map.

2.1.3.2 Align Paired-End Reads to Reference Transcriptome with bowtie2

This step is the most resource-consuming process; the processing time depends especially on the number of sequences in the transcriptome. In the following step, we used an array-like job to submit one alignment (align all reads for a given sample to the reference transcriptome) per node, using 20 CPUs on each. The following script contains instructions to copy fastq files to the scratch partition, run RSEM, and retrieve .isoforms.results and .genes.results files.

First, prepare a folder containing all fastq files (SRRxxxxx_1.fastq and SRRxxxxxx_2.fastq) and a "pe_sample" file containing accession names (SRRxxxxx). In the present example, we had 12 paired-end runs and each will be treated in a separate job.

```
#!/bin/bash

### This is a SLURM submission file.
```

```
#SBATCH -J b2_alignreads
#SBATCH -o out_alignreads.%A_%a
#SBATCH -e err_alignreads.%A_%a
#SBATCH --ntasks=1
#SBATCH --cpus-per-task=20
#SBATCH --tasks-per-node=1
#SBATCH --array=1-12
#SBATCH -t 96:00:00
#SBATCH -p kernel3

export PATH=$BOWTIE_DIR/bowtie2-2.2.4:$PATH

#load required modules
module purge
module load shared
module load gcc/4.9.0
module load slurm/14.03.0

echo Start time: $(date)

number=$SLURM_ARRAY_TASK_ID
paramfile=pe_samples
in=‘sed -n ${number}p $paramfile | awk '{print $1}'‘

if [ -e /scratch/align/CDF97_allcontigs_$in ]; then
  rm -fr /scratch/align/CDF97_allcontigs_$in
fi

mkdir /scratch/align/CDF97_allcontigs_$in
cp /path/to/fastq_files/${in}_*.fastq /scratch/tduge/CDF97_allcontigs_
$in/
cd /scratch/tduge/CDF97_allcontigs_$in

$RSEM_DIR/rsem-calculate-expression --bowtie2 -p 20 --bowtie-
chunkmbs 1024 --paired-end ${in}_1.fastq ${in}_2.fastq /scratch/ref/
CDF97_allcontigs $in

cd $HOME/CDF97_allcontigs/
cp /scratch/align/CDF97_allcontigs_$in/*.results .
rm -fr /scratch/align/CDF97_allcontigs_$in

echo End time: $(date)
```

Submit job to your computing grid. Modify if required to suit scheduler (LSF, SGE, etc.) specificities. Each array job will produce two files named

".isoforms.results" and ".genes.results" according to the gene map that has been provided to build the reference.

TIP: To analyze single-end reads, just modify the rsem-calculate-expression parameters to –single-end and indicate the corresponding single fastq file. Using a similar array task, provide a "se_sample file" containing single-end accession.

2.1.3.3 Combine RSEM Results Files to Generate Raw Count or FPKM Matrix

```
#Rscript
    library(parallel)

    table.list<-list.files(pattern=".isoforms.results")

    #get raw counts
    #read tables and store count columns
    listing.tpm<-mclapply(table.list, function(x){
         tmp<-read.table(x, header=T)
         tmp[,6]}, mc.cores=20)

    #temporary table
    big.table<-do.call("cbind", listing.tpm)
    rm(listing.tpm)

    #attribute sample names
    tmp<-read.table(table.list[1], header=T)
    colnames(big.table)<-gsub(".isoforms.results", "", table.list)
    rownames(big.table)<-as.vector(tmp[,1])

    #write to external file
    write.table(big.table, "fpkm_table")

    #get fpkm
    listing.fpkm<-mclapply(table.list, function(x){
         tmp<-read.table(x, header=T)
         tmp[,7]}, mc.cores=20)
    big.table<-do.call("cbind", listing.fpkm)
    rm(listing.fpkm)
    tmp<-read.table(table.list[1], header=T)
    colnames(big.table)<-gsub(".isoforms.results", "", table.list)
    rownames(big.table)<-as.vector(tmp[,1])
    write.table(big.table, "fpkm_table")
```

TIP: Comparison of qPCR data with in silico digital expression can be useful to assess both quality of transcript reconstruction and abundance estimation.

2.2 Transcriptome Postassembly Analysis

Once assembly and abundance estimation procedures completed, the dataset can be used to identify candidate genes associated with a specific biosynthetic pathway. To this aim, the postassembly analysis may be directed to cluster genes with or without a priori. It is often appropriated to combine these strategies to identify candidate genes. The dataset is presented as a matrix where rows are genes and columns are samples. In the following section, we will describe how to apply such procedures with R and specific bioconductor packages.

```
#read the data
tpm.table<-read.table("tpm_table", header=T, row.names=1)
fpkm.table<-read.table("fpkm_table", header=T, row.names=1)

#load annotations
annot<-read.delim("annotation/Trinotate_output.xls", header=T)
annot[,1]<-gsub("|", "_", as.vector(annot[,1]), fixed=T)
annot.ok<-cbind(as.vector(annot[,1]) ,do.call("c", lapply(strsplit
(as.vector(annot[,3]), ";", fixed=T), function(x)x[1])), as.vector
(annot[,10]))

transcrit.1.orf<-annot[which(duplicated(annot[,2])==FALSE),]
rownames(transcrit.1.orf )<-transcrit.1.orf[,1]
transcrit.1.orf<-transcrit.1.orf[which(rownames(transcrit.1.
orf ) %in% rownames(fpkm.table)),]

#prepare objects to be used in GO term enrichment analysis
#the main objective is to get all represented GO terms in the
transcriptome
#and their effective (number of transcripts with a given annotation)
GO.tot<-lapply(1:nrow(transcrit.1.orf ),
function(x)unlist(strsplit(as.vector(transcrit.1.orf[x,14]), "`")))
names(GO.tot)<-rownames(transcrit.1.orf )
all.GO<-unique(unlist(GO.tot))

#GO annotations in Trinotate xls output are combined and separated by a
"`" character
#we used this to separate individual terms
```

```
all.GO.2<-do.call("rbind", strsplit(all.GO, "~", fixed=T))
rownames(all.GO.2)<-all.GO.2[,1]
all.GO.2[,3]<-do.call("rbind", lapply(strsplit(all.GO.2[,3], " "),
function(x)ifelse(length(x)==1, x[1],
ifelse(length(x)==2, paste(x[1], x[2]),
ifelse(length(x)==3, paste(x[1], x[2], x[3]),
ifelse(length(x)>3, paste(x[1], x[2], x[3], x[4])))))))

all.GO.factor<-factor(unlist(GO.tot, use.names=FALSE))
length.GO.terms<-summary(all.GO.factor, maxsum=length(all.GO))
#we can remove GO terms with very few genes in order to have a more
general annotation
length.GO.terms<-length.GO.terms[which(length.GO.terms>4)]
```

2.2.1 Differential Expression

Basically, identifying differentially expressed genes in RNA-seq datasets is similar to the strategy applied for microarray analysis. The objective is to perform a statistical analysis of the transcript mean or median expression to identify those which are significantly more or less expressed in comparison to the remaining transcripts in an experimental condition. Such analysis can be conducted using common R packages including edgeR (Robinson, McCarthy, & Smyth, 2010) and DESeq (Anders & Huber, 2010), available at Bioconductor (http://bioconductor.org/). The resulting *P*-values obtained from linear models are then adjusted by False Discovery Rate (FDR) or family-wise error rate (Bonferroni correction) and allow comparing gene expression. The sets of differentially expressed genes may next be used in gene set enrichment analysis to have better insights in the biological processes found in a given sample, provided that sufficient annotation information has been attributed to each transcript. Although the statistical analysis may be conducted by comparing single samples, it is more appropriated to have two or more biological replicates to strengthen the different tests.

Here is an example using edgeR (freely available at Bioconductor; http://bioconductor.org/packages/release/bioc/html/edgeR.html) to compare expression levels obtained in two conditions.

```
#Rscript
#edgeR
#load packages
library(edgeR)
```

```
rnaseqMatrix <- fpkm.table[rowSums(fpkm.table)>=2,]
#declare the number of biological repeats.
#although not recommended, this works for only one replicate per condition
#be sure that expression matrix columns are ordered accordingly
Conditions <- factor(c(rep("Condition1", 1), rep("Condition2", 1)))

exp_study <- DGEList(counts=rnaseqMatrix, group=conditions)
exp_study <- calcNormFactors(exp_study)
et <- exactTest(exp_study, dispersion=0.1)
tTags <- topTags(et,n=NULL)

#get genes significantly (FDR corrected p-value<0.05) up-regulated in condition 1
DE.geneList<-rownames(tTags$table)[which(tTags$table[,4]<0.05 & tTags$table[,1]<0)]
```

For automatic multiple pairwise comparisons, we recommend to use a Trinity wrapper script named run_DE_analysis.pl. as follow:

```
#calculate enrichment of GO terms from differentially expressed gene lists
GO.tot.DEG<-GO.tot[DE.geneList]
GO.factor.DEG<<-factor(unlist(GO.tot.DEG, use.names=FALSE))
length.GO.a.terms<-summary(GO.factor.DEG, maxsum=length(all.GO))
table.count<-matrix(nrow=nlevels(GO.factor.DEG), ncol=2)
for (i in 1:length(length.GO.a.terms)){
   i.tmp<-names(length.GO.a.terms[i])
   table.count[i, 1]<-length.GO.terms[i.tmp]
   table.count[i, 2]<-length.GO.a.terms[i.tmp]
}
rownames(table.count)<-names(length.GO.a.terms)
res.pval<-sapply(1:nrow(table.count), function(y){
   q=table.count[y,2]
   m=table.count[y,1]
   n=nrow(fpkm.table)-table.count[y,1]
   k=length(DE.geneList)
   phyper(q, m, n, k, lower.tail=F)})
res.pval<-p.adjust(res.pval, method="BH")
GO.table<-cbind(table.count, "Pval"=res.pval)
```

TIP: this procedure may be used for any gene list stored in a vector object (such as those that may be obtained in the following procedures).

2.2.2 Correlation Analysis

Identification of candidate genes from a biosynthetic pathway can be carried out by correlation analyses. This procedure aims to identify a set of genes with the highest level of coexpression with a list of previously characterized genes of the corresponding pathway, used as baits. Correlation may be calculated from linear models or following the Pearson statistic (Pearson correlation coefficient, PCC). Calculating correlations for a transcript with the entire transcriptome may be computationally intensive. In biosynthetic or signaling pathways, it may be useful to (i) calculate PCC of each candidate gene with all other genes and (ii) determine intersections between lists of coexpressed genes to identify missing elements. In such a case, the lists of coexpressed genes are determined by setting an appropriate PCC threshold.

```
#Rscript
#collect candidate names stored in one given file, one name per row
candidates.list<-scan("genes_of_interest", what="character")
correlation.list<-lapply(candidates.list,
  function(x){sapply(1:nrow(fpkm.table),
    function(y)cor(as.numeric(fpkm.table[x,]), as.numeric(fpkm.
table[y,]))})
####use mclapply from 'parallel' package when several CPUs are
available
library('parallel')
cpu.numbers<-detectCores()
correlation.list<-mclapply(candidates.list,
  function(x){sapply(1:nrow(fpkm.table),
    function(y)cor(as.numeric(fpkm.table[x,]), as.numeric(fpkm.
table[y,]))}, mc.cores=cpu.numbers)

#get co-expressed gene list for r>0.8
best.correlated.list<-lapply(correlation.list,
function(x)names(which(x>0.8)))
names(best.correlated.list)<-candidates.list

#determine and visualize intersections
#for 2 intersections
library(gplots)
intersection.res<-venn(list("geneA"=best.correlated.list
[["geneA"]],
        "geneB"=best.correlated.list[["geneB"]]))
```

```
#for 3 intersections
intersection.res<-venn(list("geneA"=best.correlated.list
[["geneA"]],
            "geneB"=best.correlated.list[["geneB"]],
            "geneC"=best.correlated.list[["geneC"]]))

plot(intersection.res)
#get intersection information
attr(intersection.res, "intersections")

#retrieve annotation for a given intersection
tmp.list<-attr(intersection.res, "intersections")[[1]]
annot.tmp<-sapply(tmp.list, function(x)which(annot.ok[,1] %in% x)
[1])
annot.tmp.2<-annot.ok[annot.tmp,]
#export as a csv file, readable in any Microsoft Office Excel or
LibreOffice Calc
write.csv(annot.tmp.2, "intersect_1.csv")
```

2.2.3 Clustering Procedures

Although correlation analyses (PCC calculations for example) constitute an efficient approach to identify candidate genes, application of gene clustering procedures renders gene discrimination even more stringent. A basic guideline of these strategies is described later.

2.2.3.1 Partitioning

The partitioning methods aim at creating groups of genes with the lowest variance within each group. The reference method is the *k*-means clustering. With a given *k* value which indicate the final number of groups, *k*-means algorithm tries to associate genes within each group such as the sum of their square value is minimized. The choice of *k* is intricate but can be guided by graphical inspection. For example, one may plot *k* values (eg, 1–100) against within group sum of squares for those different *k* values. The R function *k*-means as well as other functions of Cluster package can be used ("fanny" function for example).

```
#Rscript
#test within sum of squares for different k values
withinss.values<-sapply(1:50, function(x)kmeans(fpkm.mat, x, nstart
=25, iter.max=25)$tot.withinss)
```

```
#plot total within sum of squares for each k value
plot(1:50, withinss.values)
#this plot gives a visual inspection of appropriate k values, which correspond to the lowest values minimizing the total within group sum of squares.

#partition dataset according to the optimal k value (may also be determined using silhouette plots)
kmeans.res<-kmeans(fpkm.mat, k, nstart=25, iter.max=100)

#be careful, cluster research is a randomized process, so clusters will have different names #but not composition if the command is re-run. One might set a set.seed() value for #reproducibility purposes.

library(ggplot2)
library(reshape2)

#plot cluster expression profiles : use cluster centers in kmeans.res object
#first reformat table for ggplot plotting
melted.kmeans.res<-melt(cbind.data.frame("Cluster"=paste("cluster", 1:k, kmeans.res$size, sep="_"), kmeans.res$centers))

ggplot(melted.kmeans.res, aes(variable, value))+geom_point()+geom_line(aes(group=1))+facet_wrap(~Cluster,scales="free_y", ncol=2)+theme(axis.text.x=element_text(angle=330, hjust=0))

#retrieve annotation for a given cluster
tmp.list<-names(which(kmeans.res$cluster==clusternumber))
annot.tmp<-sapply(tmp.list, function(x)which(annot.ok[,1] %in% x)[1])
annot.tmp.2<-annot.ok[annot.tmp,]
#export as a csv file, readable in any Microsoft Office Excel or LibreOffice Calc
write.csv(annot.tmp.2, "annotation_genes_clusternumber.csv")
```

The MB-RNA-seq cluster package (Si, Liu, Li, & Brutnell, 2014) was specifically designed for RNA-seq data. It adapts a model-based (non binomial or negative Poisson) distribution to an initial *k*-means partitioning.

2.2.3.2 Hierarchical Clustering

Given a dissimilarity matrix (computed by euclidean distances, for example), a hierarchical clustering procedure acts iteratively to cluster similar individuals (genes) by joining most similar individuals. New dissimilarity measures (Ward, UPGMA, Lance-William, etc.) are calculated at each iteration between the new formed cluster and the remaining genes. In the final tree, genes with similar expression patterns are grouped together.

```
#Rscript
#calculate dissimilarity matrix
d.mat<-dist(fpkm.table, method="euclidean")
#cluster genes
hclust.dmat<-hclust(d.mat, method="ward.D2")
#plot tree
plot(hclust.dmat)
#create cluster by cutting tree
#first, observe how the tree is cut for different thresholds; try different k values
rect.hclust(hclust.dmat, k=5)
#get cluster composition and size
cluster.hclust.dmat<-cutree(hclust.dmat, k=5)
cluster.size<-sapply(levels(as.factor(cluster.hclust.dmat)), function(x)length(which(cluster.hclust.dmat==x)))
names(cluster.size)<-paste("Cluster", levels(as.factor(cluster.hclust.dmat)), sep="")

#plot cluster expression profiles
mean.clust<-lapply(1:nlevels(as.factor(cluster.hclust.dmat)), function(x){
 cbind.data.frame("Mean"=apply(fpkm.table[which(cluster.hclust.dmat==x),], 2, mean),"Sample"=colnames(fpkm.table), "Cluster"=rep(paste(x, cluster.size[x], sep="_"), ncol(fpkm.table)))})
mean.clust.table<-do.call("rbind", mean.clust)

p<-ggplot(mean.clust.table, aes(x=Sample, y=Mean))
p+geom_point()+geom_line(aes(group=Cluster))+facet_wrap(~Cluster, ncol=4)

#retrieve annotation for a given cluster
tmp.list<-names(which(cluster.hclust.dmat==clusternumber))
annot.tmp<-sapply(tmp.list, function(x)which(annot.ok[,1]%in%x)[1])
annot.tmp.2<-annot.ok[annot.tmp,]
#export as a csv file, readable in any Microsoft Office Excel or LibreOffice Calc
write.csv(annot.tmp.2, "annotation_genes_clusternumber.csv")
```

In addition, agglomerative clustering may also be tested for investigation purposes with the "agnes" function from the Cluster package. Many dissimilarity measures are available with parameters (arguments to the dissimilarity methods) that may be fine-tuned to improve clustering.

2.2.3.3 HOPACH

This function associates an initial *k*-means based partitioning and a final hierarchical clustering procedure to order similar genes within a sum cluster (van der Laan & Pollard, 2003). Clusters are used to construct tree branches. The final order of genes is used to group genes displaying very correlated expression levels.

```
#Rscript
library(hopach)
gene.dist<-distancematrix(fpkm.mat,"cosangle")
gene.hobj<-hopach(fpkm.mat.sorted,dmat=gene.dist)
rm(gene.dist)akeoutput(fpkm.mat.sorted, gene.hobj, bootobj=NULL,
file="HOPACH.out",
        gene.names=rownames(fpkm.mat.sorted))

res.hopach<-read.table("HOPACH.out", header=T)
rownames(res.hopach)<-as.vector(res.hopach[,2])
#in this object, we have to look at the order of genes: this corresponds
to the order in the final tree, and genes with close expression patterns are
found near each other.
position.of.interest<-res.hopach["GeneOfInterest",1]
#this returns the position of GeneOfInterest; the further step is to
analyse functions of #genes located in the neighborhood of this position
(eg, +/- 100)
names(res.hopach[(position.of.interest-100): (position.of.interest
+100),1])
```

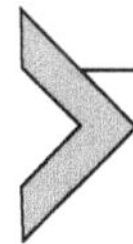

3. VALIDATION OF CANDIDATE GENE FUNCTION BY BIOLISTIC-MEDIATED VIGS

Whatever the efficiency of the procedures of candidate gene identification, functional validation of each candidate gene is required before considering their transfer into heterologous organisms to reconstitute the biosynthetic pathway of a desired valuable compound. While direct functional approaches including protein expression and biochemical assays are still frequently undertaken, rapid and potent screening of multiple candidates performed through transient gene invalidations mediated by VIGS are increasingly popular. Based on reverse genetic principles, this approach aims at transiently silencing a specific gene in planta, by using the RNA degradation system that plants deploy to respond to viral infections, and at studying its consequence on multiple biological processes such as the biosynthesis of a specialized metabolites. Over the last 15 years, plenty protocols of VIGS have been described for several plants. However, with only few exceptions, all these protocols rely on the inoculation of the viral genome through *Agrobacterium tumefaciens*-mediated transformations. However, this biological delivery strategy is subject to host specificity restriction as well as the induction of plant defense responses, which are commonplace for medicinal plants. For instance, in *C. roseus*, such type of reactions has precluded, for long, the development of an efficient procedure of plant agrotransformation. To date, three VIGS protocols have been described for *C. roseus*, which are all based on the inoculation of tobacco rattle virus (TRV) vectors using *Agrobacterium* by mechanical inoculation through piercing or pinching the stem below the meristem or by seedling infiltration (De Luca, Salim, Levac, Atsumi, & Yu, 2012; Liscombe & O'Connor, 2011; Sung, Lin, & Chen, 2014). Recently, we described a distinct delivery method relying on the transfer of the TRV vector (pTRV1 and pTRV2) by a biolistic-mediated transformation of *C. roseus* plantlets (Carqueijeiro et al., 2015). By eliminating *Agrobacterium* as shuffling vector avoiding thus host specificity problems, this strategy potentially constitutes a transferable tool for other plant specifies recalcitrant to *Agrobacterium*.

3.1 Plant Material and Growth Condition Pretransformation

Seeds of *C. roseus* (Little Bright Eye or Apricot sunstorm) were germinated and cultivated at 23°C using loam as substrate, in a green house, under a 16-h light/8-h dark cycle, with white fluorescent light (maximum intensity of 70 μmol m^{-2} s^{-1}). At the cotyledon stage, plants were individually potted and grown until the first leaf pair reached full development and the second pair just emerged (Fig. 2A).

TIP: To allow an accurate analysis of the silencing results, we recommend preparing around 10 control plants (transformed with empty vectors), 10 plants per silenced candidate gene, and 10 plants transformed with constructs generating easily identifiable phenotype modifications. Monitoring the kinetic of phenotype appearance allows determining the optimal period for sample harvesting. Silencing of genes encoding phytoene desaturase

Fig. 2 Virus-induced gene silencing in *Catharanthus roseus* by biolistic transformation (VIGS). 2 weeks old *C. roseus* plantlets presenting one pair of fully expanded leaves were used to perform the particle bombardment. (A) Time table representing the development of the leaves following transformation of VIGS vectors (B) prebombardment (0 dpb), 7 (7 dpb), and 20 (20 dpb) days after bombardment, as compared with a 20 days old nontransformed leaf (wild type). (C–F) Phenotypic aspect of *C. roseus* plants portraying different conditions including wild-type plants (C), plants transformed with empty vector pTRV2-EV (D), protoporphyrin IX magnesium chelatase pTRV2-ChlH depicting the characteristic yellow pigmentation (E), or pTRV2-PDS exhibiting the bleaching of the leaves (F), at 30 dpb. (See the color plate.)

(PDS; De Luca et al., 2012) or protoporphyrin IX magnesium chelatase (ChlH; Liscombe & O'Connor, 2011) usually leads to useful results.

3.2 Silencing Constructs for VIGS

pTRV vectors (pTRV1 and pTRV2-MCS) expressing the two components of the TRV genome were obtained from the Arabidopsis Biological Resource Center (ABRC) and were used to propagate the virus within plantlets. Fragment of 200–400 bp of the target genes is usually cloned into appropriated restriction sites of the pTRV2 multiple cloning site using classical endonuclease-based DNA manipulation. Supercoiled plasmids used for particle bombardment were isolated from *E. coli* cultures using Nucleospin Plasmid kit (Macherey-Nagel) following manufacturer's instructions.

TIP: Before each VIGS experiment, we advise to confirm plasmid integrity by basic analytical electrophoresis gel. Avoid proceed in case of plasmid degradation.

3.3 Biolistic-Mediated Transformation of *C. roseus*

3.3.1 Particle Preparation

1. Weigh 30 mg of 1 μm gold particles (Bio-Rad) in a glass tube and dry heat at 180°C for 8 h.
2. Wash the gold particles, 5 min with 1 mL of fresh 70% ethanol using vortex and sonication bath. Transfer to 2 mL sterile microcentrifuge tube.
3. Centrifuge the gold particles at 16,000 × *g* for 5 s.
4. Remove the supernatants; wash the pellets three times with 1 mL of sterilized milliQ water.
5. Centrifuge the gold particles at 16,000 × *g* for 5 s and resuspend the gold particles in 500 μL of 50% glycerol (p/v) sterile.

3.3.2 Coating of Plasmids onto Particles

1. For 10 bombardments, 10 μg of each plasmid are coated and precipitated onto 6.25 mg of gold particles. Plasmid solutions usually displayed a 1 μg/μL final concentration.
2. Mix 50 μL of 0,1 *M* spermidine with 100 μL of glycerol stocked gold particles in a sterilized tube and homogenize simultaneously with vortex and short pulses of ultrasounds with sonication bath.
3. Add the appropriated volume of purified DNA plasmid (not exceeding 10 μL for each plasmid) and mix. Maximize the coating efficiency by allowing the binding for 3 min and by vortexing each minute.
4. While homogenizing the solution with vortex, add 60 μL of 2.5 *M* $CaCl_2$ and mix for additional 15 min with a vortex at constant speed (around 1000 rpm).
5. Spin the tubes for 2 s at 16,000 × *g* to pellet gold particles coated with plasmids and remove the supernatant.
6. Wash plasmid coated gold particle pellets successively with 500 μL of 70% ethanol and 500 μL of 100% ethanol without resuspending gold particles.
7. Remove all supernatant and resuspend particles in 100 μL of 100% ethanol.
8. Spread 10 μL of coated gold particles onto each macrocarrier (Bio-Rad) to allow drying before transformation.

3.3.3 Particle Bombardment Procedure

Transformations of *C. roseus* plantlets are performed with the Bio-Rad PDS1000/He delivery system according to manufacturer's recommendations, using 1100 psi rupture disks (Bio-rad) under a vacuum pressure of 28 in. of Hg, at a stopping-screen-to-target distance of 9 cm with a 1-cm distance-of-flight of the macrocarriers. A single potted plant is placed in the biolistic device and a unique bombardment is achieved.

3.4 Posttransformation Treatments and Analysis

Following bombardment, plants are replaced in greenhouse and are cultivated under similar conditions until appearance of the phenotype of the ChlH- or PDS-silenced plants used as positive controls of gene silencing. Using these transformation conditions, up to 90–100% of the bombarded periwinkle plants display gene silencing. While limited variations arise, leaf photobleaching or yellowing typically begins around 7–10 days after bombardment while fully bleached neo-formed leaves can be retrieved 21–25 days posttransformation (Fig. 2). Leaves of silenced candidate genes can be thus harvested in the same time laps for further analysis including evaluation of gene silencing by quantitative PCR as well as measurement of alkaloid contents by HPLC analysis (Besseau et al., 2013) in order to identify candidate gene function.

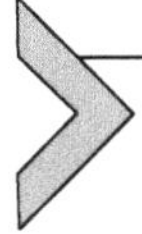

4. STUDYING THE SUBCELLULAR LOCALIZATION OF BIOSYNTHETIC PATHWAY ENZYMES IN PLANT AND YEAST CELLS TO ALLEVIATE BOTTLENECKS IN BIOENGINEERING APPROACHES

Most of the biosynthetic pathways of specialized metabolites, particularly in plants, exhibit a complex intracellular compartmentalization relying on the targeting of their enzymes to distinct organelles. An overview of the level of complexity that this type of organization can reach has been recently depicted in *C. roseus* (Courdavault et al., 2014). In this plant, the distribution of the alkaloid biosynthetic pathways in numerous subcellular compartments involves, as a corollary, manifold transmembrane transports of metabolic intermediates that could impact biosynthetic fluxes. While plants can deploy these exchanges to fine-tune the regulation of metabolites biosynthesis, a negative impact can be engendered in metabolic engineering applications based on the reconstitution of biosynthetic pathways in heterologous organisms. Such reconstitution and more generally, the understanding of the general plant physiology thus require a complete knowledge of pathway organization. In the following sections, we describe a procedure of protein subcellular localization study based on the creation of fusions with FPs and on their expression in planta through transient biolistic-mediated transformation. This procedure allows the quick and standardized obtainment of robust results of localization, which can be useful when a biosynthetic pathway is composed of numerous enzymes. Given the differences between some of the plant and yeast targeting sequences, we also present a similar strategy allowing the validation of protein subcellular localizations in yeast that is required to avoid enzyme mislocalization and the inherent disruption of metabolite biosynthesis following pathway reconstitution. The detailed protocols are illustrated with localization of enzymes of the MIA biosynthetic pathway of *C. roseus*, including strictosidine synthase (STR) and strictosidine β-D-glucosidase (SGD), two enzymes acting consecutively in the pathway in distinct subcellular compartments (Guirimand et al., 2010).

4.1 Protein Subcellular Localization in *C. roseus* Cells

While protein subcellular localization can be analyzed by biochemical approaches involving assay for enzyme activity/protein immunodetection following cell fractionation or observed directly with electron microscopy of immunogold-labeled sections, expression of fusions with FPs is the most rapid and popular approach. Following stable and/or transient expression of these proteins in plant cells, protein targeting can be determined by visualization of the subcellular fluorescent profiles using epifluorescence or confocal microscopy and simultaneously compared with the fluorescent profiles of markers for each subcellular compartment.

4.1.1 Constructs Expressing Fusions with Fluorescent Proteins in Plant Cells

A myriad of color variants of FPs is now available. The choice of the FPs used to generate fusions can be guided by the intrinsic properties of each variant (pK_a, brightness, etc.) but also by the microscopy equipment utilized for fluorescence visualization. Since the capacities of epifluorescence microscopes (eFM) to discriminate the different FPs are usually lower than those of confocal microscopes, we recommend combining FPs with no overlapping excitation and emission spectra when eFM are used. In such a case, combination of yellow FP (YFP) with cyan FP (CFP) as well as green FP (GFP) with red FP (RFP) or mcherrry provides the best results. We have developed a set of plasmids with each variant color, based on the pSCA-YFP scaffold

(Guirimand et al., 2009), allowing overexpression of fusion proteins in plant cell under the dependence of the constitutive CaMV 35S promoter. These plasmids, available upon request, display multiple cloning sites at both the 5′ and 3′ ends of the FP coding sequence, with sites of compatible restriction enzymes, enabling cloning of the coding sequence of the studied enzymes at each end of the FP with the same cDNA.

One of the main pitfalls in localization studies with fusion proteins is the masking of the targeting sequence in the studied protein by the FP. For instance, if a protein possesses a plastid transit peptide at its N-terminal end, fusion with the C-terminal end of the FP (yielding FP-protein orientation) have to be avoided. To prevent artifactual localizations, a careful analysis of the putative targeting/anchoring sequences of each protein should be performed to select the most suitable orientation.

1. Predictions of the putative targeting sequence are routinely carried out with PSORT, TargetP, Predotar, MitoProt, PredPlant PTS1, NLS mapper, TMHMM algorithms, for instance. *When applied to STR and SGD, this analysis led to the identification of a putative vacuolar/secretory signal peptide at the N-terminal end of STR (1-MANFSESKSM-MAVFFMFFLLLLSSSSSSSSSSPIL-35) and of a putative bipartite nuclear localization sequence (NLS) located at the C-terminal end of SGD (537-KKRFREEDKL-VELVKKQKY-555).*
2. Amplify the coding sequence of the studied proteins with high fidelity DNA polymerases and introduce appropriated restriction sites at both ends of the cDNA to allow cloning in the pSCA-vectors. *In our example, STR coding sequence was amplified with primers STR-for (5′-CTGAGA ACTAGTATGGCAAACTTTTCTGAATCTAAA-3′) and STR-rev (5′-CTGAGAACTAGTGCTAGAAACATAAGAATTTCCCTT-3′) that introduce the SpeI restriction site at both extremities to allow cloning into either the SpeI or NheI sites of the pSCA-CFP vector in order to express STR-CFP and CFP-STR fusions, respectively. Similarly, SGD was amplified with primers SGD-for (5′-CTGAGATCTAGAATGGGATCTAAAG ATGATCAGTCC-3′) and SGD-rev (5′-CTGAGATCTAGATTAGT ATTTTTGCTTCTTGACTAACTCAACT-3′) introducing a XbaI site (compatible with SpeI and NheI) to express the SGD-YFP and YFP-SGD fusion proteins.*
3. Following cloning into the pSCA-YFP vectors, extract and sequence the recombinant plasmids to ensure that constructs are exempt from mutations. For plant cell transformations, concentrate plasmids to final 1 $\mu g\,\mu L^{-1}$ concentration before transformation. *Note that supercoiled*

plasmids freshly extracted from E. coli cultures (using Nucleospin Plasmid kit for example) usually provide the best expression levels.

4.1.2 Cell Culture and Plating

Expression of the fusion proteins can be achieved in *C. roseus* cells and we preconize the use of the *C. roseus* C20 strain cell that is suitable for subcellular localization of proteins from the periwinkle or from other plant species.

1. *C. roseus* C20 cell suspensions are cultivated in the dark at 24°C under continuous shaking (100 rpm) for 7 days as previously described (Mérillon, Doireau, Guillot, Chénieux, & Rideau, 1986).
2. At the third day of culture, pour 4 mL of the homogenized cell suspension onto a circular piece of filter paper (45 mm diameter—Fisherbrand A70.70000) in a filtration funnel and apply weak air suction.
3. Transfer plated cells onto solid Gamborg B5 medium (8 g L^{-1} agar) (Gamborg, Miller, & Ojima, 1968) supplemented with 10 μ*M* naphthalene acetic acid (NAA) in a 45-mm Petri dish and cultivate at 24°C for 48 h in the dark before transformation.

4.1.3 Transient Cell Transformation by Biolistic

Studies of protein subcellular localizations can be carried out following stable and/or transient transformations of plant cells. While stable transformation allows performing long-term studies on selected and propagated transformed cells, this approach is more time consuming in particular because the analysis of several transformed cell lines is required to validate localization results. By contrast, transient transformations rapidly generate localization results (within 1–2 days) and allow the observation of thousands of independent transformation events in a single cell plate. As a consequence, transient transformations are suitable to characterize the localization of enzymes from biosynthetic pathways of specialized metabolites. For such transformations, numerous protocols using protoplasts have been described and notably in *C. roseus* (Duarte, Memelink, & Sottomayor, 2010). However, to avoid artifacts of localization caused by the cell stress induced by cell wall removal, we preconize the use of particle bombardments that results in the entry of a single small particle for most of the transformed cells. This constitutes a weakly traumatic situation adapted for localization studies. The protocol described hereafter has been developed according to Guirimand et al. (2009).

1. For each bombardment, 400 ng (or up to 1 μg) of purified plasmid is coated and precipitated onto 500 μg of gold particles prepared as

described in Section 3.3.1. When plasmid cotransformations are performed (with plasmids encoding subcellular markers for instance), prepare a stoichiometric mix of each plasmid. A useful set of plasmids encoding markers of each plat cell compartments has been described in Nelson, Cai, and Nebenführ (2007).

2. Mix 5 μL of 0.1 *M* spermidine with 100 μL of glycerol stocked gold particles in a sterilized tube and homogenize simultaneously with vortex and short pulses of ultrasounds (15 s; 40 W) with sonication bath.
3. Add the appropriate volume of purified DNA plasmid (usually not exceeding 3 μL for each plasmid) and mix.
4. While homogenizing the solution with vortex, add 5 μL of 2.5 *M* $CaCl_2$ and mix for additional 15 min.
5. Spin the tubes for 2 s at 16,000 × *g* to pellet gold particles coated with plasmids and remove the supernatant.
6. Wash plasmid coated gold particles pellets successively with 150 μL of 70% ethanol and 150 μL of 100% ethanol without resuspending gold particles.
7. Remove all supernatant and resuspend particles in 8 μL of 100% ethanol.
8. Spread the 8 μL of coated gold particles on each macrocarrier (Bio-Rad) and allow drying before transformation.
9. Transformations are performed with the Bio-Rad PDS100/He delivery system according to Section 3.3.3, using 1100 psi rupture disks, under a vacuum pressure of 28 in. of Hg, at a stopping-screen-to-target distance of 6 cm with a 1-cm distance-of-flight of the macrocarrier.
10. A single transformation per Petri dish of plated cells is performed and cells are cultivated for 16 h in the dark at 24°C before observation.

4.1.4 Fluorescent Protein Imaging and Epifluorescence Microscopy

Fluorescence profiles of the fusion proteins can be usually observed from 16 to 72 h postbombardment by harvesting transformed cells from the Petri dish and mounting them between slide and cover. Evolution of protein localizations has to be checked during all this period since the kinetic of targeting to each subcellular compartment greatly differs. When a fusion protein is coexpressed with a subcellular compartment fluorescent marker (a protein known to be targeted to a specific subcellular compartment and fused to a different FP) or with a second protein fused to a distinct FP, superimposition of the two distinct fluorescent signals has to be evaluated in order to definitely establish the localization of the studied protein. In the example described later, image captures of *C. roseus* transiently transformed cells expressing FP-fused proteins are performed with an Olympus BX51 eFM equipped with the Olympus DP71 digital camera with CellD imaging software (Soft Imaging System, Olympus). The YFP and CFP fluorescence signals emitted from fusion proteins are visualized using a YFP filter set (Chroma#31040, 500–520 nm excitation filter, 540–580 nm emission filter) and a Cyan GFP filter set (Chroma#31044v2, 426–446 excitation filter, 460–500 nm band pass emission filter), respectively. YFP and CFP fluorescence are successively acquired and merged with the CellD imaging software while the morphology of transformed cells is observed with differential interference contrast (DIC). Fig. 3 illustrates the sequential image capture process carried out for *C. roseus* cells cotransformed with plasmids expressing the YFP-SGD (Fig. 3A) and STR-CFP (Fig. 3B) fusion proteins,

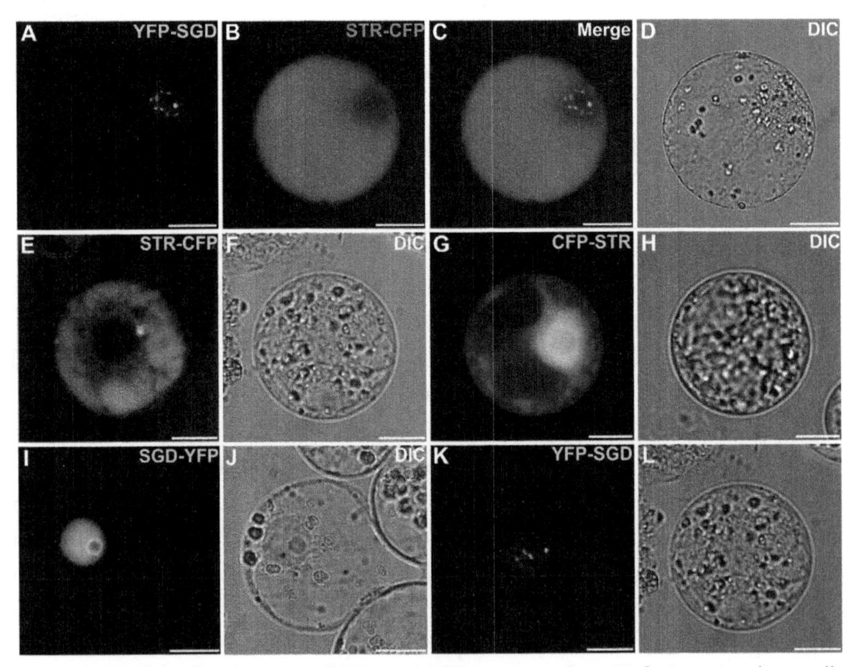

Fig. 3 Subcellular localization of STR and SGD expressed as FP fusions in plant cells. *C. roseus* cells were transiently cotransformed (A–D) or transformed (E–L) with constructs expressing YFP-SGD (A; K), STR-CFP (B; E), CFP-STR (G) of SGD-YFP (I). For cotransformation, superimposition of the two fluorescence signals appears on the merged image (C). Cell morphology (D; F; H; J; L) was observed with differential interference contrast (DIC). Bars = 10 μm. (See the color plate.)

their numerical superimposition (Fig. 3C), and the observation of cell morphology by DIC (Fig. 3D).

Depending on the cell strain used for transformation, cell aggregation can occur after platting on solid medium rendering more difficult mounting between slide and cover. In such a case, protoplasts of transformed cells can be prepared before observation as described later. Such treatments performed posttransformation and during a short period are not likely to induce modifications of localization.

1. Harvest around 250 mg of transformed cells with an inoculation loop in a tube containing 800 μL of MM Buffer (MES 20 m*M*, Mannitol 0,4 *M*).
2. Allow to decant for 5 min at room temperature, remove supernatant and resuspend cell in 800 μL of MM Buffer. Repeat this washing step twice.
3. Resuspend cell pellet in 1 mL of digestion MM Buffer (MM Buffer containing cellulase R-10 2%, Macerozyme r-10 0.3%, pectolyase 0.2%).
4. Transfer in a Petri dish (diameter 45 mm) and incubate for 2 h in the dark with slow agitation to generate protoplasts.
5. Harvest solution in a 1.7-mL Eppendorf tube, allow decanting for 5 min and remove supernatant.
6. Wash carefully protoplasts as described in step 2 and gently resuspend in a final volume of 400 μL of MM Buffer before mounting.

4.1.5 The Importance of Being Correctly Fused

As mentioned earlier, masking of targeting signals by improper orientation of fusion with FP is a common pitfall observed during subcellular localization studies and may result in artifactual protein targeting. As an illustration of this difficulty, Fig. 3 shows the analysis of the localization of STR that displays a predicted targeting sequence at its N-terminal end. In *C. roseus* transiently transformed cells, the STR-CFP fusion protein is efficiently targeted to the vacuole (Fig. 3E–F) while expression of the CFP-STR fusion protein results in protein mislocalization in the cytosol (Fig. 3G–H) probably caused by the masking of the targeting sequence that renders it non accessible and/or nonfunctional. Furthermore, the effects of an incorrect orientation of fusion with FP can be more tenuous and thus less easily identifiable. For SGD that bears a bipartite NLS at the C-terminal end, fusion with the FP does not inactivate this sequence that is still able to direct the SGD-YFP fusion in the nucleus as a soluble protein (Fig. 3I–J). However, such orientation of fusion annihilates the propensity of SGD to self-interact and to form high molecular weight complexes (appearing as small dots in the nucleus), which can be only observed when the C-terminal end of SGD is free such as in the YFP-SGD fusion protein (Fig. 3K–L). More complex situations can be also encountered when the studied protein possesses targeting sequences at both extremities as recently depicted for the *C. roseus* isopentenyl diphosphate isomerase (IDI). This protein is characterized by the presence of a N-terminal dual plastid/mitochondria targeting peptide and by a C-terminal type 1 peroxisome targeting sequence (PTS1) requiring the expression of an YFP internal fusion protein to observe the triple localization to plastid, mitochondria, and peroxisome (Guirimand, Guihur, et al., 2012; Guirimand, Simkin, et al., 2012).

4.2 Protein Subcellular Localization in Yeast Cells

While the study of the subcellular localization in the plant cells of multiple enzymes of a biosynthetic pathway of interest allows identifying potential bottlenecks associated to transmembrane exchanges of intermediates, a subsequent confirmation of the correct protein localization can be required following the transfer of the whole biosynthetic pathway in heterologous organisms. Indeed, specific plant-targeting sequences (eg, plastid targeting peptide) are inoperative in yeasts or bacteria and common localization sequences can also be subjected to misinterpretations leading to undesired protein targeting. Such heterologous mistargeting has been depicted for the expression of STR in yeast that undergone a massive secretion in the medium instead of an expected vacuolar localization, due to the promiscuity of vacuolar/secretion targeting sequences (Geerlings et al., 2001). As a consequence, we recommend validating protein localization and, if needed, to replace inefficient plant localization peptides by sequences adapted to the host organisms. Our protocol of protein subcellular localization in yeast is described later.

4.2.1 Constructs Expressing Fusions with Fluorescent Proteins in Yeast Cells

Numerous plasmids dedicated to the expression of FP fusion proteins in yeast are now available. Most of these plasmids harbor codon optimized FP coding sequences under the control of strong constitutive promoters such as ACT1 or TEF1. To avoid potential problems caused by the continuous expression of FP fusions and a possible mislocalization caused by a massive protein overexpression, plasmids bearing inducible promoters should preferentially be used. This type of plasmids can be easily generated by using skeleton of commercial plasmids such as those of the pESC series (Agilent

Technologies) possessing the pGAL1/pGAL10 inducible promoters but are also available upon request.

1. Amplify yeast codon optimized coding sequences of FP with high fidelity DNA polymerases. For instance, yeYFP and yeCFP coding sequences can be amplified with a similar primer couple composed of FLUSC1 (5′-CTGAGGTCTAGA*AGATCT*ACTAGTATGTCTAAAGGTGAAGAATTAT-3′ introducing *Xba*I, *Bgl*II, and *Spe*I restriction sites) and FLUSC2 (5′-CTGAGAGGATCCTTACCTAGGTTTGTACAATTCATCCATACCA-3′ introducing *BamHI* and *Avr*II restriction sites).
2. Digest PCR products by *Xba*I and *BamHI* and clone into pESC-LEU and/or pESC-TRP linearized by *Spe*I and *Bgl*II that generate compatible extremities with *Xba*I and *BamHI*, respectively. *In the following example, we cloned yeYFP and yeCFP coding sequences into pESC-LEU and pESC-TRP to generate pESC-LEU-YFP and pESC-TRP-CFP, respectively.*
3. Extract plasmids and sequence. The *Spe*I and *Bgl*II restriction sites initially present in the plasmid sequence have been disrupted through their annealing with *Xba*I and *BamHI* compatible extremities. The coding sequences of the studied proteins can now be cloned into the *Bgl*II or *Spe*I restriction sites (introduced by PCR) to generate a protein fused to the N-terminus of YFP/CFP or into *Avr*II to express a protein fused to the C-terminal end of YFP/CFP. *These restriction sites have been selected for compatibility with cloning strategy used to study the protein localization in plant cells using the pSCA-YFP vectors. The coding sequences of STR and SGD (Section 4.1.1) have thus been cloned in pESC-LEU-YFP and pESC-TRP-CFP, accordingly.*

4.2.2 Preparation of Yeast Competent Cells

The following protocol of yeast competent cell preparation is restricted to transformation by electroporation as described in the subsequent section.

1. Streak a *S. cerevisiae* reference strain (WT303 for instance, auxotroph to leucine and tryptophan) on solid YPD medium (10 g L^{-1} yeast extract, 20 g L^{-1} peptone, 20 g L^{-1} dextrose, and 15 g L^{-1} agar) and grow at 30° C for 48 h.
2. Pick a single colony, inoculate 10 mL of liquid YPD medium and grow at 30°C overnight.
3. Inoculate 50 mL of liquid YPD medium with 500 μL of the preculture and grow around 4 h at 30°C until reaching the end of the exponential growth (absorbance at 600 nm of 0.6–1.5).

4. Chill the yeast culture 15 min on ice before transferring into sterile 50 mL Falcon tubes. Centrifuge at 3000 × *g* for 10 min at 4°C to pellet yeast.
5. Remove the supernatant and resuspend yeast pellet in 40 mL of ice-cold DTT-supplemented lithium acetate solution (lithium acetate 100 m*M*, DTT 10 m*M*). Incubate at 28°C for 1 h.
6. Centrifuge at 3000 × *g* for 10 min at 4°C in sterile centrifuge tubes, discard the supernatant, and wash the pellet twice with 20 mL sorbitol 1 *M*.
7. Resuspend yeast pellet in 5 mL of 1 *M* sorbitol and distribute in pre-chilled eppendorf tubes before flash freezing in liquid nitrogen. Store at −80°C.

4.2.3 Protocol of Yeast Cell Transformation

Several protocols of yeast transformation have been described either by heat shock treatment or by electroporation. We recommend using this last one due to a higher efficiency of transformation.

1. Mix 0.5–2 μg of DNA Purified Plasmid (pESC-LEU-YFP and/or pESC-LEU-TRP) with 200 μL of yeast competent cells in a sterile Eppendorf tube.
2. Incubate 10 min on ice and transfer into a 0.2-cm gap width electroporation cuvette.
3. Perform electroporation by applying a 5-ms electric pulse of 1.5 kV (Bio-Rad MicroPulser Electroporator—program Sc2).
4. Plate transformed cells on CSM-LEU (YNB with ammonium 6.7 g L^{-1}, dextrose 20 g L^{-1}, DOB-LEU 500 mg L^{-1}, agar 20 g L^{-1}), CSM-TRP (YNB with ammonium 6.7 g L^{-1}, dextrose 20 g L^{-1}, DOB-TRP 500 mg L^{-1}, agar 20 g L^{-1}), or CSM-LEU-TRP (YNB with ammonium 6.7 g L^{-1}, dextrose 20 g L^{-1}, DOB-LEU-TRP 500 mg L^{-1}, agar 20 g L^{-1}) selective solid medium, depending on the combination of selection marker used in the cloning vectors and incubate at 30°C for 3–5 days.
5. Streak independently transformed yeast colonies onto CSM-LEU, CSM-TRP, or CSM-LEU-TRP containing 2% galactose to induce protein expression and grow two additional days at 30°C.
6. Resuspend transformed yeast independently in 100 μL of water and analyze fluorescence using eFM in the conditions described in Section 4.1.4.

4.2.4 *Correct and Incorrect Plant Protein Targeting in Yeast*

Since STR and SGD display a complete sequestration at the subcellular level, being targeted to vacuole and nucleus, respectively, in plant cells, we illustrated the validation of the localization of both enzymes in yeast though their expression as fusions with FP (Section 4.2.1). As compared to nonfused FP exhibiting a classical nucleocytosolic localization (Fig. 4A–B), STR-CFP was efficiently targeted to the vacuole in our experimental conditions, with a negligible secretion in the medium (Fig. 4C–D). Such localization is thus consistent with that observed in plant cells. By contrast, SGD localization in yeast produced a more complex situation. For the SGD-YFP fusion, a unique and unexpected targeting to the vacuole was observed (Fig. 4E–F) as revealed with cotransformation with STR-CFP (Fig. 5A–D). Furthermore, we observed that YFP-SGD fusions were targeted, in similar proportions, to the nucleus (Figs. 4G–H and 5E–H) or to the vacuole (Figs. 4I–J and 5I–L) and less frequently to both compartments (Figs. 4K–L and 5M–P). In this case, targeting of STR and SGD to the

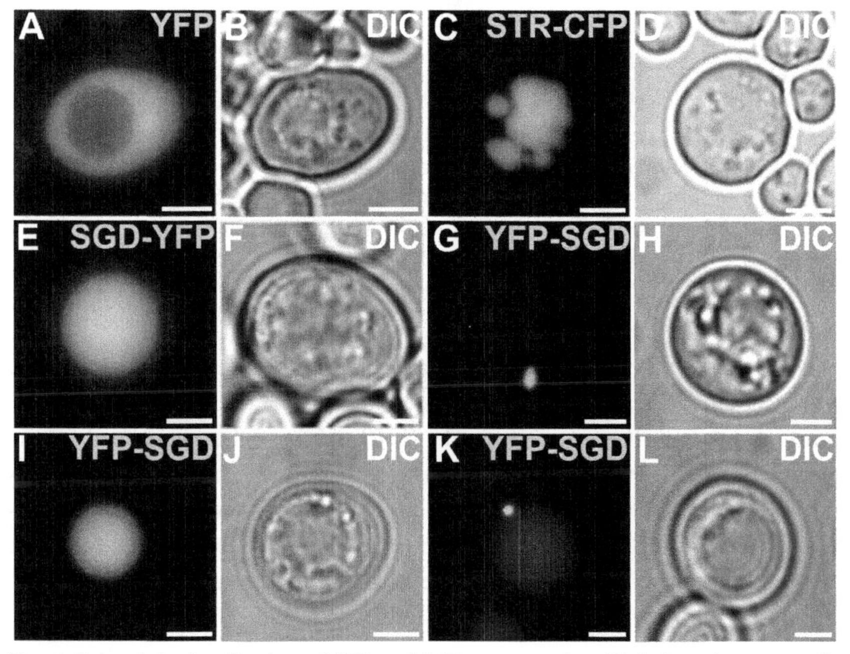

Fig. 4 Subcellular localization of STR and SGD expressed as FP fusions in yeast cells. Yeast cells were transformed with constructs expressing unfused YFP (A), STR-CFP (C), SGD-YFP (E), or YFP-SGD (G; I; K). Cell morphology (B; D; F; H; J; L) was observed with differential interference contrast (DIC). Bars = 2 μm. (See the color plate.)

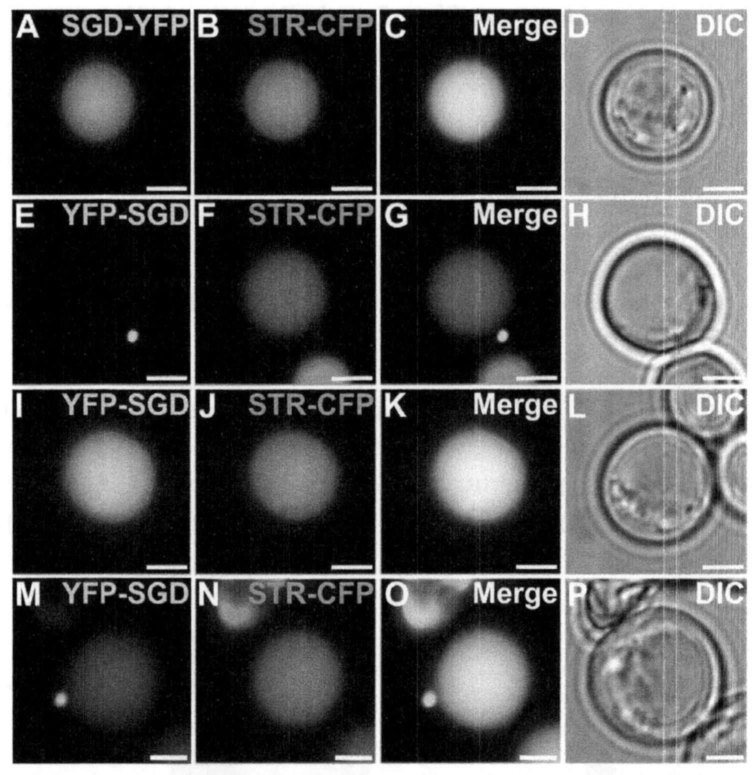

Fig. 5 Colocalization of STR and SGD expressed as FP fusions in yeast cells. Yeast cells were co transformed with constructs expressing SGD-YFP and STR-CFP (A–D) or YFP-SGD and STR-CFP (E–P). Colocalization of the two fluorescence signals appears on the merged images (C; G; K; O). Cell morphology (D; H; L; P) was observed with differential interference contrast (DIC). Bars = 2 μm. (See the color plate.)

vacuole can result in potential undesirable effects in yeast since it can lead to a massive deglucosylation of stricosidine that might be toxic as a result of protein reticulation (Guirimand et al., 2010). Albeit these mislocalizations potentially result from the fusion with FPs that could alter the functionality of the targeting sequences, it also highlights the differences of protein behavior in plant and yeast cells and the importance to validate protein. It should also be taken into consideration when small tags are added to proteins to monitor their expression in heterologous organisms.

5. CONCLUDING REMARKS

While synthetic biology now technically offers the possibility of transferring and controlling the whole biosynthetic pathway of a valuable compound in heterologous organisms, identification of all the enzymes of the pathway and of its subcellular organization still constitutes essential prerequisites to the achievement of bioengineered productions. With the recent availability of massive transcriptomics and genomics data concerning organisms producing metabolites of interest, notably for plants, and with the development of efficient tools for candidate gene prediction, validation, and characterization, such complete and intensive deciphering of pathways are more than ever right at our fingertips. As such, the protocols described earlier are part and parcel of the technical arsenal that can be deployed to attain pathway characterization. Albeit being initially developed for the Madagascar periwinkle, most of them are applicable to other plant species. We expect for instance that direct-transformation of vectors for VIGS will expand the availability of this powerful functional approach outside the host range of Agrobacterium-delivered VIGS. By showing unforeseen localization obtained with enzymes of the MIA pathway, we also highlighted the importance to take account of the subcellular localization in metabolic engineering pathways, in particular following transfer to microbial systems which may not properly process plant-targeting signals. Thereby, the study of protein subcellular localization still constitutes a milestone in the early steps of synthetic biology approaches dedicated to metabolic engineering.

ACKNOWLEDGMENTS

We gratefully acknowledge support from the Région Centre (France, ABISAL grant, Doctoral Fellowship to F.L. and Post-Doctoral Fellowship to I.C.). E.F. was financed by a fellowship from the Ministère de l'Enseignement Supérieur et de la Recherche (France). We also thank M.A. Marquet, E. Danos, and E. Marais for maintenance of cell cultures. We also acknowledge the Cascimodot Fédération (CCSC, Orléans) for access to the Région Centre computing grid.

REFERENCES

Anders, S., & Huber, W. (2010). Differential expression analysis for sequence count data. *Genome Biology*, *11*, R106.

Besseau, S., Kellner, F., Lanoue, A., Thamm, A. M. K., Salim, V., Schneider, B., et al. (2013). A pair of tabersonine 16-hydroxylases initiates the synthesis of vindoline in an organ-dependent manner in *Catharanthus roseus*. *Plant Physiology*, *163*, 1–12.

Brown, S., Clastre, M., Courdavault, V., & O'Connor, S. E. (2015). De novo production of the plant-derived alkaloid strictosidine in yeast. *Proceedings of the National Academy of Sciences of the United States of America*, *112*, 3205–3210.

Carqueijeiro, I., Masini, E., Foureau, E., Sepúlveda, L. J., Marais, E., Lanoue, A., et al. (2015). Virus-induced gene silencing in *Catharanthus roseus* by biolistic inoculation of tobacco rattle virus vectors. *Plant Biology*, *17*, 1242–1246.

Cassels, K. B., Asencio, M., Conget, P., Speisky, H., Videla, A. L., & Lissi, A. E. (1995). Structure-antioxidative activity relationships in benzylisoquinoline alkaloids. *Pharmacology Research*, *31*, 103–107.

Chang, M. C., Eachus, R. A., Trieu, W., Ro, D. K., & Keasling, J. D. (2007). Engineering *Escherichia coli* for production of functionalized terpenoids using plant P450s. *Nature Chemical Biology*, *3*, 274–277.

Courdavault, V., Papon, N., Clastre, M., Giglioli-Guivarc'h, N., St-Pierre, B., & Burlat, V. (2014). A look inside an alkaloid multisite plant: The Catharanthus logistics. *Current Opinion in Plant Biology*, *19*, 43–50.

De Luca, V., Salim, V., Levac, D., Atsumi, S. M., & Yu, F. (2012). Discovery and functional analysis of monoterpenoid indole alkaloid pathways in plants. In D. A. Hopwood (Ed.), *Methods in enzymology: Vol. 515. Natural product biosynthesis by microorganisms and plants, part A* (pp. 207–229). Waltham: Academic Press—Elsevier.

Drewes, S. E., George, J., & Khan, F. (2003). Recent findings on natural products with erectile-dysfunction activity. *Phytochemistry*, *62*, 1019–1025.

Duarte, P., Memelink, J., & Sottomayor, M. (2010). Fusion with fluorescent proteins for subcellular localization of enzymes involved in plant alkaloid biosynthesis. In E. Fett-Neto & A. Germano (Eds.), *Methods in molecular biology: Vol. 643. Plant secondary metabolism engineering* (pp. 275–290). New York: Humana Press.

Dugé de Bernonville, T., Clastre, M., Besseau, S., Oudin, A., Burlat, V., Glévarec, G., et al. (2015). Phytochemical genomics of the Madagascar periwinkle: Unravelling the last twists of the alkaloid engine. *Phytochemistry*, *113*, 9–23.

Dugé de Bernonville, T., Foureau, E., Parage, C., Lanoue, A., Clastre, M., Londono, M. A., et al. (2015). Characterization of a second secologanin synthase isoform producing both secologanin and secoxyloganin allows enhanced de novo assembly of a Catharanthus roseus transcriptome. *BMC Genomics*, *16*, 619.

Gamborg, O. L., Miller, R. A., & Ojima, K. (1968). Nutrient requirements of suspension cultures of soybean root cells. *Experimental Cell Research*, *50*, 151–158.

Geerlings, A., Redondo, F., Contin, A., Memelink, J., van Der Heijden, R., & Verpoorte, R. (2001). Biotransformation of tryptamine and secologanin into plant terpenoid indole alkaloids by transgenic yeast. *Applied Microbiology and Biotechnology*, *56*, 420–424.

Góngora-Castillo, E., Fedewa, G., Yeo, Y., Chappell, J., DellaPenna, D., & Buell, C. R. (2012). Genomic approaches for interrogating the biochemistry of medicinal plant species. *Methods in Enzymology*, *517*, 139–159.

Guirimand, G., Burlat, V., Oudin, A., Lanoue, A., St-Pierre, B., & Courdavault, V. (2009). Optimization of the transient transformation of Catharanthus roseus cells by particle bombardment and its application to the subcellular localization of hydroxymethylbutenyl 4-diphosphate synthase and geraniol 10-hydroxylase. *Plant Cell Reports*, *28*, 1215–1234.

Guirimand, G., Courdavault, V., Lanoue, A., Mahroug, S., Guihur, A., Blanc, N., et al. (2010). Strictosidine activation in Apocynaceae: Towards a "nuclear time bomb"? *BMC Plant Biology*, *10*, 182.

Guirimand, G., Guihur, A., Phillips, M. A., Oudin, A., Glévarec, G., Melin, C., et al. (2012). A single gene encodes isopentenyl diphosphate isomerase isoforms targeted to plastids, mitochondria and peroxisomes in *Catharanthus roseus*. *Plant Molecular Biology*, *79*, 443–459.

Guirimand, G., Simkin, A. J., Papon, N., Besseau, S., Burlat, V., St-Pierre, B., et al. (2012). Cycloheximide as a tool to investigate protein import in peroxisomes: A case study of the subcellular localization of isoprenoid biosynthetic enzymes. *Journal of Plant Physiology*, *169*, 825–829.

Haas, B. J., Papanicolaou, A., Yassour, M., Grabherr, M., Blood, P. D., Bowden, J., et al. (2013). De novo transcript sequence reconstruction from RNA-seq using the Trinity platform for reference generation and analysis. *Nature Protocols*, *8*, 1494–1512.

Hanson, J. R. (2003). Natural products: Secondary metabolites. In E. W. Abel (Ed.), *Tutorial chemistry texts: Vol. 17* (pp. 3–18). London: The Royal Society of Chemistry.

Hawkins, K. M., & Smolke, C. D. (2008). Production of benzylisoquinoline alkaloids in Saccharomyces cerevisiae. *Nature Chemical Biology*, *4*, 564–573.
Huang, Y., Niu, B., Gao, Y., Fu, L., & Li, W. (2010). CD-HIT suite: A web server for clustering and comparing biological sequences. *Bioinformatics*, *26*, 680–682.
Langmead, B., & Salzberg, S. L. (2012). Fast gapped-read alignment with Bowtie 2. *Nature Methods*, *9*, 357–359.
Langmead, B., Trapnell, C., Pop, M., & Salzberg, S. L. (2009). Ultrafast and memory-efficient alignment of short DNA sequences to the human genome. *Genome Biology*, *10*, R25.
Lee, M. (2011). The history of Ephedra (ma-huang). *Journal of the Royal College of Physicians of Edinburgh*, *41*, 78–84.
Li, B., & Dewey, C. N. (2011). RSEM: Accurate transcript quantification from RNA-Seq data with or without a reference genome. *BMC Bioinformatics*, *12*, 323.
Li, H., & Durbin, R. (2009). Fast and accurate short read alignment with Burrows-Wheeler transform. *Bioinformatics*, *25*, 1754–1760.
Li, L. P., Liu, W., Liu, H., Zhu, F., Zhang, D. Z., Shen, H., et al. (2015). Synergistic antifungal activity of berberine derivative B-7b and fluconazole. *PLoS One*, *10*, e0126393.
Liscombe, D. K., & O'Connor, S. E. (2011). A virus-induced gene silencing approach to understanding alkaloid metabolism in Catharanthus roseus. *Phytochemistry*, *72*, 1969–1977.
Mano, M. (2006). Vinorelbine in the management of breast cancer: New perspectives, revived role in the era of targeted therapy. *Cancer Treatment Reviews*, *32*, 106–118.
Martin, J. A., & Wang, Z. (2011). Next-generation transcriptome assembly. *Nature Reviews Genetics*, *12*, 671–682.
Mérillon, J. M., Doireau, P., Guillot, A., Chénieux, J. C., & Rideau, M. (1986). Indole alkaloid accumulation and tryptophan decarboxylase activity in *Catharanthus roseus* cells cultured in three different media. *Plant Cell Reports*, *5*, 23–26.
Miller, M. L., & Ojima, I. (2001). Chemistry and chemical biology of taxane anticancer agents. *Chemical Records*, *1*, 195–211.
Nelson, B. K., Cai, X., & Nebenführ, A. (2007). A multicolored set of in vivo organelle markers for co-localization studies in Arabidopsis and other plants. *The Plant Journal*, *51*, 1126–1136.
Newman, D. J., & Cragg, G. M. (2012). Natural products as sources of new drugs over the 30 years from 1981 to 2010. *Journal of Natural Products*, *75*, 311–335.
Paddon, C. J., Westfall, P. J., Pitera, D. J., Benjamin, K., Fisher, K., McPhee, D., et al. (2013). High-level semi-synthetic production of the potent antimalarial artemisinin. *Nature*, *496*, 528–532.
Pertea, G., Huang, X., Liang, F., Antonescu, V., Sultana, R., Karamycheva, S., et al. (2003). TIGR Gene Indices clustering tools (TGICL): A software system for fast clustering of large EST datasets. *Bioinformatics*, *19*, 651–652.
Pertea, M., Pertea, G. M., Antonescu, C. M., Chang, T. C., Mendell, J. T., & Salzberg, S. L. (2015). StringTie enables improved reconstruction of a transcriptome from RNA-seq reads. *Nature Biotechnology*, *33*, 290–295.
Pollier, J., Moses, T., & Goossens, A. (2011). Combinatorial biosynthesis in plants: A (p)review on its potential and future exploitation. *Natural Product Reports*, *28*, 1897–1916.
Ragauskas, A. J., Williams, C. K., Davison, B. H., Britovsek, G., Cairney, J., Eckert, C. A., et al. (2006). The path forward for biofuels and biomaterials. *Science*, *311*, 484–489.
Ro, D. K., Paradise, E. M., Ouellet, M., Fisher, K. J., Newman, K. L., Ndungu, J. M., et al. (2006). Production of the antimalarial drug precursor artemisinic acid in engineered yeast. *Nature*, *440*, 940–943.

Roberts, A., Feng, H., & Pachter, L. (2013). Fragment assignment in the cloud with eXpress-D. *BMC Bioinformatics*, *14*, 358.

Robinson, M. D., McCarthy, D. J., & Smyth, G. K. (2010). edgeR: A bioconductor package for differential expression analysis of digital gene expression data. *Bioinformatics*, *26*, 139–140.

Schulz, M. H., Zerbino, D. R., Vingron, M., & Birney, E. (2012). Oases: Robust de novo RNA-seq assembly across the dynamic range of expression levels. *Bioinformatics*, *28*, 1086–1092.

Si, Y., Liu, P., Li, P., & Brutnell, T. P. (2014). Model-based clustering for RNA-seq data. *Bioinformatics*, *30*, 197–205.

Stermitz, F. R., Lorenz, P., Tawara, J. N., Zenewicz, L. A., & Lewis, K. (2000). Synergy in a medicinal plant: Antimicrobial action of berberine potentiated by 5′-methoxyhydnocarpin, a multidrug pump inhibitor. *Proceedings of the National Academy of Sciences of the United States of America*, *15*, 1433–1437.

Sung, Y. C., Lin, C. P., & Chen, J. C. (2014). Optimization of virus-induced gene silencing in *Catharanthus roseus*. *Plant Pathology*, *63*, 1159–1167.

Tarselli, M. A., Raehal, K. M., Brasher, A. K., Groer, C. E., Cameron, M. D., Bohn, L. M., et al. (2011). Synthesis of conolidine, a potent non-opioid analgesic for tonic and persistent pain. *Nature Chemistry*, *3*, 449–453.

Thomas, C. J., Rahier, N. J., & Hecht, S. M. (2004). Camptothecin: Current perspectives. *Bioorganic & Medicinal Chemistry*, *12*, 1585–1604.

Trapnell, C., Roberts, A., Goff, L., Pertea, G., Kim, D., Kelley, D. R., et al. (2012). Differential gene and transcript expression analysis of RNA-seq experiments with TopHat and Cufflinks. *Nature Protocols*, *7*, 562–578.

Trapnell, C., Williams, B. A., Pertea, G., Mortazavi, A., Kwan, G., van Baren, M. J., et al. (2010). Transcript assembly and quantification by RNA-Seq reveals unannotated transcripts and isoform switching during cell differentiation. *Nature Biotechnology*, *28*, 511–515.

van der Laan, M. J., & Pollard, K. S. (2003). Hybrid clustering of gene expression data with visualization and the bootstrap. Journal of Statistical Planning and Inference, 117, 275–303.

Wink, M. (1999). Plant secondary metabolism: Biochemistry, function, and biotechnology. In M. Wink (Ed.), *Biochemistry of plant secondary metabolism* (pp. 1–16). Shefield: Shefield Academic Press.

Zhao, J., & Dixon, R. A. (2009). MATE transporters facilitate vacuolar uptake of epicatechin 3′-O-glucoside for proanthocyanidin biosynthesis in Medicago truncatula and Arabidopsis. *Plant Cell Reports*, *21*, 2323–2340.

CHAPTER NINE

Functional Expression and Characterization of Plant ABC Transporters in *Xenopus laevis* Oocytes for Transport Engineering Purposes

D. Xu*, D. Veres*, Z.M. Belew*, C.E. Olsen†, H.H. Nour-Eldin*, B.A. Halkier*,[1]

*DynaMo Center, Faculty of Science, University of Copenhagen, Frederiksberg C, Denmark
†Faculty of Science, University of Copenhagen, Frederiksberg C, Denmark
[1]Corresponding author: e-mail address: bah@plen.ku.dk

Contents

1. Introduction 208
2. Preparation of cDNA of Plant ABC Transporter Genes by In Planta "Exon Engineering" 211
3. ABC Transporter Expression in *Xenopus* Oocytes 215
4. Optimization of Transport Assay for Diffusible ABA in *Xenopus* Oocytes 217
5. Case Study: Characterization of the ABA Exporter at ABCG25 in *Xenopus* Oocytes 219
6. Conclusions 220
Acknowledgments 221
References 221

Abstract

Transport engineering in bioengineering is aimed at efficient export of the final product to reduce toxicity and feedback inhibition and to increase yield. The ATP-binding cassette (ABC) transporters with their highly diverse substrate specificity and role in cellular efflux are potentially suitable in transport engineering approaches, although their size and high number of introns make them notoriously difficult to clone. Here, we report a novel in planta "exon engineering" strategy for cloning of full-length coding sequence of ABC transporters followed by methods for biochemical characterization of ABC exporters in *Xenopus* oocytes. Although the *Xenopus* oocyte expression system is particularly suitable for expression of membrane proteins and powerful in screening for novel transporter activity, only few examples of successful expression of ABC transporter has been reported. This raises the question whether the oocytes system is suitable to express and characterize ABC transporters. Thus we have selected AtABCG25, previously characterized in insect cells as the exporter of commercially valuable abscisic acid—as

Methods in Enzymology, Volume 576
ISSN 0076-6879
http://dx.doi.org/10.1016/bs.mie.2016.04.001

case study for optimizing of characterization in *Xenopus* oocytes. The tools provided will hopefully contribute to more successful transport engineering in synthetic biology.

1. INTRODUCTION

Synthetic biology aims at engineering new biological processes for specific industrial applications such as microbial production of valuable plant products. In general, low yield can be caused by toxicity and feedback inhibition of the final products (Ikeda, 2006; Jones & Woods, 1986), and implementation of transport engineering to export specialized metabolites out of the cell may enable higher yield production (Ikeda, 2006; Ochoa-Villarreal et al., 2015; Young, Lee, & Alper, 2010) (Fig. 1). A major challenge for implementing transport engineering technology in synthetic biology is that exporters for plant natural products are generally unknown.

The ATP-binding cassette (ABC) transport family comprises both primary exporters and importers while most of them are implicated in cellular efflux. It is a ubiquitous and diverse group of proteins, whose direct utilization of ATP for energizing their transport processes allows them to transport substrates across a multitude of biological membranes (for review, see

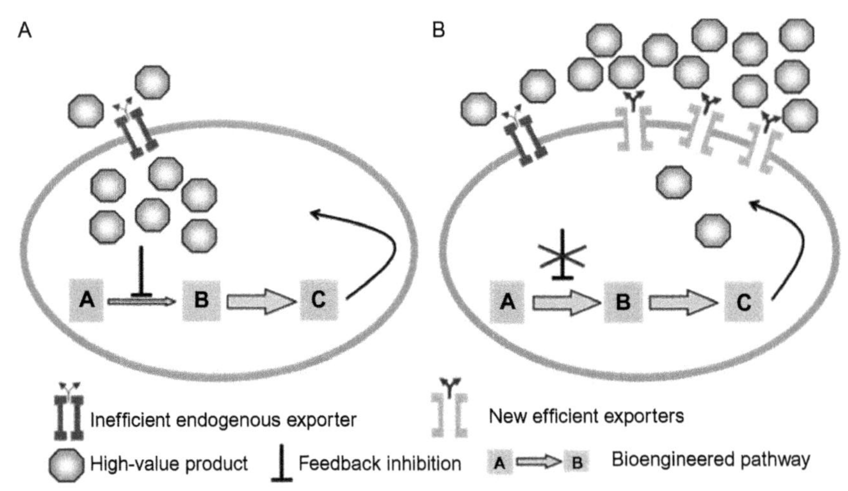

Fig. 1 Schematic illustration of the potential impact of transport engineering. (A) Pathway engineered yeast strain with low yield due to end product feedback inhibition. Purification from cell lysates. (B) High-yielding pathway and transport engineered yeast strain with no feedback inhibition and purification from growth medium. (See the color plate.)

Hwang et al., 2016; Kang et al., 2011). ABC transporters are often associated with detoxification processes. The impressive list of substrates that can be transported by ABC transporters includes peptides, carbohydrates, lipids, heavy metal chelates, inorganic acids, steroids, cell wall monomers, and xenobiotics (Theodoulou & Kerr, 2015; Kang et al., 2011). The diverse functions and substrate specificities assigned to ABC transporters and the large size of these families in the plant kingdom (eg, ~130 ABCs in *Arabidopsis*) create a rich gene pool, which may provide transport proteins of relevance for synthetic biology approaches.

Phylogenetic analysis according to domain organization has defined eight major subfamilies (ABCA to ABCH) of eukaryotic ABC proteins (Hwang et al., 2016). The structure of a prototypical ABC transporter includes four core domains, two nucleotide-binding folds (NBFs) and two transmembrane domains (TMDs) (Beis, 2015). The functional unit is either composed from two "half-transporters" with its own TMD and NBF, or one full-length protein that includes all four domains. Many biochemical functions of the long ABC transporters in plants remain unexplored (Lefevre, Baijot, & Boutry, 2015), due to difficulties in cloning of full-length CDS and/or heterologous expression for functional characterization.

The successful heterologous expression of plant ABC transporters is likely dependent on the protein and host tested (for summary, see Table 1). For example, *Saccharomyces cerevisiae* appears to be suitable for expression of plant ABCB and ABCC transporters, but not ABCG families (Chen, Sanchez-Fernandez, Lyver, Dancis, & Rea, 2007; Geisler et al., 2005; Kamimoto et al., 2012; Nagy et al., 2009; Ruzicka et al., 2010; Lefevre et al., 2015). *Xenopus laevis* oocytes are widely used to express and characterize membrane proteins, including carriers, channels, and aquaporins (Feng, Xia, Fan, Xu, & Miller, 2013). Furthermore, due to their large size and low background transport activity, the *Xenopus* oocyte expression system is well suited for characterizing transport proteins via biochemical and biophysical (electrophysiology) techniques as well as for reverse genetic screening (Nour-Eldin et al., 2012). To date, only two plant ABC proteins—a berberine transporter from *Coptis japonica,* MDR1 (CjMDR1) (Shitan et al., 2003) and a bacterial-type UDP-glucose ABC transporter OsSTAR1/2 from rice (Huang et al., 2009)—have been expressed successfully in this system. AtABCG37 and OsABCB25 were reported to be unsuccessfully expressed in *Xenopus* oocytes, probably due to their failed targeting to plasma membranes. This has resulted in a dogma

Table 1 Nonplant Expression System Used for Characterization of Plant ABC Transporters

Host Expression System	Gene Name	Substrates	References
Escherichia coli	*AtABCB14*	Malic acid	Lee et al. (2008)
Lactococcus lactis	*AtABCB25*	Glutathione polysulfides	Schaedler et al. (2014)
Schizosaccharomyces pombe	*AtABCG37*	Indole-3-butyric acid	Ruzicka et al. (2010)
	AtABCB4	Indole-3-acetic acid	Yang and Murphy (2009)
	ABCB19	Indole-3-acetic acid	Yang and Murphy (2009)
	AtABCB1	Indole-3-acetic acid	Yang and Murphy (2009)
HeLa cells	*AtABCB14*	Malic acid	Lee et al. (2008)
	AtABCG37	Indole-3-butyric acid	Ruzicka et al. (2010)
	AtABCB1	Indole-3-acetic acid	Geisler et al. (2005)
	AtABCB4		Terasaka et al. (2005)
Xenopus laevis oocytes	*CjMDR1*	Berberine	Shitan et al. (2003)
	OsSTAR1/2	UDP-glucose	Huang et al. (2009)
Saccharomyces cerevisiae	*LjABCB1*	Indole-3-acetic acid	Takanashi, Sugiyama, Sato, Tabata, and Yazaki (2012)
	AtABCB4	Indole-3-acetic acid	Santelia et al. (2005)
	AtABCC1/2	Phytochelatins	Song et al. (2010)
	AtABCC3	Phytochelatins–cadmium complex	Brunetti et al. (2015) Tommasini et al. (1998)
	AtABCC5	Inositol hexakisphosphate	Nagy et al. (2009)
	AtABCC4	Methotrexate	Klein et al. (2004)
	VvABCC1	Anthocyanidin 3-O-glucosides	Francisco et al. (2013)
	AtABCB21	Indole-3-acetic acid	Kamimoto et al. (2012)
	AtABCB23	Fe	Chen et al. (2007)
	AtABCB25	Fe	Chen et al. (2007)
	AtABCB1	Indole-3-acetic acid	Geisler et al. (2005)
	AtABCG37	Indole-3-butyric acid	Ruzicka et al. (2010)
	AtABCG29	Monolignol	Alejandro et al. (2012)
	AtABCG40	Abscisic acid	Kang et al. (2010)
Spodoptera frugiperda	*ABCG25*	Abscisic acid	Kuromori et al. (2010)

that *Xenopus* oocytes are not suited for expression and characterization of ABC transporters. However, a variety of other intracellular transporters have been successfully expressed and characterized in oocytes (Feng et al., 2013), arguing that besides plasma membrane mistargeting, the failure to functionally express ABC proteins in oocytes is case-dependent like in any other systems. Here, we set out to develop a method to clone full-length cDNAs of long ABC transporter genes using AtABCG40 (CDS approx. 4.2 kb) as an example. The method is called in planta "exon engineering" and utilizes transient expression (Voinnet, Rivas, Mestre, & Baulcombe, 2003) of long genomic fragments in leaves of *Nicotiana benthamiana* to obtain full-length cDNA. Second, we use *Arabidopsis* abscisic acid (ABA) transporter AtABCG25, which has previously been characterized only using vesicles from *Spodoptera frugiperda* cells (Kuromori et al., 2010), as a case study for functional characterization of an ABC transporter in *Xenopus* oocytes.

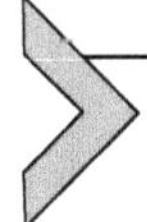

2. PREPARATION OF cDNA OF PLANT ABC TRANSPORTER GENES BY IN PLANTA "EXON ENGINEERING"

Cloning the coding sequence of full-length ABC transporter from plant cDNA is challenging for several reasons. Many ABC transporters exhibit conditional spatial and temporal expression patterns, which require optimization of growth conditions and induction and prior knowledge on transcription patterns. Moreover, even if isolating RNA from appropriate tissues and conditions, low expression levels coupled with large size have so far hindered cloning of the majority of the long full-length ABC genes from the *Arabidopsis* model plant. This difficulty is evident by the disproportional underrepresentation of full-length ABC genes in cDNA repositories such as RIKEN BRC and ABRC (full-length cDNA clones). Furthermore, many full-length ABCs are composed of many exons and introns (eg, many *Arabidopsis* ABCG family genes have more than 20 exons). This complicates alternative cloning approaches where exons could be amplified by PCR and fused via DNA assembly techniques (eg, User fusion (Geu-Flores, Nour-Eldin, Nielsen, & Halkier, 2007) or Gibson Assembly (Gibson, Smith, Hutchison, Venter, & Merryman, 2010)). Moreover, the many introns increase the risk of misannotation, which in turn complicates de novo gene synthetic strategies especially in nonmodel organism where transcriptomic data is not available.

Here, we develop an in planta "exon engineering" method as a powerful tool for cloning of cDNAs of full-length ABC transporters (Fig. 2). Briefly, we clone the genomic sequence of the given gene into a binary plant expression vector under the control of a strong constitutive promoter (eg, 35S or Ubiquitin promoter). The gene is then transiently expressed in *N. benthamiana* leaves wherein the tobacco leaves' splicing machinery correctly assembles the many exons of the heterologous DNA fragment during transcription. After 2–4 days, RNA is extracted for reverse transcription to generate preparative amounts of full-length cDNAs. We use AT1G04120

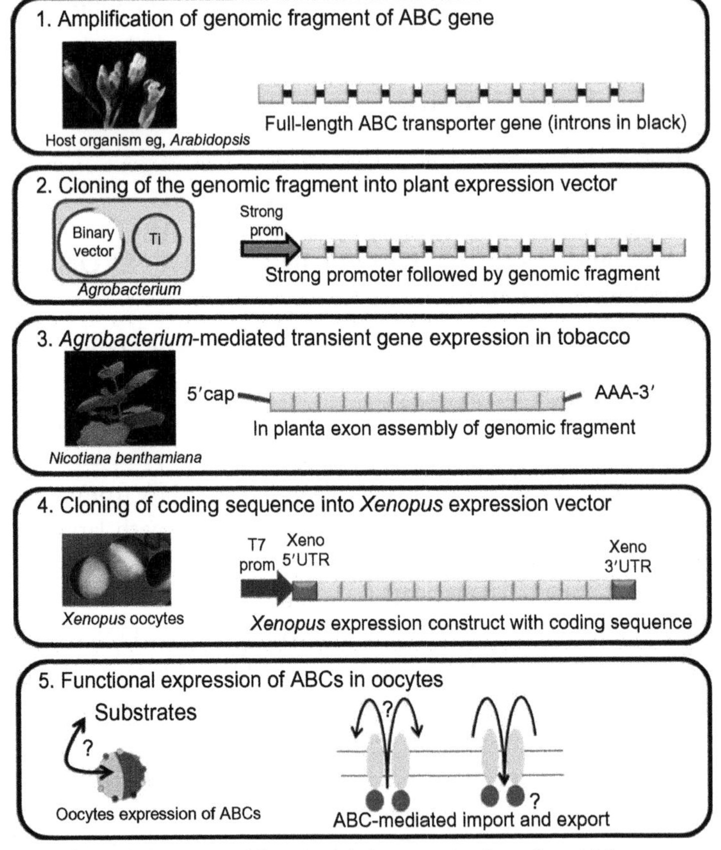

Fig. 2 Work flow of cloning and functional characterization of an ABC transporter gene from a given host with "exon engineering" strategy. Diagram representing the scalable process for characterization an ABC protein of interest from its coding genomic DNA sequence in five steps with two heterologous systems: (1) transient overexpression of gDNA in *Nicotiana benthamiana* by *Agrobacterium* infiltration for cDNA preparation; (2) overexpression of cRNA in the *Xenopus laevis* oocyte system. (See the color plate.)

(ABCG40) from the model plant *Arabidopsis* as an example for introduction of the workflow for "exon engineering" (Fig. 2).

Protocol:

1. Extract genomic DNA from flower buds of 5-week-old bolting *Arabidopsis thaliana* (Col-0) with the DNeasy Plant Mini Kit (Qiagen) according to the manufacturer's instructions. Elute the DNA from binding column with DNase-free water instead of elution buffer at 60°C for 10 min, to elute with maximum yield and to avoid EDTA, or other components interfering with the downstream reactions.

   Note: *Arabidopsis* flower buds but not flowers exhibit higher yield and quality for extraction of genomic DNA among all the tissues of *Arabidopsis*.
2. Prepare binary plant–bacteria shuttle plasmid vectors for cloning gene of interest. Here, we use the USER-compatible vector pCambia 2300-35Su as an example.

   Note: pCambia 2300-35Su is a USER™ cloning compatible in planta overexpression vector. For USER™ cloning technique and construction of the vectors, please refer to Nour-Eldin, Hansen, Norholm, Jensen, and Halkier (2006). Briefly, 5 μg plasmid was digested with 60 U *Pac*I (NEB) overnight at 37°C in a total volume of 200 μL. 20 U of Nt.BbvCI (NEB) and additional 20 U *Pac*I were added the next day, and the digestion was incubated for 1 h at 37°C. The linearized vector was purified with the Qiagen PCR purification kit.

   Note: Prior to cloning genes of interest, the success of linearization was tested by transforming *E. coli* and verifying a low background colony number (<50 colony/100 ng linearized plasmid).
3. Design 5′ primer containing a sequence coding an epitope tag (eg, Flag) after ATG code of the CDS (see sequence of primers later). N-terminal fusion with epitope tag (eg, Flag) of ABC gene will allow to analysis expression of full-length protein in different heterologous host by Western blotting. The Flag-genomic fragment was amplified from genomic DNA template and cloned into a pCambia 2300-35Su vector (Nour-Eldin et al., 2006). Amplify genomic fragment with USER-compatible primers with Phusion U Hot Start DNA Polymerase (Thermo F555S). GGCTTAAU and GGTTTAAU are the sequences in the forward and reverse primers, respectively; that are complementary to the single-stranded overhangs in the digested USER cassette. More details for design USER-compatible primers please refer to Nour-Eldin et al. (2006).

Note: The primers for cloning *Flag-AtABCG40* into pCambia 2300-35Su are

Forward:

5′**GGCTTAAU**ATGGACTACAAAGACGATGACGACAAG ATGGAGGGAACTAGTTTTCACC

Reverse:

5′**GGTTTAAU**CTATCGTTTTTGGAAATTGAAAC.

The USER cloning overhangs are in bold, the sequence encoding the FLAG tag is underlined.

Note: In order to reduce unspecific binding of primer, the PCR reaction mix should be kept on ice until it is put in preheated PCR machine.

4. Determine the ratio between PCR product and linearized plasmid based on the intensity of the bands from the gel electrophoresis. Make a mixture with molar ratio of PCR fragment to linearized plasmids approx. 3:5μL in a final volume of 10–19 μL. Add 1 μL of USER enzyme mix to the PCR tubes. Incubate the tube for 30 min at 37°C followed by 30 min at 20°C.

   Note: Do not purify PCR products for USER reaction when desired bands are present after checking by gel electrophoresis. If gel purification is required, elute with elution buffer and not ddH_2O for USER reaction.

5. Mix the USER reaction mixture with NEB10 chemical competent cells with a volume ratio (1:10) and transform by heat shock.

   Note: If cloning by USER cloning, do not use electroporation for transformation. Strain NEB10 was used because of its reduced recombination of cloned DNA and it has been successful in amplifying plasmids containing ABC transporters.

6. Validate successful cloning by sequencing and use the correct plasmids to transform *Agrobacterium tumefaciens (eg, GV3850)* by electroporation. The settings of the electroporation machine were: 400 **Ω**, 2.5 kV, and 25 μFD.

   Note: Also prepare an Agrobacterium strain carrying the silencing suppressor protein p19 protein of tomato bushy stunt virus (Voinnet et al., 2003), which is required in step 9.

7. Pick one Agrobacterium colony and grow it overnight at 28°C in YEP media supplement with appropriate antibiotics. 1 mL of the ON culture is then used to inoculate 10 mL (YEP + antibiotics) and grown overnight at 28°C.

Note: For *Agrobacterium GV3850*-containing plasmid pCambia 2300-35Su, use YEP supplemented with 50 μg/mL rifampicin plus 50 μg/mL kanamycin. It is crucial to ensure air supply to the bacteria during the entire incubation period. The O/N Agrobacterium culture should reach an OD600 between 1 and 6.

8. Prepare the 10 mL infiltration solution: 10 mM MES (pH 5.6), 10 mM $MgCl_2$ and 100 μ*M* acetosyringone in sterile H_2O for each *Agrobacterium* culture. Precipitate bacteria by centrifugation in 50 mL Falcon tubes at 4000 × *g* for 10 min. Resuspend the bacterial pellet in the solution prepared above to reach a final OD600 around 0.5–0.8. Incubate the suspension on the bench at room temperature for 1–2 h. 1 mL of resuspended solution can be infiltrated into two fully expanded tobacco leaves.
9. Perform Agrobacterium infiltration of *N. benthamiana*. More details for Agrobacterium infiltration please refer to Voinnet et al. (2003). Always coinfiltratc Agrobacterium with p19. The suspensions are mixed as follows 1/3 p19 and 2/3 of your suspension. Perform the infiltration with a 1 mL syringe. Press the syringe (no needle) on the underside of the leaf and exert a counter-pressure with your finger on the other side. Infiltrate the maximum number of leaves and avoid cotyledons.

   Note: *N. benthamiana* plants are grown in soil in small pots with a diameter of 5.5 cm in green house or climate chamber at 24°C (day) and 17°C (night) with 80–90% humidity for approximately 3–4 weeks (to 4–6 leaves stage).
10. 3–5 days after infiltration, collect 100 mg leaf samples (avoid the main vein) from infiltrated tobacco leaves for RNA extraction.
11. Extract RNA from the tobacco samples with Spectrum™ Plant Total RNA kit (Sigma) or other methods according to manufacturer's instruction. Elute with nuclease-free water to a final volume of 25 μL.
12. Generate single-stranded cDNA from 1 to 1.5 μg RNA using the SuperScript® III First-Strand Synthesis System (Thermo Fisher) with oligo dT primers. The temperature program is according to manufacturer's instructions, albeit with a 3-h 50°C incubation.

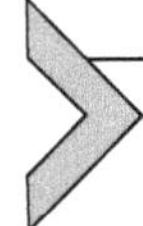

3. ABC TRANSPORTER EXPRESSION IN *XENOPUS* OOCYTES

As an example, we used the above "exon engineering" strategy to successfully clone the coding sequence of *AtABCG40* (genomic fragment approx. 6.5 kb with 21 introns, coding sequence approx. 4.2 kb) into a

Xenopus oocytes expression vector. Following RNA extraction from tobacco leaves and cDNA generation, the coding sequence of ABCG40 was amplified by PCR using primers that added the epitope tag (eg, Flag) sequence to the 5′ end of the gene and USER cloning sequences to both ends. After verification of the *Xenopus* expression construct by sequencing, cRNA was generated via in vitro transcription and was microinjected into oocytes for expression of ABCG40. As a prerequisite, we verified that the majority of cRNA of *Flag-ABCG40* was full-length (Fig. 3A). After expression for 4 days in oocytes, we performed Western blotting and detected strong expression of full size Flag-ABCG40 protein in the oocyte extracts (Fig. 3B).

Protocol:

1. The Flag-CDS fragment was amplified from cDNA template with same primers used for cloning pCambia2300Su-Flag-AtABCG40 and cloned into a *Xenopus* expression vector (eg, pNBIu) (Nour-Eldin et al., 2006).
2. Carry out in vitro transcription of cRNA on the purified PCR products (concentration approx. 100 ng/μL) using the mMESSAGE Kit

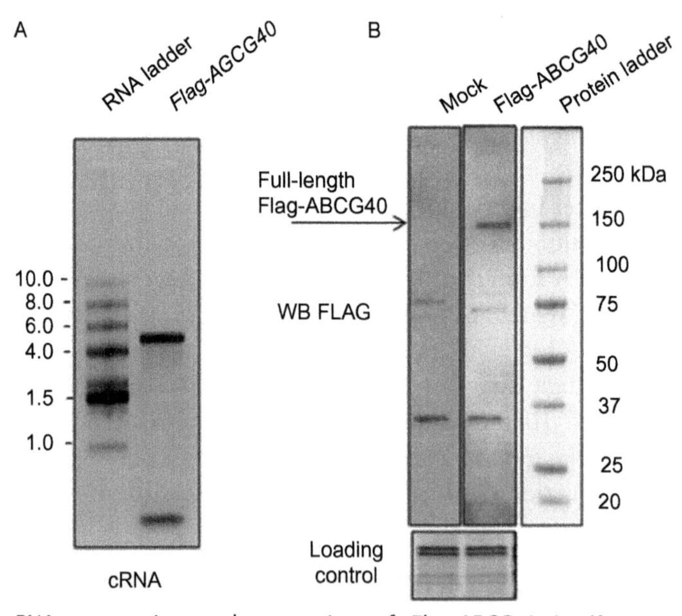

Fig. 3 cRNA preparation and expression of Flag-ABCG40 in *Xenopus* oocytes. (A) Validation of the length and integrity cRNA of Flag-AGCG40 after in vitro transcription. (B) Western blotting detection of Flag-ABCG40. The intact Flag-ABCG40 is shown by *arrow*.

(Jorgensen, Nour-Eldin, & Halkier, 2016) according to manufacturer's instruction (https://www.thermofisher.com/order/catalog/product/AM1344?ICID=search-product).

Note: There is a trade-off between RNA yield and good capping efficiency. For in vitro transcription of long, full-length cRNA fragments (CDS $\geq$4 kb), a 2-h incubation time during in vitro transcription with addition of 0.5 μL of GTP (30 mM) gave the best yield and quality of cRNA.

3. Analyze cRNA by gel electrophoresis using denaturing formaldehyde agarose gel to check the yield and quality of cRNA (Fig. 3A).
4. Preparation of oocytes and cRNA injection was carried out according to previous detailed description (Jorgensen et al., 2016).
5. To verify expression, we analyzed the expression level of full-length ABC protein with SDS-PAGE and Western blotting (Fig. 3B) according to previous description with minor modification (see note) (Jorgensen et al., 2016).

   Note: To detect the FLAG tag, we used mouse Monoclonal ANTI-FLAG® M2 antibody (F3165 SIGMA) (1:2500). We blocked PVDF membrane overnight in PBST with 3% skimmed milk at 4°C and with mild agitation and incubate the primary and secondary antibody in PBST with 3% skimmed milk at room temperature for 1 h.

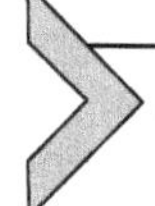

4. OPTIMIZATION OF TRANSPORT ASSAY FOR DIFFUSIBLE ABA IN *XENOPUS* OOCYTES

ABA is a weak organic acid with a pK_a of 4.8. At lower pH, the uncharged form can diffuse across membranes. Consequently, the rate of diffusion is dependent on the pH of the solution: at pH 5.8 the diffusion will be higher than at pH 7.4 (Fig. 4A). By comparing ABA concentration in transporter-expressing oocytes with mock-injected oocytes, we can characterize either import or export activity of a given transporter.

In order to investigate background uptake of ABA in *Xenopus* oocytes, we measured ABA uptake over time in control oocytes with and without adding ATP into the media (100 μ*M* ABA) (Fig. 4B). No significant difference was observed. This indicates that externally added ATP does not influence the background uptake of ABA in oocytes. In addition, ABC transporter inhibitor glibenclamide have no additional effect when ATP is present (Fig. 4B), which indicates background ABA uptake is not due to endogenous ABC importers.

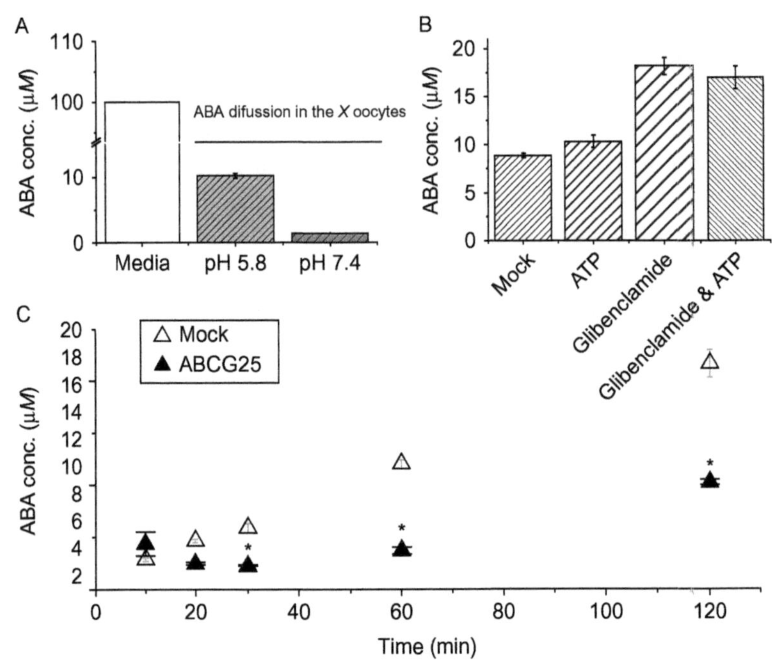

Fig. 4 Optimization of ABA transport assay for characterization of exporter AtABCG25. (A) Diffusion of ABA in the transport buffer at different pH for 60 min. Partially diffusion of ABA at pH 5.8 allows for both import and export assay. (B) Assay for detection of background transport activity by endogenous (ABC) transporters in oocytes. Mg-ATP (1 mM, pH 7), ABC transporter inhibitor glibenclamide (100 μ*M*, DMSO), and their combination were added into the media (pH 5.8) containing 100 μ*M* ABA for 60 min. No significant difference of ABA content between mock with ATP treatment, as well as glibenclamide with glibenclamide and ATP (*T*-test, $P < 0.05$). (C) Time course of export of ABA by ABCG25 in the oocytes at pH 5.8. Media ABA concentration is 100 μ*M*. For all experiments, each data points contain at least three repeats, each data point is the average from 5 oocytes. Significant differences in ABA content between mock (uninjected) oocytes and ABCG25-expressing oocytes are indicated by asterisks (*$P < 0.05$).

Protocol:

1. Prepare ABA-containing assay media in either Kulori buffer pH 5.8 or pH 7.4 (90 mM NaCl, 1 mM KCl, 1 mM $MgCl_2$, 5 mM MES (pH 5.8)/ HEPES (pH 7.4)). Adjust pH using TRIS buffer. Add ABA to a final concentration of 100 μM.

   Note: The pH of the assay media must be measured and adjusted if necessary after the addition of the substrate and/or inhibitors.

2. Subsequently, incubate the oocytes in 1 mL of the assay media for a given time point at room temperature.

Note: We use 24-well plates for assays and preincubate oocytes in "empty" buffer solutions to allow adaptation of oocytes to assay pH and inhibitors before transferring them to substrate containing solutions. We typically add 15 oocytes to each well. Preincubation time for pH adaptation is approx. 5 min.

3. Stop the assay by adding an equal amount of Kulori buffer pH 7.4 and transfer the oocytes as quickly as possible to washing buffer (Kulori buffer pH 7.4).
4. Wash the oocytes in washing buffer (Kulori buffer pH 7.4) at least 3 times, then transfer intact oocytes into 3 × 1.5-mL tubes (5 oocytes per tube). Remove the excess washing buffer from the tubes and add 50 μL 50% methanol. Homogenize by pipetting.
5. Centrifuge at 10,000 × *g* for 10 min at 4°C and transfer the supernatant into new tube and incubate at −20°C for at least 1 h. Centrifuge the samples again at 21,000 × *g* for 15 min at 4°C and transfer the supernatant into HPLC vials.
6. Analyze the ABA content by liquid chromatography–mass spectrometry (LC–MS).

   LC–MS was carried out using an Agilent 1100 Series LC (Agilent Technologies, Germany) coupled to a Bruker HCT-Ultra ion trap mass spectrometer (Bruker Daltonics, Bremen, Germany). A Zorbax SB-C18 column (Agilent; 1.8 μm, 2.1 × 50 mm) was used at a flow rate of 0.2 mL/min. The oven temperature was maintained at 35 degree. The mobile phases were: A, water with 0.1% (v/v) HCOOH and 10 μ*M* NaCl; B, acetonitrile with 0.1% (v/v) HCOOH. The gradient program was: 0–0.5 min, isocratic 6% B; 0.5–12.5 min, linear gradient 6–55% B; 12.5–13.1 min, linear gradient 55–95% B; 13.1–15.5 isocratic 95% B; 15.60–20 min, isocratic 6% B. The flow rate was increased to 0.3 mL/min in the interval 15.2–17.5 min. The mass spectrometer was run in positive electrospray mode.
7. Process the data with Data Analysis (Bruker Daltonics) according to the manufacturer's instruction. Make a standard dilution curve for quantification of ABA concentration in oocytes.

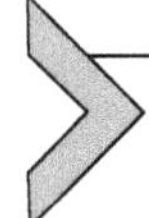

5. CASE STUDY: CHARACTERIZATION OF THE ABA EXPORTER AT ABCG25 IN *XENOPUS* OOCYTES

AtABCG25 is an *Arabidopsis* ABA exporter characterized in *S. frugiperda* vesicles (Kuromori et al., 2010). The efflux activity can be detected as ABA uptake of the regenerated membrane vesicles, because the regenerated membrane includes inside-out vesicles (Kuromori et al., 2010). However, the possibility that the observed uptake is due to import activity using vesicle-mediated transport assay cannot be excluded. In order to further validate the characterization of ABCG25, we measured uptake of ABA by AtABCG25 in oocytes over time. At different time points (after 20 min), the ABCG25-expressing oocytes accumulated ABA to significantly lower levels than control oocytes, showing that the ABA is likely exported out of the oocyte by ABCG25 (Fig. 4C). The biggest difference in ABA levels between AtABCG25-expressing oocytes and control oocytes was seen between 60 and 120 min. The functional expression of AtABCG25 in oocytes as an ABA exporter indicates that besides plant B-type ABC, *Xenopus* oocytes are able to express plant G-type ABC exporter, which has not been shown before.

Protocol:

1. Three days after injection, the oocytes injected with either RNA or water, are first preincubated in "empty" Kulori buffer pH 5.8 for 5 min.
2. Subsequently, transfer the oocytes to 1 mL assay volume consisting of Kulori buffer pH 5.8 and 100 μ*M* ABA.
3. Incubate assay for given time points at room temperature.
4. Stop the assay by adding an equal amount of Kulori buffer pH 7.4 and transfer the oocytes as quickly as possible to washing buffer (Kulori buffer pH 7.4).
5. Wash the oocytes 3 times in Kulori buffer pH 7.4 and then transfer intact oocytes into 1.5 mL Eppendorf tubes, 5 oocytes into each tube and three tubes for each type of RNA.
6. Remove the excess washing buffer from the tubes and add 50 μL 50% methanol. Homogenize by pipetting.
7. Centrifuge at 10,000 × *g* for 10 min at 4°C and transfer the supernatant into new tube and incubate at −20°C for at least 1 h. Centrifuge the samples again at 21,000 × *g* for 15 min at 4°C and transfer the supernatant into HPLC vials.
8. Analyze the ABA content by LC–MS.

6. CONCLUSIONS

In planta "exon engineering" provides a reliable and unique tool for getting fast access to preparative amounts of long full-length cDNAs (eg, ABC transporters) regardless of its length and expression level from a given plant genomic template. In contrast to de novo DNA synthesis technology, in planta "exon engineering" requires neither a template coding sequence nor optimization of chemical synthesis and gene fusion to clone full-length coding sequence. Instead, the conserved mechanism of transcription and splicing within eukaryotes enables this strategy for cloning any eukaryotic coding sequence. Furthermore, Agro-infiltration in *N. benthamiana* leaf usually is easily established. Besides the cloning of ABCG40 described here, we have in a pilot experiment cloned 5 out of 12 other full-length ABCG genes using this technique.

By implementing in planta "exon engineering" strategy, we successfully cloned the coding sequence of intron-rich, long ABC transporters. Additionally, this study shows that the *Xenopus* oocyte can serve as a new platform for functional assessment of plasma membrane ABC protein for export of specialized metabolites, as demonstrated using *Arabidopsis* abscisic acid exporter ABCG25. Successful cloning and expression of ABC transporters is a prerequisite for implementing ABC transporters in transport engineering strategies in synthetic biology. We hope that the presented cloning strategy as well as the expression and export assay methods can be used as a tool package for transport engineering of commercial valuable compounds.

ACKNOWLEDGMENTS

The work was supported by the Danish National Research Foundation (Grant no. DNRF99).

REFERENCES

Alejandro, S., Lee, Y., Tohge, T., Sudre, D., Osorio, S., Park, J., ... Martinoia, E. (2012). AtABCG29 is a monolignol transporter involved in lignin biosynthesis. *Current Biology*, *22*(13), 1207–1212. Retrieved from <Go to ISI>://WOS:000306379600025.

Beis, K. (2015). Structural basis for the mechanism of ABC transporters. *Biochemical Society Transactions*, *43*, 889–893. Retrieved from <Go to ISI>://WOS:000363759700019, http://www.biochemsoctrans.org/content/ppbiost/43/5/889.full.pdf.

Brunetti, P., Zanella, L., De Paolis, A., Di Litta, D., Cecchetti, V., Falasca, G., et al. (2015). Cadmium-inducible expression of the ABC-type transporter AtABCC3 increases phytochelatin-mediated cadmium tolerance in Arabidopsis. *Journal of Experimental Botany*, *66*, 3815–3829.

Chen, S., Sanchez-Fernandez, R., Lyver, E. R., Dancis, A., & Rea, P. A. (2007). Functional characterization of AtATM1, AtATM2, and AtATM3, a subfamily of Arabidopsis half-molecule ATP-binding cassette transporters implicated in iron homeostasis. *Journal of Biological Chemistry*, *282*(29), 21561–21571. Retrieved from <Go to ISI>://WOS:000248047500083.

Feng, H., Xia, X., Fan, X., Xu, G., & Miller, A. J. (2013). Optimizing plant transporter expression in Xenopus oocytes. *Plant Methods*, *9*(1), 48. http://dx.doi.org/10.1186/1746-4811-9-48.

Francisco, R. M., Regalado, A., Ageorges, A., Burla, B. J., Bassin, B., Eisenach, C., et al. (2013). ABCC1, an ATP binding cassette protein from grape berry, transports anthocyanidin 3-O-glucosides. *Plant Cell*, *25*, 1840–1854.

Geisler, M., Blakeslee, J. J., Bouchard, R., Lee, O. R., Vincenzetti, V., Bandyopadhyay, A., ... Martinoia, E. (2005). Cellular efflux of auxin catalyzed by the Arabidopsis MDR/PGP transporter AtPGP1. *Plant Journal*, *44*(2), 179–194. Retrieved from <Go to ISI>://WOS:000232885700001.

Geu-Flores, F., Nour-Eldin, H. H., Nielsen, M. T., & Halkier, B. A. (2007). USER fusion: A rapid and efficient method for simultaneous fusion and cloning of multiple PCR products. *Nucleic Acids Research*, *35*(7), e55. http://dx.doi.org/10.1093/nar/gkm106.

Gibson, D. G., Smith, H. O., Hutchison, C. A., 3rd., Venter, J. C., & Merryman, C. (2010). Chemical synthesis of the mouse mitochondrial genome. *Nature Methods*, 7(11), 901–903. http://dx.doi.org/10.1038/nmeth.1515.

Huang, C. F., Yamaji, N., Mitani, N., Yano, M., Nagamura, Y., & Ma, J. F. (2009). A bacterial-type ABC transporter is involved in aluminum tolerance in rice. *Plant Cell*, *21*(2), 655–667. Retrieved from <Go to ISI>://WOS:000264706500025, http://www.ncbi.nlm.nih.gov/pmc/articles/PMC2660611/pdf/tpc2100655.pdf.

Hwang, J. U., Song, W. Y., Hong, D., Ko, D., Yamaoka, Y., Jang, S., et al. (2016). Plant ABC transporters enable many unique aspects of a terrestrial plant's lifestyle. *Molecular Plant*, *9*, 338–355.

Ikeda, M. (2006). Towards bacterial strains overproducing L-tryptophan and other aromatics by metabolic engineering. *Applied Microbiology and Biotechnology*, *69*(6), 615–626. http://dx.doi.org/10.1007/s00253-005-0252-y.

Jones, D. T., & Woods, D. R. (1986). Acetone-butanol fermentation revisited. *Microbiological Reviews*, *50*(4), 484–524. Retrieved from, http://www.ncbi.nlm.nih.gov/pubmed/3540574.

Jorgensen, M. E., Nour-Eldin, H. H., & Halkier, B. A. (2016). A western blot protocol for detection of proteins heterologously expressed in Xenopus laevis oocytes. *Methods in Molecular Biology*, *1405*, 99–107. http://dx.doi.org/10.1007/978-1-4939-3393-8_10.

Kamimoto, Y., Terasaka, K., Hamamoto, M., Takanashi, K., Fukuda, S., Shitan, N., ... Yazaki, K. (2012). Arabidopsis ABCB21 is a facultative auxin importer/exporter regulated by cytoplasmic auxin concentration. *Plant and Cell Physiology*, *53*(12), 2090–2100. Retrieved from <Go to ISI>://WOS:000312105400011.

Kang, J., Hwang, J. U., Lee, M., Kim, Y. Y., Assmann, S. M., Martinoia, E., & Lee, Y. (2010). PDR-type ABC transporter mediates cellular uptake of the phytohormone abscisic acid. *Proceedings of the National Academy of Sciences of the United States of America*, *107*(5), 2355–2360. http://dx.doi.org/10.1073/pnas.0909222107.

Kang, J., Park, J., Choi, H., Burla, B., Kretzschmar, T., Lee, Y., & Martinoia, E. (2011). Plant ABC transporters. *Arabidopsis book*, *Vol. 9*, e0153. American Society of Plant Biologists: Rockville, MD. http://dx.doi.org/10.1199/tab.0153.

Klein, M., Geisler, M., Suh, S. J., Kolukisaoglu, H. U., Azevedo, L., Plaza, S., ... Martinoia, E. (2004). Disruption of AtMRP4, a guard cell plasma membrane ABCC-type ABC transporter, leads to deregulation of stomatal opening and increased drought susceptibility. *Plant Journal*, *39*(2), 219–236. http://dx.doi.org/10.1111/j.1365-313X.2004.02125.x.

Kuromori, T., Miyaji, T., Yabuuchi, H., Shimizu, H., Sugimoto, E., Kamiya, A., ... Shinozaki, K. (2010). ABC transporter AtABCG25 is involved in abscisic acid transport and responses. *Proceedings of the National Academy of Sciences of the United States of America*,

107(5), 2361–2366. Retrieved from <Go to ISI>://WOS:000274296300099, http://www.ncbi.nlm.nih.gov/pmc/articles/PMC2836683/pdf/pnas.200912516.pdf.

Lee, M., Choi, Y., Burla, B., Kim, Y. Y., Jeon, B., Maeshima, M., ... Lee, Y. (2008). The ABC transporter AtABCB14 is a malate importer and modulates stomatal response to CO2. *Nature Cell Biology*, *10*(10), 1217–1223. Retrieved from <Go to ISI>://WOS:000259682900016, http://www.nature.com/ncb/journal/v10/n10/pdf/ncb1782.pdf.

Lefevre, F., Baijot, A., & Boutry, M. (2015). Plant ABC transporters: Time for biochemistry? *Biochemical Society Transactions*, *43*, 931–936. Retrieved from <Go to ISI>://WOS:000363759700026. http://www.biochemsoctrans.org/content/ppbiost/43/5/931.full.pdf.

Nagy, R., Grob, H., Weder, B., Green, P., Klein, M., Frelet-Barrand, A., ... Martinoia, E. (2009). The Arabidopsis ATP-binding cassette protein AtMRP5/AtABCC5 is a high affinity inositol hexakisphosphate transporter involved in guard cell signaling and phytate storage. *Journal of Biological Chemistry*, *284*(48), 33614–33622. Retrieved from <Go to ISI>://WOS:000272028500065.

Nour-Eldin, H. H., Andersen, T. G., Burow, M., Madsen, S. R., Jorgensen, M. E., Olsen, C. E., ... Halkier, B. A. (2012). NRT/PTR transporters are essential for translocation of glucosinolate defence compounds to seeds. *Nature*, *488*(7412), 531–534. http://dx.doi.org/10.1038/nature11285.

Nour-Eldin, H. H., Hansen, B. G., Norholm, M. H., Jensen, J. K., & Halkier, B. A. (2006). Advancing uracil-excision based cloning towards an ideal technique for cloning PCR fragments. *Nucleic Acids Research*, *34*(18), e122. http://dx.doi.org/10.1093/nar/gkl635.

Ochoa-Villarreal, M., Howat, S., Hong, S., Jang, M. O., Jin, Y. W., Lee, E. K., & Loake, G. J. (2015). Plant cell culture strategies for the production of natural products. *BMB Reports*, *49*(3), 149–158. Retrieved from, http://www.ncbi.nlm.nih.gov/pubmed/26698871.

Ruzicka, K., Strader, L. C., Bailly, A., Yang, H., Blakeslee, J., Langowski, L., ... Friml, J. (2010). Arabidopsis PIS1 encodes the ABCG37 transporter of auxinic compounds including the auxin precursor indole-3-butyric acid. *Proceedings of the National Academy of Sciences of the United States of America*, *107*(23), 10749–10753. http://dx.doi.org/10.1073/pnas.1005878107.

Santelia, D., Vincenzetti, V., Azzarello, E., Bovet, L., Fukao, Y., Duchtig, P., ... Geisler, M. (2005). MDR-like ABC transporter AtPGP4 is involved in auxin-mediated lateral root and root hair development. *FEBS Letter*, *579*(24), 5399–5406. http://dx.doi.org/10.1016/j.febslet.2005.08.061.

Schaedler, T. A., Thornton, J. D., Kruse, I., Schwarzlander, M., Meyer, A. J., van Veen, H. W., & Balk, J. (2014). A conserved mitochondrial ATP-binding cassette transporter exports glutathione polysulfide for cytosolic metal cofactor assembly. *Journal of Biological Chemistry*, *289*(34), 23264–23274. Retrieved from <Go to ISI>://WOS:000341505000001, http://www.jbc.org/content/289/34/23264.full.pdf.

Shitan, N., Bazin, I., Dan, K., Obata, K., Kigawa, K., Ueda, K., ... Yazaki, K. (2003). Involvement of CjMDR1, a plant multidrug-resistance-type ATP-binding cassette protein, in alkaloid transport in Coptis japonica. *Proceedings of the National Academy of Sci of the United States of America*, *100*(2), 751–756. http://dx.doi.org/10.1073/pnas.0134257100.

Song, W. Y., Park, J., Mendoza-Cozatl, D. G., Suter-Grotemeyer, M., Shim, D., Hortensteiner, S., et al. (2010). Arsenic tolerance in Arabidopsis is mediated by two ABCC-type phytochelatin transporters. *Proceedings of the National Academy of Sciences of the United States of America*, *107*, 21187–21192.

Takanashi, K., Sugiyama, A., Sato, S., Tabata, S., & Yazaki, K. (2012). LjABCB1, an ATP-binding cassette protein specifically induced in uninfected cells of Lotus japonicus nodules. *Journal of Plant Physiology*, *169*, 322–326.

Terasaka, K., Blakeslee, J. J., Titapiwatanakun, B., Peer, W. A., Bandyopadhyay, A., Makam, S. N., et al. (2005). PGP4, an ATP binding cassette P-glycoprotein, catalyzes auxin transport in Arabidopsis thaliana roots. *Plant Cell*, *17*, 2922–2939.

Theodoulou, F. L., & Kerr, I. D. (2015). ABC transporter research: Going strong 40 years on. *Biochemical Society Transactions*, *43*, 1033–1040. Retrieved from <Go to ISI>://WOS:000363759700039, http://www.ncbi.nlm.nih.gov/pmc/articles/PMC4652935/pdf/bst0431033.pdf.

Tommasini, R., Vogt, E., Fromenteau, M., Hortensteiner, S., Matile, P., Amrhein, N., et al. (1998). An ABC-transporter of Arabidopsis thaliana has both glutathione-conjugate and chlorophyll catabolite transport activity. *Plant Journal*, *13*, 773–780.

Voinnet, O., Rivas, S., Mestre, P., & Baulcombe, D. (2003). An enhanced transient expression system in plants based on suppression of gene silencing by the p19 protein of tomato bushy stunt virus. *Plant Journal*, *33*(5), 949–956. Retrieved from, http://www.ncbi.nlm.nih.gov/pubmed/12609035.

Yang, H., & Murphy, A. S. (2009). Functional expression and characterization of Arabidopsis ABCB, AUX 1 and PIN auxin transporters in Schizosaccharomyces pombe. *Plant Journal*, *59*(1), 179–191. http://dx.doi.org/10.1111/j.1365-313X.2009.03856.x.

Young, E., Lee, S. M., & Alper, H. (2010). Optimizing pentose utilization in yeast: the need for novel tools and approaches. *Biotechnology for Biofuels*, *3*, 24. https://biotechnologyforbiofuels.biomedcentral.com/articles/10.1186/1754-6834-3-24.

CHAPTER TEN

Quantifying the Metabolites of the Methylerythritol 4-Phosphate (MEP) Pathway in Plants and Bacteria by Liquid Chromatography–Triple Quadrupole Mass Spectrometry

D. González-Cabanelas*, A. Hammerbacher†, B. Raguschke*, J. Gershenzon*[,1], L.P. Wright*

*Max Planck Institute for Chemical Ecology, Jena, Germany
†Forestry and Agricultural Biotechnology Institute (FABI), University of Pretoria, Pretoria, South Africa
[1]Corresponding author: e-mail address: gershenzon@ice.mpg.de

Contents

1. Introduction 226
2. Preparation of Stable Isotope-Labeled Internal Standards 229
   2.1 Growing *E. coli* with Labeled Glucose as a Carbon Source 230
   2.2 Isolation of MEcDP by Export from Transgenic *E. coli* 232
3. Extraction of Methylerythritol Phosphate Pathway Intermediates from Biological Sources 234
   3.1 Extraction of Methylerythritol Phosphate Pathway Intermediates from Plant Materials 234
   3.2 Extraction of Methylerythritol Phosphate Pathway Intermediates from Bacterial Cultures 235
4. Analysis of Methylerythritol Phosphate Pathway Metabolites by LC–MS/MS 236
   4.1 Separation by HILIC 237
   4.2 Detection by Tandem Mass Spectrometry 240
   4.3 Quantification 242
5. Discussion and Summary 244
References 245

Abstract

The 2-C-methyl-D-erythritol 4-phosphate (MEP) pathway occurs in the plastids of higher plants and in most economically important prokaryotes where it is responsible for the biosynthesis of the isoprenoid building blocks, isopentenyl diphosphate and dimethylallyl diphosphate. These five-carbon compounds are the substrates for the enormous variety of terpenoid products, including many essential metabolites and substances of commercial value. Increased knowledge of the regulation of the MEP

Methods in Enzymology, Volume 576
ISSN 0076-6879
http://dx.doi.org/10.1016/bs.mie.2016.02.025

pathway is critical to understanding many aspects of plant and microbial metabolism as well as in developing biotechnological platforms for producing these commercially valuable isoprenoids. To achieve this goal, researchers must have the ability to investigate the in vivo kinetics of the pathway by accurately measuring the concentrations of MEP pathway metabolites. However, the low levels of these metabolites complicate their accurate determination without suitable internal standards. This chapter describes a sensitive method to accurately determine the concentrations of MEP pathway metabolites occurring at trace amounts in biological samples using liquid chromatography coupled to triple quadrupole mass spectrometry. In addition, simple protocols are given for producing stable isotope-labeled internal standards for these analyses.

1. INTRODUCTION

Isoprenoids, also known as terpenoids, are a very diverse group of compounds present in all living organisms with a multitude of different functions (Croteau, Kutchan, & Lewis, 2000; McGarvey & Croteau, 1995). Despite the structural diversity of this group, isoprenoids are all derived from the same five-carbon building blocks, isopentenyl diphosphate (IDP) and its isomer dimethylallyl diphosphate (DMADP). Isoprenoids have important functions in primary metabolism as electron carriers in respiration and photosynthesis (ubiquinone, plastoquinone), membrane components (sterols, cholesterol), photosynthetic pigments (carotenoids, chlorophylls), and plant growth regulators (abscisic acid, brassinosteroids, cytokinins, strigolactones). In secondary metabolism, an enormous variety of structures play a multitude of roles in defense against enemies, attraction of mutualists, and internal signaling (Bouvier, Rahier, & Camara, 2005; Croteau et al., 2000; Gershenzon & Dudareva, 2007; Rodriguez-Concepcion, 2006).

Isoprenoid biosynthesis is unusual in the fact that two completely different pathways each produce the basic five-carbon precursors IDP and DMADP (Eisenreich et al., 1998; Rohmer, Knani, Simonin, Sutter, & Sahm, 1993; Rohmer, Seemann, Horbach, BringerMeyer, & Sahm, 1996). The mevalonate (MVA) pathway was once thought to be only responsible for the biosynthesis of these precursors, but in the 1990s the methylerythritol 4-phosphate (MEP) pathway was discovered as an alternative route to IDP and DMADP (Cordoba, Salmi, & Leon, 2009; Eisenreich et al., 1998). Most organisms use only one of the two pathways for the biosynthesis of the isoprenoid precursors. The MVA pathway is exclusive to archaea, fungi, and animals, while the MEP pathway is present in most eubacteria and apicomplexan protozoa (like *Plasmodium falciparum*).

By contrast, plants, green algae, and some bacteria use both pathways (Lichtenthaler, Rohmer, & Schwender, 1997; Rodriguez-Concepcion & Boronat, 2002 Rohmer, 1999).

In contrast to the MVA pathway that involves sequential additions of acetyl-CoA units followed by reduction and phosphorylation, the MEP pathway starts with the synthesis of 1-deoxy-D-xylulose 5-phosphate (DXP) by condensation of glyceraldehyde 3-phosphate (G3P) and pyruvate (Pyr) via DXP synthase (DXS). DXP is then reduced by DXP reductoisomerase (DXR) to 2-C-methyl-D-erythitol 4-phosphate (MEP) that is converted into 2-C-methyl-D-erythritol-2,4-cyclodiphosphate (MEcDP) in three steps via the cytidyl conjugates, 4-(cytidine 5′-diphospho)-2-C-methyl-D-erythritol (ME-CDP) and ME-CDP phosphate (MEP-CDP) as intermediates. Finally, MEcDP is successively reduced to (*E*)-4-hydroxy-3-methylbut-2-enyl 4-diphosphate (HMBDP), catalyzed by HMBDP synthase (HDS), and then to a mixture of IDP and DMADP, catalyzed by HMBDP reductase (HDR), to complete the pathway (Hunter, 2007; Phillips, Leon, Boronat, & Rodriguez-Concepcion, 2008) (Fig. 1). Since the discovery of the MEP pathway, increasing attention has been paid to the individual intermediates both because of interest in pathway regulation and the finding that in plants some intermediates are also involved in complex signaling cascades from plastids (site of the MEP pathway in plants) to the nucleus (Gonzalez-Cabanelas et al., 2015; Ward, Baker, Llewellyn, Hawkins, & Beale, 2011; Xiao et al., 2012). This highlights the importance of developing analytical tools to detect and quantify MEP intermediates to investigate these and other possible functions.

The absence of the MEP pathway in animals, including humans, makes it an attractive target for the development of compounds, such as antibiotics and herbicides, targeted at organisms possessing the pathway (Testa & Brown, 2003). In addition, the commercial value of some isoprenoids as pigments (carotenoids), industrial materials (resins and rubbers), flavor and fragrances (monoterpenes), drugs (taxol, artemisinin), and even biofuels has triggered interest in enhancing their production (Rodriguez-Concepcion, 2006; Tippmann, Chen, Siewers, & Nielsen, 2013). During the last decade many attempts at engineering the MEP pathway have been made in order to increase the supply of IPP and DMAPP for the biosynthesis of valuable isoprenoids, but with few exceptions the results have been disappointing (Ajikumar et al., 2010; Klein-Marcuschamer, Ajikumar, & Stephanopoulos, 2007). In fact, when metabolic engineering of isoprenoid biosynthesis has proved successful in *Escherichia coli*, the native MEP pathway

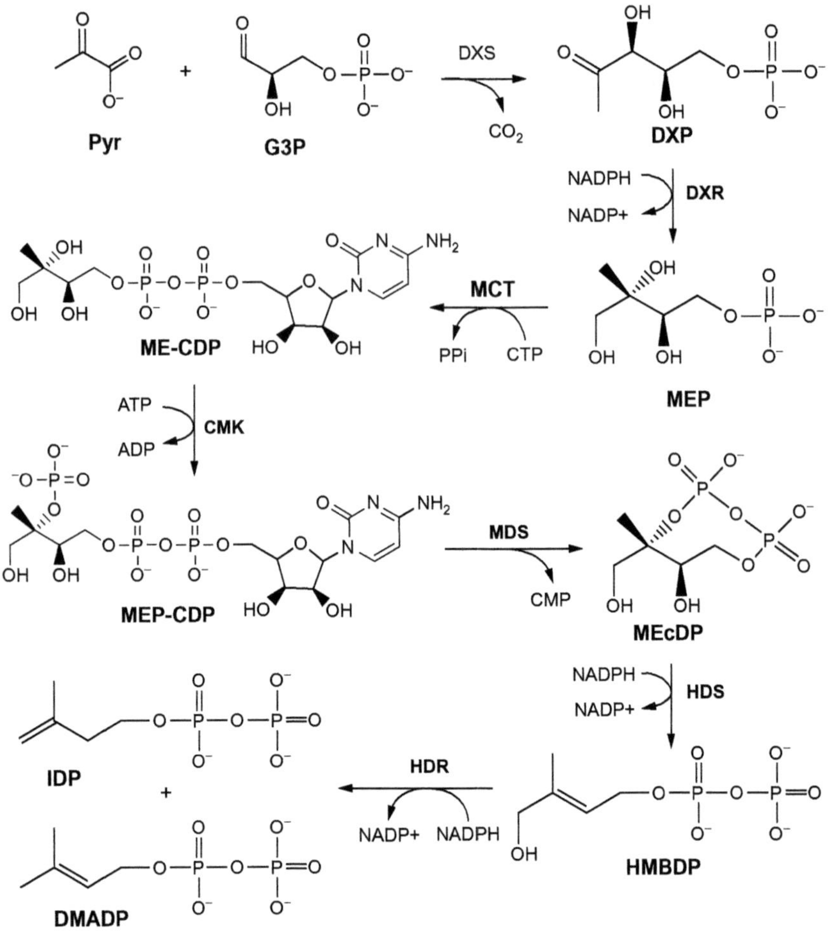

Fig. 1 The 2-C-methyl-D-erythritol 4-phosphate (MEP) pathway. Pyruvate (Pyr) and glyceraldehyde 3-phosphate (G3P) are condensed into 1-deoxy-D-xylulose 5-phosphate (DXP) by DXP synthase (DXS) with a loss of carbon dioxide. DXP is reduced to MEP by DXP reductoisomerase (DXR) and the product reacts with cytidine triphosphate (CTP) to produce 4-(cytidine 5′-diphospho)-2-C-methyl-D-erythritol (ME-CDP) with a loss of a diphosphate (PPi) moiety. ME-CDP is converted to ME-CDP phosphate (MEP-CDP) by ME-CDP kinase (CMK) using adenosine triphosphate (ATP), which loses a phosphate to form adenosine diphosphate (ADP). MEP-CDP is cyclized via a loss of cytidine monophosphate (CMP) to 2-C-methyl-D-erythritol 2,4-cyclodiphosphate (MEcDP) by MEcDP synthase (MDS). MEcDP is reduced to (*E*)-4-hydroxy-3-methylbut-2-enyl diphosphate (HMBDP) by HMBDP synthase (HDS). HMBDP is reduced to a mixture of isopentenyl diphosphate (IDP) and dimethylallyl diphosphate (DMADP) by HMBDP reductase (HDR).

has been ignored and the MVA pathway ectopically expressed (Cervin, Whited, Chotani, Valle, & Fioresi, 2009; Tsuruta et al., 2009). However, it would actually be better to employ the MEP pathway for this purpose since it has a lower energy requirement than the MVA pathway when using only glucose and glycerol as carbon sources (Ajikumar et al., 2010; Steinbüchel, 2003). The lack of success in engineering the MEP pathway is often attributed to insufficient knowledge of its control (Wright et al., 2014).

The majority of the work done on the regulation of the MEP pathway has focused on the regulation of gene expression. However, some studies indicated that posttranscriptional and especially posttranslational regulation are also important in MEP pathway regulation (Flores-Perez, Sauret-Gueto, Gas, Jarvis, & Rodriguez-Concepcion, 2008; Guevara-Garcia et al., 2005; Kobayashi et al., 2007). In this context, measurements of enzyme activities and the accompanying pools of metabolites are needed to understand the operation of the pathway. Equally important is the in vivo measurement of flux. Since the MEP pathway leads to several major products, the only suitable approach to measure flux is to follow the incorporation rate of stable isotopic label into metabolites (Wright et al., 2014). Here we describe extraction, analysis, and quantification methods to measure intermediates of the MEP pathway, including their amounts of stable isotopic label. These procedures will further our understanding how the pathway works in different organisms, providing a foundation for increased production of isoprenoids with high commercial values.

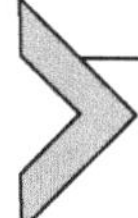

2. PREPARATION OF STABLE ISOTOPE-LABELED INTERNAL STANDARDS

Stable isotope-labeled internal standards have identical physical and chemical properties to their target analytes but a different mass. Hence such standards show the same separation characteristics as the analyte during chromatography, but can be distinguished by mass spectrometry. The addition of stable isotope-labeled internal standards to a sample therefore facilitates the identification and absolute quantification of metabolites. Knowing the amount of standard added allows the measurement of the quantity of the analyte (recovery), the ion suppression effect of the matrix components, and mass detector fluctuations (Stokvis, Rosing, & Beijnen, 2005). However, stable isotope-labeled internal standards are very expensive and often, as is the case for the MEP pathway metabolites, not commercially available.

We therefore developed a method to produce such standards for MEP pathway intermediates by growing *E. coli* BL21(DE3) cells in M9 minimal medium using universally labeled ^{13}C-D-glucose as the carbon source.

2.1 Growing *E. coli* with Labeled Glucose as a Carbon Source

A survey was first conducted of *E. coli* cultures at various growth stages to determine at what stage MEP pathway metabolites are at the highest levels. A 5-mL starter culture of *E. coli* BL21(DE3) was grown in 50 mL M9 minimal medium and the MEP pathway metabolites were measured at regular intervals over 14 h at 37°C on a shaker set at 220 revolutions per minute (rpm). The highest concentrations of the metabolites were detected after 5–6 h of growth when the optical density at 600 nm (OD_{600}) was between 1.3 and 1.6 (Fig. 2). Harvesting cells at this stage grown with ^{13}C-labeled glucose as the carbon source, we obtained high levels of MEP pathway metabolites with an excess of 80% fully labeled ^{13}C-compounds and the remainder partially labeled; unlabeled metabolites were not detectable (Fig. 3).

1. *E. coli* BL21(DE3) was grown overnight in 5 mL LB medium at 37°C in a shaking incubator at 220 rpm. The next day the cells were centrifuged at 4200 × *g* for 5 min and the supernatant was removed. Removal of the LB medium is necessary to prevent the formation of unlabeled MEP pathway metabolites.
2. Resuspend cells in 50 mL M9 minimal medium containing [$^{13}C_6$]-D-glucose as carbon source and grow cells at 37°C in a shaking incubator at 220 rpm until an OD_{600} value between 1.3 and 1.6 is reached (5–6 h).
3. Centrifuge cells for 5 min at 4200 × *g* and 4°C, remove the supernatant, and add 5 mL ice-cold extraction solvent consisting of acetonitrile/methanol/water (2:2:1, v/v/v) containing 0.1% ammonium hydroxide to stop the metabolism and extract the MEP pathway intermediates. Vortex the cell suspension thoroughly and incubate on ice for at least 15 min with occasional vortexing.
4. Transfer the liquid phase to a new tube after centrifuging 5 min at 20,000 × *g* and 4°C and dry in a rotary evaporator. Alternatively the solvent can be separated into aliquots and evaporated under a stream of nitrogen gas at 40°C in a heating block or centrifugal vacuum evaporator at 45°C.
5. Dissolve residue in 500 μL 10 m*M* ammonium acetate, adjusted to pH 9 with ammonium hydroxide, centrifuge 5 min at 20,000 × *g* and 4°C, and transfer supernatant to a new 1.5-mL safe-lock microcentrifuge tube kept on ice.

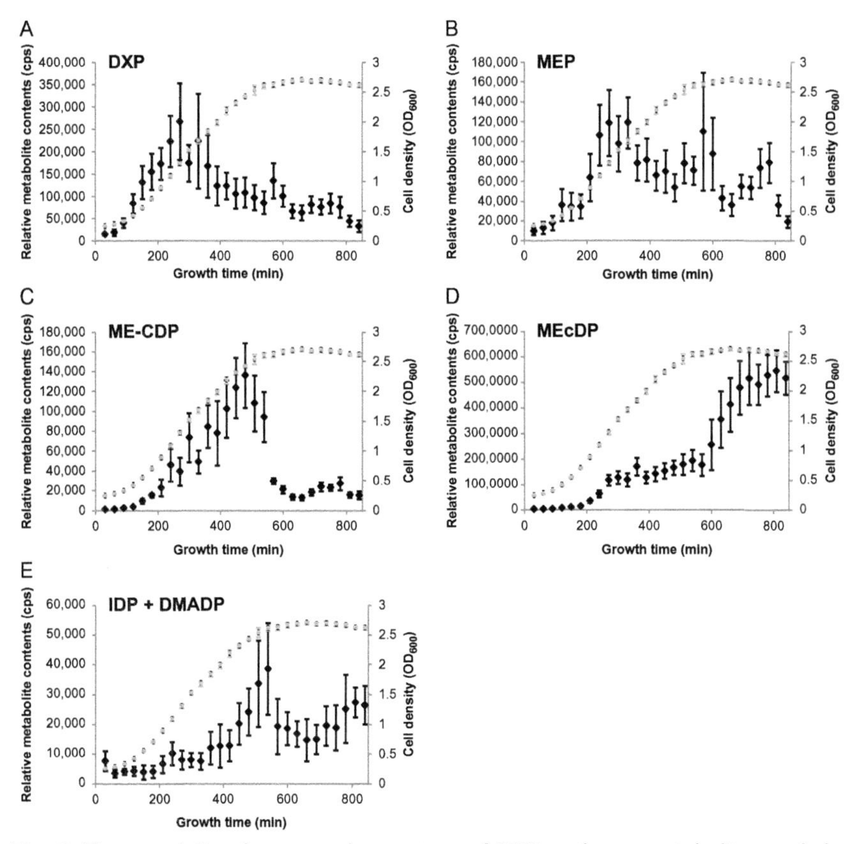

Fig. 2 The correlation between the content of MEP pathway metabolites and the growth phase of an *E. coli* culture. Shown are the average amount of metabolites from three different experiments (*solid symbols*), expressed as peak area of a 5 μL sample extract. The metabolites were extracted and analyzed as described in the text from 1 mL *E. coli* grown in M9 minimal medium incubated at 37°C and harvested at 30 min intervals. The bacterial cell density (*open symbols*) was measured by determining the optical density at 600 nm. Shown are the relative contents of (A) DXP, (B) MEP, (C) ME-CDP, (D) MEcDP, and (E) IDP + DMADP.

6. Extract with 500 μL chloroform by vortexing thoroughly. Centrifuge 5 min at 20,000 × *g* at 4°C, and remove the upper aqueous phase containing the stable isotope-labeled internal standards. Although the standards are stable for at least a month when stored at 4°C, we recommend making aliquots and drying under a steam of nitrogen gas for longer term storage. We typically dilute the preparation 1:9 (v/v) with extraction solvent or with the MEP pathway metabolite extract for absolute quantification.

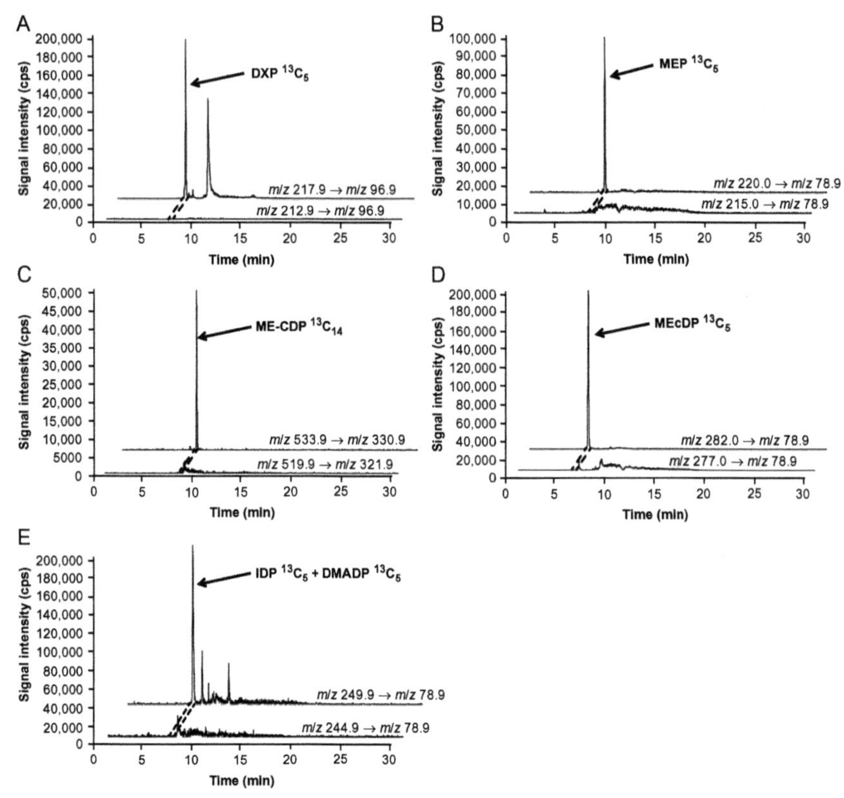

Fig. 3 Stable isotope-labeled MEP pathway intermediates extracted from *E. coli* grown at 37°C in M9 minimal medium with universally labeled glucose $^{13}C_6$ used as carbon source. The metabolites were analyzed with LC–MS/MS using electrospray ionization after separation on a HILIC column as described in the text. The selected reaction monitoring (SRM) of the fully labeled (A) DXP, (B) MEP, (C) ME-CDP, (D) MEcDP, and (E) IDP + DMADP are shown in the rear traces with the SRM of the unlabeled compound in the front traces.

2.2 Isolation of MEcDP by Export from Transgenic *E. coli*

Recent research has shown that the MEP pathway intermediate MEcDP also functions in stress responses in *Arabidopsis thaliana* (Gonzalez-Cabanelas et al., 2015; Xiao et al., 2012). However, research on the additional roles of MEcDP is hampered by the excessive cost of this metabolite from commercial suppliers. Thus we have developed a rapid and simple method to produce MEcDP at low cost that exploits the tendency of transgenic *E. coli* to export MEcDP to the medium with increased flux through the MEP pathway (Zhou, Zou, Stephanopoulos, & Too, 2012). The first enzyme in the MEP pathway, DXS, is the major flux controlling enzyme (Wright et al., 2014). By overexpressing DXS in *E. coli* BL21(DE3) cells, we increased pathway flux and export of MEcDP to the medium. This results not only in greater amounts of MEcDP but its occurrence in the medium simplifies its purification. After concentrating the MEcDP through freeze-drying, this metabolite was purified using chromatography on microcrystalline cellulose.

1. Start a small 5 mL culture of transformed *E. coli* BL21(DE3) cells overexpressing DXS in LB medium and grow it overnight at 37°C shaking at 220 rpm.
2. Add the 5 mL starter culture to 45 mL LB medium and grow at 37°C and 220 rpm shaking until an OD_{600} between 0.4 and 0.5 is reached when the cells are in exponential growth. Induce the expression of DXS cloned in the pH 9 plasmid (Yu & Liu, 2006) by adding isopropyl β-D-1-thiogalactopyranoside (IPTG) to a final concentration of 1 m*M* and let the cells grow under the same conditions for another 3 h. IPTG induction is necessary due to the T7 promoter used on the pH 9 plasmid.
3. Replace the LB medium with M9 minimal medium by centrifuging the cells for 5 min at 4200 × *g* at 28°C, removing the old medium and resuspending the cells in M9 medium with universally labeled ^{13}C-D-glucose as carbon source. Incubate the cells at 28°C and 220 rpm.
4. Lyophilize the supernatant after removing the cells by centrifuging for 5 min at 4°C at 4200 × *g*. Resuspend the pellet in 5 mL of isopropanol/methanol/water (2:2:1, v/v/v) containing 0.1% formic acid. Centrifuge at 4200 × *g* for 5 min at 4°C to remove any precipitate.
5. Load the sample on a microcrystalline cellulose column with a diameter of 10 mm and a length of 600 mm and elute with isopropanol/methanol/water (2:2:1, v/v/v) containing 0.1% formic acid at a flow rate of 1.0 mL min^{-1}. Fractions of 4 mL can then be collected and analyzed for MEcDP content using direct injection into the triple quadrupole mass spectrometer (MS/MS). Inject a 1-μL portion of the collected fraction into a solvent flow of 1 mL min^{-1} connected to the ion source of the MS/MS. The solvent used is the same as was used for the running solvent on the cellulose column. The parameters used for detection with the mass spectrometer are described in Section 4.2. Combine the fractions containing MEcDP and evaporate the solvent using a rotary evaporator to obtain the purified MEcDP. Using this approach we could purify approximately 2 mg MEcDP with an NMR determined purity of 80%. If more MEcDP is needed, the culture volume can be scaled up.

3. EXTRACTION OF METHYLERYTHRITOL PHOSPHATE PATHWAY INTERMEDIATES FROM BIOLOGICAL SOURCES

As a primary pathway with a direct connection to photosynthesis (Li & Sharkey, 2013), the MEP pathway has a rapid flux rate and so care must be taken to quench metabolic activity as soon as possible after harvest in order to obtain accurate measurements of metabolic pools. For example, wild-type *A. thaliana* has a MEP pathway flux of 3.44 pmol mg^{-1} min^{-1} when grown at 21°C at a light intensity of 140 μmol m^{-2} s^{-1} photosynthetic photon flux density. Hence to have an effect of less than 5% on a typical amount of ~20 pmol mg^{-1} DXP (Wright et al., 2014), metabolism should be halted within 20 s. This is usually accomplished by freezing in liquid nitrogen.

Another consideration in the extraction of MEP pathway metabolites is choosing a buffer to maximize their stability. Since nearly all of these metabolites are phosphorylated, a slightly alkaline pH of 9 was selected to maintain ionization for increased stability and solubility. At higher pH values, there is a risk of degrading MEcDP by cleaving the cyclodiphosphate linkage (Ge, d'Avignon, Ackerman, & Sammons, 2012).

3.1 Extraction of Methylerythritol Phosphate Pathway Intermediates from Plant Materials

1. For fast quenching of metabolism material was directly immersed in liquid nitrogen after harvest. For typical pathway flux (Wright et al., 2014), quenching within 20 s should be adequate for accurate, reproducible analyses. However with more rapid flux or smaller pool sizes, the time window necessary to deactivate metabolism might need to be reduced further. In such cases, it may be advisable to flash-freeze the plant material while still under the light source (Arrivault et al., 2009). Once frozen, the next step is to homogenize the plant material to obtain a fine powder. This can be done using a mortar and pestle precooled in liquid nitrogen. An additional method is to first freeze-dry the material and then homogenize it in a vertical oscillating ball mill. Most important is that the homogenized material be a fine powder without any distinguishable particles to assure full extraction of the metabolites.
2. For extraction, weigh approximately 5 mg of dried plant material into a 1.5-mL safe-lock microcentrifuge tube, taking care to note the exact weight. This weight will be used later to normalize the quantified

compound amounts to the weight of plant material extracted. A weight within 0.5 mg of the intended 5 mg gives maximum extraction efficiency with the amounts of extraction solvents listed. Add 250 μL of an acetonitrile–10 m*M* ammonium acetate pH 9 mixture (1:1, v/v) and vortex thoroughly. The acetonitrile in the extraction solvent helps to dissolve membranes and is needed for maximum extraction of the metabolites from chloroplasts.

3. After centrifugation for 5 min at 20,000 × *g* in a microcentrifuge, 200 μL of the supernatant is transferred to a new 1.5-mL safe-lock microcentrifuge tube. For maximum efficiency, the pellet should be extracted with another 250 μL of extraction solvent as described in step 2.
4. The supernatants are then combined and the extract evaporated under a stream of nitrogen in a 40°C heating block. Alternatively the supernatant can be dried in a centrifugal vacuum evaporator at 45°C.
5. Dissolve the residue in 100 μL of 10 m*M* ammonium acetate, adjusted to pH 9 with ammonium hydroxide, by thorough vortexing and add 100 μL of chloroform to extract the hydrophobic compounds from the sample. This is to prevent unnecessary contamination of the hydrophilic interaction liquid chromatography (HILIC) column and mass spectrometer. The extract is then centrifuged for 5 min at 20,000 × *g* to facilitate fast phase separation.
6. Transfer 50 μL of the aqueous phase to a new 1.5-mL safe-lock microcentrifuge tube; add 50 μL acetonitrile, vortex, and centrifuge 5 min at 20,000 × *g* to remove any precipitate. The supernatant is then analyzed by liquid chromatography-triple quadrupole mass spectrometry (LC–MS/MS) as described in Section 4 or stored at −20°C for later analysis. In the presence of isotopically labeled internal standards, the extract can be stored for an extended period without detectable losses.

3.2 Extraction of Methylerythritol Phosphate Pathway Intermediates from Bacterial Cultures

The methylerythritol phosphate pathway is present in most eubacteria, including *E. coli*, as an essential biosynthetic route. Due to the absence of a rigid cell wall, we extract the bacteria cells with a mixture of ice-cold acetonitrile, methanol, and water to break the cell membrane and quench metabolic processes. The *E. coli* cultures were grown in 50 mL M9 minimal medium. Extraction proceeds as follows:

1. Collect *E. coli* cells from a 1 mL growth culture and centrifuge at 20,000 × *g* for 5 min at 4°C to precipitate the cell pellet.

2. Freeze the sample in liquid nitrogen and store at –20°C for later extraction, or extract the cells directly by adding 1 mL of ice-cold acetonitrile/methanol/water (2:2:1, v/v/v) containing 0.1% ammonium hydroxide. Incubate on ice for at least 15 min with occasional vortexing to facilitate membrane breakage and maximum extraction efficiency.
3. Transfer the liquid phase to a new 1.5-mL safe-lock microcentrifuge tube after centrifuging for 5 min at 20,000 × *g*. Dry under a stream of nitrogen gas at 40°C in a heating block or centrifugal vacuum evaporator at 45°C.
4. Dissolve the residue in 100 μL 10 m*M* ammonium acetate, adjusted to pH 9 with ammonium hydroxide, by thorough vortexing and centrifuge 5 min at 20,000 × *g*. Transfer 90 μL of the supernatant to a new 1.5-mL safe-lock microcentrifuge tube.
5. Extract with an equal volume of chloroform and separate the phases by centrifugation at 20,000 × *g*. Transfer 50 μL of the upper aqueous phase to a new 1.5-mL safe-lock microcentrifuge tube; add 50 μL acetonitrile, vortex, and centrifuge at 20,000 × *g* for 5 min to remove any suspended particles. Transfer the supernatant to an HPLC vial for LC–MS/MS analysis as described in Section 4 or store at –20°C for later analysis. In the presence of isotopically labeled internal standards, the extract can be stored for an extended period without detectable losses.

4. ANALYSIS OF METHYLERYTHRITOL PHOSPHATE PATHWAY METABOLITES BY LC–MS/MS

Mass spectrometry furnishes a very sensitive, information-rich platform for analyzing most metabolites, but to take full advantage it is necessary to optimize the chromatographic separation. This is particularly true for phosphorylated analytes, such as the MEP pathway metabolites, not only to improve sensitivity but also to minimize overlap with other phosphorylated compounds with similar masses that are also likely to experience similar fragmentations due to the facile loss of the phosphate groups. Separation of highly hydrophilic phosphorylated compounds is typically performed using counter ion liquid chromatography (Luo, Groenke, Takors, Wandrey, & Oldiges, 2007) or ion exchange liquid chromatography (Lunn et al., 2006). However, in counter ion chromatography mass spectrometry signals are reduced due to ion suppression and contamination of the mass spectrometer by the counter ion (Rutters, Mohring, Rullkotter, Griep-Raming, & Metyger, 2000). Ion exchange chromatography requires a membrane-based electrolytic suppressor when coupled to a mass spectrometer to continuously neutralize pH and remove salts prior to infusion (Rabin, Stillian, Barreto, Friedman, & Toofan, 1993), which is not practical for multipurpose instruments and entails additional costs.

HILIC is an attractive alternative to separate small polar compounds (Li & Sharkey, 2013). HILIC employs polar stationary phases (Olsen, 2001; Oyler et al., 1996) and mobile phase solvents containing a water miscible organic solvent with a small amount of water. The low water content in the mobile phase results in a liquid–liquid extraction system with an aqueous layer around the surface of the polar stationary phase and an organic layer elsewhere. Increasing the aqueous content of the mobile phase causes the elution of compounds that had been retained in the water layer associated with the stationary phase (Buszewski & Noga, 2012; Mateos-Vivas, Rodriguez-Gonzalo, Garcia-Gomez, & Carabias-Martinez, 2015) through a partitioning effect. Silica is the most widely used chromatographic support in HILIC and can be modified by the binding of a variety of functional groups, such as aminopropyl, amide, or cyano moieties (Ikegami, Tomomatsu, Takubo, Horie, & Tanaka, 2008). A typical mobile phase employed in HILIC retention contains between 70% and 90% of acetonitrile (Alpert, 1990; Naidong, 2003), which provides enough water for the creation of the aqueous layer and the complete solubilization of the hydrophilic compounds from the sample.

4.1 Separation by HILIC

To choose the best HILIC stationary phase to separate and detect the MEP pathway metabolites, we tested columns of Nucleodex β-OH (5 μm, 200 × 4.0 mm; Macherey-Nagel) (Pulido, Toledo-Ortiz, Phillips, Wright, & Rodriguez-Concepcion, 2013), Kinetex HILIC (2.6 μm, 150 × 2.1 mm; Phenomenex), Zorbax RX-SIL (5 μm, 150 × 2.1 mm; Agilent), Atlantis HILIC (3 μm, 150 × 2.1 mm; Waters) (Ghirardo et al., 2014), ZIC-pHILIC (5 μm, 150 × 2.1 mm, Merck) (Wright et al., 2014), XBridge BEH Amide (3.5 μm, 150 × 2.1 mm; Waters) (Gonzalez-Cabanelas et al., 2015; Saladie, Wright, Garcia-Mas, Rodriguez-Concepcion, & Phillips, 2014; Wright & Phillips, 2014), and XBridge BEH Amide XP (2.5 μm, 150 × 2.1 mm; Waters). The best results were obtained with the XBridge BEH Amide XP column which uses an ethylene-bridged hybrid silica gel support prepared from tetraethoxysilane and bis(triethoxysilyl)ethane with trifunctional-bonded amide as functional group.

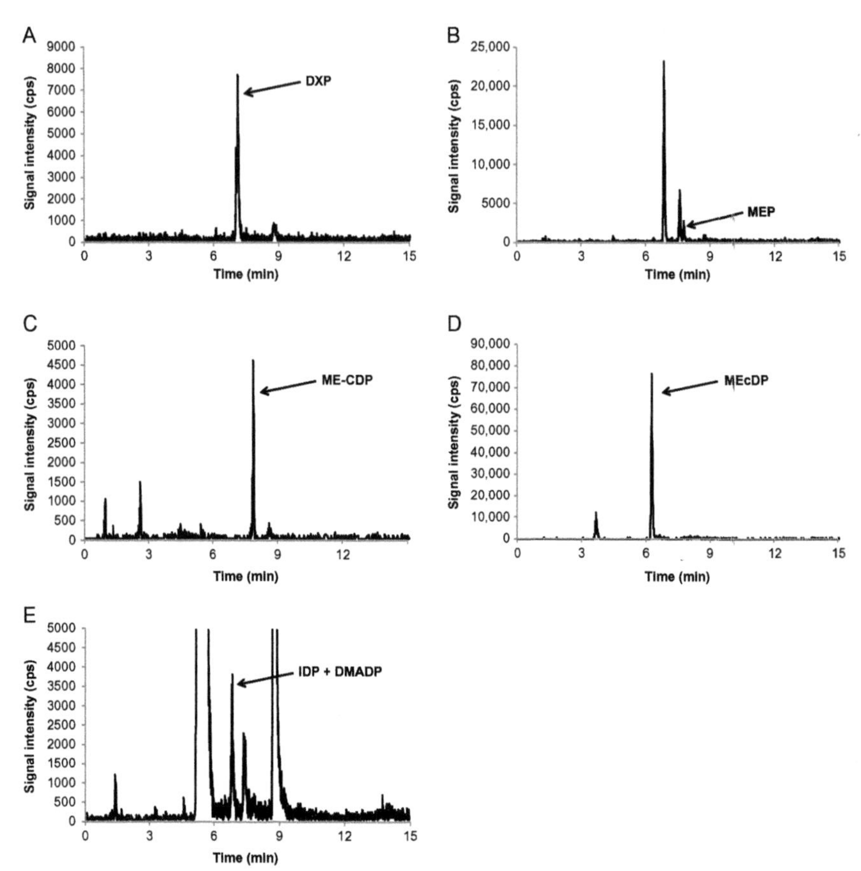

Fig. 4 MEP pathway intermediates extracted from a mature *A. thaliana* rosette. The metabolites were analyzed with LC–MS/MS using electrospray ionization after separation on a HILIC column as described in the text. The selected reaction monitoring (SRM) of (A) DXP, (B) MEP, (C) ME-CDP, (D) MEcDP, and (E) IDP + DMADP with the peaks representing the MEP pathway metabolites indicated by an *arrow*.

The version with 2.5 μm resin particle size was chosen to better separate the metabolic intermediate MEP from various coeluting compounds (Fig. 4). The column was run at a high flow rate of 500 μL min^{-1} relative to the small 2.1-mm column diameter (van Deemter, Zuiderweg, & Klinkenberg, 1956) to improve separation efficiency. This leads to column back pressures of up to 500 bar, which are too high for standard HPLC systems. In the absence of a high-pressure HPLC instrument, we recommend using the XBridge BEH Amide column with a 3.5-μm particle size which runs at a maximum back pressure of around 220 bar under similar flow rates.

The mobile phase was selected to reduce peak broadening and tailing. The diphosphorylated compounds, IDP and DMADP, showed especially pronounced peak tailing caused by hydrogen bonding between the phosphate groups and free silanol groups of the silica in the resin (Gritti, Kazakevich, & Guiochon, 2007) as well as interaction with metals in the LC–MS/MS system (Tuytten et al., 2006). Hydrogen bonding was minimized by using an end-capped stationary phase of the XBridge BEH Amide XP column. Interaction with the metal contained in the instrumental setup was reduced by using ammonium bicarbonate as a buffer additive in the mobile phase (Asakawa et al., 2008; Guo & Gaiki, 2005), and by increasing the pH (Preinerstorfer, Schiesel, Lammerhofer, & Lindner, 2010). Increasing the pH by more than 1 unit above the pK_a of the analytes has the additional benefit of stabilizing their charge (Boersema, Mohammed, & Heck, 2008; Buszewski & Noga, 2012). However, the high pH led to decreased retention due to repulsion between the negatively charged phosphates and the residual negative charge of the stationary phase (Alpert, 1990). To increase the retention, we increased the ammonium bicarbonate to 20 m*M* (the maximum advisable concentration for liquid chromatography with mass spectrometry detection) to decrease the repulsion between analytes and the stationary phase (Asakawa et al., 2008) and to increase the volume or hydrophilicity of the aqueous layer (Mateos-Vivas et al., 2015). To increase analyte retention further, a polar cationic base was added to the mobile phase to form hydrophilic ion pairs with the phosphate groups and so further reduce electrostatic repulsion and increase retention (Mateos-Vivas et al., 2015). This was ammonium hydroxide added until a pH of 10.5 was reached.

Another advantage of HILIC separation in comparison with other chromatographic methods is the generally small effect of temperature on retention properties (Buszewski & Noga, 2012). Our analysis did not show marked differences in the retention times or in the quality of the chromatograms under different temperatures so to promote column lifetime we used a temperature of 25°C.

The optimized separation conditions used the XBridge BEH Amide XP Column (2.5 μm, 150 × 2.1 mm; Waters) together with a guard column containing the same sorbent (2.5 μm, 10 × 2.1 mm) and a SSI™ high-pressure precolumn filter (Sigma-Aldrich). The mobile phases were 20 m*M* ammonium bicarbonate adjusted to pH 10.5 with LC–MS grade ammonium hydroxide as solvent A, and 80% acetonitrile containing 20 m*M* ammonium bicarbonate adjusted to pH 10.5 with LC–MS grade ammonium hydroxide

Table 1 Liquid Chromatography Solvent Gradient Profile for Separation of Methylerythritol Phosphate Pathway Intermediates on the XBridge BEH Amide XP HILIC Column

Time (min)	% Solvent A[a]	% Solvent B[b]
0	0	100
5	16	84
10	16	84
11	40	60
15	40	60
15.10	0	100
30	0	100

[a]Solvent A is 20 m*M* ammonium bicarbonate, pH 10.5, adjusted with LC–MS grade ammonium hydroxide
[b]Solvent B is 80% acetonitrile with 20 m*M* ammonium bicarbonate adjusted to pH 10.5 with LC–MS grade ammonium hydroxide.
The column temperature was 25°C with a flow rate of 500 μL min^{-1} and 5 μL sample injected.

for solvent B. The solvent gradient profile is described in Table 1, with steps of elution (0–10 min), column wash (10–15 min), and column equilibration (15–30 min). Separation was performed at 25°C with a flow rate of 500 μL min^{-1}, and the volume injected was 5 μL. Fig. 4 shows a typical chromatogram of MEP metabolites extracted from *A. thaliana* in which six of the eight metabolites of the MEP pathway are detected and quantified. Of the metabolites that cannot be quantified, MEP-CDP is very labile and spontaneously breaks down to ME-CDP, while HMBDP occurs at levels below our limit of quantification (LOQ).

4.2 Detection by Tandem Mass Spectrometry

The mobile phase used for HILIC is especially suitable for standard electrospray ionization mass spectrometry because of its high volatility (Ikegami et al., 2008). The use of mass spectrometry detection has long been appreciated for metabolite analyses because of its high sensitivity and ability to separate coeluting compounds based on mass measurements. For pathway studies, mass spectrometric detection also facilitates measurements of stable isotope-labeled compounds which can be used for absolute quantification of intermediate pools and for label incorporation experiments for determination of flux. Tandem mass spectrometry in particular is extremely useful for detection against the complexity of plant or

Table 2 Ion Source Parameter Settings Used for Ionizing the MEP Pathway Metabolites in the Ion Source of an API 5000 Tandem Mass Spectrometer (AB Sciex)

Ion Source Parameter	Optimized Setting
Ion spray voltage	–4500 V
Ion source temperature	700°C
Nebulizer gas	70 psi
Heating gas	30 psi
Curtain gas	30 psi

bacterial matrices. Using the widely available API 5000 triple quadrupole mass spectrometer (AB Sciex), the ion source parameters were optimized for all of the MEP pathway intermediates under investigation simultaneously (Table 2). The ion source was used in the negative ionization mode with nitrogen gas for the nebulizing, heating, and curtain gas. The results should be applicable to all mass spectrometers from this manufacturer, since all these use the same ion source.

The individual intermediates were quantified on a triple quadrupole mass analyzer by multiple reaction monitoring using specific precursor ion—product ion reactions for each metabolite. The optimized parameters for each reaction are given in Table 3. We also give parameters for quantifying stable isotope-labeled standards of each metabolite. With the molecular ion as a precursor, most of the MEP pathway metabolites gave either *m/z* 96.9 or *m/z* 78.9 as the most sensitively detected product ion after fragmentation, and these were used for quantification. These masses both correspond to a fragment from the phosphate group of the precursor ion, with *m/z* 96.9 representing [phosphate–H]$^-$ and *m/z* 78.9 representing [phosphate–H_2O–H]$^-$. These phosphate fragment ions are the only detectable product ions for MEP, MEcDP, and IDP/DMADP and the most sensitively detected for DXP. Use of the phosphate fragment as product ions for quantification is also advantageous when measuring the incorporation of ^{13}C because of the reduced scan combinations necessary. Since the product ion does not contain any carbons, only the mass increase in the precursor ion needs to be measured for determining the amount of ^{13}C in the metabolites. For the MEP pathway intermediate ME-CDP, pronounced background noise was observed when monitoring the phosphate product ion, and therefore we used the *m/z* 321.9 ion as product ion for quantification which represents the cytidine diphosphate fragment.

Table 3 Parameters for Quantifying Individual MEP Pathway Metabolites and Their Respective Stable Isotopically Labeled Internal Standards by Tandem Mass Spectrometry with a Triple Quadrupole Analyzer

Compound	Retention Time (min)	Precursor Ion → Product Ion	Declustering Potential (V)	Collision Energy (V)	Cell Exit Potential (V)
DXP	7.1	212.9 → 96.9	–60	–16	–15
DXP-$^{13}C_5$	7.1	217.9 → 96.9	–60	–16	–15
MEP	7.8	215.0 → 78.9	–70	–36	–11
MEP-$^{13}C_5$	7.8	220.0 → 78.9	–70	–36	–11
ME-CDP	7.9	519.9 → 321.9	–105	–30	–9
ME-CDP-$^{13}C_{14}$	7.9	533.9 → 330.9	–105	–30	–9
MEcDP	6.3	277.0 → 78.9	–50	–38	–11
MEcDP-$^{13}C_5$	6.3	282.0 → 78.9	–50	–38	–11
IDP and DMADP	6.8	244.9 → 78.9	–45	–24	–11
IDP-$^{13}C_5$ and DMADP-$^{13}C_5$	6.8	249.9 → 78.9	–45	–24	–11

The retention times for the different metabolites as well as the masses of the precursor and product ions are also shown. Nitrogen was used as collision gas at a CAD setting of 10 and the entrance potential was set at –10 V for all metabolites.

4.3 Quantification

The accurate determination of analyte abundance by liquid chromatography-tandem mass spectrometry is affected by a range of factors, such as the stability of the molecule, recovery from the sample, degree of ion suppression, and limits of detection and quantification. The sample matrix is frequently observed to reduce the ionization efficiency of specific analytes due to other molecules that coelute with the compound of interest (Tang & Kebarle, 1993). In the analysis of MEP pathway intermediates, the use of a HILIC column should diminish ion suppression because of the high content of organic solvents which evaporate faster and so promote ionization of solvated analytes. The use of the negative ionization mode might also decrease ion suppression since fewer compounds are ionized under this mode than under positive ionization. Other strategies to reduce ion suppression include sample dilution or a decrease in injection volume to reduce the amount of interfering compounds.

To compensate for ion suppression, authentic standards are necessary. Stable isotope-labeled standards are ideal for this purpose because they behave similarly to the unlabeled analytes during the sample preparation, chromatography, and the ionization process (Van Eeckhaut, Lanckmans, Sarre, Smolders, & Michotte, 2009), and can be distinguished from the sample compounds based on their mass. Yet their expense or availability is often a major barrier to use. We have described methods for producing affordable stable isotope-labeled standards of MEP metabolites in Section 2.1 (Fig. 3). Their addition allows us to compensate not only for ion suppression but also for the loss of analytes during extraction and variations in instrument performance (Table 4) (Birkemeyer, Luedemann, Wagner, Erban, & Kopka, 2005). Optimization of the concentrations of stable isotopically labeled internal standards themselves is also necessary to avoid possible ion suppression effects. The concentrations of such standards should be kept in the same range as the natural occurrence of the sample compounds and in the linear range of the calibration curve (Bakhtiar & Majumdar, 2007; Freitas, Gotz, Ruff, Singer, & Muller, 2004). To be able to fully exploit an analytical method, it is necessary to know the limits of detection (LOD) and LOQ (Table 4). LOD is the lowest analyte concentration that can be detected, but not necessarily quantified, while the LOQ is the lowest concentration of an analyte that can be determined with acceptable precision and accuracy (Armbruster & Pry, 2008).

Table 4 Extraction Efficiency, Ion Suppression, Limit of Detection (LOD) and Limit of Quantification (LOQ) for the Analysis of the MEP Pathway Intermediates DXP, MEP, ME-CDP, MEcDP, and IDP/DMADP in *Arabidopsis thaliana* Leaf Extracts as Described in This Chapter

	DXP	MEP	ME-CDP	MEcDP	IDP/DMADP
Extraction efficiency (%)	74.5 ±3.0	62.9 ±1.9	78.0 ±2.5	83.9 ±3.2	24.2 ±1.9
Ion suppression (%)	65.7 ±2.7	64.5 ±2.9	75.1 ±4.0	55.3 ±3.5	58.3 ±2.3
LOD (pmol mL^{-1})	4.73 ±0.52	19.83 ±1.38	1.61 ±0.01	27.17 ±0.42	0.78 ±0.02
LOQ (pmol mL^{-1})	7.88 ±0.86	33.05 ±2.3	2.68 ±0.01	45.28 ±0.71	1.29 ±0.03

The extraction efficiency and ion suppression values represents the average ± SE ($n=10$). The LOD was taken as the amount of metabolite detected that gave a signal-to-noise (S/N) ratio of 3 and the LOQ was taken as the amount necessary to give an S/N ratio of 5.

5. DISCUSSION AND SUMMARY

Other methods have been described for analyzing the MEP pathway metabolites from biological samples, most of which also use a sensitive mass spectrometer for detection. A notable exception is the measurement of MEcDP in plants under stress conditions with ^{31}P-NMR spectroscopy. Oxidative stress caused by high temperature and light intensities inhibits HDS that results in the drastic accumulation of the HDS substrate MEcDP to levels detectable by NMR (Rivasseau et al., 2009). When MEP pathway metabolites have been analyzed from crude extracts of biological samples, liquid chromatography separations are usually done by either counter ion liquid chromatography (Balcke, Bennewitz, Zabel, & Tissier, 2014; Zhang et al., 2011; Zhou et al., 2012) or HILIC (Baidoo, Xiao, Dehesh, & Keasling, 2014; Li & Sharkey, 2013). The mass analyzers include TOF (Baidoo et al., 2014; Perez-Gil et al., 2012; Zhou et al., 2012), quadrupole MS (Baidoo et al., 2014), MS/MS (Li & Sharkey, 2013; Zhang et al., 2011), and tripleTOF mass spectrometers (MS^2-QTOF) (Balcke et al., 2014). MS^2-QTOF instruments are a hybrid between MS/MS and TOF, where the third quadrupole is replaced with a TOF. This configuration allows for the accurate mass detection of the precursor ion as well as untargeted mass spectra determination of compounds through the sequential window acquisition of all theoretical mass spectra (SWATH) technique (Balcke et al., 2014).

However, none of these previous methods directly measure the recovery efficiency and ion suppression during analysis. Technical variation during extraction procedures and ion suppression fluctuations due to slight differences in the matrix of different samples can lead to decreased accuracy and precision. The addition of stable isotopically labeled internal standards is an excellent remedy for these problems, more effective than the use of external standard curves. Although protocols have been published for producing isotopically labeled standards of MEP metabolites (Hecht et al., 2001; Illarionova et al., 2006; Schuhr et al., 2001), they require tackling the intermediates one at a time using some sophisticated synthetic protocols. The protocol furnished here provides a method for producing stable isotope-labeled internal standards of the MEP pathway using only equipment and resources normally present in biochemical laboratories. These standards have facilitated the quantification of six of the eight MEP pathway intermediates in the model plant *A. thaliana*, a species that accumulates only low quantities of specialized isoprenoids. Thus the methods should be applicable to a range of plant and microbial taxa regardless of their level of isoprenoid production.

REFERENCES

Ajikumar, P. K., Xiao, W. H., Tyo, K. E., Wang, Y., Simeon, F., Leonard, E., et al. (2010). Isoprenoid pathway optimization for Taxol precursor overproduction in *Escherichia coli*. *Science*, *330*, 70–74.

Alpert, A. J. (1990). Hydrophilic-interaction chromatography for the separation of peptides, nucleic acids and other polar compounds. *Journal of Chromatography A*, *499*, 177–196.

Armbruster, D. A., & Pry, T. (2008). Limit of blank, limit of detection and limit of quantitation. *Clinical Biochemistry Reviews*, *29*(Suppl 1), S49–S52.

Arrivault, S., Guenther, M., Ivakov, A., Feil, R., Vosloh, D., van Dongen, J. T., et al. (2009). Use of reverse-phase liquid chromatography, linked to tandem mass spectrometry, to profile the Calvin cycle and other metabolic intermediates in Arabidopsis rosettes at different carbon dioxide concentrations. *Plant Journal*, *59*, 826–839.

Asakawa, Y., Tokida, N., Ozawa, C., Ishiba, M., Tagaya, O., & Asakawa, N. (2008). Suppression effects of carbonate on the interaction between stainless steel and phosphate groups of phosphate compounds in high-performance liquid chromatography and electrospray ionization mass spectrometry. *Journal of Chromatography A*, *1198–1199*, 80–86.

Baidoo, E. E., Xiao, Y., Dehesh, K., & Keasling, J. D. (2014). Metabolite profiling of plastidial deoxyxylulose-5-phosphate pathway intermediates by liquid chromatography and mass spectrometry. *Methods in Molecular Biology*, *1153*, 57–76.

Bakhtiar, R., & Majumdar, T. K. (2007). Tracking problems and possible solutions in the quantitative determination of small molecule drugs and metabolites in biological fluids using liquid chromatography-mass spectrometry. *Journal of Pharmacological and Toxicological Methods*, *55*, 262–278.

Balcke, G. U., Bennewitz, S., Zabel, S., & Tissier, A. (2014). Isoprenoid and metabolite profiling of plant trichomes. *Methods in Molecular Biology*, *1153*, 189–202.

Birkemeyer, C., Luedemann, A., Wagner, C., Erban, A., & Kopka, J. (2005). Metabolome analysis: The potential of in vivo labeling with stable isotopes for metabolite profiling. *Trends in Biotechnology*, *23*, 28–33.

Boersema, P. J., Mohammed, S., & Heck, A. J. (2008). Hydrophilic interaction liquid chromatography (HILIC) in proteomics. *Analytical and Bioanalytical Chemistry*, *391*, 151–159.

Bouvier, F., Rahier, A., & Camara, B. (2005). Biogenesis, molecular regulation and function of plant isoprenoids. *Progress in Lipid Research*, *44*, 357–429.

Buszewski, B., & Noga, S. (2012). Hydrophilic interaction liquid chromatography (HILIC)—A powerful separation technique. *Analytical and Bioanalytical Chemistry*, *402*, 231–247.

Cervin, M.A., Whited, G.M., Chotani, G.K., Valle, F., Fioresi, C., Sanford, K.J., et al. (2009). Compositions and methods for producing isoprene. United States patent application. US 2009/0203102 A1.

Cordoba, E., Salmi, M., & Leon, P. (2009). Unravelling the regulatory mechanisms that modulate the MEP pathway in higher plants. *Journal of Experimental Botany*, *60*, 2933–2943.

Croteau, R., Kutchan, T. M., & Lewis, N. G. (2000). Natural products (secondary metabolites). In W. Gruissem, B. Buchanan, & R. Jones (Eds.), *Biochemistry and molecular biology of plants* (pp. 1250–1268). Rockville, MD: American Society of Plant Biologists.

Eisenreich, W., Schwarz, M., Cartayrade, A., Arigoni, D., Zenk, M. H., & Bacher, A. (1998). The deoxyxylulose phosphate pathway of terpenoid biosynthesis in plants and microorganisms. *Chemical Biology*, *5*, R221–R233.

Flores-Perez, U., Sauret-Gueto, S., Gas, E., Jarvis, P., & Rodriguez-Concepcion, M. (2008). A mutant impaired in the production of plastome-encoded proteins uncovers a mechanism for the homeostasis of isoprenoid biosynthetic enzymes in Arabidopsis plastids. *Plant Cell*, *20*, 1303–1315.

Freitas, L. G., Gotz, C. W., Ruff, M., Singer, H. P., & Muller, S. R. (2004). Quantification of the new triketone herbicides, sulcotrione and mesotrione, and other important herbicides and metabolites, at the ng/l level in surface waters using liquid chromatography-tandem mass spectrometry. *Journal of Chromatography A*, *1028*, 277–286.

Ge, X., d'Avignon, D. A., Ackerman, J. J. H., & Sammons, R. D. (2012). Observation and identification of 2-C-methyl-d-erythritol-2,4-cyclopyrophosphate in horseweed and ryegrass treated with glyphosate. *Pesticide Biochemistry and Physiology*, *104*, 187–191.

Gershenzon, J., & Dudareva, N. (2007). The function of terpene natural products in the natural world. *Nature Chemical Biology*, *3*, 408–414.

Ghirardo, A., Wright, L. P., Bi, Z., Rosenkranz, M., Pulido, P., Rodriguez-Concepcion, M., et al. (2014). Metabolic flux analysis of plastidic isoprenoid biosynthesis in poplar leaves emitting and nonemitting isoprene. *Plant Physiology*, *165*, 37–51.

Gonzalez-Cabanelas, D., Wright, L. P., Paetz, C., Onkokesung, N., Gershenzon, J., Rodriguez-Concepcion, M., et al. (2015). The diversion of 2-C-methyl-D-erythritol-2,4-cyclodiphosphate from the 2-C-methyl-D-erythritol 4-phosphate pathway to hemiterpene glycosides mediates stress responses in Arabidopsis thaliana. *Plant Journal*, *82*, 122–137.

Gritti, F., Kazakevich, Y. V., & Guiochon, G. (2007). Effect of the surface coverage of endcapped C18-silica on the excess adsorption isotherms of commonly used organic solvents from water in reversed phase liquid chromatography. *Journal of Chromatography A*, *1169*, 111–124.

Guevara-Garcia, A. A., San Roman, C., Arroyo, A., Cortes, M. E., Gutierrez-Nava, M. L., & Leon, P. (2005). The characterization of the Arabidopsis clb6 mutant illustrates the importance of post-transcriptional regulation of the methyl-D-erythritol 4-phosphate pathway. *Plant Cell*, *17*, 628–643.

Guo, Y., & Gaiki, S. (2005). Retention behavior of small polar compounds on polar stationary phases in hydrophilic interaction chromatography. *Journal of Chromatography A*, *1074*, 71–80.

Hecht, S., Kis, K., Eisenreich, W., Amslinger, S., Wungsintaweekul, J., Herz, S., et al. (2001). Enzyme-assisted preparation of isotope-labeled 1-deoxy-d-xylulose 5-phosphate. *Journal of Organic Chemistry*, *66*, 3948–3952.

Hunter, W. N. (2007). The non-mevalonate pathway of isoprenoid precursor biosynthesis. *Journal of Biological Chemistry*, *282*, 21573–21577.

Ikegami, T., Tomomatsu, K., Takubo, H., Horie, K., & Tanaka, N. (2008). Separation efficiencies in hydrophilic interaction chromatography. *Journal of Chromatography A*, *1184*, 474–503.

Illarionova, V., Kaiser, J., Ostrozhenkova, E., Bacher, A., Fischer, M., Eisenreich, W., et al. (2006). Nonmevalonate terpene biosynthesis enzymes as antiinfective drug targets: Substrate synthesis and high-throughput screening methods. *Journal of Organic Chemistry*, *71*, 8824–8834.

Klein-Marcuschamer, D., Ajikumar, P. K., & Stephanopoulos, G. (2007). Engineering microbial cell factories for biosynthesis of isoprenoid molecules: Beyond lycopene. *Trends in Biotechnology*, *25*, 417–424.

Kobayashi, K., Suzuki, M., Tang, J., Nagata, N., Ohyama, K., Seki, H., et al. (2007). Lovastatin insensitive 1, a Novel pentatricopeptide repeat protein, is a potential regulatory factor of isoprenoid biosynthesis in Arabidopsis. *Plant and Cell Physiology*, *48*, 322–331.

Li, Z., & Sharkey, T. D. (2013). Metabolic profiling of the methylerythritol phosphate pathway reveals the source of post-illumination isoprene burst from leaves. *Plant, Cell and Environment*, *36*, 429–437.

Lichtenthaler, H. K., Rohmer, M., & Schwender, J. (1997). Two independent biochemical pathways for isopentenyl diphosphate and isoprenoid biosynthesis in higher plants. *Physiologia Plantarum*, *101*, 643–652.

Lunn, J. E., Feil, R., Hendriks, J. H., Gibon, Y., Morcuende, R., Osuna, D., et al. (2006). Sugar-induced increases in trehalose 6-phosphate are correlated with redox activation of ADPglucose pyrophosphorylase and higher rates of starch synthesis in Arabidopsis thaliana. *Biochemistry Journal*, *397*, 139–148.

Luo, B., Groenke, K., Takors, R., Wandrey, C., & Oldiges, M. (2007). Simultaneous determination of multiple intracellular metabolites in glycolysis, pentose phosphate pathway and tricarboxylic acid cycle by liquid chromatography-mass spectrometry. *Journal of Chromatography A*, *1147*, 153–164.

Mateos-Vivas, M., Rodriguez-Gonzalo, E., Garcia-Gomez, D., & Carabias-Martinez, R. (2015). Hydrophilic interaction chromatography coupled to tandem mass spectrometry in the presence of hydrophilic ion-pairing reagents for the separation of nucleosides and nucleotide mono-, di- and triphosphates. *Journal of Chromatography A*, *1414*, 129–137.

McGarvey, D. J., & Croteau, R. (1995). Terpenoid metabolism. *Plant Cell*, 7, 1015–1026.

Naidong, W. (2003). Bioanalytical liquid chromatography tandem mass spectrometry methods on underivatized silica columns with aqueous/organic mobile phases. *Journal of Chromatography B*, *796*, 209–224.

Olsen, B. A. (2001). Hydrophilic interaction chromatography using amino and silica columns for the determination of polar pharmaceuticals and impurities. *Journal of Chromatography A*, *913*, 113–122.

Oyler, A. R., Armstrong, B. L., Cha, J. Y., Zhou, M. X., Yang, Q., Robinson, R. I., et al. (1996). Hydrophilic interaction chromatography on amino-silica phases complements reversed-phase high-performance chromatography and capillary electrophoresis for peptide analysis. *Journal of Chromatography A*, *724*, 378–383.

Perez-Gil, J., Uros, E. M., Sauret-Gueto, S., Lois, L. M., Kirby, J., Nishimoto, M., et al. (2012). Mutations in *Escherichia coli* aceE and ribB genes allow survival of strains defective in the first step of the isoprenoid biosynthesis pathway. *PLoS One*, 7(8), e43775. http://dx.doi.org/10.1371/journal.pone.0043775.

Phillips, M. A., Leon, P., Boronat, A., & Rodriguez-Concepcion, M. (2008). The plastidial MEP pathway: Unified nomenclature and resources. *Trends in Plant Science*, *13*, 619–623.

Preinerstorfer, B., Schiesel, S., Lammerhofer, M., & Lindner, W. (2010). Metabolic profiling of intracellular metabolites in fermentation broths from beta-lactam antibiotics production by liquid chromatography-tandem mass spectrometry methods. *Journal of Chromatography A*, *1217*, 312–328.

Pulido, P., Toledo-Ortiz, G., Phillips, M. A., Wright, L. P., & Rodriguez-Concepcion, M. (2013). Arabidopsis J-protein J20 delivers the first enzyme of the plastidial isoprenoid pathway to protein quality control. *Plant Cell*, *25*, 4183–4194.

Rabin, S., Stillian, J., Barreto, V., Friedman, K., & Toofan, M. (1993). New membrane-based electrolytic suppressor device for suppressed conductivity detection in ion chromatography. *Journal of Chromatography*, *640*, 97–109.

Rivasseau, C., Seemann, M., Boisson, A. M., Streb, P., Gout, E., Douce, R., et al. (2009). Accumulation of 2-C-methyl-D-erythritol 2,4-cyclodiphosphate in illuminated plant leaves at supraoptimal temperatures reveals a bottleneck of the prokaryotic methylerythritol 4-phosphate pathway of isoprenoid biosynthesis. *Plant, Cell and Environment*, *32*, 82–92.

Rodriguez-Concepcion, M. (2006). Early steps in isoprenoid biosynthesis: Multilevel regulation of the supply of common precursors in plant cells. *Phytochemistry Reviews*, *5*, 1–15.

Rodriguez-Concepcion, M., & Boronat, A. (2002). Elucidation of the methylerythritol phosphate pathway for isoprenoid biosynthesis in bacteria and plastids. A metabolic milestone achieved through genomics. *Plant Physiology*, *130*, 1079–1089.

Rohmer, M. (1999). The discovery of a mevalonate-independent pathway for isoprenoid biosynthesis in bacteria, algae and higher plants. *Natural Product Reports*, *16*, 565–574.

Rohmer, M., Knani, M., Simonin, P., Sutter, B., & Sahm, H. (1993). Isoprenoid biosynthesis in bacteria: A novel pathway for the early steps leading to isopentenyl diphosphate. *Biochemistry Journal, 295*(Pt 2), 517–524.

Rohmer, M., Seemann, M., Horbach, S., BringerMeyer, S., & Sahm, H. (1996). Glyceraldehyde 3-phosphate and pyruvate as precursors of isoprenic units in an alternative non-mevalonate pathway for terpenoid biosynthesis. *Journal of the American Chemical Society, 118*, 2564–2566.

Rutters, H., Mohring, T., Rullkotter, J., Griep-Raming, J., & Metyger, J. O. (2000). The persistent memory effect of triethylamine in the analysis of phospholipids by liquid chromatography/mass spectrometry. *Rapid Communications in Mass Spectrometry, 14*, 122–123.

Saladie, M., Wright, L. P., Garcia-Mas, J., Rodriguez-Concepcion, M., & Phillips, M. A. (2014). The 2-C-methylerythritol 4-phosphate pathway in melon is regulated by specialized isoforms for the first and last steps. *Journal of Experimental Botany, 65*, 5077–5092.

Schuhr, C. A., Hecht, S., Kis, K., Eisenreich, W., Wungsintaweekul, J., Bacher, A., et al. (2001). Studies on the non-mevalonate pathway—Preparation and properties of isotope-labeled 2C-methyl-D-erythritol 2,4-cyclodiphosphate. *European Journal of Organic Chemistry, 2001*, 3221–3226.

Steinbüchel, A. (2003). Production of rubber-like polymers by microorganisms. *Current Opinion in Microbiology, 6*, 261–270.

Stokvis, E., Rosing, H., & Beijnen, J. H. (2005). Stable isotopically labeled internal standards in quantitative bioanalysis using liquid chromatography/mass spectrometry: Necessity or not? *Rapid Communications in Mass Spectrometry, 19*, 401–407.

Tang, L., & Kebarle, P. (1993). Dependence of ion intensity in electrospray mass-spectrometry on the concentration of the analytes in the electrosprayed solution. *Analytical Chemistry, 65*, 3654–3668.

Testa, C. A., & Brown, M. J. (2003). The methylerythritol phosphate pathway and its significance as a novel drug target. *Current Pharmaceutical Biotechnology, 4*, 248–259.

Tippmann, S., Chen, Y., Siewers, V., & Nielsen, J. (2013). From flavors and pharmaceuticals to advanced biofuels: Production of isoprenoids in Saccharomyces cerevisiae. *Biotechnology Journal, 8*, 1435–1444.

Tsuruta, H., Paddon, C. J., Eng, D., Lenihan, J. R., Horning, T., Anthony, L. C., et al. (2009). Precursor of the antimalarial agent artemisinin, in *Escherichia coli*. *PLoS One, 4*, e4489.

Tuytten, R., Lemiere, F., Witters, E., Van Dongen, W., Slegers, H., Newton, R. P., et al. (2006). Stainless steel electrospray probe: A dead end for phosphorylated organic compounds? *Journal of Chromatography A, 1104*, 209–221.

van Deemter, J. J., Zuiderweg, F. J., & Klinkenberg, A. (1956). Longitudinal diffusion and resistance to mass transfer as causes of nonideality in chromatography. *Chemical Engineering Science, 5*, 271–289.

Van Eeckhaut, A., Lanckmans, K., Sarre, S., Smolders, I., & Michotte, Y. (2009). Validation of bioanalytical LC-MS/MS assays: Evaluation of matrix effects. *Journal of Chromatography B, 877*, 2198–2207.

Ward, J. L., Baker, J. M., Llewellyn, A. M., Hawkins, N. D., & Beale, M. H. (2011). Metabolomic analysis of Arabidopsis reveals hemiterpenoid glycosides as products of a nitrate ion-regulated, carbon flux overflow. *Proceedings of the National Academy of Science of the United States America, 108*, 10762–10767.

Wright, L. P., & Phillips, M. A. (2014). Measuring the activity of 1-deoxy-D-xylulose 5-phosphate synthase, the first enzyme in the MEP pathway, in plant extracts. *Methods in Molecular Biology, 1153*, 9–20.

Wright, L. P., Rohwer, J. M., Ghirardo, A., Hammerbacher, A., Ortiz-Alcaide, M., Raguschke, B., et al. (2014). Deoxyxylulose 5-phosphate synthase controls flux through the methylerythritol 4-phosphate pathway in Arabidopsis. *Plant Physiology*, *165*, 1488–1504.

Xiao, Y. M., Savchenko, T., Baidoo, E. E. K., Chehab, W. E., Hayden, D. M., Tolstikov, V., et al. (2012). Retrograde signaling by the plastidial metabolite MEcPP regulates expression of nuclear stress-response genes. *Cell*, *149*, 1525–1535.

Yu, X. H., & Liu, C. J. (2006). Development of an analytical method for genome-wide functional identification of plant acyl-coenzyme A-dependent acyltransferases. *Analytical Biochemistry*, *358*, 146–148.

Zhang, B., Watts, K. M., Hodge, D., Kemp, L. M., Hunstad, D. A., Hicks, L. M., et al. (2011). A second target of the antimalarial and antibacterial agent fosmidomycin revealed by cellular metabolic profiling. *Biochemistry*, *50*, 3570–3577.

Zhou, K., Zou, R., Stephanopoulos, G., & Too, H. P. (2012). Metabolite profiling identified methylerythritol cyclodiphosphate efflux as a limiting step in microbial isoprenoid production. *PLoS One*, 7, e47513.

CHAPTER ELEVEN

Establishing the Architecture of Plant Gene Regulatory Networks

F. Yang*[,†], W.Z. Ouma*[,†], W. Li*[,‡], A.I. Doseff*[,‡], E. Grotewold*[,†,1]
*The Ohio State University, Columbus, OH, United States
[†]Center for Applied Sciences (CAPS), The Ohio State University, Columbus, OH, United States
[‡]Heart and Lung Research Institute, The Ohio State University, Columbus, OH, United States
[1]Corresponding author: e-mail address: grotewold.1@osu.edu

Contents

1. Introduction 252
2. The *cis*-Regulatory Apparatus 254
   2.1 Establishing TSSs 254
   2.2 Promoters and Enhancers 257
   2.3 *cis*-Regulatory Variation 259
3. The *Trans*-Acting Factors 260
   3.1 Transcription Factors 261
   3.2 Cofactors 266
4. Transcription Factor Centered Approaches 266
   4.1 ChIP Approaches 266
   4.2 In Vitro TF–DNA Interaction Approaches 269
   4.3 In Silico TF-Location Prediction Approaches 273
5. Gene-Centered Approaches 274
   5.1 Y1H Approaches 274
   5.2 Electrophoretic Mobility Shift Assay (EMSA) Based Methods 279
6. Resources for Studying Plant GRNs 281
   6.1 Plant Transcription Factor ORFeome Collections (TFomes) 281
   6.2 Databases 283
7. Conclusions 286
Acknowledgment 287
References 287

Abstract

Gene regulatory grids (GRGs) encompass the space of all the possible transcription factor (TF)–target gene interactions that regulate gene expression, with gene regulatory networks (GRNs) representing a temporal and spatial manifestation of a portion of the GRG, essential for the specification of gene expression. Thus, understanding GRG architecture provides a valuable tool to explain how genes are expressed in an organism, an important aspect of synthetic biology and essential toward the development of the "in silico" cell. Progress has been made in some unicellular model systems (eg, yeast),

Methods in Enzymology, Volume 576
ISSN 0076-6879
http://dx.doi.org/10.1016/bs.mie.2016.03.003

but significant challenges remain in more complex multicellular organisms such as plants. Key to understanding the organization of GRGs is therefore identifying the genes that TFs bind to, and control. The application of sensitive and high-throughput methods to investigate genome-wide TF–target gene interactions is providing a wealth of information that can be linked to important agronomic traits. We describe here the methods and resources that have been developed to investigate the architecture of plant GRGs and GRNs. We also provide information regarding where to obtain clones or other resources necessary for synthetic biology or metabolic engineering.

1. INTRODUCTION

Control of gene expression is central to all cellular processes. Gene expression can be controlled at multiple levels, including RNA processing, mRNA transport, stability, and translation, as well as protein modification and stability. Transcription is a highly regulated process controlled in large part by transcription factors (TFs), proteins that specify when and how eukaryotic genes are expressed. TFs are operationally defined here as proteins that bind to DNA in a sequence-specific fashion to small motifs that function as *cis*-regulatory element (CRE, often 6–8 bps long). CREs are often enriched at regions just upstream of transcription start sites (TSSs) but can be also located anywhere in the gene, or even function at long distances when forming part of enhancers (ENCODE Project Consortium, 2012). However, the interaction of a single TF with a CRE is insufficient to provide the affinity and specificity necessary for the exquisite gene control found in vivo. Thus, complexes of TFs and coactivator/repressor proteins are recruited to specific *cis*-regulatory modules (CRMs) in the DNA regulatory region of genes. CRMs are usually formed by multiple CREs, and each CRM is responsible for a specific gene expression output (eg, temporal or spatial). Overall gene expression is thus provided by the combined action of all the modules, which may operate through defined biological steps (Yuh, Bolouri, & Davidson, 1998, 2001).

TFs are often modular in structure, combining the DNA-binding domain (DBD) with activation or repression domains that interact with components of the basal transcriptional machinery to modulate the activity of RNA Polymerase II (RNP-II). Protein–protein interactions and other modifications play key roles in the in vivo interaction of TFs with DNA, resulting in the specificity that TFs display in vivo. Superimposed on the activity of TFs is chromatin structure, which can affect the accessibility of the DNA to TFs, or to components of the basal transcriptional machinery, and histone modifications play a central role in regulation of gene expression (Mellor, Dudek, & Clynes, 2008; Peterson & Laniel, 2004).

TFs often control the expression of genes encoding other TFs, resulting in the formation of complex interaction webs. We previously defined gene regulatory grids (GRGs) as the space of all the possible TF–target gene interactions, with gene regulatory networks (GRNs) representing a temporal and spatial manifestation of a portion of the GRG (Mejia-Guerra, Pomeranz, Morohashi, & Grotewold, 2012). According to this definition, GRGs are static and provide an organismal blueprint for gene regulation, while GRNs are dynamic. Establishing GRGs requires a good understanding of the parts that conform the grid, including the CREs, the TFs, and their interactions. Placing CREs in a genic context entails empirically determining TSSs. Establishing GRGs can be approached from a gene-centered (ie, a gene is know, but not its regulators) or a TF-centered (the TF is known, but not its targets) perspective, each requiring a different toolbox and providing a different perspective of the GRG (Fig. 1).

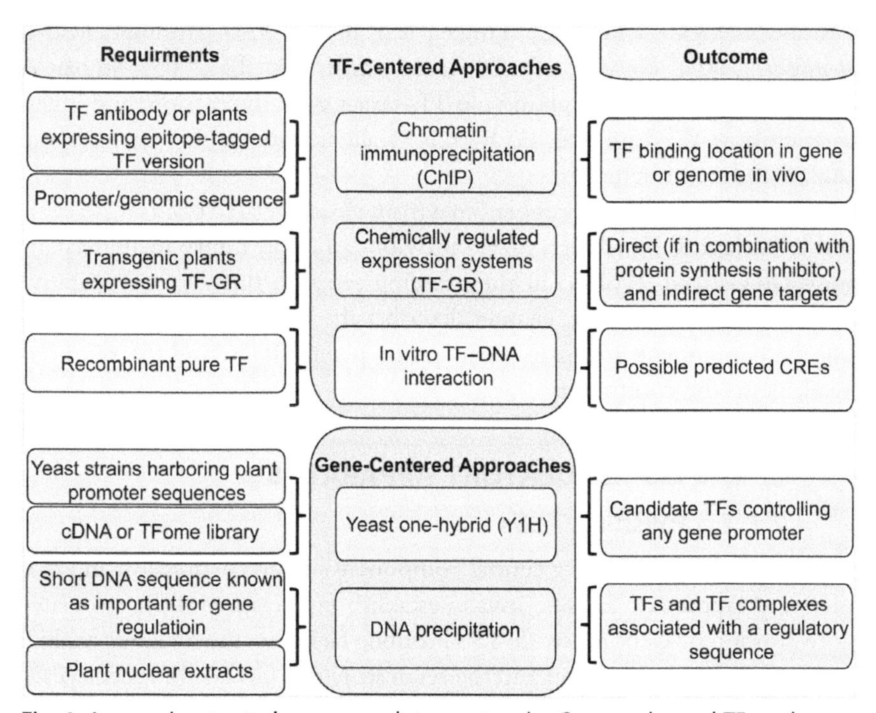

Fig. 1 Approaches to study gene regulatory networks. Commonly used TF- and gene-centered approaches to establish gene regulatory networks with their requirements and expected outcomes. (See the color plate.)

The evolution of morphological features in animals is frequently a consequence of distinctive changes in CREs and in the associated TFs (Davidson, 2001; Gibson & Honeycutt, 2002; Wray et al., 2003). Similarly, alterations in plant form are often associated with variations in CREs and TFs (Costa, Fox, Hanna, Baxter, & Coen, 2005; Doebley & Lukens, 1998), best illustrated by the molecular events associated with crop domestication (Doebley, Stec, & Hubbard, 1997; Li, Zhou, & Sang, 2006; White & Doebley, 1998, 1999). Of similar agronomic significance is the increasing awareness that many QTLs are associated with natural variation associated with the *cis*- or *trans*-acting regulatory apparatus (McMullen et al., 1998; Salvi et al., 2007). Studies have established the central role of TFs in providing tolerance to many abiotic stresses such as drought, salt, or cold (Kasuga, Liu, Miura, Yamaguchi-Shinozaki, & Shinozaki, 1999; Yamaguchi-Shinozaki & Shinozaki, 1994, 2005; Zhu, 2002). Heterosis, or hybrid vigor, representing the increased performance of a hybrid progeny when compared with the inbred parents, is likely to be related in part to allelic variations in gene expression (Birchler, Riddle, Auger, & Veitia, 2005; Brunner, Fengler, Morgante, Tingey, & Rafalski, 2005; Brunner, Pea, & Rafalski, 2005; Guo et al., 2004; Morgante et al., 2005; Stupar & Springer, 2006). Identifying in vivo TF–target gene interactions and investigating how they are affected by CRM allelic variants will increase our understanding of heterosis.

Finally, the sequencing of genomes from closely related plants (Cao et al., 2011; Weigel & Nordborg, 2015) provides an opportunity to understand how GRGs have evolved in the growing research thrust of comparative regulatory genomics. This chapter describes the methods used for identifying the components necessary for building plant GRGs, and the resources available to the community.

2. THE *CIS*-REGULATORY APPARATUS

2.1 Establishing TSSs

Transcription initiation is a central component of gene expression and regulation. Mapping of TSSs permits to establish the position of the associated core promoters. Genome-wide TSS profiling facilitates the dynamic analysis of transcriptional regulation and the reconstruction GRNs. In recent years, two 5′-cap selection techniques were developed for the efficient selection of the 5′-end of capped transcripts and applied to genome-wide TSS identification (Batut, Dobin, Plessy, Carninci, & Gingeras, 2013; Hashimoto et al.,

2004; Morton et al., 2014; Takahashi, Lassmann, Murata, & Carninci, 2012). Cap-trapper (Carninci et al., 1997) selectively captures 5′-capped RNA by chemical biotinylation of the 5′-cap and provides the basis for the cap analysis of gene expression (CAGE) and RNA annotation and mapping of promoters for analysis of gene expression (RAMPAGE) techniques (Batut & Gingeras, 2013). Oligo-capping (Maruyama & Sugano, 1994) replaces the mRNA cap structure with an oligoribonucleotide (r-oligo), and this provides the base for 5′-end serial analysis of gene expression (5′ SAGE) (Hashimoto et al., 2004) and paired-end analysis of transcription (PEAT) start sites (Ni et al., 2010). We describe these methods in some detail later.

2.1.1 Cap Analysis of Gene Expression

CAGE, a 5′ cap-trapper and deep sequencing-based technique, allows mapping of all the TSS of both capped coding and noncoding RNAs (Shiraki et al., 2003). CAGE has been utilized in genome-wide mapping of specific TSSs in eukaryotes (Hoskins et al., 2011; Kawaji et al., 2006; Mejia-Guerra et al., 2015; Ponjavic et al., 2006), revealing the existence of alternatively regulated TSSs (Carninci et al., 2006; Gustincich et al., 2006; Mejia-Guerra et al., 2015) and motifs associated with transcriptional initiation (Frith et al., 2008).

Detailed CAGE protocols have been developed (Takahashi, Kato, Murata, & Carninci, 2012). To generate a CAGE library, cDNA is first reverse transcribed using a random primer including the *Eco*P15I sequence with total RNA as template. Cap and 3′-ends are then biotinylated, and after RNase digestion of nonhybridized single-stranded RNA, 5′ complete cDNAs hybridized to biotinylated capped RNAs are captured by streptavidin-coated magnetic beads. Next, the cDNA is released from the RNA and ligated to a 5′-linker containing a barcode and *Eco*P15I sequence. The double-stranded 5′-linkers are then denatured to allow the biotin-modified second primer to anneal to the single-stranded cDNA, and prime second-strand cDNA synthesis. Subsequently, the cDNA is digested with *Eco*P15I, which cleaves 27 bp inside the 5′-end of the cDNA. Next, a 3′-linker containing the 3′ Illumina primer sequence is ligated to the 3′-end. The 96-bp CAGE tags are amplified with the primers, which are both compatible with the Illumina flow cell surface. Using this protocol, we recently carried out genome-wide mapping of TSSs in two widely used maize inbred lines and in two plant tissues, roots and shoots. These studies not only provided important information regarding maize core promoter architecture but also contributed significantly to improving maize genome annotation (Mejia-Guerra et al., 2015).

2.1.2 Other Methods

RAMPAGE is a cap-trapper method based on the synthesis of 5′-complete cDNAs from total RNA (Batut & Gingeras, 2013), thereby achieving increased accuracy in TSS mapping. This approach combines two orthogonal enrichment strategies, namely template switching (Hirzmann, Luo, Hahnen, & Hobom, 1993) and cap trapping (Carninci et al., 1996). Template switching makes use of unique properties of certain reverse-transcriptase enzymes to add adaptor sequences to the end of 5′-complete cDNAs. This method has so far been applied successfully to humans and *Drosophila* samples (Batut et al., 2013).

The other 5′-cap selection technique, the oligo-capping method, consists of three steps of enzymatic reactions (Bai et al., 2014; Maruyama & Sugano, 1994). First, the phosphate residue in the 5′-ends of truncated (eg, partially degraded) mRNAs that are noncapped is hydrolyzed by bacterial alkaline phosphatase. The cap structure remains intact during this reaction. Second, tobacco acid pyrophosphatase (TAP) cleaves the cap structure, leaving a phosphate at the 5′-end. Then, the synthetic r-oligo is ligated to the 5′-end phosphate by T4 RNA ligase. First-strand cDNA is synthesized using a random hexamer adaptor primer for 5′-end enriched cDNA. Primers that contain part of the adaptor primer sequence and the cap-replaced oligonucleotide sequence are then used to amplify the 5′-end enriched cDNA fragment.

In contrast to cap-trapper, RNA ligase may show some sequence preference (Hestand et al., 2010). Both 5′ SAGE (Hashimoto et al., 2004) and PEAT (Ni et al., 2010) are based on this method, although PEAT captures not only 5′- but also 3′-ends of mRNAs. 5′ SAGE has so far been used to identify TSSs in humans (Hodo et al., 2010) and *Apis mellifera* (honeybees) (Zheng et al., 2011), but not in plants. PEAT, however, was applied to *Arabidopsis thaliana* (Morton et al., 2014). PEAT is a powerful tool for measuring coupled initiation–termination events, and thus used for defining gene borders. However, it includes a large number of steps, comprising ligation to plasmids and a plasmid amplification step, which can result in both size and representational bias in comparison with CAGE (Ni et al., 2010; Takahashi, Lassmann, et al., 2012).

2.2 Promoters and Enhancers

The role of a gene promoter is to directly interact with basal transcription machinery components, including RNP-II and TFs to control DNA transcription (Smale & Kadonaga, 2003). Promoters are classified into core, proximal, and distal. Once TSSs have been determined, the core promoter, corresponding to DNA region in the immediate vicinity of the TSS, can be identified. The proximal promoter corresponds to the region immediately upstream of the core promoter that tends to contain primary regulatory elements bound by specific TFs. The distal promoter comprises sequences farther upstream of the core promoter.

Initiation of transcription often occurs at multiple TSSs, constituting a TSS cluster. Each TSS is linked to its own promoter, resulting in promoter clusters (Carninci et al., 2006). Although the CREs participating in core promoter function vary between genes, some core promoter general features are reasonably well established, primarily in yeast and metazoans. For example, metazoan promoters can be classified into three major types (Lenhard, Sandelin, & Carninci, 2012). Type I promoters display sharp TSS clusters, are enriched in TATA motifs, associate with disordered nucleosomes, and express in a tissue-specific pattern. Type II promoters display broad TSS clusters, are enriched in CpG islands, associate with ordered nucleosomes, and are expressed ubiquitously. Type III promoters usually have large CpG islands extending into the gene body, characterized by the recruitment of Polycomb repressive complexes and trimethylated histone 3 Lys 27 (H3K27m3) repressive marks, and correspond to genes that control multicellular development and differentiation (Juven-Gershon & Kadonaga, 2010; Lenhard et al., 2012). However, in plants, high-resolution genome-scale TSS information is only available for *Arabidopsis* (Morton et al., 2014) and maize (Mejia-Guerra et al., 2015). Plant and metazoan promoters share the characteristic that the TATA motif is enriched in sharp promoters but otherwise are largely different from the perspective of CREs (Mejia-Guerra et al., 2015).

In zebra fish, two different transcription initiation grammars set apart the transition from maternal to zygotic early embryonic development, highlighting the importance of genome-wide core promoter selection during developmental switches (Haberle et al., 2014). In mammals, more than half of the protein-coding transcriptional units had two or more alternative promoters (Carninci et al., 2006). We determined a similar situation in maize, where extensive alternative promoter usage was found between tissues or genotypes (Mejia-Guerra et al., 2015). Thus, it is important to remember that the TSS is a static feature, but rather is affected by a number of different factors. Furthermore, alternative promoter usage in some protein-coding genes generates alternative N-termini, impacting protein sequence and function (Carninci et al., 2006; Mejia-Guerra et al., 2015).

Unlike promoters, enhancers increase the rate of transcription independent of their position and orientation with respect to target genes (Bulger & Groudine, 2011). Enhancers act through a variety of TFs that ensure their correct match with target promoters and consequent gene activation (Erokhin, Vassetzky, Georgiev, & Chetverina, 2015). In multicellular organisms, enhancers are primarily responsible for the precise control of spatiotemporal patterns of gene expression (Kim & Shiekhattar, 2015). Historically, enhancers were experimentally identified by fusing the candidate DNA segment to a minimal promoter upstream of a reporter gene. In plants, a number of enhancers, often located in intergenic or intronic regions, have been identified (Clark, Wagler, Quijada, & Doebley, 2006; Krizek, 2015; McGarry & Ayre, 2008; Raatz et al., 2011; Schauer et al., 2009; Singer, Cox, & Liu, 2010; Stam et al., 2002; Yang et al., 2005). For instance, one of the first reported long-distance regulatory elements in plants is located ~100 kb upstream of the maize *b1* transcribed region (Stam et al., 2002), and the AINTEGUMENTA-LIKE6/PLETHORA3 (AIL6/PLT3) intronic sequences serve as transcriptional enhancer elements that confer expression in young flowers (Krizek, 2015). More recently, several different approaches have been combined for the unbiased discovery of tissue- and cell-specific enhancers (Bowman, 2015). Active enhancers generate a chromatin environment characterized by low nucleosome occupancy reflected in DNaseI hypersensitivity (DHS) (Gross & Garrard, 1988). Nucleosome destabilization at enhancer areas is facilitated by the binding of transcriptional coactivators and by the presence of a highly dynamic monomethylated histone H3 at lysine 4 (H3K4me1) and acetylated histone H3 at lysine 27 (H3K27ac) (Birney et al., 2007). Cell or tissue of interest isolated by laser capture or automated cell sorting can be used in a variety of assays to detect all of these molecular signatures and create a genome-wide map of candidate enhancers (Bowman, 2015). Results from such experiments have changed our views on the prevalence of enhancers. For example, it was predicted that the human genome contains approximately one million enhancers (ENCODE Project Consortium, 2012; Heintzman & Ren, 2009). In plants, enhancer prediction based exclusively on DHSs has been achieved but is so far restricted to *Arabidopsis* (Sullivan et al., 2014; Zhu, Zhang, Zhang,

Liu, & Jiang, 2015). The implementation of chromosome conformation capture (3C) permitted the in vivo identification of physical interactions between promoters and enhancers (Hovel, Louwers, & Stam, 2012; Louwers, Bader, et al., 2009; Louwers, Splinter, van Driel, de Laat, & Stam, 2009). Despite the recent progress, significant challenges remain in plant enhancer identification and elucidate the interaction mechanism between promoters and enhancers.

2.3 *cis*-Regulatory Variation

Genetic variation provides the basis for plant breeding. Natural variation in gene regulatory sequences is the major regulatory force behind expression differences in human and animals (Brem, Yvert, Clinton, & Kruglyak, 2002; Crowley et al., 2015; West et al., 2007; Yan, Yuan, Velculescu, Vogelstein, & Kinzler, 2002). In plants, the importance of *cis*-regulatory variation for genetic adaptation includes, for instance, polymorphisms in a *cis*-regulatory region of the *teosinte branched1* gene have been related with maize domestication (Wang, Stec, Hey, Lukens, & Doebley, 1999), a *cis*-regulatory element associated with maize flowering time QTL (Salvi et al., 2007), and heterosis (Guo et al., 2006, 2004).

However, different from the protein-coding lexicon, so far there is no regulatory lexicon that allows prediction of CREs important for gene expression. Allele-specific expression (ASE) provides an indirect measure for quantifying *cis*-regulatory effects by determining the relative proportions of alleles present in the transcript pool of hybrids (von Korff et al., 2009). Since both alleles in a hybrid are expressed in the same cell and are exposed to common regulatory factors, genes exhibiting asymmetric allele expression are likely controlled by *cis*-regulatory variation. Detection of ASE in hybrids offers the advantage that the two alleles are compared under identical conditions within a single individual genotype, providing an internal control for confounding factors such as differences in mRNA preparation and quality, and environmental and *trans*-acting factors (Stupar et al., 2008; von Korff et al., 2009). With a list of genes of interest and known single nucleotide polymorphisms, reverse-transcription PCR (RT-PCR) and genomic PCR analysis or single base primer extension assay for both parental and hybrid lines can be used to detect allelic variation (Guo et al., 2004; Zhuang & Adams, 2007). For the genome-wide profiling of allelic expression in hybrids, RNA-Seq or microarrays have been used (Stupar, Hermanson, & Springer, 2007; Stupar & Springer, 2006). In addition, allelic chromatin immunoprecipitation (ChIP) has been utilized to determine epigenetic marks and how TFs recognize CRE allelic variants (Bryzgalov et al., 2013; Chen et al., 2013; Ni, Hall, Battenhouse, & Iyer, 2012).

Previous studies quantifying *cis*-regulatory variation in plants have focused on outbreeding species such as poplar (Zhuang & Adams, 2007) and, in particular, maize, which has high levels of genetic diversity (Birchler, Yao, & Chudalayandi, 2006; Guo et al., 2006, 2004; Stupar & Springer, 2006). Approximately 80% of the differentially expressed maize genes display additive expression patterns in the hybrids relative to the inbred parents, indicating that intraspecific variation in gene expression levels is largely a consequence of *cis*-regulatory variation (Stupar & Springer, 2006). Allele-specific gene expression can be tissue specific (Guo et al., 2004; Stupar & Springer, 2006; Zhuang & Adams, 2007), can result in differential stress responses (Guo et al., 2004), or control different developmental processes (Adams & Wendel, 2005; Salvi et al., 2007). Knowledge of the frequency of *cis*-regulatory variation in crop plants has important implications for understanding how plants adapt to new environments and the molecular bases of heterosis. The identification of specific nucleotide changes that underlie differences in gene expression will also provide important tools for plant breeders to develop cultivars better adapted to changing climate conditions, and high yield hybrids.

3. THE *TRANS*-ACTING FACTORS

Control of gene transcription is an integrated mechanism involving *cis*-acting sequences and *trans*-acting factors. *Trans*-acting factors are proteins that bind to the *cis*-acting sequences in DNA to direct gene expression. They play important roles in regulating gene expression, such as chromatin remodeling and recruitment/stabilization of the RNP-II transcription–initiation complex. *Trans*-acting factors are classified into different functional groups. The first major class of *trans*-acting factors comprises TFs that can function as activators or repressors of transcription. The second class is formed by cofactors that include coactivators and corepressors, which usually do not bind to DNA directly, but mediate the transcriptional effects of TFs through protein–protein interactions. The third class corresponds to RNP-II subunits. Each of these subunits plays specific roles in TSS selection, alteration of elongation rates, interaction with activators, and stability of RNP-II. The fourth class corresponds to general transcription factors

(GTF), important components of the RNP-II transcription initiation complex. Different GTFs interact with the promoter at different regions, thereby exhibiting various functions (Karin, 1990; Latchman, 1997; Lee & Young, 2000; Nikolov & Burley, 1997; Roeder, 1996). Therefore, a TF controls gene transcription by interacting with components of the basal transcriptional machinery to facilitate or inhibit its association with the basal promoter, resulting in an increase or a decrease in overall transcription rates. Although several types of *trans*-acting factors exist, our discussion focuses on TFs and cofactors.

3.1 Transcription Factors

3.1.1 Functional Assays to Determine Transcriptional Regulatory Activity

TFs have a modular structure consisting of domains performing two or more functions, including DNA binding and transcription activation or repression. The DBD recognizes specific DNA target sequences. Plant TFs contain a variety of DNA-binding motifs allowing them to bind to specific DNA sequences. Most of these DBDs were originally identified in TFs from yeast or animals. In addition, TFs are often categorized into 50–60 classes (eg, MYB, bZIP, AP2/EREBP, WRKY, and NAC) based on structural features of the DBD, or of protein–protein interaction domains essential for TF function (Grotewold & Gray, 2009; Pabo & Sauer, 1992; Riechmann & Ratcliffe, 2000).

ChIP and yeast one-hybrid (Y1H) assays coupled to qPCR or deep sequencing are two major methods used to examine binding of TFs to target genes (see Sections 4.1 and 5.1). Alternatively, chimeric constructs containing TF fused to GAL4 activation domain ($GAL4^{AD}$) transformed in plant transient expression assay can be used to test TF binding to target gene promoter, helping reveal the promoter–DNA interaction (PDI) space. Transcriptional activation of the target gene by the chimeric protein provides evidence to support a particular PDI.

The transcription activation domain (TAD) interacts with other proteins, including coactivators and GTFs, to modulate transcription from the RNP-II complex bound to the core promoter. TADs are generally classified as acidic, proline, or glutamine-rich domains (Roberts, 2000). To determine whether TFs harbor a TAD, a GAL4 DNA-binding domain ($GAL4^{DBD}$) yeast transactivation assay is usually performed. $GAL4^{DBD}$-TF fused constructs are transformed into yeast strain containing a selectable gene. The positive growth of yeast cells on selective media indicates that the TF is capable of activating transcription of selective marker gene through its GAL4^{DBD} recruitment to the upstream activating sequence (UAS) containing GAL4-binding site (GAL4^{BS}). This demonstrates that the TF harbors a TAD active in yeast. A potential limitation of this approach is that TADs defined in yeast may not function in plants, since yeast recognizes only acidic TADs. Thus, this approach is usually followed by validation in plants.

One of the most common methods for characterizing transcriptional regulatory element activity in plants is the reporter gene assay. For this, the promoter region is cloned into a plasmid upstream of an easily detectable reporter, such as luciferase, β-galactosidase, or green fluorescent protein (GFP). The resulting construct is then cotransformed with a plasmid overexpressing a TF into protoplast or plant cells, and the activity of the reporter measured to determine if the TFs are capable of regulating transcription (Feller, Hernandez, & Grotewold, 2006; Hernandez, Feller, Morohashi, Frame, & Grotewold, 2007; Morohashi et al., 2012). Alternatively, the TF-GAL4^{DBD} construct is cotransformed with a reporter plasmid harboring GAL4^{BS} upstream of a minimal plant promoter (eg, from the constitutive viral *CaMV 35S*) to test for activation. To determine repression, full *CaMV 35S* promoter can be placed upstream of GAL4^{BS} driving luciferase. Establishing which region of the protein corresponds to the TAD often involves testing multiple mutation and deletion constructs (Ohta, Matsui, Hiratsu, Shinshi, & Ohme-Takagi, 2001; Sainz, Goff, & Chandler, 1997).

3.1.2 TFs Function as Activators and/or Repressors of Transcription

Through protein–protein interactions, TFs can activate or repress transcription (Glass, Rose, & Rosenfeld, 1997; Triezenberg, 1995), in a gene target-specific manner. Repressor domains often function by interacting with other TFs, such as basal TFs, coactivator proteins, or corepressor proteins (Hanna-Rose & Hansen, 1996; Licht, Ro, English, Grossel, & Hansen, 1993; Um, Li, & Manley, 1995). Other repressor domains function as histone-modifying factors that can participate in altering chromatin structure (Yang, Inouye, Zeng, Bearss, & Seto, 1996; Yang, Vickers, Brehm, Kouzarides, & Sharrocks, 2001). Among the best-characterized plant transcriptional repression motifs is the EAR domain (ERF-associated amphiphilic repression), associated with stress and defense functions (Kazan, 2006). The EAR domain is sufficient to convert activators into repressors (Hiratsu, Matsui, Koyama, & Ohme-Takagi, 2003), providing a powerful biotechnological tool. Indeed, the minimal functional region of the EAR domain of the SUPERMAN zinc-finger transcription factor, which has been optimized for specificity, if often used in what is known as the SRDX motif. This is a short 12 amino acid region with the sequence LDLDLELRLGFA (Hiratsu et al., 2003). Several TFs have been fused the SRDX, in a technique referred to as chimeric repressor silencing technology (CRES-T) (Hiratsu et al., 2003; Ohta et al., 2001). Another important plant transcriptional repression region consists of the BRD domain (B3 repression domain), which contains the R/KLFGV core motif (Ikeda & Ohme-Takagi, 2009).

3.1.3 Functional Methods to Identify TFs Controlling Agronomic Traits

A major long-term challenge in agriculture is to breed ideal crop varieties with enhanced agronomic traits, such as increased crop production and plant resistance to stress and pathogens. Most agronomic traits are quantitative and exhibit complex genetic control, resulting to slow progress in conventional crop improvement (Yang, Vanderbeld, Wan, & Huang, 2010). Although genetically engineered crops are available, they are mainly based on single-gene traits, such as expression of the Cry proteins derived from *Bacillus thuringiensis* for insect resistance (Romeis, Meissle, & Bigler, 2006). In order to improve crop agronomic traits efficiently, it is imperative to understand regulatory mechanisms involved in different plant responses, thereby enabling manipulation of key regulators.

TFs that are important for regulating plant responses to biotic and abiotic stresses, pathogens, changes in nitrate availability, plant metabolic pathway, and secondary wall formation (Doebley & Lukens, 1998; Gray, Caparros-Ruiz, & Grotewold, 2012; Rushton & Somssich, 1998; Singh, Foley, & Onate-Sanchez, 2002; Vidal, Alvarez, Moyano, & Gutierrez, 2015; Zhong & Ye, 2015). Thus, TFs are good candidates for manipulation because of their role as major regulators in controlling metabolic pathways and modifying complex traits in crop plants (Century, Reuber, & Ratcliffe, 2008; Grotewold, 2008). Members of large TF families, such as MYB, NAC, AP2/ERF, bZIP, and WRKY, are involved in regulating plant growth and development responses to various stresses, and many have been applied toward the improvement of plant agronomic traits (Mizoi, Shinozaki, & Yamaguchi-Shinozaki, 2012; Nakashima, Takasaki, Mizoi, Shinozaki, & Yamaguchi-Shinozaki, 2012; Rushton, Somssich, Ringler, & Shen, 2010) (Table 1). There is an increasing interest in employing TFs as tools for engineering the next generation of crops with increased yields, or more resistant to biotic/abiotic stress conditions (Grotewold, 2008). Natural variation, the analysis of mutants, and overexpression have been some of the main strategies

Table 1 Representative Examples of Genetic Engineering Using TFs to Improve Agronomic Traits

Gene	Gene Source	Transgenic Host	Approach	Agronomic Traits	References
OsWRK30	Rice (*Oryza sativa*)	Rice (*Oryza sativa*)	OE	Enhanced drought tolerance	Shen et al. (2012)
OsbZIP16	Rice (*Oryza sativa*)	Rice (*Oryza sativa*)	OE	Enhanced drought tolerance	Chen et al. (2012)
OsWRKY01	Rice (*Oryza sativa*)	Rice (*Oryza sativa*)	OE	Enhanced drought tolerance	Berri et al. (2009)
OsSNAC1	Rice (*Oryza sativa*)	Rice (*Oryza sativa*)	OE	Enhanced drought tolerance	Hu et al. (2006)
OsSNAC2	Rice (*Oryza sativa*)	Rice (*Oryza sativa*)	OE	Enhanced cold and salinity tolerance	Hu et al. (2008)
OsNAC5	Rice (*Oryza sativa*)	Rice (*Oryza sativa*)	OE	Enhanced drought and salinity tolerance	Takasaki et al. (2010)
OsbZIP23	Rice (*Oryza sativa*)	Rice (*Oryza sativa*)	OE	Enhanced drought and salinity tolerance	Xiang, Tang, Du, Ye, and Xiong (2008)
OsWRKY11	Rice (*Oryza sativa*)	Rice (*Oryza sativa*)	OE	Enhanced heat and drought tolerance	Wu, Shiroto, Kishitani, Ito, and Toriyama (2009)
OsMYB2	Rice (*Oryza sativa*)	Rice (*Oryza sativa*)	OE	Enhanced salt, cold, and drought tolerance	Yang, Dai, and Zhang (2012)

OsMYB55	Rice (*Oryza sativa*)	Rice (*Oryza sativa*)	OE	Enhanced heat stress tolerance	El-Kereamy et al. (2012)
AtMYB12	*Arabidopsis thaliana*	Tomato (*Solanum lycopersicum*)	OE	Enhanced phenylpropanoid production	Luo et al. (2008); Zhang et al. (2015)
SlMADS1	Tomato (*Solanum lycopersicum*)	Tomato (*Solanum lycopersicum*)	RNAi	Shorten ripening time	Dong et al. (2013)
SlERF5	Tomato (*Solanum lycopersicum*)	Tomato (*Solanum lycopersicum*)	OE	Enhanced drought and salinity tolerance	Pan et al. (2012)
StMYB1R-1	Potato (*Solanum tuberosum*)	Potato (*Solanum tuberosum*)	OE	Enhanced drought tolerance	Shin et al. (2011)
GhDREB	Cotton (*Gossypium hirsutum*)	Wheat (*Triticum aestivum* L.)	OE	Enhanced tolerance to drought, high salt, and freezing stresses	Gao et al. (2009)

OE, overexpression.

used in the past, but novel genome engineering tools including zinc-finger nucleases (ZFNs), transcription activator-like effector nucleases (TALENs), and more recently, clustered regularly interspaced short palindromic repeats (CRISPR) technologies provide powerful tools for TF function manipulation moving forward (Jirschitzka, Mattern, Gershenzon, & D'Auria, 2013; Rabara, Tripathi, & Rushton, 2014; Shan et al., 2013; Sornaraj, Luang, Lopato, & Hrmova, 2015; Tripathi, Rabara, & Rushton, 2014).

3.2 Cofactors

We define coactivators and corepressors as important players in the control of gene expression that are not necessarily involved in binding DNA, but which are tethered to regulatory complexes by interactions with TFs (Maston, Evans, & Green, 2006). An example is provided by the Mediator complex, which modulates all stages of transcription (initiation, elongation, and termination) by providing a bridge between TFs, GTFs, and RNP-II (Conaway & Conaway, 2011; Lee & Young, 2013). The plant Mediator complex has so far been characterized from *Arabidopsis* and rice (Backstrom, Elfving, Nilsson, Wingsle, & Bjorklund, 2007; Mathur, Vyas, Kapoor, & Tyagi, 2011), and Mediator components have been implicated in numerous plant biological processes, including flowering time control (Backstrom et al., 2007; Cerdan & Chory, 2003), pathogen resistance (Dhawan et al., 2009; Kidd et al., 2009), cold tolerance (Boyce et al., 2003), microRNA production (Kim et al., 2011), the regulation of cell division and patterning (Autran et al., 2002), and controlling lignin biosynthesis (Bonawitz et al., 2012). Cofactors and coactivators are often identified as part of phenotypic screens, or through protein–protein interaction experiments with TFs.

4. TRANSCRIPTION FACTOR CENTERED APPROACHES

4.1 ChIP Approaches

Establishing the architecture of plant GRNs and GRGs begins with the assembly of the network components that consist of nodes (for instance TFs, CREs, and target genes), followed by determination of regulatory relationships between components (Mejia-Guerra et al., 2012). Relationships between these components can be represented graphically as a directed network whose node interactions correspond to physical interactions determined by in vivo or in vitro experiments. One of the in vivo techniques,

ChIP, relies on the ability to immunoprecipitate (IP) protein–DNA complexes using an antibody specific to the TF of interest (Kuo & Allis, 1999). A variation of ChIP that enables determination of TF cobinding events, termed serial ChIP (sChIP), can be used to determine whether binding of two TFs to the same promoter is correlated (Kinoshita et al., 2004; Xie & Grotewold, 2008). Additionally, ChIP methods that make use of antibodies against certain histone protein modifications have been used to decipher epigenomic landscapes (Margueron, Trojer, & Reinberg, 2005; Pfluger & Wagner, 2007; Tariq & Paszkowski, 2004). However, determination of genome-wide TF-binding sites (TFBSs) using ChIP necessitated incorporation of additional high-throughput techniques, namely ChIP-Seq and ChIP-chip (Kim & Ren, 2006).

A general ChIP procedure begins with cross-linking the TF–DNA complexes in the harvested tissue using a fixative such as formaldehyde (Blecher-Gonen et al., 2013; Johnson, Mortazavi, Myers, & Wold, 2007; Kaufmann et al., 2010; Morohashi, Xie, & Grotewold, 2009). This is followed by nuclei isolation, IP using an antibody specific to the TF of interest, and isolation of the antibody–TF–DNA complexes. DNA is then recovered from the complex by reverse cross-linking and used for the generation of ChIP-Seq or ChIP-chip libraries. To determine whether there is significant enrichment, additional libraries are made in parallel to serve as negative controls. These libraries can be generated from one or more of the following: (i) plants that do not express the TF of interest; (ii) non-IPed genomic DNA, otherwise referred to as input; or (iii) genomic DNA obtained from mock IP using an idiotypic antibody such as IgG. In ChIP-chip, prepared libraries are hybridized to a microarray that has oligonucleotide probes representing part of or the entire genome. Arrays are then scanned to generate signal intensities, followed by normalization and comparison of intensities between the "ChIPed" and the control array. The goal of the analysis is to identify TFBSs, represented as "peaks" in the array signals. To this end, several statistical models for comparing "peak" signal intensities between the ChIPed and the control arrays have been developed (Buck, Nobel, & Lieb, 2005; Johnson et al., 2006; Zheng, Barrera, Ren, & Wu, 2007). In plants, this contributed to elucidation of histone modification in *Arabidopsis* seedlings and cell cultures (Benhamed et al., 2008; Bernatavichute, Zhang, Cokus, Pellegrini, & Jacobsen, 2008; Turck et al., 2007). Additionally, ChIP-chip has been applied in the construction of *Arabidopsis* GRNs (Lee et al., 2007; Morohashi & Grotewold, 2009; Oh et al., 2009).

ChIP-Seq, unlike ChIP-chip, furnishes large numbers of short sequence reads from next-generation sequencing (NGS) that are subsequently aligned to the reference genome for identification of TFBSs. Read alignment is a crucial step in analysis of ChIP-Seq data (Blecher-Gonen et al., 2013; Johnson et al., 2007; Kaufmann et al., 2010; Ouma et al., 2015). Failure to align large proportions of reads results in loss of meaningful biological information (Ouma et al., 2015). After read alignment, potential TFBSs can be identified by comparing the "ChIPed" to the control alignment profiles (He et al., 2015; Kaufmann et al., 2010). The result is genomic locations referred to as "peaks" indicating potential TFBSs. Several statistical and probabilistic models have been developed for comparison of ChIPed vs control alignment profiles for the identification of TFBSs (Cairns et al., 2011; He et al., 2015; Houlès, Rodier, Le Cam, Sardet, & Kirsh, 2015; Kaufmann et al., 2010; Mo, 2012; Muiño, Kaufmann, van Ham, Angenent, & Krajewski, 2011; Nikolayeva & Robinson, 2014; Shin, Liu, Duan, Zhang, & Liu, 2013; Spyrou, Stark, Lynch, & Tavaré, 2009; Zhang et al., 2008), designed to compare the number of uniquely aligned reads at each genomic location of the ChIPed and the control alignment profiles. More sophisticated statistical models that take into account reproducibility of ChIP signal enrichment have also been developed, for instance the irreproducibility discovery rate (IDR) proposed by ENCODE and modENCODE consortia (Landt et al., 2012). TF target genes are then identified from the vicinity of the TFBSs.

Unlike ChIP-chip, ChIP-Seq has been applied more extensively in identification of genome-wide TFBSs, largely due to its comparatively high resolution. In plants for instance, genome-wide TFBSs of APETALA1 were identified (Kaufmann et al., 2010). APETALA1 is a MADS-domain master regulator of *Arabidopsis* flowering time genes that orchestrates initiation of floral growth by integrating pathways involved in plant growth, patterning, and hormone regulation. DNA-binding landscape of another MADS TF, SEPALLATA3, was deciphered by both ChIP-chip and ChIP-Seq (Kaufmann et al., 2009). In addition, characterization of SEP3 TF target genes revealed its role in development of floral organs: regulation of growth and auxin pathways, either independently or combinatorially with other MADS-domain proteins. Similarly, ChIP-Seq was employed in identification of binding sites of another floral growth regulator, LEAFY (Moyroud et al., 2011). In maize, ChIP-Seq has contributed to the characterization of direct targets for several TFs (Bolduc et al., 2012; Eveland et al., 2014; Li et al., 2015).

As previously mentioned, the success with which reads are aligned to the reference genome impacts successful identification of TFBSs. Analyses of plant ChIP-Seq data from *Arabidopsis* and maize revealed that a large proportion (as high as 80%) of reads fails to align uniquely to the reference genome (Ouma et al., 2015), albeit using a popular short reads alignment software Bowtie (Langmead, Trapnell, Pop, & Salzberg, 2009). We found that this problem of a majority of ChIP-Seq reads not aligning was not limited to plant ChIP-Seq results, but common to other organism analyzed (Ouma et al., 2015). The observed "read unalignment" phenomenon prompted us to perform an in-depth investigation to uncover provenance of unaligned reads in ChIP-Seq datasets. A taxonomic classification of unaligned reads revealed two-pronged provenance of reads: (i) presence of sequences of bacterial and metazoan origin, likely contaminant sequences introduced during sample preparation stages; and (ii) presence of potentially legitimate reads belonging to the same taxonomic units as the source organisms. This intriguing observation of potentially legitimate reads led us to further investigate their provenance, by first determining whether they exhibited "alignable" properties. Interestingly, a large proportion (between 40% and 50% of unaligned reads) was realigned to their respective reference genomes using SHRiMP, a short read sequence alignment software capable of handling indels (Rumble et al., 2009). We further observed that the realigned reads harbor vital biological information (Ouma et al., 2015). Altogether, these observations suggest that unaligned reads from previous ChIP-Seq studies can still be recovered and used for enhancing construction of GRNs and GRGs.

4.2 In Vitro TF–DNA Interaction Approaches

4.2.1 SELEX and SELEX-Seq

Systematic evolution of ligands by exponential enrichment (SELEX) is an in vitro approach for determining binding affinities of proteins, peptides, drugs, and small molecules to either DNA or RNA (Ellington & Szostak, 1990; Tuerk & Gold, 1990; Yang, Yang, Schluesener, & Zhang, 2007). SELEX is an iterative screening process that selects RNA, single-stranded or double-stranded DNA (dsDNA) molecules from a large pool of oligonucleotides of variant sequences, followed by multiple rounds of enrichment of bound oligonucleotides. The iterative enrichment process includes elution of bound oligonucleotides and their subsequent amplification by PCR, resulting in a final product of nucleic acid aptamers. Oligonucleotides are designed in such a way that the variable region is of a constant length flanked by constant regions in the 5′- and 3′-ends. These flanking regions serve as primers for the amplification stage. Aptamers generated from SELEX can be sequenced and used to develop computational models, such as position-specific weight matrix (PWM), and subsequently employed in identification of TFBSs and their putative target genes, as done for a number of plant TFs (Chai, Xie, & Grotewold, 2011; Grotewold, Drummond, Bowen, & Peterson, 1994; Romero et al., 1998; Sainz, Grotewold, & Chandler, 1997; Williams & Grotewold, 1997). To improve affinity purification of the TF-oligonucleotide complexes, a SELEX variation was developed that included a TF fused to His_6-tagged cellulose D (CELD) (Xue, 2005). The CELD-based DNA-binding assay enabled quantitation of DNA-binding activity using CELD reporter-based assays. The technique was used to simultaneously identify DNA-binding sites of nine plant DNA-binding proteins belonging to the AP2, bHLH, NAC, and MYB TF super-families (Xue, 2005).

Several variations of the conventional SELEX now are available. Capillary electrophoresis SELEX (CE-SELEX) was developed to improve the efficiency of selection by reducing the number of iterations needed for selecting high-affinity aptamers (Dong et al., 2015; Eaton et al., 2015; Kasahara, Irisawa, Ozaki, Obika, & Kuwahara, 2013; Mosing & Bowser, 2009). Simply, the incubated oligonucleotide-target sample is injected through a free solution capillary electrophoresis. Separation of bound from unbound oligonucleotide is easily achieved since bound oligonucleotides travel at different velocities compared to unbound. Subsequent steps of PCR amplification are however similar to the conventional SELEX. The advantage of CE-SELEX over conventional SELEX was demonstrated by identifying high-affinity aptamers bound by IgE requiring just four rounds of selection, compared to 8–12 rounds required in a conventional SELEX (Mosing & Bowser, 2009). Other SELEX variations are (and not limited to) tailored (or primer-free) SELEX (Vater, Jarosch, Buchner, & Klussmann, 2003), toggle-SELEX (White et al., 2001), expression cassette SELEX (Martell, Nevins, & Sullenger, 2002), photo-SELEX (Golden, Collins, Willis, & Koch, 2000), cell-SELEX (Avci-Adali, Metzger, Perle, Ziemer, & Wendel, 2010; Paul, Avci-Adali, Ziemer, & Wendel, 2009), SELEX-SAGE (Roulet et al., 2002), and SELEX with NGS (SELEX-Seq) (Gu, Wang, Yang, Xu, & Wang, 2013; Jolma et al., 2010; Riley et al., 2014; Zykovich, Korf, & Segal, 2009).

With the exception of SELEX-SAGE, which is very labor intensive, one limitation of the aforementioned SELEX variations is the low-throughput.

SELEX coupled to next-generation DNA sequencing platforms like Illumina (SELEX-Seq) was subsequently developed. An improved multiplexed SELEX-Seq (Jolma et al., 2010) included barcoded oligonucleotides and affinity-tagged proteins that enabled generation of DNA-binding models (PWM) for 19 TFs in parallel. SELEX-Seq superiority over conventional SELEX methods in generating in vitro DNA-binding models is evident in its increased throughput, which is as much as 1000-fold higher when compared to SELEX-SAGE (Jolma et al., 2010; Riley et al., 2014).

4.2.2 Protein Binding Microarrays (PBMs)

PBMs, also known as dsDNA microarrays, are one of the high-throughput in vitro techniques employed in determining TF DNA-binding profiles (DBPs) using high-density dsDNA (Bulyk, Gentalen, Lockhart, & Church, 1999). PBMs provide DBPs used in construction of PWMs and subsequent identification of TFBSs and target genes. The first step in studying in vitro DBPs involves preparation of high-density dsDNA microarrays. First, complex probes are designed (Berger & Bulyk, 2006; Berger et al., 2006; Mintseris & Eisen, 2006; Philippakis, Qureshi, Berger, & Bulyk, 2008), followed by the manufacture of a high-density single-stranded DNA (ssDNA) microarray using any of the available platforms (eg, Agilent, Nimblegen, Affymetrix). The ssDNA microarray is then converted to dsDNA microarray by use of different techniques that include constant primer elongation (Bulyk et al., 1999), hairpin primer elongation (Mukherjee et al., 2004), and hairpin formation (Warren et al., 2006). The second step of determining DBPs PBMs involves the actual binding of an epitope-tagged TF of interest to the dsDNA microarray. Finally, the TF–dsDNA interactions are reported using a variety of ways, such as use of a fluoro-labeled antitag antibody (Berger & Bulyk, 2009). This is followed by fluoroscanning of dsDNA microarrays with a genechip scanner (Berger & Bulyk, 2009). An important consideration of the PBMs is to have dsDNA microarrays in replicates to confirm the reproducibility of the assays. To account for the influence of the background signal density of dsDNA probes on the signal of TF binding, an additional staining of dsDNA using a fluorophore such as SYBR green dye followed by scanning is carried out in parallel (Berger & Bulyk, 2006; Berger et al., 2006; Egener, Roulet, Zehnder, Bucher, & Mermod, 2005; Mukherjee et al., 2004). The background signal is then subtracted from the TF-binding signal, averaged (or median values obtained), and normalized (Berger & Bulyk, 2009; Berger et al., 2006). The resulting signal intensities are used in a wide range of bioinformatics analyses that include determination of the following: TF specificity and affinity on various DNA sequences, DBP motifs, interdependence of nucleotides in a binding site, and prediction TFBSs and target genes.

PBMs have been employed in the construction of plant GRNs and GRGs. For example, a PBM containing all possible 11-mer dsDNA probes (termed PBM11) was designed and used to determine binding specificities of *Arabidopsis* MYC2 and ERF1 (Godoy et al., 2011). In addition, presence of all possible combinations of 11-mer probes enabled identification of variants of G-box that are bound by MYC2 with medium and high affinity, while GCC variants were bound by ERF1, albeit with low affinity. This demonstrated that MYC2 is more "promiscuous" in binding compared to ERF1. Identifying relative binding affinities of a TF to all possible sequences of a given length is important for determining all functional TFBS, especially low-affinity sites, since both low- and high-affinity sites have been shown to be biologically relevant (Farley et al., 2015; Jiang & Levine, 1993; Tuupanen et al., 2009).

Similarly, a TF–target gene identification workflow was developed that determined binding specificities of 12 *Arabidopsis* NAC TFs using gapped and ungapped 8-mers (Lindemose et al., 2014). Clustering the 8-mers revealed three distinct classes of TFs based on their binding specificities that were comparable to the TFs phylogenetic relationship (Lindemose et al., 2014). As a result, potential TFBSs of the 12 NAC TFs were identified by mapping high-scoring 8-mers to promoters of *Arabidopsis* genes. A more detailed and high-throughput analysis of DNA-binding specificities 63 *Arabidopsis* TFs representing 23 families was performed (Franco-Zorrilla et al., 2014). Their PBM analysis revealed that about half of the TFs studied contained secondary DNA-interacting motifs.

4.2.3 Chemically Regulated Gene Expression Systems

Transgenic approaches of studying regulation of gene expression have traditionally relied on expression of transgenes driven by strong promoters, such as *CaMV 35S*. However, more flexible chemical inducible systems, usually containing nonplant sources, have been developed (Aoyama & Chau, 1997; Bohner, Lenk, Rieping, Herold, & Gatz, 1999; Gatz, 1997; Padidam, 2003; Zuo & Chua, 2000). Such systems are composed of two main units: (i) a transcriptional unit, usually a TF that responds to a chemical (ligand); and (ii) a response unit that facilitates binding of the expressed TF to the genes of interest (Padidam, 2003). The three main receptors used in the inducible systems are tetracyclin (TetR), estrogen (ER), and glucocorticoid (GR). The GR fused to plant TFs inducible system (TF-GR) is a powerful posttranscriptional regulatory tool that responds to dexamethasone (DEX) treatment, which binds to the GR transcriptional activator with high specificity (Evans, 1988). Expression of the TF–GR fusion protein can be controlled by the native TF promoter or by a constitutive promoter. In the absence of the inducer (DEX), the TF is prevented from entering the nucleus by HSP90 and therefore TF–GR fails to bind the target gene promoter. However, treatment with DEX frees the TF–GR fusion protein, enabling it to bind the target gene promoter, thereby initiating transcription (Gatz & Lenk, 1998). This approach however does not distinguish direct from indirect TF targets. Indirect targets can include TF target genes that code for TFs that are downstream regulators. Indirect regulatory effects can be ruled out by inhibiting translation of downstream regulatory targets using a translation inhibitor such as cycloheximide (Bargmann et al., 2013; Eklund et al., 2010). TF GR steroid-inducible systems have widely been used in plants to study TFs involved in nutrient signaling (Para et al., 2014), trichome development (Lloyd, Schena, Walbot, & Davis, 1994; Morohashi & Grotewold, 2009), floral initiation and development (Busch, Bomblies, & Weigel, 1999; Sablowski & Meyerowitz, 1998; Samach et al., 2000), and leaf cell fate determination (Aoyama et al., 1995).

4.3 In Silico TF-Location Prediction Approaches

Prediction of TFBSs using computational means is a fast but often imprecise way of inferring regulatory relationships. However, it can be helpful in cases where in vivo or in vitro approaches are not feasible, for instance in resource or time-limited scenarios. Presence of fully sequenced genomes and conservation of DNA sequences (motifs) bound by evolutionary-related TF, coupled with PWMs (position weight matrix), enables prediction of TFBSs based on sequence alone. A collection of available PWMs, usually obtained from in vitro techniques such as SELEX-Seq and PMBs (see Sections 4.2.1 and 4.2.2), can be obtained from databases such as TRANSFAC (Wingender et al., 2001) and JASPER (http://jaspar.genereg.net/). Examples of web-based software for TFBSs prediction relying on classical PWMs are PROMO (Farré et al., 2003), INSECT (Rohr, Parra, Yankilevich, & Perez-Castro, 2013), TESS (Stoeckert, Salas, Brunk, & Overton, 1999), MATRIX SEARCH (Chen, Hertz, & Stormo, 1995), SIGNAL SCAN (Prestridge, 1996), and Match™ (Kel et al., 2003). However, a limitation of classical TFBSs predictions algorithms is that the dependence on only PWMs fails to account for nucleotide position interdependencies that have been observed in biochemical assays (Berger & Bulyk, 2006; Bulyk, Johnson, & Church, 2002; Man & Stormo, 2001) and quantitative analysis of PBM data (Mukherjee et al., 2004; Weirauch et al., 2013; Zhao, Ruan, Pandey, & Stormo, 2012). Improved algorithms whose binding models incorporate dinucleotide dependencies, composition, and dinucleotide interactions between all possible pairwise positions have been shown to be predictively superior compared to classical PWM algorithms (Weirauch et al., 2013; Zhao et al., 2012). A further variation of TFBS prediction algorithms incorporates flexible statistical models that can predict binding sites of variable lengths. Such models include Bayesian hierarchical hidden Markov models (HMMs) and models based on Boltzmann chain (a generalization of HMMs) (Bulyk et al., 2002; Mathelier & Wasserman, 2013; Mehta, Schwab, & Sengupta, 2011; Moyroud et al., 2011; Wasson & Hartemink, 2009). Incorporation of TF structural information results in improvement of predictive power of algorithms. For example, a predictive biophysical model containing structural information was capable of detecting LEAFY binding sites in *Arabidopsis*, and LEAFY orthologs binding sites in maize, sorghum, rice, and *Brachypodium distachyon* (Moyroud et al., 2011). An experimental validation of the predicted TFBSs by ChIP-Seq revealed a high correlation between predicted and in vivo sites.

5. GENE-CENTERED APPROACHES

5.1 Y1H Approaches

The Y1H screening system is a powerful and efficient genetic method for a gene-centered approach to identify TF–target gene interactions, derived from the yeast two-hybrid technique (Fields & Song, 1989). It is based on genetic selection in yeast for the specific interaction between a TF (protein prey) and a DNA sequence (DNA bait) resulting in activation of a reporter gene for selection. A schematic overview of the workflow of the Y1H screening is presented in Fig. 2.

5.1.1 Principle and Work Flow of Y1H

The first step of Y1H involves cloning the DNA bait into a vector to drive the expression of selectable markers (eg, *LacZ*) and/or reporter genes (eg, *HIS3*) (Fields & Song, 1989). Traditionally, bait and prey constructs are generated by cloning with restriction enzymes and ligation, until

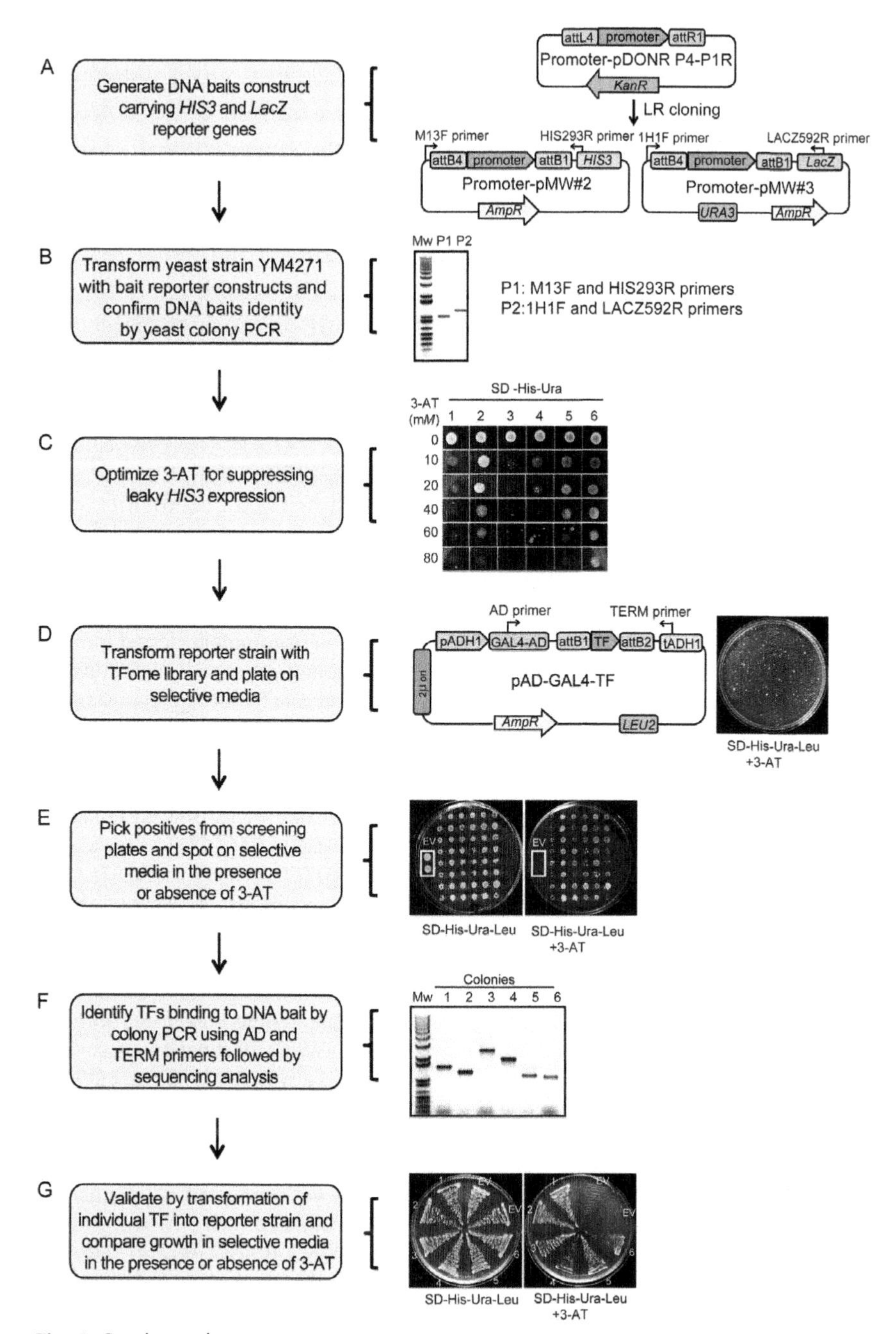

Fig. 2 See legend on next page.

recombination cloning was developed (Li & Shapiro, 1993). Recombination-based cloning systems, such as Gateway, enabled the high-throughput cloning of DNA baits and preys in a fraction of the time, allowing for the efficient establishment of gene-centered GRNs (Deplancke, Dupuy, Vidal, & Walhout, 2004).

To promote recombination between the reporter construct and the yeast genomic DNA, bait-reporter constructs are linearized and integrated into the yeast genome by individual transformation. The resulting transformed yeast cells harboring the integrated reporter construct are grown in selective media. To decrease false positives, two selective markers or reporter genes (eg, *HIS3* and *LacZ*) can be integrated into the same yeast genome at different locations. As an example, a DNA bait construct can harbor the promoter of interest driving *HIS3*. The linearized plasmid is integrated into *his3-200 locus* by homologous recombination, thereby allowing transformants to be selected on synthetic media lacking histidine by activation of *HIS3* reporter gene. If the screen relies also on the activation of *LacZ*, the construct

Fig. 2 Y1H flowchart to identify maize PDIs. (A) DNA bait promoters cloned into the pDONR P4-P1R entry vector were subsequently subcloned into the Y1H destination vectors pMW#2 and pMW#3 containing the *HIS3* and *LacZ* reporter genes by LR Gateway cloning. (B) Y1H strains were generated by genome integration of the destination constructs into the YM4271 yeast strain (*MATa, ura3-52, his3-D 200, ade2-101, ade5, lys2-801, leu2-3,112, trp1-901, tyr1-501, gal4 D, gal80 D, ade5::hisG*). Genomic integration was tested by colony PCR using the primer pair P1 for the *His* construct (M13F: 5′-GTAAAACGACGGCCAGT-3′ and HIS293R: 5′-GGGACCACCCTTTAAAGAGA-3′) and primer pair P2 for the *LacZ* construct (1H1F: 5′-GTTCGGAGATTACCGAATCAA-3′ and LACZ592R, 5′-ATGCGCTCAGGTCAAATTCAGA-3′) (Reece-Hoyes & Walhout, 2012). (C) The generated DNA bait yeast strains were tested for autoactivation by growing colonies onto synthetic defined (SD) media lacking His and Ura with increasing 3-AT concentrations, as shown for six independent colonies (#1–6). (D) Transformation of a yeast reporter strain with a $GAL4^{AD}$-TFome library. The TFome library (Burdo et al., 2014) comprises TFs ORFs cloned into the pAD-GAL4-GW-C1 destination vector and PDIs were selected by growth on SD–His–Ura–Leu media with 3-AT. (E) Selected colonies were picked and spotted onto SD–His–Ura–Leu and SD–His–Ura–Leu with 3-AT plates. Colonies representing the yeast strain used for the screen transformed with the pAD-GAL4-GW-C1 empty vector (EV) were used as a negative control (white squares). (F) The identification of the TFs binding to the DNA baits was conducted by yeast colony PCR of DNA isolated from colonies selected in SD–His–Ura–Leu with 3-AT plates using primers AD and TERM primers (AD: 5′-CGCGTTTGGAATCACTACAGGG-3′ and TERM: 5′-GGAGACTTGACCAAACCTCTGGCG-3′) (Walhout & Vidal, 2001) followed by DNA sequencing. (G) Validation of Y1H interactions was conducted by transforming individual TFs into the yeast strain used for the original screening followed by selection onto SD–His–SD–His–Ura–Leu + 3-AT. TFs #1, 2, 3, 4, 5, and 6 represent confirmed Promoter–TF interactions. EV was used as a negative control. (See the color plate.)

harboring promoter::*LacZ* contains, for example, the *URA3* marker and is integrated into the mutant *ura3-52* locus, resulting in the positive growth in synthetic media lacking uracil. Integrations of double selection constructs are generally performed sequentially. However, integrating both constructs simultaneously by cotransformation is feasible with a decrease in transformation efficiency. The obtained strains should be confirmed for correct integration of the reporter construct by colony PCR.

After identifying positive yeast strains, autoactivation tests are required to test background reporter gene expression. Background *HIS3* expression conferred by the bait DNA can enable growth on media lacking histidine, and therefore 3-amino-1,2,4-triazole (3-AT), a competitive inhibitor of the *HIS3* enzyme, is commonly applied (at concentrations ranging 5–100 m*M*) to the media when performing the screen. Thus, the growth of the yeast cells harboring a TF that recognizes promoter relies on the activation of *HIS3* expression that overcomes inhibitory 3-AT concentrations. If the transformed yeast strain continues to grow at high levels of 3-AT, shortening the DNA bait sequence and generating a new transformed strain are usually a convenient starting point.

The Y1H assay screening is done by transformation of the yeast strains with the DNA bait with a cDNA or TFome library (discussed in Section 5.1.2). The vector used for expressing the TF is usually designed to produce a translational fusion with a strong TAD (such as $GAL4^{AD}$). This converts most DNA-binding proteins into transcriptional activators. DNA-binding proteins that recognize the DNA bait will result in the activation of the selectable marker, and therefore growth in selective media. To determine the identity of the selected TFs, PCR can be performed on the colonies, followed by sequencing and BLAST analysis. Subsequently, promising clones are transformed to the bait-reporter strain individually to validate the interaction in yeast.

5.1.2 Comparing cDNA and TFome Libraries for Y1H

Libraries containing cDNA fused to the $GAL4^{AD}$ are commonly used for Y1H screens (Arda & Walhout, 2010; Deplancke et al., 2004; Deplancke, Mukhopadhyay, et al., 2006; Deplancke, Vermeirssen, Arda, Martinez, & Walhout, 2006; Martinez et al., 2008; Vermeirssen et al., 2007). However, there are two major limitations in using cDNA libraries for Y1H. First, it is often difficult to get full-length cDNA clones, due to limitations in processivity of the reverse-transcriptase enzyme. Second, TFs are general expressed at low levels and constitute ~7% of all coding genes

(Mejia-Guerra et al., 2012; Reece-Hoyes et al., 2005; Wilson et al., 2008), resulting in underrepresentation of TFs in the cDNA library, unless they are first normalized (Reece-Hoyes & Marian Walhout, 2012; Walhout, 2011).

A solution to the limitations of using cDNA libraries for gene-centered approaches is to generate TF open reading frame libraries (TFome), in which all clones encode full-length TFs fused to a TAD in destination vector (TAD-TFome) by Gateway cloning. Such TFome collections exist for plants, such as maize and *Arabidopsis* (Burdo et al., 2014; Pruneda-Paz et al., 2014), and can be obtained at the Arabidopsis Biological Resource Center (ABRC). Screening TFome libraries helps with the detection of TFs underrepresented in nonnormalized cDNA libraries and significantly enhances the establishment of GRNs (Taylor-Teeples et al., 2015). The generation of TFome libraries requires a complete TF inventory (Burdo et al., 2014; Pruneda-Paz et al., 2014). This raises a major problem, since transcriptional regulators lacking a typical DBD (Deplancke, Mukhopadhyay, et al., 2006; Hu et al., 2009) might be missed.

5.1.3 Comparison of Transformation and Mating Y1H Strategies

As described in Section 5.1.2, Y1H screens can be achieved by transforming the TAD-TFome library into a haploid bait yeast strain that harbors the bait constructs. Alternatively, individual TAD-TF clones can be introduced first into a haploid yeast strain of the opposite mating type by transformation, and then the bait and prey strains mated and positive interactions examined in the resulting diploids (Gaudinier et al., 2011; Vermeirssen et al., 2007). The mating method is generally used for array-based screens with robotic assistance for transferring of yeast between media plates, whereas direct transformation is mainly applied for library-based screens. The mating method is therefore faster and less labor-intensive because when an interaction is obtained, the corresponding TF is immediately identified. In contrast, in the case of the transformation method, the TF responsible for growth in the selective media needs to be identified either by isolating the plasmid and rescuing it in *E. coli*, followed by sequencing, or by direct yeast colony PCR followed by sequencing. In the presence of low copy Y1H prey expression vectors, more TFs are generally detected by transformation rather than by mating. However, the protein–DNA interactions are often more reproducible by mating than by transformation (Vermeirssen et al., 2007). Conversely, application of the high copy of the Y1H prey expression vector on mating detects most of these missing PDIs. The reason is not quite clear, possibly due to the increasing levels of the prey protein affecting diploid yeast in some aspect (Reece-Hoyes & Marian Walhout, 2012). An enhanced Y1H system was developed (Gaudinier et al., 2011; Reece-Hoyes & Marian Walhout, 2012) and recently successfully applied toward the identification of protein–DNA interactions associated with secondary cell wall formation in *A. thaliana* (Taylor-Teeples et al., 2015).

5.1.4 Limitations of Y1H-Based Screens

5.1.4.1 False Negatives

One of the limitations of the Y1H approach is the inability to uncover all possible TF–promoter interactions. Possible reasons include: (i) TFs that bind DNA as obligatory heterodimers with other TFs or cofactors cannot be retrieved, (ii) TFs that depend on specific posttranslational modifications absent in yeast might not be identified, and (iii) lowly abundant TFs in cDNA libraries are difficult to be recovered, although this can be solved by using TFome libraries (Arda & Walhout, 2010; Burdo et al., 2014; Pruneda-Paz et al., 2014; Walhout, 2011).

5.1.4.2 False Positives

Y1H may provide technical and biological false positives. Technical false positives may represent PDI that lack reproducibility, TFs that activate the reporters even while binding outside the promoter region or by mutations that arise in the bait strain during screening (Walhout & Vidal, 1999). Technical false positives can be reduced by using more stringent screening conditions, using freshly grown yeast DNA bait strains, by validating of all PDI identified, and by using multiple reporter assays such as combination of growth in the absence of histidine and ability to induce β-galactosidase. TFs retrieved from cDNA library screen are required to be in frame. Interactions identified at high 3-AT concentrations may represent less specific and reproducible PDI (Arda & Walhout, 2010; Walhout, 2011). Biological false positives may represent artificial PDI that are observed in yeast, but not in the organism studied. These interactions can be discarded upon subsequent validation by other methods. Important to keep in mind is that the Y1H technique provides a powerful technique to narrow down possible PDIs but cannot be used in itself as proof of the interactions.

5.2 Electrophoretic Mobility Shift Assay (EMSA) Based Methods

EMSA is extensively used to analyze nucleic acid–protein interactions (Fried & Crothers, 1981; Garner & Revzin, 1981). EMSA is based on the principle that DNA–protein complexes are larger and move slowly when subjected to nondenaturing polyacrylamide gel electrophoresis (PAGE), compared to the unbound (free) DNA probe. Since the rate of DNA migration is shifted or retarded when bound to protein, the assay is also referred to as a gel shift or gel retardation assay. DNA probes used in EMSA are typically double-stranded oligonucleotides of 20–25 bp containing a response element and can be radio-, fluoro-, or hapten-labeled. The DNA and crude nuclear extract or recombinant TF are incubated together in a binding reaction and separated by PAGE. A supershift assay can be performed to specifically assert the DNA–protein interactions by using an antibody specific to the TF of interest. However, when the binding factor is unknown, two-dimensional-EMSA (2D-EMSA) can be used to identify sequence-specific DNA-binding proteins in crude nuclear extracts (Stead & McDowall, 2007). 2D-EMSA utilizes the resolving power of SDS-PAGE to enable identification of proteins that, in a prior EMSA step, have altered mobility as a consequence of binding a probe. After purifying the protein band (spot) from the gel, matrix-assisted laser desorption/ionization time-of-flight mass spectrometry (MALDI-TOF MS) is used to identify the protein. This approach has been successfully employed for a bacterial lysate (Stead & McDowall, 2007), but the technique is challenging in cases of low-abundance TFs in a complex nuclear extract.

The success of 2D-EMSA justified the development of another technique, 3D-EMSA (Jiang, Jarrett, & Haskins, 2009). 3D-EMSA merges nondenaturing EMSA with 2-D electrophoresis (2-DE) to purify TFs. This technique is performed by cutting TF–CRE complex bands from PAGE and then further resolving them by 2-DE. Since 3D-EMSA combines the high specificity of EMSA and the high resolution of 2-DE, it is possible to purify TFs from crude nuclear extracts. For example, this method was used to successfully purify a GFP–CCAAT enhancer binding protein (CEBP) fusion protein constructed to bind CEBP's canonical sequence (CAAT), from bacterial crude extracts (Jiang et al., 2009; Jiang, Jia, & Jarrett, 2011). Although 3D-EMSA provides the highest selectivity for binding CREs, some contaminating proteins are still observed. This is likely due to protein–protein interactions; hence 3D-EMSA can be combined with the southwestern technique to solve more challenging TF purification problems (Jiang et al., 2011). TFs that bind a specific CRE are recognized as spots at particular positions and can be identified by high-performance liquid chromatography-electrospray ionization-tandem mass spectrometry (LC-ESI-MS). These techniques enabled successful characterization of the activator protein1 (AP1) and CEBP complexes from a human embryonic kidney cell line (HEK293) nuclear extract (Jiang et al., 2011). Overall, these techniques provide powerful and robust methods for investigating the low-abundance TF proteome.

The most significant benefit of EMSA, compared to Y1H, is its ability to resolve complexes of different stoichiometry or conformation. Another major advantage is that the source of the DNA-binding protein may be crude nuclear or whole-cell extract, in vitro transcription product, or a purified preparation. In addition, the relatively low ionic strength of the electrophoresis buffer helps to stabilize transient interactions, permitting even labile complexes to be resolved and analyzed by this method (Dey et al., 2012; Fried & Crothers, 1984; Stockley, 2009).

In plants, EMSA with crude nuclear extracts was used to identify interactions of known motifs with nuclear proteins in different genotypes (delViso, Casaretto, & Quatrano, 2007), tissues (Gomez-Maldonado et al., 2004; Rawat, Xu, Yao, & Chye, 2005), or treatments (Comelli, Viola, & Gonzalez, 2009; Rawat et al., 2005), followed by Y1H (Comelli et al., 2009), phage display (Korfhage, Trezzini, Meier, Hahlbrock, & Somssich, 1994), or directly verified with predicted TFs (Gomez-Maldonado et al., 2004). Nevertheless, 2-DE/3-DE TF purification and characterization methods have not yet been reported for plant TFs, likely because the techniques are laborious, technically challenging and exhibit low sensitivity.

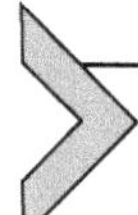

6. RESOURCES FOR STUDYING PLANT GRNs

6.1 Plant Transcription Factor ORFeome Collections (TFomes)

Several TFome collections have been developed for *Arabidopsis*, maize, and rice (Table 2). All these collections are in recombination-ready vectors of the Gateway system, allowing the rapid transfer of the ORFs to an increasing number of Gateway compatible vectors, such as those suitable for expression in yeast, bacteria, or plants (Curtis & Grossniklaus, 2003; Deplancke et al., 2004; Earley et al., 2006; Karimi, Inze, & Depicker, 2002). These collections have been transferred to yeast expression vectors and utilized to identify regulators of the *Arabidopsis* circadian clock (Pruneda-Paz, Breton, Para, & Kay, 2009), components of the *Arabidopsis* Mediator complex (Ou et al., 2011), and genes involved in the abiotic stress response in protoplast transactivation studies (Wehner et al., 2011). Some of these collections have also been used for the development of *Arabidopsis* overexpression lines (Coego et al., 2014; Weiste, Iven, Fischer, Onate-Sanchez, & Droge-Laser, 2007).

Table 2 Plant TFome Collections Available

Institution	Species	Number of TFs Included	Vectors	Availability of Collection	References
REGIA	*Arabidopsis thaliana*	884	pDONR™201 pDEST™22	ABRC	Castrillo et al. (2011); Paz-Ares (2002)
PKU-Yale	*Arabidopsis thaliana*	1282	pENTR™TOPO pDEST™22	ABRC	Gong et al. (2004); Ou et al. (2011)
RBC	*Arabidopsis thaliana*	1498	pDONR™207 pDEST™GAD424	RBC	Mitsuda et al. (2010)
REGULATORS	*Arabidopsis thaliana*	288	pDONR™221 pDEST™22	ABRC	Castrillo et al. (2011)
NPGRCC	*Arabidopsis thaliana*	96	pENTR™/D-TOPO pDEST™22	ABRC	Ou et al. (2011)
OSU-UT	*Zea mays*	2017	pENTR™/SD/D-TOPO pAD-GAL4-GWC1	GRASSIUS/ ABRC	Burdo et al. (2014)
OSU-UT	*Oryza sativa*	62	pENTR™/SD/D-TOPO	GRASSIUS/ Addgene	Unpublished

NPGRCC, The National Plant Gene Research Center of China; OSU-UT, The Ohio State University-University of Toledo; PKU-Yale, Peking University-Yale; RBC, RIKEN BioResource Center; REGIA, REgulatory Gene Initiative in Arabidopsis.

6.2 Databases

Databases play a key role in studying architecture and functions of GRNs and GRGs by providing resources necessary for construction and elucidation of network dynamics. To this end, plant gene regulatory databases can be classified into three major groups based on the information they host: (i) exclusively TF-based databases (TFDBs); (ii) exclusively sequence-based databases (*cis*-DBs); and (iii) TF-target sequence-based databases (TF-*cis*-DBs) (Fig. 3 and Table 3). TFDBs host information related exclusively to TFs, which include TF structures (primary, secondary, and tertiary structures), as well as their protein and DBDs, family classifications, expression data, and evolutionary information. *cis*-DBs host predominantly

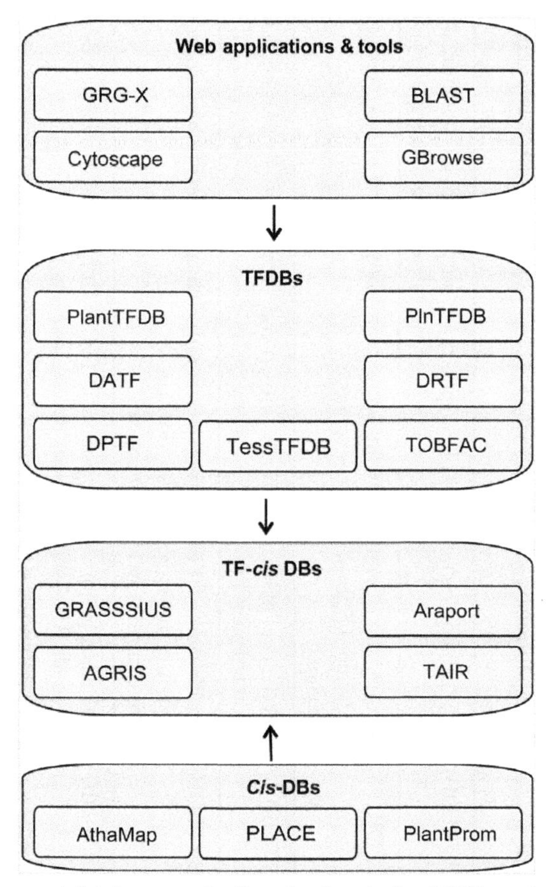

Fig. 3 Databases and database applications for the study of GRNs and grids. TF-*cis*-DBs host information contained in both TFDBs (TF-related information) and *cis*-DBs (*cis*-regulatory element-related information). (See the color plate.)

Table 3 Databases Hosting Information for Plant GRNs and GRGs

Database (Database Type)	Information Hosted	URL	References
PlantTFDB (TFDB)	TF evolutionary and functional information	http://planttfdb.cbi.pku.edu.cn/	Guo et al. (2008); Zhang et al. (2011)
DATFAP (TFDB)	EST TF sequences for 13 plant species	http://cgi-www.daimi.au.dk/cgi-chili/datfap/frontdoor.py	Fredslund (2008)
DPTF (TFDB)	A collection of experimentally known and predicted TF from poplar (*Populus trichocarpa*)	http://dptf.cbi.pku.edu.cn	Zhu et al. (2007)
DRTF (TFDB)	A collection of experimentally known and predicted rice TFs	http://drtf.cbi.pku.edu.cn/	Gao et al. (2006)
PlnTFDB (TFDB)	A collection of TFs and their descriptions (protein families, structures, domain architectures) for 19 plant species	http://plntfdb.bio.uni-potsdam.de/v3.0/	Pérez-Rodríguez et al. (2010)
AthaMap (*cis*-DB)	*A. thaliana* genome-wide potential TFBSs and small RNA-binding sites determined from published PWMs	http://www.athamap.de/	Bülow et al. (2010)
PLACE (*cis*-DB)	Published plant *cis*-regulatory elements/motifs	https://sogo.dna.affrc.go.jp/cgi-bin/sogo.cgi?lang=en&pj=640&action=page&page=newplace	Higo, Ugawa, Iwamoto, and Korenaga (1999)

PlantProm (*cis*-DB)	A collection of plant promoter sequences for RNA Pol II and TSS	http://linux1.softberry.com/berry.phtml?topic=plantprom&group=data&subgroup=plantprom	Shahmuradov, Gammerman, Hancock, Bramley, and Solovyev (2003)
GRASSIUS (TF-*cis* DB)	Three interlinked databases (GrassTFDB, GrassPROMDB, GrassCoRegDB); and computational tools for studying transcriptional gene regulation in *Z. mays*, *O. sativa*, *S. bicolor*, *S. officinarum*, and *Brachypodium sylvaticum*	http://grassius.org/	Yilmaz et al. (2009)
AGRIS (TF-*cis* DB)	Three interlinked *A. thaliana* databases: AtCisDB, AtTFDB, and AtRegNet	http://arabidopsis.med.ohio-state.edu/	Yilmaz et al. (2011)
SoyDB (TF-*cis* DB)	A knowledgebase for *Glycine max* TFs	http://casp.rnet.missouri.edu/soydb/	Wang et al. (2010)
LegumeTFDB (TF-*cis* DB)	Predicted TF-encoding genes for *G. max*, *Lotus japonicus*, *Medicago truncatula*	http://legumetfdb.psc.riken.jp/	Moolhuijzen et al. (2006)

cis-regulatory information such as promoter sequences, CREs, small RNA-binding sites, enhancer sequences, and TSSs. The TF-*cis*-DBs incorporate both TF and DNA sequence information, either separately or in an integrated manner such as TF–target gene relationships.

Our group hosts two highly visited plant TF-*cis*-DB databases that provide regulatory information, GRASSIUS (Yilmaz et al., 2009) and AGRIS (Yilmaz et al., 2011). GRASSIUS is an information knowledgebase and computational tools database for transcriptional gene regulation in maize, rice, sorghum, sugarcane, and *B. distachyon*. GRASSIUS hosts three interlinked databases: GrassTFDB (contains TF collection for these grass species, classified into different TF families); GrassPROMDB (a collection of promoter sequences); and GrassCoRegDB (contains a collection of transcriptional coregulators). Additionally, GRASSIUS hosts tools that aid the study and visualization of regulatory relationships. These include the gene regulatory grid explorer (GRG-X), a web application for visualizing regulatory networks, GRASSIUS Blast, and a GBrowse genome browser. In addition, GRASSIUS serves as a portal for access to the grasses TFome collections (Burdo et al., 2014). AGRIS is an information resource comprising of three interlinked *Arabidopsis* databases hosting approximately 33,000 promoter regions and descriptions of experimentally validated CREs (AtCisDB), information on approximately 1770 TFs (AtTFDB), and 17,694 manually curated regulatory interactions (AtRegNet). Similarly, AGRIS hosts GRG-X for visualization of regulatory interactions. Other notable plant databases with regulatory information include SoyDB, a knowledgebase for soybean (*Glycine max*) TFs (Wang et al., 2010), and LegumeTFDB, which provides predicted TF-encoding genes for several legumes (Moolhuijzen et al., 2006). Other databases that host plant genetic and molecular biology data such as TAIR and Araport (Krishnakumar et al., 2015) contain regulatory information that contribute to deciphering of plant GRNs and GRGs (see Table 3 for a complete list of database resources for studying GRNs and GRGs).

7. CONCLUSIONS

From all the above, it is clear that the time is ripe for exploring and dissecting plant GRGs and GRNs. The principal components of these grids, TFs and CREs, provide essential ingredients for synthetic biology and rational plant metabolic engineering that will lead to plant improvement and agronomic advances. Information available on how TFs interact with specific CREs provides an opportunity for building bioinspired synthetic biology circuits. The plant community is very unique with regard to sharing methods and resources, and as described in this chapter, there are a large number of public web-based resources for accessing information, as well as for requesting clones and other components of the toolbox that the synthetic biologists or metabolic engineer requires.

ACKNOWLEDGMENT

This research was funded by Grant NSF IOS-1125620 to E.G. and A.I.D.

REFERENCES

Adams, K. L., & Wendel, J. F. (2005). Allele-specific, bidirectional silencing of an alcohol dehydrogenase gene in different organs of interspecific diploid cotton hybrids. *Genetics*, *171*(4), 2139–2142.

Aoyama, T., & Chau, N. H. (1997). A glucocorticoid-mediated transcriptional induction system in transgenic plants. *The Plant Journal*, *11*(3), 605–612.

Aoyama, T., Dong, C. H., Wu, Y., Carabelli, M., Sessa, G., Ruberti, I., et al. (1995). Ectopic expression of the *Arabidopsis* transcriptional activator Athb-1 alters leaf cell fate in tobacco. *The Plant Cell*, 7(11), 1773–1785.

Arda, H. E., & Walhout, A. J. (2010). Gene-centered regulatory networks. *Briefings in Functional Genomics*, *9*(1), 4–12.

Autran, D., Jonak, C., Belcram, K., Beemster, G. T., Kronenberger, J., Grandjean, O., et al. (2002). Cell numbers and leaf development in *Arabidopsis*: A functional analysis of the *STRUWWELPETER* gene. *The EMBO Journal*, *21*(22), 6036–6049.

Avci-Adali, M., Metzger, M., Perle, N., Ziemer, G., & Wendel, H. P. (2010). Pitfalls of cell-systematic evolution of ligands by exponential enrichment (SELEX): Existing dead cells during in vitro selection anticipate the enrichment of specific aptamers. *Oligonucleotides*, *20*(6), 317–323.

Backstrom, S., Elfving, N., Nilsson, R., Wingsle, G., & Bjorklund, S. (2007). Purification of a plant mediator from *Arabidopsis thaliana* identifies PFT1 as the Med25 subunit. *Molecular Cell*, *26*(5), 717–729.

Bai, L., Wang, Q., Li, H., Cheng, M., Zhang, N., & Li, H. (2014). 5′-Cap selection methods and their application in full-length cDNA library construction and transcription start site profiling. *Journal of Shanghai Jiaotong University (Science)*, *19*(5), 7.

Bargmann, B. O. R., Marshall-Colon, A., Efroni, I., Ruffel, S., Birnbaum, K. D., Coruzzi, G. M., et al. (2013). TARGET: A transient transformation system for genome-wide transcription factor target discovery. *Molecular Plant*, *6*(3), 978–980.

Batut, P., Dobin, A., Plessy, C., Carninci, P., & Gingeras, T. R. (2013). High-fidelity promoter profiling reveals widespread alternative promoter usage and transposon-driven developmental gene expression. *Genome Research*, *23*(1), 169–180.

Batut, P., & Gingeras, T. R. (2013). RAMPAGE: Promoter activity profiling by paired-end sequencing of 5'-complete cDNAs. *Current Protocols in Molecular Biology*, *104*, Unit 25B.11.

Benhamed, M., Martin-Magniette, M.-L. L., Taconnat, L., Bitton, F., Servet, C., De Clercq, R., et al. (2008). Genome-scale *Arabidopsis* promoter array identifies targets of the histone acetyltransferase GCN5. *The Plant Journal*, *56*(3), 493–504.

Berger, M. F., & Bulyk, M. L. (2006). Protein binding microarrays (PBMs) for rapid, high-throughput characterization of the sequence specificities of DNA binding proteins. *Methods in Molecular Biology (Clifton, NJ)*, *338*, 245–260.
Berger, M. F., & Bulyk, M. L. (2009). Universal protein-binding microarrays for the comprehensive characterization of the DNA-binding specificities of transcription factors. *Nature Protocols*, *4*(3), 393–411.
Berger, M. F., Philippakis, A. A., Qureshi, A. M., He, F. S., Estep, P. W., & Bulyk, M. L. (2006). Compact, universal DNA microarrays to comprehensively determine transcription-factor binding site specificities. *Nature Biotechnology*, *24*(11), 1429–1435.
Bernatavichute, Y. V., Zhang, X., Cokus, S., Pellegrini, M., & Jacobsen, S. E. (2008). Genome-wide association of histone H3 lysine nine methylation with CHG DNA methylation in *Arabidopsis thaliana*. *PLoS One*, *3*(9), e3156.
Berri, S., Abbruscato, P., Faivre-Rampant, O., Brasileiro, A. C., Fumasoni, I., Satoh, K., et al. (2009). Characterization of *WRKY* co-regulatory networks in rice and *Arabidopsis*. *BMC Plant Biology*, *9*, 120.
Birchler, J. A., Riddle, N. C., Auger, D. L., & Veitia, R. A. (2005). Dosage balance in gene regulation: Biological implications. *Trends in Genetics*, *21*(4), 219–226.
Birchler, J. A., Yao, H., & Chudalayandi, S. (2006). Unraveling the genetic basis of hybrid vigor. *Proceedings of the National Academy of Sciences of the United States of America*, *103*(35), 12957–12958.
Birney, E., Stamatoyannopoulos, J. A., Dutta, A., Guigo, R., Gingeras, T. R., Margulies, E. H., et al. (2007). Identification and analysis of functional elements in 1% of the human genome by the ENCODE pilot project. *Nature*, *447*(7146), 799–816.
Blecher-Gonen, R., Barnett-Itzhaki, Z., Jaitin, D., Amann-Zalcenstein, D., Lara-Astiaso, D., & Amit, I. (2013). High-throughput chromatin immunoprecipitation for genome-wide mapping of in vivo protein-DNA interactions and epigenomic states. *Nature Protocols*, *8*(3), 539–554.
Bohner, S., Lenk, I., Rieping, M., Herold, M., & Gatz, C. (1999). Technical advance: Transcriptional activator TGV mediates dexamethasone-inducible and tetracycline-inactivatable gene expression. *The Plant Journal*, *19*(1), 87–95.
Bolduc, N., Yilmaz, A., Mejia-Guerra, M. K., Morohashi, K., O'Connor, D., Grotewold, E., et al. (2012). Unraveling the KNOTTED1 regulatory network in maize meristems. *Genes & Development*, *26*(15), 1685–1690.
Bonawitz, N. D., Soltau, W. L., Blatchley, M. R., Powers, B. L., Hurlock, A. K., Seals, L. A., et al. (2012). REF4 and RFR1, subunits of the transcriptional coregulatory complex mediator, are required for phenylpropanoid homeostasis in Arabidopsis. *The Journal of Biological Chemistry*, *287*(8), 5434–5445.
Bowman, S. K. (2015). Discovering enhancers by mapping chromatin features in primary tissue. *Genomics*, *106*(3), 140–144.
Boyce, J. M., Knight, H., Deyholos, M., Openshaw, M. R., Galbraith, D. W., Warren, G., et al. (2003). The *sfr6* mutant of *Arabidopsis* is defective in transcriptional activation via CBF/DREB1 and DREB2 and shows sensitivity to osmotic stress. *The Plant Journal*, *34*(4), 395–406.
Brem, R. B., Yvert, G., Clinton, R., & Kruglyak, L. (2002). Genetic dissection of transcriptional regulation in budding yeast. *Science*, *296*(5568), 752–755.
Brunner, S., Fengler, K., Morgante, M., Tingey, S., & Rafalski, A. (2005). Evolution of DNA sequence nonhomologies among maize inbreds. *The Plant Cell*, *17*(2), 343–360.
Brunner, S., Pea, G., & Rafalski, A. (2005). Origins, genetic organization and transcription of a family of non-autonomous helitron elements in maize. *The Plant Journal*, *43*(6), 799–810.
Bryzgalov, L. O., Antontseva, E. V., Matveeva, M. Y., Shilov, A. G., Kashina, E. V., Mordvinov, V. A., et al. (2013). Detection of regulatory SNPs in human genome using ChIP-seq ENCODE data. *PLoS One*, *8*(10), e78833.

Buck, M. J., Nobel, A. B., & Lieb, J. D. (2005). ChIPOTle: A user-friendly tool for the analysis of ChIP-chip data. *Genome Biology*, *6*(11), R97.
Bulger, M., & Groudine, M. (2011). Functional and mechanistic diversity of distal transcription enhancers. *Cell*, *144*(3), 327–339.
Bülow, L., Brill, Y., & Hehl, R. (2010). AthaMap-assisted transcription factor target gene identification in *Arabidopsis thaliana*. *Database*, *2010*. baq034.
Bulyk, M. L., Gentalen, E., Lockhart, D. J., & Church, G. M. (1999). Quantifying DNA-protein interactions by double-stranded DNA arrays. *Nature Biotechnology*, *17*(6), 573–577.
Bulyk, M. L., Johnson, P. L., & Church, G. M. (2002). Nucleotides of transcription factor binding sites exert interdependent effects on the binding affinities of transcription factors. *Nucleic Acids Research*, *30*(5), 1255–1261.
Burdo, B., Gray, J., Goetting-Minesky, M. P., Wittler, B., Hunt, M., Li, T., et al. (2014). The Maize TFome—Development of a transcription factor open reading frame collection for functional genomics. *The Plant Journal*, *80*, 356–366.
Busch, M. A., Bomblies, K., & Weigel, D. (1999). Activation of a floral homeotic gene in *Arabidopsis*. *Science*, *285*, 585–587.
Cairns, J., Spyrou, C., Stark, R., Smith, M. L., Lynch, A. G., & Tavaré, S. (2011). BayesPeak—An R package for analysing ChIP-seq data. *Bioinformatics*, *27*(5), 713–714.
Cao, J., Schneeberger, K., Ossowski, S., Gunther, T., Bender, S., Fitz, J., et al. (2011). Whole-genome sequencing of multiple *Arabidopsis thaliana* populations. *Nature Genetics*, *43*(10), 956–963.
Carninci, P., Kvam, C., Kitamura, A., Ohsumi, T., Okazaki, Y., Itoh, M., et al. (1996). High-efficiency full-length cDNA cloning by biotinylated CAP trapper. *Genomics*, *37*(3), 327–336.
Carninci, P., Sandelin, A., Lenhard, B., Katayama, S., Shimokawa, K., Ponjavic, J., et al. (2006). Genome-wide analysis of mammalian promoter architecture and evolution. *Nature Genetics*, *38*(6), 626–635.
Carninci, P., Westover, A., Nishiyama, Y., Ohsumi, T., Itoh, M., Nagaoka, S., et al. (1997). High efficiency selection of full-length cDNA by improved biotinylated cap trapper. *DNA Research*, *4*(1), 61–66.
Castrillo, G., Turck, F., Leveugle, M., Lecharny, A., Carbonero, P., Coupland, G., et al. (2011). Speeding *cis-trans* regulation discovery by phylogenomic analyses coupled with screenings of an arrayed library of Arabidopsis transcription factors. *PLoS One*, *6*(6), e21524.
Century, K., Reuber, T. L., & Ratcliffe, O. J. (2008). Regulating the regulators: The future prospects for transcription-factor-based agricultural biotechnology products. *Plant Physiology*, *147*(1), 20–29.
Cerdan, P. D., & Chory, J. (2003). Regulation of flowering time by light quality. *Nature*, *423*(6942), 881–885.
Chai, C., Xie, Z., & Grotewold, E. (2011). SELEX (Systematic Evolution of Ligands by EXponential Enrichment), as a powerful tool for deciphering the protein-DNA interaction space. *Methods in Molecular Biology*, *754*, 249–258.
Chen, H., Chen, W., Zhou, J., He, H., Chen, L., Chen, H., et al. (2012). Basic leucine zipper transcription factor OsbZIP16 positively regulates drought resistance in rice. *Plant Science*, *193–194*, 8–17.
Chen, Q. K., Hertz, G. Z., & Stormo, G. D. (1995). MATRIX SEARCH 1.0: A computer program that scans DNA sequences for transcriptional elements using a database of weight matrices. *Computer Applications in the Biosciences: CABIOS*, *11*(5), 563–566.
Chen, C. C., Xiao, S., Xie, D., Cao, X., Song, C. X., Wang, T., et al. (2013). Understanding variation in transcription factor binding by modeling transcription factor genome-epigenome interactions. *PLoS Computational Biology*, *9*(12), e1003367.

Clark, R. M., Wagler, T. N., Quijada, P., & Doebley, J. (2006). A distant upstream enhancer at the maize domestication gene tb1 has pleiotropic effects on plant and inflorescent architecture. *Nature Genetics*, *38*(5), 594–597.

Coego, A., Brizuela, E., Castillejo, P., Ruiz, S., Koncz, C., del Pozo, J. C., et al. (2014). The TRANSPLANTA collection of *Arabidopsis* lines: A resource for functional analysis of transcription factors based on their conditional overexpression. *The Plant Journal*, 77(6), 944–953.

Comelli, R. N., Viola, I. L., & Gonzalez, D. H. (2009). Characterization of promoter elements required for expression and induction by sucrose of the Arabidopsis *COX5b-1* nuclear gene, encoding the zinc-binding subunit of cytochrome *c* oxidase. *Plant Molecular Biology*, *69*(6), 729–743.

Conaway, R. C., & Conaway, J. W. (2011). Function and regulation of the Mediator complex. *Current Opinion in Genetics & Development*, *21*(2), 225–230.

Costa, M. M., Fox, S., Hanna, A. I., Baxter, C., & Coen, E. (2005). Evolution of regulatory interactions controlling floral asymmetry. *Development*, *132*(22), 5093–5101.

Crowley, J. J., Zhabotynsky, V., Sun, W., Huang, S., Pakatci, I. K., Kim, Y., et al. (2015). Analyses of allele-specific gene expression in highly divergent mouse crosses identifies pervasive allelic imbalance. *Nature Genetics*, *47*(4), 353–360.

Curtis, M. D., & Grossniklaus, U. (2003). A gateway cloning vector set for high-throughput functional analysis of genes in planta. *Plant Physiology*, *133*(2), 462–469.

Davidson, E. H. (2001). Gene regulatory functions in development. *Genomic regulatory systems: Development and evolution* (pp. 11–23). San Diego: Academic Press.

del Viso, F., Casaretto, J. A., & Quatrano, R. S. (2007). 14-3-3 Proteins are components of the transcription complex of the *ATEM1* promoter in *Arabidopsis*. *Planta*, *227*(1), 167–175.

Deplancke, B., Dupuy, D., Vidal, M., & Walhout, A. J. (2004). A gateway-compatible yeast one-hybrid system. *Genome Research*, *14*(10B), 2093–2101.

Deplancke, B., Mukhopadhyay, A., Ao, W., Elewa, A. M., Grove, C. A., Martinez, N. J., et al. (2006). A gene-centered *C. elegans* protein-DNA interaction network. *Cell*, *125*(6), 1193–1205.

Deplancke, B., Vermeirssen, V., Arda, H. E., Martinez, N. J., & Walhout, A. J. (2006). Gateway-compatible yeast one-hybrid screens. *CSH Protocols*, *2006*(5).

Dey, B., Thukral, S., Krishnan, S., Chakrobarty, M., Gupta, S., Manghani, C., et al. (2012). DNA-protein interactions: Methods for detection and analysis. *Molecular and Cellular Biochemistry*, *365*(1–2), 279–299.

Dhawan, R., Luo, H., Foerster, A. M., Abuqamar, S., Du, H. N., Briggs, S. D., et al. (2009). HISTONE MONOUBIQUITINATION1 interacts with a subunit of the mediator complex and regulates defense against necrotrophic fungal pathogens in *Arabidopsis*. *The Plant Cell*, *21*(3), 1000–1019.

Doebley, J., & Lukens, L. (1998). Transcriptional regulators and the evolution of plant form. *The Plant Cell*, *10*, 1075–1082.

Doebley, J., Stec, A., & Hubbard, L. (1997). The evolution of apical dominance in maize. *Nature*, *386*, 485–488.

Dong, T., Hu, Z., Deng, L., Wang, Y., Zhu, M., Zhang, J., et al. (2013). A tomato MADS-box transcription factor, SlMADS1, acts as a negative regulator of fruit ripening. *Plant Physiology*, *163*(2), 1026–1036.

Dong, L., Tan, Q., Ye, W., Liu, D., Chen, H., Hu, H., et al. (2015). Screening and identifying a novel ssDNA aptamer against alpha-fetoprotein using CE-SELEX. *Scientific Reports*, *5*, 15552.

Earley, K. W., Haag, J. R., Pontes, O., Opper, K., Juehne, T., Song, K., et al. (2006). Gateway-compatible vectors for plant functional genomics and proteomics. *The Plant Journal*, *45*(4), 616–629.

Eaton, R. M., Shallcross, J. A., Mael, L. E., Mears, K. S., Minkoff, L., Scoville, D. J., et al. (2015). Selection of DNA aptamers for ovarian cancer biomarker HE4 using CE-SELEX and high-throughput sequencing. *Analytical and Bioanalytical Chemistry*, *407*(23), 6965–6973.

Egener, T., Roulet, E., Zehnder, M., Bucher, P., & Mermod, N. (2005). Proof of concept for microarray-based detection of DNA-binding oncogenes in cell extracts. *Nucleic Acids Research*, *33*(8), e79.

Eklund, D. M., Ståldal, V., Valsecchi, I., Cierlik, I., Eriksson, C., Hiratsu, K., et al. (2010). The *Arabidopsis thaliana* STYLISH1 protein acts as a transcriptional activator regulating auxin biosynthesis. *The Plant Cell*, *22*(2), 349–363.

El-Kereamy, A., Bi, Y. M., Ranathunge, K., Beatty, P. H., Good, A. G., & Rothstein, S. J. (2012). The rice R2R3-MYB transcription factor OsMYB55 is involved in the tolerance to high temperature and modulates amino acid metabolism. *PLoS One*, 7(12), e52030.

Ellington, A. D., & Szostak, J. W. (1990). *In vitro* selection of RNA molecules that bind specific ligands. *Nature*, *346*(6287), 818–822.

ENCODE Project Consortium (2012). An integrated encyclopedia of DNA elements in the human genome. *Nature*, *489*(7414), 57–74.

Erokhin, M., Vassetzky, Y., Georgiev, P., & Chetverina, D. (2015). Eukaryotic enhancers: Common features, regulation, and participation in diseases. *Cellular and Molecular Life Sciences*, *72*(12), 2361–2375.

Evans, R. M. (1988). The steroid and thyroid hormone receptor superfamily. *Science*, *240*(4854), 889–895.

Eveland, A. L., Goldschmidt, A., Pautler, M., Morohashi, K., Liseron-Monfils, C., Lewis, M. W., et al. (2014). Regulatory modules controlling maize inflorescence architecture. *Genome Research*, *24*, 431–443.

Farley, E. K., Olson, K. M., Zhang, W., Brandt, A. J., Rokhsar, D. S., & Levine, M. S. (2015). Suboptimization of developmental enhancers. *Science*, *350*(6258), 325–328.

Farré, D., Roset, R., Huerta, M., Adsuara, J. E. E., Roselló, L., Albà, M. M., et al. (2003). Identification of patterns in biological sequences at the ALGGEN server: PROMO and MALGEN. *Nucleic Acids Research*, *31*(13), 3651–3653.

Feller, A., Hernandez, J. M., & Grotewold, E. (2006). An ACT-like domain participates in the dimerization of several plant bHLH transcription factors. *The Journal of Biological Chemistry*, *281*, 28964–28974.

Fields, S., & Song, O. (1989). A novel genetic system to detect protein-protein interactions. *Nature*, *340*(6230), 245–246.

Franco-Zorrilla, J. M. M., López-Vidriero, I., Carrasco, J. L. L., Godoy, M., Vera, P., & Solano, R. (2014). DNA-binding specificities of plant transcription factors and their potential to define target genes. *Proceedings of the National Academy of Sciences of the United States of America*, *111*(6), 2367–2372.

Fredslund, J. (2008). DATFAP: A database of primers and homology alignments for transcription factors from 13 plant species. *BMC Genomics*, *9*, 140.

Fried, M., & Crothers, D. M. (1981). Equilibria and kinetics of lac repressor-operator interactions by polyacrylamide gel electrophoresis. *Nucleic Acids Research*, *9*(23), 6505–6525.

Fried, M. G., & Crothers, D. M. (1984). Kinetics and mechanism in the reaction of gene regulatory proteins with DNA. *Journal of Molecular Biology*, *172*(3), 263–282.

Frith, M. C., Valen, E., Krogh, A., Hayashizaki, Y., Carninci, P., & Sandelin, A. (2008). A code for transcription initiation in mammalian genomes. *Genome Research*, *18*(1), 1–12.

Gao, S. Q., Chen, M., Xia, L. Q., Xiu, H. J., Xu, Z. S., Li, L. C., et al. (2009). A cotton (*Gossypium hirsutum*) DRE-binding transcription factor gene, *GhDREB*, confers enhanced tolerance to drought, high salt, and freezing stresses in transgenic wheat. *Plant Cell Reports*, *28*(2), 301–311.

Gao, G., Zhong, Y., Guo, A., Zhu, Q., Tang, W., Zheng, W., et al. (2006). DRTF: A database of rice transcription factors. *Bioinformatics (Oxford, England)*, *22*(10), 1286–1287.

Garner, M. M., & Revzin, A. (1981). A gel electrophoresis method for quantifying the binding of proteins to specific DNA regions: Application to components of the *Escherichia coli* lactose operon regulatory system. *Nucleic Acids Research*, *9*(13), 3047–3060.

Gatz, C. (1997). Chemical control of gene expression. *Annual Review of Plant Physiology and Plant Molecular Biology*, *48*, 89–108.

Gatz, C., & Lenk, I. (1998). Promoters that respond to chemical inducers. *Trends in Plant Science*, *3*, 352–358.

Gaudinier, A., Zhang, L., Reece-Hoyes, J. S., Taylor-Teeples, M., Pu, L., Liu, Z., et al. (2011). Enhanced Y1H assays for *Arabidopsis*. *Nature Methods*, *8*(12), 1053–1055.

Gibson, G., & Honeycutt, E. (2002). The evolution of developmental regulatory pathways. *Current Opinion in Genetics & Development*, *12*(6), 695–700.

Glass, C. K., Rose, D. W., & Rosenfeld, M. G. (1997). Nuclear receptor coactivators. *Current Opinion in Cell Biology*, *9*(2), 222–232.

Godoy, M., Franco-Zorrilla, J. M. M., Pérez-Pérez, J., Oliveros, J. C., Lorenzo, O., & Solano, R. (2011). Improved protein-binding microarrays for the identification of DNA-binding specificities of transcription factors. *The Plant Journal*, *66*(4), 700–711.

Golden, M. C., Collins, B. D., Willis, M. C., & Koch, T. H. (2000). Diagnostic potential of PhotoSELEX-evolved ssDNA aptamers. *Journal of Biotechnology*, *81*(2–3), 167–178.

Gomez-Maldonado, J., Avila, C., Torre, F., Canas, R., Canovas, F. M., & Campbell, M. M. (2004). Functional interactions between a glutamine synthetase promoter and MYB proteins. *The Plant Journal*, *39*(4), 513–526.

Gong, W., Shen, Y. P., Ma, L. G., Pan, Y., Du, Y. L., Wang, D. H., et al. (2004). Genome-wide ORFeome cloning and analysis of *Arabidopsis* transcription factor genes. *Plant Physiology*, *135*(2), 773–782.

Gray, J., Caparros-Ruiz, D., & Grotewold, E. (2012). Grass phenylpropanoids: Regulate before using! *Plant Science*, *184*, 112–120.

Gross, D. S., & Garrard, W. T. (1988). Nuclease hypersensitive sites in chromatin. *Annual Review of Biochemistry*, *57*, 159–197.

Grotewold, E. (2008). Transcription factors for predictive plant metabolic engineering: Are we there yet? *Current Opinion in Biotechnology*, *19*(2), 138–144.

Grotewold, E., Drummond, B. J., Bowen, B., & Peterson, T. (1994). The *myb*-homologous *P* gene controls phlobaphene pigmentation in maize floral organs by directly activating a flavonoid biosynthetic gene subset. *Cell*, *76*(3), 543–553.

Grotewold, E., & Gray, J. (2009). Maize transcription factors. In S. Hake & J. Bennetzen (Eds.), *The maize handbook: Vol. 2* (pp. 693–714). New York: Springer.

Gu, G., Wang, T., Yang, Y., Xu, X., & Wang, J. (2013). An improved SELEX-Seq strategy for characterizing DNA-binding specificity of transcription factor: NF-κB as an example. *PLoS One*, *8*(10), e76109.

Guo, A.-Y. Y., Chen, X., Gao, G., Zhang, H., Zhu, Q.-H. H., Liu, X.-C. C., et al. (2008). PlantTFDB: A comprehensive plant transcription factor database. *Nucleic Acids Research*, *36*(Database issue), D966–D969.

Guo, M., Rupe, M. A., Yang, X., Crasta, O., Zinselmeier, C., Smith, O. S., et al. (2006). Genome-wide transcript analysis of maize hybrids: Allelic additive gene expression and yield heterosis. *Theoretical and Applied Genetics*, *113*(5), 831–845.

Guo, M., Rupe, M. A., Zinselmeier, C., Habben, J., Bowen, B. A., & Smith, O. S. (2004). Allelic variation of gene expression in maize hybrids. *The Plant Cell*, *16*(7), 1707–1716.

Gustincich, S., Sandelin, A., Plessy, C., Katayama, S., Simone, R., Lazarevic, D., et al. (2006). The complexity of the mammalian transcriptome. *The Journal of Physiology*, *575*(Pt. 2), 321–332.

Haberle, V., Li, N., Hadzhiev, Y., Plessy, C., Previti, C., Nepal, C., et al. (2014). Two independent transcription initiation codes overlap on vertebrate core promoters. *Nature*, *507*(7492), 381–385.
Hanna-Rose, W., & Hansen, U. (1996). Active repression mechanisms of eukaryotic transcription repressors. *Trends in Genetics*, *12*(6), 229–234.
Hashimoto, S., Suzuki, Y., Kasai, Y., Morohoshi, K., Yamada, T., Sese, J., et al. (2004). 5′-End SAGE for the analysis of transcriptional start sites. *Nature Biotechnology*, *22*(9), 1146–1149.
He, X., Cicek, A. E., Wang, Y., Schulz, M. H., Le, H.-S. S., & Bar-Joseph, Z. (2015). *De novo* ChIP-seq analysis. *Genome Biology*, *16*(1), 205.
Heintzman, N. D., & Ren, B. (2009). Finding distal regulatory elements in the human genome. *Current Opinion in Genetics & Development*, *19*(6), 541–549.
Hernandez, J. M., Feller, A., Morohashi, K., Frame, K., & Grotewold, E. (2007). The basic helix loop helix domain of maize R links transcriptional regulation and histone modifications by recruitment of an EMSY-related factor. *Proceedings of the National Academy of Sciences of the United States of America*, *104*(43), 17222–17227.
Hestand, M. S., Klingenhoff, A., Scherf, M., Ariyurek, Y., Ramos, Y., van Workum, W., et al. (2010). Tissue-specific transcript annotation and expression profiling with complementary next-generation sequencing technologies. *Nucleic Acids Research*, *38*(16), e165.
Higo, K., Ugawa, Y., Iwamoto, M., & Korenaga, T. (1999). Plant cis-acting regulatory DNA elements (PLACE) database: 1999. *Nucleic Acids Research*, *27*(1), 297–300.
Hiratsu, K., Matsui, K., Koyama, T., & Ohme-Takagi, M. (2003). Dominant repression of target genes by chimeric repressors that include the EAR motif, a repression domain, in *Arabidopsis*. *The Plant Journal*, *34*, 733–739.
Hirzmann, J., Luo, D., Hahnen, J., & Hobom, G. (1993). Determination of messenger RNA 5′-ends by reverse transcription of the cap structure. *Nucleic Acids Research*, *21*(15), 3597–3598.
Hodo, Y., Hashimoto, S., Honda, M., Yamashita, T., Suzuki, Y., Sugano, S., et al. (2010). Comprehensive gene expression analysis of 5'-end of mRNA identified novel intronic transcripts associated with hepatocellular carcinoma. *Genomics*, *95*(4), 217–223.
Hoskins, R. A., Landolin, J. M., Brown, J. B., Sandler, J. E., Takahashi, H., Lassmann, T., et al. (2011). Genome-wide analysis of promoter architecture in *Drosophila melanogaster*. *Genome Research*, *21*(2), 182–192.
Houlès, T., Rodier, G., Le Cam, L., Sardet, C., & Kirsh, O. (2015). Description of an optimized ChIP-seq analysis pipeline dedicated to genome wide identification of E4F1 binding sites in primary and transformed MEFs. *Genomics Data*, *5*, 368–370.
Hovel, I., Louwers, M., & Stam, M. (2012). 3C technologies in plants. *Methods*, *58*(3), 204–211.
Hu, H., Dai, M., Yao, J., Xiao, B., Li, X., Zhang, Q., et al. (2006). Overexpressing a NAM, ATAF, and CUC (NAC) transcription factor enhances drought resistance and salt tolerance in rice. *Proceedings of the National Academy of Sciences of the United States of America*, *103*(35), 12987–12992.
Hu, S., Xie, Z., Onishi, A., Yu, X., Jiang, L., Lin, J., et al. (2009). Profiling the human protein-DNA interactome reveals ERK2 as a transcriptional repressor of interferon signaling. *Cell*, *139*(3), 610–622.
Hu, H., You, J., Fang, Y., Zhu, X., Qi, Z., & Xiong, L. (2008). Characterization of transcription factor gene *SNAC2* conferring cold and salt tolerance in rice. *Plant Molecular Biology*, *67*(1–2), 169–181.
Ikeda, M., & Ohme-Takagi, M. (2009). A novel group of transcriptional repressors in Arabidopsis. *Plant & Cell Physiology*, *50*(5), 970–975.
Jiang, D., Jarrett, H. W., & Haskins, W. E. (2009). Methods for proteomic analysis of transcription factors. *Journal of Chromatography. A*, *1216*(41), 6881–6889.

Jiang, D., Jia, Y., & Jarrett, H. W. (2011). Transcription factor proteomics: Identification by a novel gel mobility shift-three-dimensional electrophoresis method coupled with southwestern blot and high-performance liquid chromatography-electrospray-mass spectrometry analysis. *Journal of Chromatography. A, 1218*(39), 7003–7015.

Jiang, J., & Levine, M. (1993). Binding affinities and cooperative interactions with bHLH activators delimit threshold responses to the dorsal gradient morphogen. *Cell, 72*(5), 741–752.

Jirschitzka, J., Mattern, D. J., Gershenzon, J., & D'Auria, J. C. (2013). Learning from nature: New approaches to the metabolic engineering of plant defense pathways. *Current Opinion in Biotechnology, 24*(2), 320–328.

Johnson, W. E., Li, W., Meyer, C. A., Gottardo, R., Carroll, J. S., Brown, M., & Liu, X. S. (2006). Model-based analysis of tiling-arrays for ChIP-chip. *Proceedings of the National Academy of Sciences of the United States of America, 103*(33), 12457–12462.

Johnson, D. S., Mortazavi, A., Myers, R. M., & Wold, B. (2007). Genome-wide mapping of in vivo protein-DNA interactions. *Science, 316*(5830), 1497–1502.

Jolma, A., Kivioja, T., Toivonen, J., Cheng, L., Wei, G., Enge, M., et al. (2010). Multiplexed massively parallel SELEX for characterization of human transcription factor binding specificities. *Genome Research, 20*(6), 861–873.

Juven-Gershon, T., & Kadonaga, J. T. (2010). Regulation of gene expression via the core promoter and the basal transcriptional machinery. *Developmental Biology, 339*(2), 225–229.

Karimi, M., Inze, D., & Depicker, A. (2002). GATEWAY vectors for *Agrobacterium*-mediated plant transformation. *Trends in Plant Science*, 7(5), 193–195.

Karin, M. (1990). Too many transcription factors: Positive and negative interactions. *The New Biologist, 2*(2), 126–131.

Kasahara, Y., Irisawa, Y., Ozaki, H., Obika, S., & Kuwahara, M. (2013). 2′,4′-BNA/LNA aptamers: CE-SELEX using a DNA-based library of full-length 2′-O,4′-C-methylene-bridged/linked bicyclic ribonucleotides. *Bioorganic & Medicinal Chemistry Letters, 23*(5), 1288–1292.

Kasuga, M., Liu, Q., Miura, S., Yamaguchi-Shinozaki, K., & Shinozaki, K. (1999). Improving plant drought, salt and freezing tolerance by gene transfer of a single stress-inducible transcription factor. *Nature Biotechnology, 17*, 287–291.

Kaufmann, K., Muino, J. M., Jauregui, R., Airoldi, C. A., Smaczniak, C., Krajewski, P., et al. (2009). Target genes of the MADS transcription factor SEPALLATA3: Integration of developmental and hormonal pathways in the *Arabidopsis* flower. *PLoS Biology*, 7(4), e1000090.

Kaufmann, K., Muiño, J. M., Østerås, M., Farinelli, L., Krajewski, P., & Angenent, G. C. (2010). Chromatin immunoprecipitation (ChIP) of plant transcription factors followed by sequencing (ChIP-SEQ) or hybridization to whole genome arrays (ChIP-CHIP). *Nature Protocols, 5*(3), 457–472.

Kawaji, H., Frith, M. C., Katayama, S., Sandelin, A., Kai, C., Kawai, J., et al. (2006). Dynamic usage of transcription start sites within core promoters. *Genome Biology*, 7(12), R118.

Kazan, K. (2006). Negative regulation of defence and stress genes by EAR-motif-containing repressors. *Trends in Plant Science, 11*(3), 109–112.

Kel, A. E., Gossling, E., Reuter, I., Cheremushkin, E., Kel-Margoulis, O. V., & Wingender, E. (2003). MATCH: A tool for searching transcription factor binding sites in DNA sequences. *Nucleic Acids Research, 31*(13), 3576–3579.

Kidd, B. N., Edgar, C. I., Kumar, K. K., Aitken, E. A., Schenk, P. M., Manners, J. M., et al. (2009). The mediator complex subunit PFT1 is a key regulator of jasmonate-dependent defense in *Arabidopsis*. *The Plant Cell, 21*(8), 2237–2252.

Kim, T. H., & Ren, B. (2006). Genome-wide analysis of protein-DNA interactions. *Annual Review of Genomics and Human Genetics*, 7, 81–102.

Kim, T. K., & Shiekhattar, R. (2015). Architectural and functional commonalities between enhancers and promoters. *Cell, 162*(5), 948–959.
Kim, Y. J., Zheng, B., Yu, Y., Won, S. Y., Mo, B., & Chen, X. (2011). The role of Mediator in small and long noncoding RNA production in *Arabidopsis thaliana. The EMBO Journal, 30*(5), 814–822.
Kinoshita, T., Miura, A., Choi, Y., Kinoshita, Y., Cao, X., Jacobsen, S. E., et al. (2004). One-way control of *FWA* imprinting in *Arabidopsis* endosperm by DNA methylation. *Science, 303*(5657), 521–523.
Korfhage, U., Trezzini, G. F., Meier, I., Hahlbrock, K., & Somssich, I. E. (1994). Plant homeodomain protein involved in transcriptional regulation of a pathogen defense-related gene. *The Plant Cell, 6*(5), 695–708.
Krishnakumar, V., Hanlon, M. R., Contrino, S., Ferlanti, E. S., Karamycheva, S., Kim, M., et al. (2015). Araport: The *Arabidopsis* information portal. *Nucleic Acids Research, 43*(Database issue), D1003–D1009.
Krizek, B. A. (2015). Intronic sequences are required for *AINTEGUMENTA-LIKE6* expression in *Arabidopsis* flowers. *BMC Research Notes, 8*, 556.
Kuo, M. H., & Allis, C. D. (1999). In vivo cross-linking and immunoprecipitation for studying dynamic protein:DNA associations in a chromatin environment. *Methods (San Diego, Calif.), 19*(3), 425–433.
Landt, S. G., Marinov, G. K., Kundaje, A., Kheradpour, P., Pauli, F., Batzoglou, S., et al. (2012). ChIP-seq guidelines and practices of the ENCODE and modENCODE consortia. *Genome Research, 22*(9), 1813–1831.
Langmead, B., Trapnell, C., Pop, M., & Salzberg, S. L. (2009). Ultrafast and memory-efficient alignment of short DNA sequences to the human genome. *Genome Biology, 10*(3), R25.
Latchman, D. S. (1997). Transcription factors: An overview. *The International Journal of Biochemistry & Cell Biology, 29*(12), 1305–1312.
Lee, J., He, K., Stolc, V., Lee, H., Figueroa, P., Gao, Y., et al. (2007). Analysis of transcription factor HY5 genomic binding sites revealed its hierarchical role in light regulation of development. *The Plant Cell, 19*(3), 731–749.
Lee, T. I., & Young, R. A. (2000). Transcription of eukaryotic protein-coding genes. *Annual Review of Genetics, 34*, 77–137.
Lee, T. I., & Young, R. A. (2013). Transcriptional regulation and its misregulation in disease. *Cell, 152*(6), 1237–1251.
Lenhard, B., Sandelin, A., & Carninci, P. (2012). Metazoan promoters: Emerging characteristics and insights into transcriptional regulation. *Nature Reviews. Genetics, 13*(4), 233–245.
Li, C., Qiao, Z., Qi, W., Wang, Q., Yuan, Y., Yang, X., et al. (2015). Genome-wide characterization of *cis*-acting DNA targets reveals the transcriptional regulatory framework of *opaque2* in maize. *The Plant Cell, 27*(3), 532–545.
Li, X.-M., & Shapiro, L. J. (1993). Three-step PCR mutagenesis for 'linker scanning'. *Nucleic Acids Research, 21*, 3745–3748.
Li, C., Zhou, A., & Sang, T. (2006). Rice domestication by reducing shattering. *Science, 311*(5769), 1936–1939.
Licht, J. D., Ro, M., English, M. A., Grossel, M., & Hansen, U. (1993). Selective repression of transcriptional activators at a distance by the *Drosophila* Kruppel protein. *Proceedings of the National Academy of Sciences of the United States of America, 90*(23), 11361–11365.
Lindemose, S., Jensen, M. K., Van de Velde, J., O'Shea, C., Heyndrickx, K. S., Workman, C. T., et al. (2014). A DNA-binding-site landscape and regulatory network analysis for NAC transcription factors in *Arabidopsis thaliana. Nucleic Acids Research, 42*(12), 7681–7693.
Lloyd, A. M., Schena, M., Walbot, V., & Davis, R. W. (1994). Epidermal cell fate determination in *Arabidopsis*: Patterns defined by a steroid-inducible regulator. *Science, 266*, 436–439.

Louwers, M., Bader, R., Haring, M., van Driel, R., de Laat, W., & Stam, M. (2009). Tissue- and expression level-specific chromatin looping at maize *b1* epialleles. *The Plant Cell, 21*(3), 832–842.

Louwers, M., Splinter, E., van Driel, R., de Laat, W., & Stam, M. (2009). Studying physical chromatin interactions in plants using chromosome conformation capture (3C). *Nature Protocols, 4*(8), 1216–1229.

Luo, J., Butelli, E., Hill, L., Parr, A., Niggeweg, R., Bailey, P., et al. (2008). AtMYB12 regulates caffeoyl quinic acid and flavonol synthesis in tomato: Expression in fruit results in very high levels of both types of polyphenol. *The Plant Journal, 56*(2), 316–326.

Man, T. K., & Stormo, G. D. (2001). Non-independence of Mnt repressor-operator interaction determined by a new quantitative multiple fluorescence relative affinity (QuMFRA) assay. *Nucleic Acids Research, 29*(12), 2471–2478.

Margueron, R., Trojer, P., & Reinberg, D. (2005). The key to development: Interpreting the histone code? *Current Opinion in Genetics & Development, 15*(2), 163–176.

Martell, R. E., Nevins, J. R., & Sullenger, B. A. (2002). Optimizing aptamer activity for gene therapy applications using expression cassette SELEX. *Molecular Therapy, 6*(1), 30–34.

Martinez, N. J., Ow, M. C., Barrasa, M. I., Hammell, M., Sequerra, R., Doucette-Stamm, L., et al. (2008). A *C. elegans* genome-scale microRNA network contains composite feedback motifs with high flux capacity. *Genes & Development, 22*(18), 2535–2549.

Maruyama, K., & Sugano, S. (1994). Oligo-capping: A simple method to replace the cap structure of eukaryotic mRNAs with oligoribonucleotides. *Gene, 138*(1–2), 171–174.

Maston, G. A., Evans, S. K., & Green, M. R. (2006). Transcriptional regulatory elements in the human genome. *Annual Review of Genomics and Human Genetics*, 7, 29–59.

Mathelier, A., & Wasserman, W. W. (2013). The next generation of transcription factor binding site prediction. *PLoS Computational Biology, 9*(9), e1003214.

Mathur, S., Vyas, S., Kapoor, S., & Tyagi, A. K. (2011). The Mediator complex in plants: Structure, phylogeny, and expression profiling of representative genes in a dicot (*Arabidopsis*) and a monocot (rice) during reproduction and abiotic stress. *Plant Physiology, 157*(4), 1609–1627.

McGarry, R. C., & Ayre, B. G. (2008). A DNA element between At4g28630 and At4g28640 confers companion-cell specific expression following the sink-to-source transition in mature minor vein phloem. *Planta, 228*(5), 839–849.

McMullen, M. D., Byrne, P. F., Snook, M. E., Wiseman, B. R., Lee, E. A., Widstrom, N. W., et al. (1998). Quantitative trait loci and metabolic pathways. *Proceedings of the National Academy of Sciences of the United States of America, 95*(5), 1996–2000.

Mehta, P., Schwab, D. J., & Sengupta, A. M. (2011). Statistical mechanics of transcription-factor binding site discovery using hidden Markov models. *Journal of Statistical Physics, 142*(6), 1187–1205.

Mejia-Guerra, M. K., Li, W., Galeano, N. F., Vidal, M., Gray, J., Doseff, A. I., et al. (2015). Core promoter plasticity between maize tissues and genotypes contrasts with predominance of sharp transcription initiation sites. *The Plant Cell, 27*(12), 3309–3320.

Mejia-Guerra, M. K., Pomeranz, M., Morohashi, K., & Grotewold, E. (2012). From plant gene regulatory grids to network dynamics. *Biochimica et Biophysica Acta, 1819*(5), 454–465.

Mellor, J., Dudek, P., & Clynes, D. (2008). A glimpse into the epigenetic landscape of gene regulation. *Current Opinion in Genetics & Development, 18*(2), 116–122.

Mintseris, J., & Eisen, M. B. (2006). Design of a combinatorial DNA microarray for protein-DNA interaction studies. *BMC Bioinformatics*, 7, 429.

Mitsuda, N., Ikeda, M., Takada, S., Takiguchi, Y., Kondou, Y., Yoshizumi, T., et al. (2010). Efficient yeast one-/two-hybrid screening using a library composed only of transcription factors in *Arabidopsis thaliana*. *Plant & Cell Physiology, 51*(12), 2145–2151.

Mizoi, J., Shinozaki, K., & Yamaguchi-Shinozaki, K. (2012). AP2/ERF family transcription factors in plant abiotic stress responses. *Biochimica et Biophysica Acta*, *1819*(2), 86–96.

Mo, Q. (2012). A fully Bayesian hidden Ising model for ChIP-seq data analysis. *Biostatistics (Oxford, England)*, *13*(1), 113–128.

Moolhuijzen, P., Cakir, M., Hunter, A., Schibeci, D., Macgregor, A., Smith, C., et al. (2006). LegumeDB1 bioinformatics resource: Comparative genomic analysis and novel cross-genera marker identification in lupin and pasture legume species. *Genome/National Research Council Canada = Génome/Conseil national de recherches Canada*, *49*(6), 689–699.

Morgante, M., Brunner, S., Pea, G., Fengler, K., Zuccolo, A., & Rafalski, A. (2005). Gene duplication and exon shuffling by helitron-like transposons generate intraspecies diversity in maize. *Nature Genetics*, *37*(9), 997–1002.

Morohashi, K., Casas, M. I., Ferreyra, L. F., Mejia-Guerra, M. K., Pourcel, L., Yilmaz, A., et al. (2012). A genome-wide regulatory framework identifies maize *pericarp color1* controlled genes. *The Plant Cell*, *24*(7), 2745–2764.

Morohashi, K., & Grotewold, E. (2009). A systems approach reveals regulatory circuitry for *Arabidopsis* trichome initiation by the GL3 and GL1 selectors. *PLoS Genetics*, *5*(2), e1000396.

Morohashi, K., Xie, Z., & Grotewold, E. (2009). Gene-specific and genome-wide ChIP approaches to study plant transcriptional networks. *Methods in Molecular Biology*, *553*, 3–12.

Morton, T., Petricka, J., Corcoran, D. L., Li, S., Winter, C. M., Carda, A., et al. (2014). Paired-end analysis of transcription start sites in *Arabidopsis* reveals plant-specific promoter signatures. *The Plant Cell*, *26*(7), 2746–2760.

Mosing, R. K., & Bowser, M. T. (2009). Isolating aptamers using capillary electrophoresis-SELEX (CE-SELEX). *Methods in Molecular Biology (Clifton, NJ)*, *535*, 33–43.

Moyroud, E., Minguet, E. G., Ott, F., Yant, L., Posé, D., Monniaux, M., et al. (2011). Prediction of regulatory interactions from genome sequences using a biophysical model for the *Arabidopsis* LEAFY transcription factor. *The Plant Cell*, *23*(4), 1293–1306.

Muiño, J. M., Kaufmann, K., van Ham, R. C., Angenent, G. C., & Krajewski, P. (2011). ChIP-seq Analysis in R (CSAR): An R package for the statistical detection of protein-bound genomic regions. *Plant Methods*, 7, 11.

Mukherjee, S., Berger, M. F., Jona, G., Wang, X. S., Muzzey, D., Snyder, M., et al. (2004). Rapid analysis of the DNA-binding specificities of transcription factors with DNA microarrays. *Nature Genetics*, *36*(12), 1331–1339.

Nakashima, K., Takasaki, H., Mizoi, J., Shinozaki, K., & Yamaguchi-Shinozaki, K. (2012). NAC transcription factors in plant abiotic stress responses. *Biochimica et Biophysica Acta*, *1819*(2), 97–103.

Ni, T., Corcoran, D. L., Rach, E. A., Song, S., Spana, E. P., Gao, Y., et al. (2010). A paired-end sequencing strategy to map the complex landscape of transcription initiation. *Nature Methods*, 7(7), 521–527.

Ni, Y., Hall, A. W., Battenhouse, A., & Iyer, V. R. (2012). Simultaneous SNP identification and assessment of allele-specific bias from ChIP-seq data. *BMC Genetics*, *13*, 46.

Nikolayeva, O., & Robinson, M. D. (2014). edgeR for differential RNA-seq and ChIP-seq analysis: An application to stem cell biology. *Methods in Molecular Biology (Clifton, NJ)*, *1150*, 45–79.

Nikolov, D. B., & Burley, S. K. (1997). RNA polymerase II transcription initiation: A structural view. *Proceedings of the National Academy of Sciences of the United States of America*, *94*(1), 15–22.

Oh, E., Kang, H., Yamaguchi, S., Park, J., Lee, D., Kamiya, Y., et al. (2009). Genome-wide analysis of genes targeted by PHYTOCHROME INTERACTING FACTOR 3-LIKE5 during seed germination in *Arabidopsis*. *The Plant Cell*, *21*(2), 403–419.

Ohta, M., Matsui, K., Hiratsu, K., Shinshi, H., & Ohme-Takagi, M. (2001). Repression domains of class II ERF transcriptional repressors share an essential motif for active repression. *The Plant Cell, 13*(8), 1959–1968.

Ou, B., Yin, K. Q., Liu, S. N., Yang, Y., Gu, T., Wing Hui, J. M., et al. (2011). A high-throughput screening system for *Arabidopsis* transcription factors and its application to Med25-dependent transcriptional regulation. *Molecular Plant, 4*(3), 546–555.

Ouma, W. Z., Mejia-Guerra, M. K., Yilmaz, A., Pareja-Tobes, P., Li, W., Doseff, A. I., et al. (2015). Important biological information uncovered in previously unaligned reads from chromatin immunoprecipitation experiments (ChIP-Seq). *Scientific Reports, 5*, 8635.

Pabo, C. O., & Sauer, R. T. (1992). Transcription factors: Structural families and principles of DNA recognition. *Annual Review of Biochemistry, 61*, 1053–1095.

Padidam, M. (2003). Chemically regulated gene expression in plants. *Current Opinion in Plant Biology, 6*(2), 169–177.

Pan, Y., Seymour, G. B., Lu, C., Hu, Z., Chen, X., & Chen, G. (2012). An ethylene response factor (ERF5) promoting adaptation to drought and salt tolerance in tomato. *Plant Cell Reports, 31*(2), 349–360.

Para, A., Li, Y., Marshall-Colon, A., Varala, K., Francoeur, N. J., Moran, T. M., et al. (2014). Hit-and-run transcriptional control by bZIP1 mediates rapid nutrient signaling in *Arabidopsis*. *Proceedings of the National Academy of Sciences of the United States of America, 111*(28), 10371–10376.

Paul, A., Avci-Adali, M., Ziemer, G., & Wendel, H. P. (2009). Streptavidin-coated magnetic beads for DNA strand separation implicate a multitude of problems during cell-SELEX. *Oligonucleotides, 19*(3), 243–254.

Paz-Ares, J. (2002). REGIA, an EU project on functional genomics of transcription factors from *Arabidopsis thaliana*. *Comparative and Functional Genomics, 3*(2), 102–108.

Pérez-Rodríguez, P., Riaño-Pachón, D. M., Corrêa, L. G. G., Rensing, S. A., Kersten, B., & Mueller-Roeber, B. (2010). PlnTFDB: Updated content and new features of the plant transcription factor database. *Nucleic Acids Research, 38*(Database issue), D822–D827.

Peterson, C. L., & Laniel, M. A. (2004). Histones and histone modifications. *Current Biology, 14*(14), R546–R551.

Pfluger, J., & Wagner, D. (2007). Histone modifications and dynamic regulation of genome accessibility in plants. *Current Opinion in Plant Biology, 10*(6), 645–652.

Philippakis, A. A., Qureshi, A. M., Berger, M. F., & Bulyk, M. L. (2008). Design of compact, universal DNA microarrays for protein binding microarray experiments. *Journal of Computational Biology, 15*(7), 655–665.

Ponjavic, J., Lenhard, B., Kai, C., Kawai, J., Carninci, P., Hayashizaki, Y., et al. (2006). Transcriptional and structural impact of TATA-initiation site spacing in mammalian core promoters. *Genome Biology*, 7(8), R78.

Prestridge, D. S. (1996). SIGNAL SCAN 4.0: Additional databases and sequence formats. *Computer Applications in the Biosciences: CABIOS, 12*(2), 157–160.

Pruneda-Paz, J. L., Breton, G., Nagel, D. H., Kang, S. E., Bonaldi, K., Doherty, C. J., et al. (2014). A genome-scale resource for the functional characterization of *Arabidopsis* transcription factors. *Cell Reports, 8*(2), 622–632.

Pruneda-Paz, J. L., Breton, G., Para, A., & Kay, S. A. (2009). A functional genomics approach reveals CHE as a component of the *Arabidopsis* circadian clock. *Science, 323*(5920), 1481–1485.

Raatz, B., Eicker, A., Schmitz, G., Fuss, E., Muller, D., Rossmann, S., et al. (2011). Specific expression of *LATERAL SUPPRESSOR* is controlled by an evolutionarily conserved 3' enhancer. *The Plant Journal, 68*(3), 400–412.

Rabara, R. C., Tripathi, P., & Rushton, P. J. (2014). The potential of transcription factor-based genetic engineering in improving crop tolerance to drought. *OMICS, 18*(10), 601–614.

Rawat, R., Xu, Z. F., Yao, K. M., & Chye, M. L. (2005). Identification of cis-elements for ethylene and circadian regulation of the *Solanum melongena* gene encoding cysteine proteinase. *Plant Molecular Biology*, *57*(5), 629–643.

Reece-Hoyes, J. S., Deplancke, B., Shingles, J., Grove, C. A., Hope, I. A., & Walhout, A. J. (2005). A compendium of *Caenorhabditis elegans* regulatory transcription factors: A resource for mapping transcription regulatory networks. *Genome Biology*, *6*(13), R110.

Reece-Hoyes, J. S., & Marian Walhout, A. J. (2012). Yeast one-hybrid assays: A historical and technical perspective. *Methods*, *57*(4), 441–447.

Reece-Hoyes, J. S., & Walhout, A. J. (2012). Gene-centered yeast one-hybrid assays. *Methods in Molecular Biology*, *812*, 189–208.

Riechmann, J. L., & Ratcliffe, O. J. (2000). A genomic perspective on plant transcription factors. *Current Opinion in Plant Biology*, *3*, 423–434.

Riley, T. R., Slattery, M., Abe, N., Rastogi, C., Liu, D., Mann, R. S., et al. (2014). SELEX-seq: A method for characterizing the complete repertoire of binding site preferences for transcription factor complexes. *Methods in Molecular Biology (Clifton, NJ)*, *1196*, 255–278.

Roberts, S. G. (2000). Mechanisms of action of transcription activation and repression domains. *Cellular and Molecular Life Sciences*, *57*(8–9), 1149–1160.

Roeder, R. G. (1996). The role of general initiation factors in transcription by RNA polymerase II. *Trends in Biochemical Sciences*, *21*(9), 327–335.

Rohr, C. O., Parra, R. G., Yankilevich, P., & Perez-Castro, C. (2013). INSECT: IN-silico SEarch for Co-occurring Transcription factors. *Bioinformatics*, *29*(22), 2852–2858.

Romeis, J., Meissle, M., & Bigler, F. (2006). Transgenic crops expressing *Bacillus thuringiensis* toxins and biological control. *Nature Biotechnology*, *24*(1), 63–71.

Romero, I., Fuertes, A., Benito, M. J., Malpica, J. M., Leyva, A., & Paz-Ares, J. (1998). More than 80 *R2R3-MYB* regulatory genes in the genome of *Arabidopsis thaliana*. *The Plant Journal*, *14*(3), 273–284.

Roulet, E., Busso, S., Camargo, A. A., Simpson, A. J., Mermod, N., & Bucher, P. (2002). High-throughput SELEX SAGE method for quantitative modeling of transcription-factor binding sites. *Nature Biotechnology*, *20*(8), 831–835.

Rumble, S. M., Lacroute, P., Dalca, A. V., Fiume, M., Sidow, A., & Brudno, M. (2009). SHRiMP: Accurate mapping of short color-space reads. *PLoS Computational Biology*, *5*(5), e1000386.

Rushton, P. J., & Somssich, I. E. (1998). Transcriptional control of plant genes responsive to pathogens. *Current Opinion in Plant Biology*, *1*(4), 311–315.

Rushton, P. J., Somssich, I. E., Ringler, P., & Shen, Q. J. (2010). WRKY transcription factors. *Trends in Plant Science*, *15*(5), 247–258.

Sablowski, R. W., & Meyerowitz, E. M. (1998). A homolog of *NO APICAL MERISTEM* is an immediate target of the floral homeotic genes *APETALA3/PISTILLATA*. *Cell*, *92*(1), 93–103.

Sainz, M. B., Goff, S. A., & Chandler, V. L. (1997). Extensive mutagenesis of a transcriptional activation domain identifies single hydrophobic and acidic amino acids important for activation in vivo. *Molecular and Cellular Biology*, *17*(1), 115–122.

Sainz, M. B., Grotewold, E., & Chandler, V. L. (1997). Evidence for direct activation of an anthocyanin promoter by the maize C1 protein and comparison of DNA binding by related Myb domain proteins. *The Plant Cell*, *9*, 611–625.

Salvi, S., Sponza, G., Morgante, M., Tomes, D., Niu, X., Fengler, K. A., et al. (2007). Conserved noncoding genomic sequences associated with a flowering-time quantitative trait locus in maize. *Proceedings of the National Academy of Sciences of the United States of America*, *104*(27), 11376–11381.

Samach, A., Onouchi, H., Gold, S. E., Ditta, G. S., Schwarz-Sommer, Z., Yanofsky, M. F., et al. (2000). Distinct roles of CONSTANS target genes in reproductive development of *Arabidopsis*. *Science*, *288*(5471), 1613–1616.
Schauer, S. E., Schluter, P. M., Baskar, R., Gheyselinck, J., Bolanos, A., Curtis, M. D., et al. (2009). Intronic regulatory elements determine the divergent expression patterns of *AGAMOUS-LIKE6* subfamily members in *Arabidopsis*. *The Plant Journal*, *59*(6), 987–1000.
Shahmuradov, I. A., Gammerman, A. J., Hancock, J. M., Bramley, P. M., & Solovyev, V. V. (2003). PlantProm: A database of plant promoter sequences. *Nucleic Acids Research*, *31*(1), 114–117.
Shan, Q., Wang, Y., Li, J., Zhang, Y., Chen, K., Liang, Z., et al. (2013). Targeted genome modification of crop plants using a CRISPR-Cas system. *Nature Biotechnology*, *31*(8), 686–688.
Shen, H., Liu, C., Zhang, Y., Meng, X., Zhou, X., Chu, C., et al. (2012). OsWRKY30 is activated by MAP kinases to confer drought tolerance in rice. *Plant Molecular Biology*, *80*(3), 241–253.
Shin, H., Liu, T., Duan, X., Zhang, Y., & Liu, X. S. (2013). Computational methodology for ChIP-seq analysis. *Quantitative Biology*, *1*(1), 54–70.
Shin, D., Moon, S. J., Han, S., Kim, B. G., Park, S. R., Lee, S. K., et al. (2011). Expression of *StMYB1R-1*, a novel potato single MYB-like domain transcription factor, increases drought tolerance. *Plant Physiology*, *155*(1), 421–432.
Shiraki, T., Kondo, S., Katayama, S., Waki, K., Kasukawa, T., Kawaji, H., et al. (2003). Cap analysis gene expression for high-throughput analysis of transcriptional starting point and identification of promoter usage. *Proceedings of the National Academy of Sciences of the United States of America*, *100*(26), 15776–15781.
Singer, S. D., Cox, K. D., & Liu, Z. (2010). Both the constitutive *cauliflower mosaic virus 35S* and tissue-specific *AGAMOUS* enhancers activate transcription autonomously in *Arabidopsis thaliana*. *Plant Molecular Biology*, *74*(3), 293–305.
Singh, K., Foley, R., & Onate-Sanchez, L. (2002). Transcription factors in plant defense and stress responses. *Current Opinion in Plant Biology*, *5*(5), 430–436.
Smale, S. T., & Kadonaga, J. T. (2003). The RNA polymerase II core promoter. *Annual Review of Biochemistry*, *72*, 449–479.
Sornaraj, P., Luang, S., Lopato, S., & Hrmova, M. (2015). Basic leucine zipper (bZIP) transcription factors involved in abiotic stresses: A molecular model of a wheat bZIP factor and implications of its structure in function. *Biochimica et Biophysica Acta*, *1860*(1 Pt. A), 46–56.
Spyrou, C., Stark, R., Lynch, A. G., & Tavaré, S. (2009). BayesPeak: Bayesian analysis of ChIP-seq data. *BMC Bioinformatics*, *10*, 299.
Stam, M., Belele, C., Ramakrishna, W., Dorweiler, J. E., Bennetzen, J. L., & Chandler, V. L. (2002). The regulatory regions required for *B'* paramutation and expression are located far upstream of the maize *b1* transcribed sequences. *Genetics*, *162*(2), 917–930.
Stead, J. A., & McDowall, K. J. (2007). Two-dimensional gel electrophoresis for identifying proteins that bind DNA or RNA. *Nature Protocols*, *2*(8), 1839–1848.
Stockley, P. G. (2009). Filter-binding assays. *Methods in Molecular Biology*, *543*, 1–14.
Stoeckert, C. J., Salas, F., Brunk, B., & Overton, G. C. (1999). EpoDB: A prototype database for the analysis of genes expressed during vertebrate erythropoiesis. *Nucleic Acids Research*, *27*(1), 200–203.
Stupar, R. M., Gardiner, J. M., Oldre, A. G., Haun, W. J., Chandler, V. L., & Springer, N. M. (2008). Gene expression analyses in maize inbreds and hybrids with varying levels of heterosis. *BMC Plant Biology*, *8*, 33.
Stupar, R. M., Hermanson, P. J., & Springer, N. M. (2007). Nonadditive expression and parent-of-origin effects identified by microarray and allele-specific expression profiling of maize endosperm. *Plant Physiology*, *145*(2), 411–425.

Stupar, R. M., & Springer, N. M. (2006). *cis*-Transcriptional variation in maize inbred lines B73 and Mo17 leads to additive expression patterns in the F_1 hybrid. *Genetics*, *173*(4), 2199–2210.
Sullivan, A. M., Arsovski, A. A., Lempe, J., Bubb, K. L., Weirauch, M. T., Sabo, P. J., et al. (2014). Mapping and dynamics of regulatory DNA and transcription factor networks in *A. thaliana*. *Cell Reports*, *8*(6), 2015–2030.
Takahashi, H., Kato, S., Murata, M., & Carninci, P. (2012). CAGE (cap analysis of gene expression): A protocol for the detection of promoter and transcriptional networks. *Methods in Molecular Biology*, *786*, 181–200.
Takahashi, H., Lassmann, T., Murata, M., & Carninci, P. (2012). 5' end-centered expression profiling using cap-analysis gene expression and next-generation sequencing. *Nature Protocols*, 7(3), 542–561.
Takasaki, H., Maruyama, K., Kidokoro, S., Ito, Y., Fujita, Y., Shinozaki, K., et al. (2010). The abiotic stress-responsive NAC-type transcription factor OsNAC5 regulates stress-inducible genes and stress tolerance in rice. *Molecular Genetics and Genomics*, *284*(3), 173–183.
Tariq, M., & Paszkowski, J. (2004). DNA and histone methylation in plants. *Trends in Genetics*, *20*(6), 244–251.
Taylor-Teeples, M., Lin, L., de Lucas, M., Turco, G., Toal, T. W., Gaudinier, A., et al. (2015). An *Arabidopsis* gene regulatory network for secondary cell wall synthesis. *Nature*, *517*(7536), 571–575.
Triezenberg, S. J. (1995). Structure and function of transcriptional activation domains. *Current Opinion in Genetics & Development*, *5*(2), 190–196.
Tripathi, P., Rabara, R. C., & Rushton, P. J. (2014). A systems biology perspective on the role of WRKY transcription factors in drought responses in plants. *Planta*, *239*(2), 255–266.
Tuerk, C., & Gold, L. (1990). Systematic evolution of ligands by exponential enrichment: RNA ligands to bacteriophage T4 DNA polymerase. *Science*, *249*(4968), 505–510.
Turck, F., Roudier, F., Farrona, S., Martin-Magniette, M.-L. L., Guillaume, E., Buisine, N., et al. (2007). Arabidopsis TFL2/LHP1 specifically associates with genes marked by trimethylation of histone H3 lysine 27. *PLoS Genetics*, *3*(6).
Tuupanen, S., Turunen, M., Lehtonen, R., Hallikas, O., Vanharanta, S., Kivioja, T., et al. (2009). The common colorectal cancer predisposition SNP rs6983267 at chromosome 8q24 confers potential to enhanced Wnt signaling. *Nature Genetics*, *41*(8), 885–890.
Um, M., Li, C., & Manley, J. L. (1995). The transcriptional repressor even-skipped interacts directly with TATA-binding protein. *Molecular and Cellular Biology*, *15*(9), 5007–5016.
Vater, A., Jarosch, F., Buchner, K., & Klussmann, S. (2003). Short bioactive Spiegelmers to migraine-associated calcitonin gene-related peptide rapidly identified by a novel approach: Tailored-SELEX. *Nucleic Acids Research*, *31*(21), e130.
Vermeirssen, V., Barrasa, M. I., Hidalgo, C. A., Babon, J. A., Sequerra, R., Doucette-Stamm, L., et al. (2007). Transcription factor modularity in a gene-centered *C. elegans* core neuronal protein-DNA interaction network. *Genome Research*, *17*(7), 1061–1071.
Vidal, E. A., Alvarez, J. M., Moyano, T. C., & Gutierrez, R. A. (2015). Transcriptional networks in the nitrate response of *Arabidopsis thaliana*. *Current Opinion in Plant Biology*, *27*, 125–132.
von Korff, M., Radovic, S., Choumane, W., Stamati, K., Udupa, S. M., Grando, S., et al. (2009). Asymmetric allele-specific expression in relation to developmental variation and drought stress in barley hybrids. *The Plant Journal*, *59*(1), 14–26.
Walhout, A. J. (2011). What does biologically meaningful mean? A perspective on gene regulatory network validation. *Genome Biology*, *12*(4), 109.

Walhout, A. J., & Vidal, M. (1999). A genetic strategy to eliminate self-activator baits prior to high-throughput yeast two-hybrid screens. *Genome Research*, *9*(11), 1128–1134.
Walhout, A. J., & Vidal, M. (2001). High-throughput yeast two-hybrid assays for large-scale protein interaction mapping. *Methods*, *24*(3), 297–306.
Wang, Z., Libault, M., Joshi, T., Valliyodan, B., Nguyen, H. T., Xu, D., et al. (2010). SoyDB: A knowledge database of soybean transcription factors. *BMC Plant Biology*, *10*, 14.
Wang, R. L., Stec, A., Hey, J., Lukens, L., & Doebley, J. (1999). The limits of selection during maize domestication. *Nature*, *398*(6724), 236–239.
Warren, C. L., Kratochvil, N. C., Hauschild, K. E., Foister, S., Brezinski, M. L., Dervan, P. B., et al. (2006). Defining the sequence-recognition profile of DNA-binding molecules. *Proceedings of the National Academy of Sciences of the United States of America*, *103*(4), 867–872.
Wasson, T., & Hartemink, A. J. (2009). An ensemble model of competitive multi-factor binding of the genome. *Genome Research*, *19*(11), 2101–2112.
Wehner, N., Hartmann, L., Ehlert, A., Bottner, S., Onate-Sanchez, L., & Droge-Laser, W. (2011). High-throughput protoplast transactivation (PTA) system for the analysis of Arabidopsis transcription factor function. *The Plant Journal*, *68*(3), 560–569.
Weigel, D., & Nordborg, M. (2015). Population genomics for understanding adaptation in wild plant species. *Annual Review of Genetics*, *23*(49), 315–338.
Weirauch, M. T., Cote, A., Norel, R., Annala, M., Zhao, Y., Riley, T. R., et al. (2013). Evaluation of methods for modeling transcription factor sequence specificity. *Nature Biotechnology*, *31*(2), 126–134.
Weiste, C., Iven, T., Fischer, U., Onate-Sanchez, L., & Droge-Laser, W. (2007). In planta ORFeome analysis by large-scale over-expression of GATEWAY-compatible cDNA clones: Screening of ERF transcription factors involved in abiotic stress defense. *The Plant Journal*, *52*(2), 382–390.
West, M. A., Kim, K., Kliebenstein, D. J., van Leeuwen, H., Michelmore, R. W., Doerge, R. W., et al. (2007). Global eQTL mapping reveals the complex genetic architecture of transcript-level variation in *Arabidopsis*. *Genetics*, *175*(3), 1441–1450.
White, S., & Doebley, J. (1998). Of genes and genomes and the origin of maize. *Trends in Genetics*, *14*(8), 327–332.
White, S. E., & Doebley, J. F. (1999). The molecular evolution of *terminal ear1*, a regulatory gene in the genus Zea. *Genetics*, *153*(3), 1455–1462.
White, R., Rusconi, C., Scardino, E., Wolberg, A., Lawson, J., Hoffman, M., et al. (2001). Generation of species cross-reactive aptamers using "toggle" SELEX. *Molecular Therapy*, *4*(6), 567–573.
Williams, C. E., & Grotewold, E. (1997). Differences between plant and animal Myb domains are fundamental for DNA-binding, and chimeric Myb domains have novel DNA-binding specificities. *The Journal of Biological Chemistry*, *272*, 563–571.
Wilson, M. D., Barbosa-Morais, N. L., Schmidt, D., Conboy, C. M., Vanes, L., Tybulewicz, V. L., et al. (2008). Species-specific transcription in mice carrying human chromosome 21. *Science*, *322*(5900), 434–438.
Wingender, E., Chen, X., Fricke, E., Geffers, R., Hehl, R., Liebich, I., et al. (2001). The TRANSFAC system on gene expression regulation. *Nucleic Acids Research*, *29*(1), 281–283.
Wray, G. A., Hahn, M. W., Abouheif, E., Balhoff, J. P., Pizer, M., Rockman, M. V., et al. (2003). The evolution of transcriptional regulation in eukaryotes. *Molecular Biology and Evolution*, *20*(9), 1377–1419.
Wu, X., Shiroto, Y., Kishitani, S., Ito, Y., & Toriyama, K. (2009). Enhanced heat and drought tolerance in transgenic rice seedlings overexpressing *OsWRKY11* under the control of *HSP101* promoter. *Plant Cell Reports*, *28*(1), 21–30.

Xiang, Y., Tang, N., Du, H., Ye, H., & Xiong, L. (2008). Characterization of OsbZIP23 as a key player of the basic leucine zipper transcription factor family for conferring abscisic acid sensitivity and salinity and drought tolerance in rice. *Plant Physiology*, *148*(4), 1938–1952.
Xie, Z., & Grotewold, E. (2008). Serial ChIP as a tool to investigate the co-localization or exclusion of proteins on plant genes. *Plant Methods*, *4*, 25.
Xue, G.-P. P. (2005). A CELD-fusion method for rapid determination of the DNA-binding sequence specificity of novel plant DNA-binding proteins. *The Plant Journal*, *41*(4), 638–649.
Yamaguchi-Shinozaki, K., & Shinozaki, K. (1994). A novel *cis*-acting element in an Arabidopsis gene is involved in responsiveness to drought, low-temperature, or high-salt stress. *The Plant Cell*, *6*, 251–264.
Yamaguchi-Shinozaki, K., & Shinozaki, K. (2005). Organization of *cis*-acting regulatory elements in osmotic- and cold-stress-responsive promoters. *Trends in Plant Science*, *10*(2), 88–94.
Yan, H., Yuan, W., Velculescu, V. E., Vogelstein, B., & Kinzler, K. W. (2002). Allelic variation in human gene expression. *Science*, *297*(5584), 1143.
Yang, A., Dai, X., & Zhang, W. H. (2012). A R2R3-type MYB gene, *OsMYB2*, is involved in salt, cold, and dehydration tolerance in rice. *Journal of Experimental Botany*, *63*(7), 2541–2556.
Yang, W. M., Inouye, C., Zeng, Y., Bearss, D., & Seto, E. (1996). Transcriptional repression by YY1 is mediated by interaction with a mammalian homolog of the yeast global regulator RPD3. *Proceedings of the National Academy of Sciences of the United States of America*, *93*(23), 12845–12850.
Yang, W., Jefferson, R. A., Huttner, E., Moore, J. M., Gagliano, W. B., & Grossniklaus, U. (2005). An egg apparatus-specific enhancer of *Arabidopsis*, identified by enhancer detection. *Plant Physiology*, *139*(3), 1421–1432.
Yang, S., Vanderbeld, B., Wan, J., & Huang, Y. (2010). Narrowing down the targets: Towards successful genetic engineering of drought-tolerant crops. *Molecular Plant*, *3*(3), 469–490.
Yang, S. H., Vickers, E., Brehm, A., Kouzarides, T., & Sharrocks, A. D. (2001). Temporal recruitment of the mSin3A-histone deacetylase corepressor complex to the ETS domain transcription factor Elk-1. *Molecular and Cellular Biology*, *21*(8), 2802–2814.
Yang, Y., Yang, D., Schluesener, H. J., & Zhang, Z. (2007). Advances in SELEX and application of aptamers in the central nervous system. *Biomolecular Engineering*, *24*(6), 583–592.
Yilmaz, A., Mejia-Guerra, M., Kurz, K., Liang, X., Welch, L., & Grotewold, E. (2011). AGRIS: Arabidopsis Gene Regulatory Information Server, an update. *Nucleic Acids Research*, *39*, D1118–D1122.
Yilmaz, A., Nishiyama, M. Y., Garcia-Fuentes, B., Souza, G. M., Janies, D., Gray, J., et al. (2009). GRASSIUS: A platform for comparative regulatory genomics across the grasses. *Plant Physiology*, *149*(1), 171–180.
Yuh, C.-H., Bolouri, H., & Davidson, E. H. (1998). Genomic cis-regulatory logic: Experimental and computational analysis of a sea urchin gene. *Science*, *279*(5358), 1896–1902.
Yuh, C. H., Bolouri, H., & Davidson, E. H. (2001). *cis*-Regulatory logic in the *endo16* gene: Switching from a specification to a differentiation mode of control. *Development*, *128*(5), 617–629.
Zhang, Y., Butelli, E., Alseekh, S., Tohge, T., Rallapalli, G., Luo, J., et al. (2015). Multi-level engineering facilitates the production of phenylpropanoid compounds in tomato. *Nature Communications*, *6*, 8635.
Zhang, H., Jin, J., Tang, L., Zhao, Y., Gu, X., Gao, G., et al. (2011). PlantTFDB 2.0: Update and improvement of the comprehensive plant transcription factor database. *Nucleic Acids Research*, *39*(Database issue), D1114–D1117.

Zhang, Y., Liu, T., Meyer, C. A., Eeckhoute, J., Johnson, D. S., Bernstein, B. E., et al. (2008). Model-based analysis of ChIP-Seq (MACS). *Genome Biology*, *9*(9), R137.

Zhao, Y., Ruan, S., Pandey, M., & Stormo, G. D. (2012). Improved models for transcription factor binding site identification using nonindependent interactions. *Genetics*, *191*(3), 781–790.

Zheng, M., Barrera, L. O., Ren, B., & Wu, Y. N. (2007). ChIP-chip: Data, model, and analysis. *Biometrics*, *63*(3), 787–796.

Zheng, H., Sun, L., Peng, W., Shen, Y., Wang, Y., Xu, B., et al. (2011). Global identification of transcription start sites in the genome of *Apis mellifera* using 5'LongSAGE. *Journal of Experimental Zoology. Part B, Molecular and Developmental Evolution*, *316*(7), 500–514.

Zhong, R., & Ye, Z. H. (2015). Secondary cell walls: Biosynthesis, patterned deposition and transcriptional regulation. *Plant & Cell Physiology*, *56*(2), 195–214.

Zhu, J.-K. (2002). Salt and drought stress signal transduction in plants. *Annual Review of Plant Biology*, *53*, 247–273.

Zhu, Q.-H. H., Guo, A.-Y. Y., Gao, G., Zhong, Y.-F. F., Xu, M., Huang, M., et al. (2007). DPTF: A database of poplar transcription factors. *Bioinformatics (Oxford, England)*, *23*(10), 1307–1308.

Zhu, B., Zhang, W., Zhang, T., Liu, B., & Jiang, J. (2015). Genome-wide prediction and validation of intergenic enhancers in *Arabidopsis* using open chromatin signatures. *The Plant Cell*, *27*(9), 2415–2426.

Zhuang, Y., & Adams, K. L. (2007). Extensive allelic variation in gene expression in populus F1 hybrids. *Genetics*, *177*(4), 1987–1996.

Zuo, J., & Chua, N. H. (2000). Chemical-inducible systems for regulated expression of plant genes. *Current Opinion in Biotechnology*, *11*(2), 146–151.

Zykovich, A., Korf, I., & Segal, D. J. (2009). Bind-n-Seq: High-throughput analysis of in vitro protein-DNA interactions using massively parallel sequencing. *Nucleic Acids Research*, *37*(22), e151.

CHAPTER TWELVE

Engineering of Tomato Glandular Trichomes for the Production of Specialized Metabolites

R.W.J. Kortbeek*, J. Xu*, A. Ramirez*, E. Spyropoulou*, P. Diergaarde[†], I. Otten-Bruggeman[†], M. de Both[†], R. Nagel[‡], A. Schmidt[‡], R.C. Schuurink*[,1], P.M. Bleeker*

*Swammerdam Institute for Life Sciences, University of Amsterdam, Amsterdam, The Netherlands
[†]Keygene N.V., Wageningen, The Netherlands
[‡]Max Planck Institute for Chemical Ecology, Jena, Germany
[1]Corresponding author: e-mail address: R.C.Schuurink@uva.nl

Contents

1. Introduction 306
  1.1 Trichomes on Plants 306
  1.2 Specialized Metabolites and Application 306
  1.3 Production of Specialized Metabolites in Tomato Trichomes 307
  1.4 Modifying Terpene Production 308
  1.5 Focus 310
2. Materials and Technology 310
  2.1 Transient Transformation of Tomato Trichomes by Microparticle Bombardment 310
  2.2 Stable Transformation of Tomato Trichomes 316
3. Proof of Concept: Targeted Expression of a Terpene Precursor Gene in Tomato Glandular Trichomes 318
  3.1 Stable Trichome-Specific Expression of FPS 318
  3.2 Functional Validation of Trichome Engineering 321
4. Summary 325
Acknowledgments 326
References 327

Abstract

Glandular trichomes are specialized tissues on the epidermis of many plant species. On tomato they synthesize, store, and emit a variety of metabolites such as terpenoids, which play a role in the interaction with insects. Glandular trichomes are excellent tissues for studying the biosynthesis of specialized plant metabolites and are especially suitable targets for metabolic engineering. Here we describe the strategy for engineering tomato glandular trichomes, first with a transient expression system to provide proof

Methods in Enzymology, Volume 576
ISSN 0076-6879
http://dx.doi.org/10.1016/bs.mie.2016.02.014

of trichome specificity of selected promoters. Using microparticle bombardment, the trichome specificity of a terpene-synthase promoter could be validated in a relatively fast way. Second, we describe a method for stable expression of genes of interest in trichomes. Trichome-specific expression of another terpene-synthase promoter driving the yellow-fluorescence protein-gene is presented. Finally, we describe a case of the overexpression of farnesyl diphosphate synthase (FPS), specifically in tomato glandular trichomes, providing an important precursor in the biosynthetic pathway of sesquiterpenoids. FPS was targeted to the plastid aiming to engineer sesquiterpenoid production, but interestingly leading to a loss of monoterpenoid production in the transgenic tomato trichomes. With this example we show that trichomes are amenable to engineering though, even with knowledge of a biochemical pathway, the result of such engineering can be unexpected.

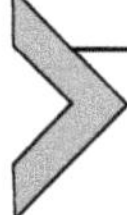

1. INTRODUCTION

1.1 Trichomes on Plants

Trichomes are specialized structures of epidermal origin found as extrusions or appendages on plant surfaces, but they are also found on lichens and even algae (Engene et al., 2012). Trichomes are very diverse in appearance and function in stress resistance, including excessive light or temperature, and insect and pathogen defence. They are usually shaped as hair-like structures, but can also appear as scales, buds or papillae and range from very small unicellular to very big multicellular structures. On tomato the trichomes are shaped as hairs and are differentiated in eight types including four that are glandular (Glas et al., 2012). Glandular trichomes are typically multicellular with one or more glandular cells that can produce, store, emit, or exude specialized compounds into the environment (volatile) or onto the epidermal surface (nonvolatile) (Schilmiller, Last, & Pichersky, 2008; Wagner, 1991). Targeting of particular biosynthetic pathways to these specialized structures is thought to be highly regulated as not to interfere with the plants' development. Glandular trichomes are well-known sources of essential oils and resins that have a widespread use in agricultural, pharmaceutical, and cosmetic industries.

1.2 Specialized Metabolites and Application

In addition to the primary metabolites that most plants have in common, plants can produce a large variety of specialized metabolites. Many of these metabolites, including terpenoids, acylsugars, indoles, phenolic compounds, and methyl ketones have a role in defence against herbivores and pathogenic organisms. All these compounds have been described to be synthesized by or present in glandular trichomes (for a review, see Glas et al., 2012). Terpenoids are the largest group of specialized metabolites with more than 30,000 terpenoids known to date to be produced by plants. Also tomato glandular trichomes produce a wide array of terpenoids, which are mostly specialized metabolites with roles in defence against pests and in attracting beneficial insects such as natural enemies of herbivores (Kappers et al., 2005; Simmons & Gurr, 2005; Weinhold & Baldwin, 2011).

1.3 Production of Specialized Metabolites in Tomato Trichomes

Terpenoids are synthesized via distinct metabolic pathways inside tomato trichome head cell(s). They are derived from rather simple universal C_5 isoprene building blocks but can result in a wide variety of decorated $(C_5)_n$ molecules. Two main pathways were identified for the production of these building blocks. Both these pathways result in the production of the intermediate C_5-molecules isopentenyl diphosphate (IPP) and its double bond isomer dimethylallyl diphosphate (DMAPP) but in different cellular compartments (Fig. 1). Both C_5 compounds are required in specific ratios by the prenyltransferases that subsequently form terpene precursors. The cytosolic mevalonate (MVA) pathway starts from acetyl-CoA and proceeds through the intermediate MVA into IPP and DMAPP and results in the biosynthesis of C_{15}-sesquiterpenes. The MVA pathway is present in all eukaryotes, but plants have an alternative biosynthetic pathway toward IPP and DMAPP called the MEP pathway (methyl-erythritol phosphate pathway) that is localized in the plastid. In tomato trichome plastids, IPP and DMAPP are then converted to C_{10}-, C_{15}-, and C_{20}-precursors that are further converted into mono-, sesqui-, and diterpenes, respectively (Fig. 1). There are indications that some intermediate molecules are able to cross the plastid membrane (Dudareva et al., 2005). The glandular trichomes of tomato are rather special in the sense that, besides canonical *trans*-precursors (GPP, geranyl diphosphate; FPP, farnesyl diphosphate; and GGPP, geranylgeranyl diphosphate) they can also convert IPP and DMAPP to *cis*-precursors in plastids. Neryl diphosphate is the *cis*-precursor for many monoterpenes in *Solanum lycopersicum* (Schilmiller et al., 2009) and in the wild tomato *Solanum habrochaites* plastidial sesquiterpenes are produced from *cis*-FPP (Sallaud et al., 2009).

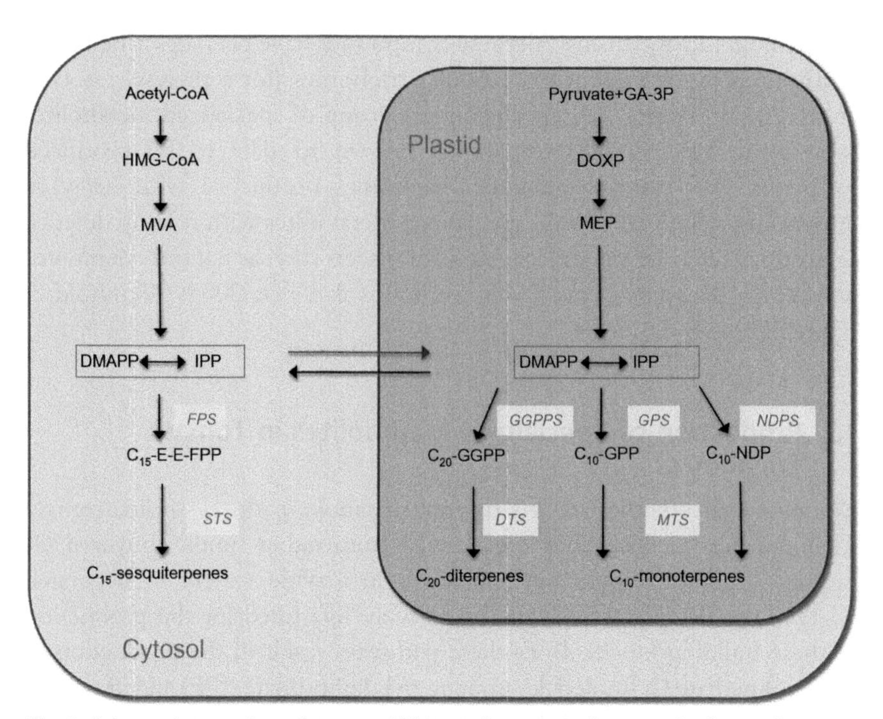

Fig. 1 Schematic overview of terpenoid biosynthesis in trichomes of cultivated tomato. The synthesis of terpenoids takes place in the cytosol through the mevalonate (MVA) pathway, in the plastid via the methyl-erythritol phosphate (MEP) pathway but also (partly) in the mitochondria, ER and/or peroxisomes (not depicted). Abbreviations: *Acetyl-CoA*, acetoacetyl-coenzyme A; *HMG-CoA*, 3-hydroxy-3-methylglutaryl-coenzyme A; *MVA*, mevalonate; *DMAPP*, dimethylallyl diphosphate; *IPP*, isopentenyl diphosphate; *E,E-FPP*, *trans*-farnesyl diphosphate; *FPS*, farnesyl diphosphate synthase; *STS*, sesquiterpene synthase; *GA-3P*, glyceraldehyde-3-phosphate; *DOXP*, 1-deoxy-D-xylulose 5-phosphate; *GGPP*, geranylgeranyl diphosphate; *GGPPS*, geranylgeranyl diphosphate synthase; *DPS*, diterpene synthase; *GPP*, geranyl diphosphate; *GPS*, geranyl diphosphate synthase; *NDP*, neryl diphosphate; *NDPS*, neryl diphosphate synthase; *MTS*, monoterpene synthase. (See the color plate.)

1.4 Modifying Terpene Production

The initial attempts to increase terpene production in plants were not done in glandular trichomes. Aharoni et al. (2003) showed that ectopic expression of a terpene synthase (FaNES, a strawberry linalool/nerolidol synthase) in *Arabidopsis thaliana* leaves was more successful when targeted to plastids than in the cytosol but did lead to growth retardation upon high expression levels. Targeting a terpene synthase to a specific cellular compartment for generating higher levels of terpenes was also pursued by Wu et al. (2006) and Huang et al. (2010). Wu et al. (2006) showed that sesquiterpene production in *Nicotiana tabacum* leaves was elevated as a consequence of directing a naturally cytosolic prenyltransferase (farnesyl diphosphate synthase, FPS) plus a sesquiterpene synthase to the plastid. It was reasoned that this increase in sesquiterpene production in the plastid was caused by the compartmentalized production. In addition to the plastid, IPP isomerase as well as FPS are also present in mitochondria (Cunillera, Boronat, & Ferrer, 1997; Phillips, D'Auria, Gershenzon, & Pichersky, 2008), providing an alternative engineering route. When FaNES was targeted to the mitochondria nerolidol levels were greatly increased compared to control, but also compared to plastid-targeted FaNES, due to the ample availability of FPP in the mitochondria (Kappers et al., 2005). Terpenoid biosynthetic pathways are under regulation of feedback inhibition. It is known that the levels of IPP and DMAPP, but also downstream metabolites such as FPP can influence enzyme activity of upstream precursors (Banerjee et al., 2013; Closa et al., 2010; Tholl & Lee, 2011).

Glandular trichomes are regarded excellent tissues for studying the biosynthesis of specialized plant metabolites, and as especially suitable targets for metabolic engineering (Ennajdaoui et al., 2010; Schilmiller et al., 2008). Using trichome-specific promoters rather than strong ubiquitous promoters has the advantage that possibly toxic products are not produced outside of these specialized tissues and that perturbation of metabolic pathways in the rest of the plant is avoided (Tissier, 2012). However, unexpected negative effects of trichome-targeted expression have been observed, though it is speculated this might be due to the use of an enhancer element in the expression cassette causing ectopic expression in meristematic or progenitor cells (Wu et al., 2012; Zhang et al., 2015).

Despite its potential, relatively few cases of trichome engineering have been reported (Tissier, 2012). Examples of the successful use of trichome-specific promoters include the targeted expression of a taxadiene synthase in tobacco under control of the cembatrienol synthase promoter (Rontein et al., 2008) and a beta-ketothiolase under control of two different cotton trichome promoters (John & Keller, 1996). Two genes from tomato, essential for the production of santalene and bergamotene, were successfully expressed in the glandular trichomes of tobacco (Sallaud et al., 2009). *Z*-abienol production was successfully engineered in the glandular trichomes of *Nicotiana sylvestris*, which normally does not produce this compound (Sallaud et al., 2012). In tomato, methylketone production was slightly increased by expressing two methylketone synthases under control of a trichome-specific promoter, while the same genes expressed under control of the strong and ubiquitous cauliflower mosaic virus 35S promoter showed lesions in their leaves (Yu & Pichersky, 2014). Two genes from wild tomato *S. habrochaites*, necessary for the production of 7-epizingiberene were introduced into cultivated tomato under different trichome-specific promoters (*ShMKS1* and *SlTPS5*, respectively) leading to production of 7-epizingiberene (Bleeker et al., 2012). Although this was sufficient to validate insect resistance, compared to the wild tomato from which these genes originate, only very low levels of this sesquiterpene were produced. This despite the fact that both the precursor and biosynthetic gene are plastid localized with the less stringently regulated MEP pathway (Bleeker et al., 2012; Sallaud et al., 2009). These examples indicate that optimizing metabolic engineering in trichomes might require additional factors, such as the upstream component of the precursor pathway (Lange et al., 2011).

1.5 Focus

Various promoters of genes expressed specifically in glandular trichomes have been identified (Table 1) which can be used to drive the expression of genes of interest. Here we focus on the use of (a number of) tomato trichome-specific promoters to achieve targeted expression. In this method, we describe the engineering of tomato glandular trichomes, first with a transient expression system to get proof of trichome specificity of promoters, and second a stable expression system for production of specialized compounds. As proof of concept, we describe a case of the latter with the specific overexpression of an FPS. Like Wu et al. (2006), we opted for avian FPS to avoid (post)transcriptional regulation and feedback inhibition, and as the cytosolic MVA flux appears to be much more regulated (Wu et al., 2012), we decided to target the expression of FPS to the plastids of glandular trichome secretory cells mediated by a transit peptide signal sequence.

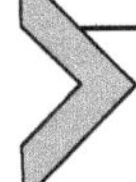

2. MATERIALS AND TECHNOLOGY

2.1 Transient Transformation of Tomato Trichomes by Microparticle Bombardment

Transient-plant transformation is a broadly applicable and relatively rapid way of checking the expression of designed plant-expression vectors in vivo. For this purpose, promoter activity is often visualized driving GUS (beta-glucuronidase) or fluorescent markers. The *Agrobacterium*-mediated transient-transformation assay (ATTA) is a well-known technique with high

Table 1 List of Trichome-Specific Promoters

Gene	Protein Description	Species	Reporter	Trichome Type	References
*Aa*ADS	Amorpha-4,11-diene synthase	*Artemisia annua*	GUS	Glandular trichomes	Wang et al. (2011)
CYP71AV1	Cytochrome P450	*Artemisia annua*	GUS	Glandular and nonglandular trichomes	Wang, Han, Kanagarajan, Lundgren, and Brodelius (2013)
*Aa*DBR2	Artemisinic aldehyde Δ11(13) reductase	*Artemisia annua*	GUS	Glandular secretory trichomes	Jiang et al. (2013)
*Aa*GL2	Transcription factor	*Artemisia annua*	GUS	Glandular and T-shaped trichomes	Jindal, Longchar, Singh, and Gupta (2015)
*Aa*MIXTA-Like1	Transcription factor	*Artemisia annua*	GUS	Glandular and T-shaped trichomes	Jindal et al. (2015)
SQAPI	Squash aspartic protease inhibitor	*Cucurbita maxima*	GUS	Long glandular trichomes in mature plant	Anandan, Gatehouse, Marshall, Murray, and Christeller (2009)
*Ns*CBTS-2a	Cembratrien-ol synthase	*Nicotiana sylvestris*	GUS	Tall glandular trichomes	Ennajdaoui et al. (2010)
CYP71D16	Cytochrome P450	*Nicotiana tabacum*	GUS	Glandular and nonglandular trichomes	Wang (2002)
T-phylloplanin	Surface-localized protein	*Nicotiana tabacum*	GFP, GUS	Short, procumbent glandular trichomes	Shepherd, Bass, Houtz, and Wagner (2005)
*Nt*LTP1	Lipid transfer protein	*Nicotiana tabacum*	GFP, GUS	Long glandular trichomes	Choi et al. (2012)

Continued

Table 1 List of Trichome-Specific Promoters—cont'd

Gene	Protein Description	Species	Reporter	Trichome Type	References
*Nt*CPS2	Copalyl diphosphate synthase	*Nicotiana tabacum*	GUS	Glandular-secreting trichomes	Sallaud et al. (2012)
*Sh*MKS1	Methylketone synthase	*Solanum habrochaites*	GFP	Type VI glandular trichomes	Akhtar et al. (2013)
*Sh*MKS2	Methylketone synthase	*Solanum habrochaites*	GFP	Strong in type VI, weak in other types	Yu and Pichersky (2014)
*Sl*ASAT4	Acylsucrose acyltransferase	*Solanum lycopersicum*	GFP	Type I and IV glandular trichomes	Schilmiller, Charbonneau, and Last (2012)
*Sl*EOT1	Transcription factor	*Solanum lycopersicum*	GUS–sYFP1	Type VI glandular trichomes	Spyropoulou, Haring, and Schuurink (2014)
*Sl*TPS5	Linalool synthase	*Solanum lycopersicum*	GUS–sYFP1	Type VI glandular trichomes	Spyropoulou et al. (2014)
*Sl*ASAT3	Acylsucrose acyltransferase	*Solanum lycopersicum*	GFP	Type I and IV glandular trichomes	Schilmiller et al. (2015)
*Sl*IPMS3	Isopropylmalate synthase	*Solanum lycopersicum*	GFP, GUS	Type I and IV glandular trichomes	Ning et al. (2015)

Abbreviations: *GUS*, β-glucuronidase; *GFP*, green fluorescent protein; *sYFP1*, yellow fluorescent protein.

transformation efficiency (Kapila, De Rycke, Van Montagu, & Angenon, 1997; Wroblewski, Tomczak, & Michelmore, 2005). However, construct delivery to glandular trichomes by *Agrobacterium* is difficult, possibly due to the trichome barrier cells that warrant specificity of transport to the stalk and head cells (Dell & McComb, 1978; Werker, 2000). Previous promoter studies in glandular trichomes have been done in stably transformed plants (Table 1), which is more time-consuming compared to transient transformation. Here we present a step-wise optimized method to transiently transform glandular trichomes on tomato stems using a microparticle bombardment system. The presented method is exemplified by transforming a GUS–sYFP marker gene driven by the *S. lycopersicum* Terpene Synthase 5 (*SlTPS5*) promoter. *SlTPS5* has previously been described and demonstrated to encode a linalool synthase with specific expression in type VI tomato trichomes (Falara et al., 2011; van Schie, Haring, & Schuurink, 2007).

2.1.1 Preparation of the pSlTPS5: GUS–sYFP1 Construct

A 1254-bp genomic sequence from *S. lycopersicum* cv. Moneymaker upstream of the start codon of *TPS5* (AY840091) was cloned between restriction sites *Sac*I (5′) and *Xba*I, replacing the 35S promoter of vector pJVII, a pMON999-based vector (Monsanto, St. Louis, MO). The *SlTPS5* promoter was cloned upstream of the start codon of *uid*A (GUS) fused to a yellow fluorescent protein (sYFP1) (Kremers, Goedhart, van Munster, & Gadella, 2006), followed by the *Nos* terminator (t*Nos*) (Fig. 2A). The final construct was sequenced and then transformed to *Escherichia coli* DH5α cells, grown overnight at 37°C in Luria Broth containing the appropriate antibiotic, after which the plasmid was isolated using a Midi Plasmid Purification Kit (Qiagen).

2.1.2 Particle Coating with DNA

Thirty-five microliter of gold particle solution (74 mg Au in 1.2 mL 100% ethanol) was placed in a 1.5-mL tube and spun shortly at maximum speed. The ethanol was carefully removed and without disturbing the pellet 1 mL of deionized water MQ was added to the tube. After spinning at 2000 rpm for 2 min, water was removed as much as possible. Next, the appropriate amount of plasmid (25 μg) was added in 250 μL of ice-cold MQ water. A quick vortexing step was followed by addition of 250 μL of ice-cold 2.5 *M* $CaCl_2$ and 50 μL of ice-cold freshly prepared 1.4 m*M* spermidine solution. Spermidine (05292 Sigma-Aldrich) was melted at 65°C and

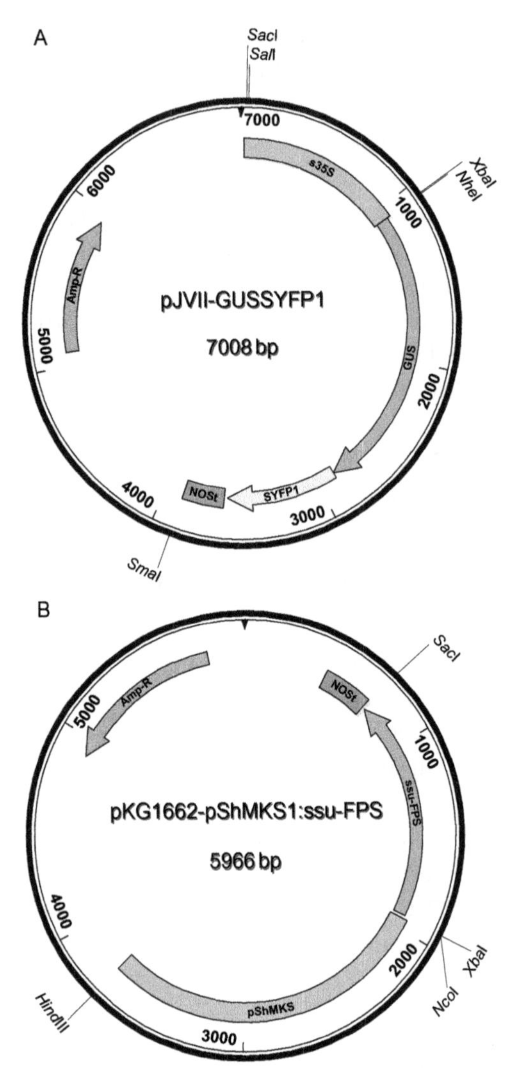

Fig. 2 Maps of plasmids used for engineering. (A) Map of the pJVII–GUS–sYFP plasmid used for transient and stable tomato transformation with trichome-specific promoters. For transient transformation by particle bombardment, the *S. lycopersicum* terpene synthase 5 (*SlTPS5*) promoter was used to drive GUS–sYFP. For stable transformation, the promoter of *S. lycopersicum* terpene synthase 9 (*SlTPS9*) was used. (B) pKG11662-pShMKS1:ssu-FPS construct used in the engineering of specialized metabolism in tomato trichomes. Abbreviations: *GUS*, β-glucuronidase; *sYFP1*, yellow fluorescent protein; *NOSt*, nopaline synthase terminator; *Amp-R*, β-lactamase ampicillin-resistance gene. (See the color plate.)

aliquoted under argon and stored at 80°C prior to use. The final mixture was vortexed for 10 min at 4°C and pelleted by spinning for 10 min at 500 rpm. After removal of the supernatant, the pellet was washed six times with 600 μL of 100% ethanol, by pipetting up and down to resuspend the pellet and spinning 1 min at maximum speed. Washing steps were repeated until a homogeneous suspension was obtained without visible gold clumps. Finally, 36 μL of 100% ethanol was added and the pellet was resuspended by vortexing. This amount of DNA-coated microparticles suspension was sufficient to perform three particle bombardments (10 μL each).

2.1.3 Plant Material and Growing Conditions

Tomato plants (*S. lycopersicum* cv. Moneymaker) were grown in soil in a greenhouse with a day/night temperature of 23/18°C and a 16/8 h light/dark regime for 4 weeks. For the transient transformation of trichomes, 3–4 cm long stem pieces were collected and distributed in the center of three Petri dishes containing 1% MS medium immediately before the bombardment.

2.1.4 Bombardment Conditions

Ten microliter of the DNA-coated gold particles (microcarrier) suspension was loaded onto the center of a macrocarrier, desiccated, and delivered according to the procedure described for the Biolistic PDS-1000/He system (https://www.bio-rad.com). Specific conditions used were rupture disks of 1100 PSI accelerating pressure, 9 cm target distance and 28 in. Hg vacuum pressure. The Petri dishes were sealed with Parafilm and bombarded stems were incubated in a growth chamber at 26°C for 48 h.

2.1.5 GUS Expression Assay

Forty-eight hours after bombardment, stem pieces were transferred to 15 mL tubes and submerged in 5-bromo-4-chloro-1*H*-indol-3yl-*b*-D-glucopyranosiduronic acid (X-Gluc) buffer containing 1 m*M* X-Gluc, 100 m*M* sodium phosphate buffer, pH 7.0, and 0.1% Triton X-100 (v/v). After 16 h of incubation at 37°C in the staining buffer, the stem pieces were washed three times with 50 m*M* sodium phosphate buffer (pH 7.2) and finally resuspended and preserved in the same buffer. Blue staining was assessed using a binocular microscope.

2.1.6 Analysis of Promoter Trichome Specificity

Here examination of the bombarded tomato stems showed that, even though the whole stem was hit by the DNA-coated particles, only trichomes

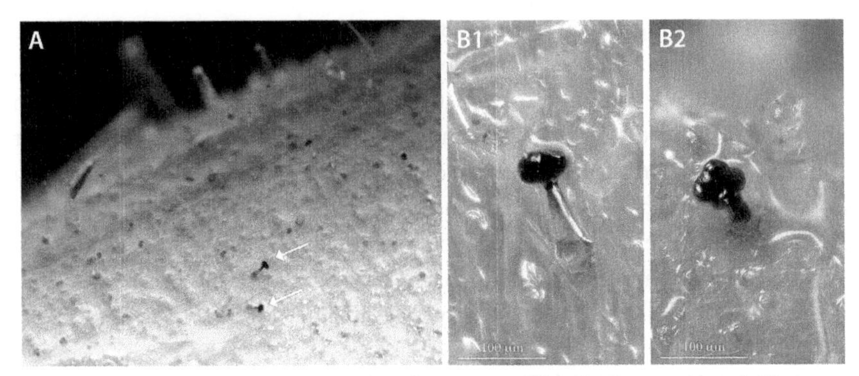

Fig. 3 GUS activity shows trichome-specific activity of the *SlTPS5* promoter. β-Glucuronidase (GUS) enzymatic activity in *S. lycopersicum* cv. Moneymaker stem trichomes transformed, by particle bombardment, with the pJVII-pSlTPS5:GUS construct containing GUS gene driven by the *S. lycopersicum* terpene synthase 5 (*SlTPS5*) promoter. Panoramic view (A) and close up (B1,2) of type VI glandular trichomes expressing SlTPS5: GUS. GUS activity was determined by incubation with X-Gluc (16 h). *Photograph by Jan van Arkel, IBED, University of Amsterdam.* (See the color plate.)

type VI expressed the GUS gene (Fig. 3A). A closer view on the stained trichomes showed that the *blue* coloring was not restricted to the four head cells but was also present in the basal stem of the type VI trichome (Fig. 3B1 and B2). This was likely due to diffusion of chloro-bromoindigo, to which GUS converts X-Gluc, from the head to adjacent cells as the staining was always found in a diffused pattern in the basal cells. However, it is also possible that the GUS protein itself has diffused (Mascarenhas & Hamilton, 1992).

Taken together, transient transformation by microparticle bombardment as presented here can be a quick and powerful way to assay the specificity of promoters in glandular trichomes. Importantly, trichomes can remain intact after delivery of the plasmid and are still able to synthesize the reporter protein. We therefore conclude this is a proper method to validate new trichome-specific promoters.

2.2 Stable Transformation of Tomato Trichomes

Although particle bombardment can indicate trichome specificity, the number of trichomes hit might not be very high and the procedure is destructive. Alternatively, for a more in-depth study, one can make stable transgenic plants expressing a promoter–reporter construct. In this section, a construct with the *SlTPS9* promoter driving a *GUS–sYFP1* fusion was introduced into *S. lycopersicum* cv. Moneymaker to verify the promoter activity of *SlTPS9* in stably transformed plants.

2.2.1 Preparation of Construct pSlTPS9: GUS–sYFP1

A 1557-bp fragment of genomic sequence upstream from the start codon of *SlTPS9* was cloned in vector pJVII between restriction sites *Sac*I (5′) and *Nhe*I as described in Section 2.1.1 (Fig. 1B). The construct was verified by sequencing and then the expression cassette was transferred to the MCS of the binary vector pBINplus (van Engelen et al., 1995) using the *Sac*I and *Sma*I restriction sites. The final construct was transformed to *Agrobacterium tumefaciens* GV3101 (pMP90).

2.2.2 Stable Tomato Transformation

Tomato (*S. lycopersicum* cv. Moneymaker) seeds were surface sterilized in 70% ethanol for 2 min followed by 20 min in 25% hypochlorite. After rinsing five times in sterile water they were placed on germination medium, which consists of 2.5 g L^{-1} Murashige and Skoog medium-including Gamborg B5 vitamins (MS + Vit B5), 10 g L^{-1} sucrose, and 0.5 g L^{-1} MES, pH 5.8 (Cortina & Culianez-Macia, 2004). Seedlings were grown at 25°C and 70% relative humidity for 10 days (90 μmol m^{-2} s^{-1}; 8 h dark, 16 h light). Cotyledons of sterile tomato seedlings were cut off, and the tips were removed and sectioned transversely with a scalpel in two fragments. Cotyledon cuts were placed adaxial-side down in 90 × 15 mm Petri dishes containing coculture medium (COM) and incubated for 1 day. The COM medium was composed of 4.5 g L^{-1} MS + Vit B5, 30 g L^{-1} sucrose, 0.5 g L^{-1} MES, 2 mg L^{-1} zeatin, 0.1 mg L^{-1} indole-3-acetic acid (IAA), 0.05 mg L^{-1} 2,4-dichlorophenoxyacetic acid (2,4-D), and 200 μ*M* acetosyringone, pH 5.8. *A. tumefaciens* strain GV3101 (pMP90) harboring the construct, which also contains the selection gene neomycin phosphotransferase (*NPTII*), was grown overnight to OD600 of 0.6–0.8 in modified Luria Bertani medium (1% Bacto Trypton, 0.5% yeast extract, and 0.25% NaCl, pH 7.0). Prior to cocultivation the culture was centrifuged (15 min at 3000 rcf) and the pellet was resuspended in liquid medium consisting of 4.5 g L^{-1} MS + Vit B5, 30 g L^{-1} sucrose, and 0.5 g L^{-1} MES, pH 5.8. Tomato cotyledon explants were removed from the COM plates and transferred to the bacterial suspension for 5 min. Next, they were placed on fresh COM plates after shortly drying on sterile filter paper. After 2 days of cultivation on COM, the explants were placed on postculture medium consisting of 4.5 g L^{-1} MS + Vit B5, 30 g L^{-1} sucrose, 0.5 g L^{-1} MES, 2 mg L^{-1} zeatin, 0.1 mg L^{-1} IAA, 200 mg L^{-1} cefotaxime, and 50 mg L^{-1} vancomycin, pH 5.8. Another 3 days later, the explants were transferred to shoot-inducing medium (SIM) composed of 4.5 g L^{-1} MS + Vit B5, 10 g L^{-1} glucose, 0.5 g L^{-1} MES, 2 mg L^{-1} zeatin, 0.1 mg L^{-1} IAA,

100 mg L^{-1} kanamycin, and 500 mg L^{-1} carbenicilin, pH 5.8. The plates were incubated at 25°C, 70% RH under fluorescent light (90 μmol m^{-2} s^{-1}; 8 h dark, 16 h light). Explants were transferred to fresh SIM every 2 weeks. Calli were removed from explants when they grew over 0.5 cm in width and transferred to fresh SIM. Emerging shoots from these calli were harvested and placed in sterile plant containers (68 × 66 mm) containing root-inducing medium (RIM). RIM consisted of 4.5 g L^{-1} MS + Vit B5, 10 g L^{-1} sucrose, 0.5 g L^{-1} MES, 0.25 mg L^{-1} indole-3-butyric acid (IBA), and 200 mg L^{-1} cefotaxime, pH 5.8. After root formation, the plants were gently removed from the containers and potted in soil.

2.2.3 Transgene Expression

Five independent transgenic primary (T0) lines were obtained after PCR verification of the presence of the *YFP* gene on genomic DNA isolated from leaves. Based on *YFP* expression in T0 stems and leaves, one transgenic line was selected for further analysis. Transgene expression was further confirmed by observing *YFP* fluorescence using an EVOSfl inverted microscope (http://www.thermofisher.com) in stems and leaves collected from the transgenic T0 plants. The selected transgenic line exhibited strong YFP fluorescence in glandular stem and leaf trichomes (Fig. 4B and D). Fig. 4 shows that fluorescence was specific to the four glandular head cells of the type VI trichomes. This result confirms the trichome-specific expression of *SlTPS9* as found by Bleeker, Diergaarde, et al. (2011) and Bleeker, Spyropoulou, et al. (2011). It furthermore shows that the method presented here is suitable to drive transgene expression specifically in the glandular trichomes of stably transformed tomatoes.

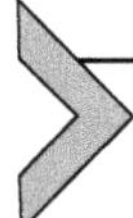

3. PROOF OF CONCEPT: TARGETED EXPRESSION OF A TERPENE PRECURSOR GENE IN TOMATO GLANDULAR TRICHOMES

3.1 Stable Trichome-Specific Expression of FPS

3.1.1 Preparation of pMKS1:ssu-SlFPS and pMKS1:ssu-GgFPS Transgenic Tomatoes

The *ShMKS1* promoter sequence was amplified from genomic DNA of the wild tomato *S. habrochaites* accession PI126449. The 1733-bp fragment was cloned in front of GUS in the pKG1662 vector (patent US2011/0113512A1), of which the 35S promoter was removed using *Hin*dIII and *Nco*I. The signal peptide of the Rubisco-small subunit (*ssu*) (168 bp)

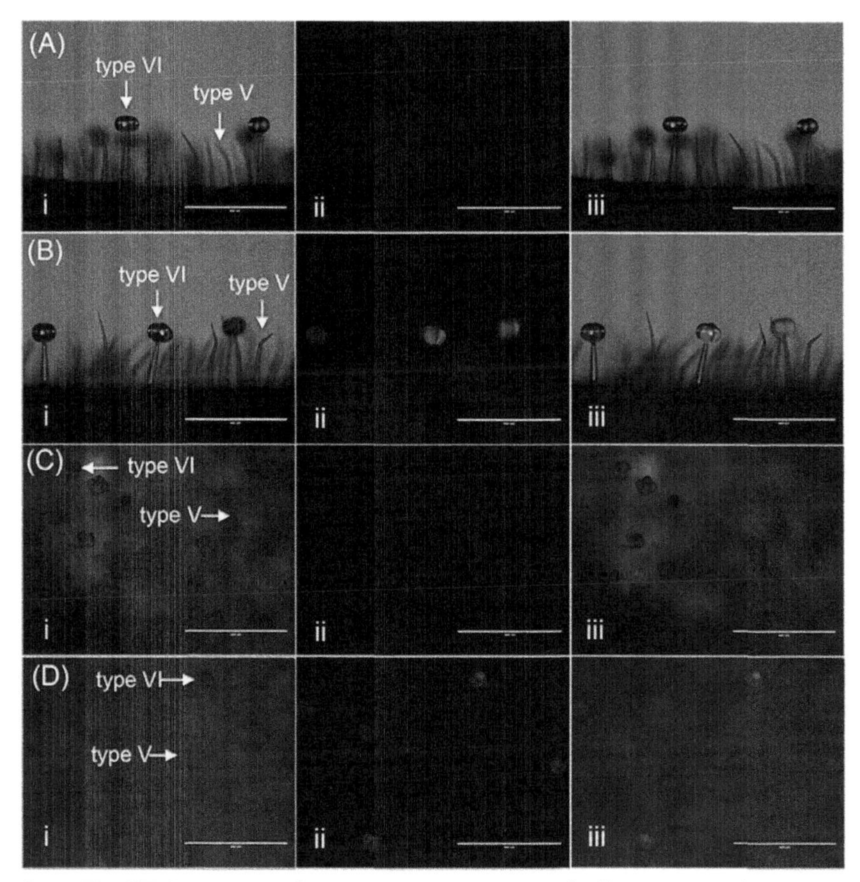

Fig. 4 The *SlTPS9* promoter drives sYFP expression specifically in type VI glandular trichomes of stable transgenic tomato plants. (A) Stem trichomes of untransformed plant. (B) Stem trichomes of tomato transformed with the *SlTPS9:GUS–sYFP* construct. (C) Leaf trichomes of untransformed plants. (D) Leaf trichomes of tomato transformed with the *SlTPS9:GUS–sYFP* construct. Images: (i) Normal light (brightfield), (ii) GFP filter, and (iii) merged image of (i) and (ii). Images were taken using the EVOSfl inverted microscope (http://www.thermofisher.com). *Arrows* indicate type VI glandular trichomes and type V nonglandular trichomes. Scale bars: 400 μm.

was amplified from *S. lycopersicum* with primers creating *Nco*I (3′) and modified *Nco*I (5′) ends. *Gallus gallus* FPS (*Gg*FPS) was amplified from chicken liver cDNA and digested with *Nco*I and *Sac*I. The two gene fragments were ligated into the *ShMKS1* promoter containing pGK1662 vector using the *Nco*I and *Sac*I restriction sites, replacing the GUS gene. This p*MKS1:ssu-GgFPS:35ST* cassette was subsequently cut out using *Hind*III/*Sac*I and ligated into pBINplus (van Engelen et al., 1995). Similar reactions were done to create the *ssu-SlFPS* constructs with the exception that FPS was amplified from *S. lycopersicum* cDNA. All constructs were checked by sequencing and then transformed to *A. tumefaciens* GV3101 and used to transform tomato *S. lycopersicum* var. Moneymaker as described in Section 2.2.2 of this chapter.

After stable plant transformation, successfully transformed plants growing on selective medium were checked for the presence of the transgene by PCR. Amplification was done on gDNA from leaf material using a *MKS1* promoter-specific primer and a gene-specific primer. In this way, at least three independent lines, containing either the *ssu-SlFPS* (*ssu-SlFPS* lines) or the *ssu-GgFPS* (*ssu-GgFPS* lines) transgene were selected for further analysis.

3.1.2 Expression Analysis of the Different FPS Transgenes

To confirm transgene expression, *FPS* transcript levels in the trichomes of three independently transformed lines were measured by quantitative real-time PCR (qRT-PCR) and compared to trichomes of untransformed plants. Glandular trichomes can contain large amounts of etheric oils and saccharides, which can make it hard to isolate RNA. Commercially available kits often suffice to obtain high quality RNA out of trichomes. However, for high RNA yield, we used an 80°C phenol extraction buffer followed by chloroform:isoamylalcohol phase separation and precipitation in 2 *M* LiCl. For a full method description, see Verwoerd, Dekker, and Hoekema (1989). After RNA isolation, DNase treatment and cDNA synthesis were done using commercially available kits from Ambion and Fermentas, respectively (www.thermofisher.com).

To quantify the FPS transcripts in the *ssu-SlFPS* and *ssu-GgFPS* lines, we used a *FPS*-specific primer pair to amplify the cDNA and normalized their levels to those of *Rubisco conjugating enzyme 1* (*RCE1*).

As shown in Fig. 5, *FPS* levels in the trichomes of three independent ssu-SlFPS lines are only moderately elevated compared to untransformed tomato ($n=3$). It must be noted that these also include basal expression levels of the native, cytosolic, *FPS*. Therefore transgene *ssu-SlFPS* transgene expression is expected to be lower than represented. Interestingly, transgene *GgFPS* expression of the three independent *ssu-GgFPS* lines showed to be much higher compared to the *ssu-SlFPS* lines (Fig. 6).

From these results it appears that *SlFPS* levels are under (post)transcriptional regulation in tomato trichomes. The *ShMKS1* promoter has been shown to be highly active in transgenic tomato trichomes when driving GFP (Akhtar et al., 2013). However, we found only a moderate increase

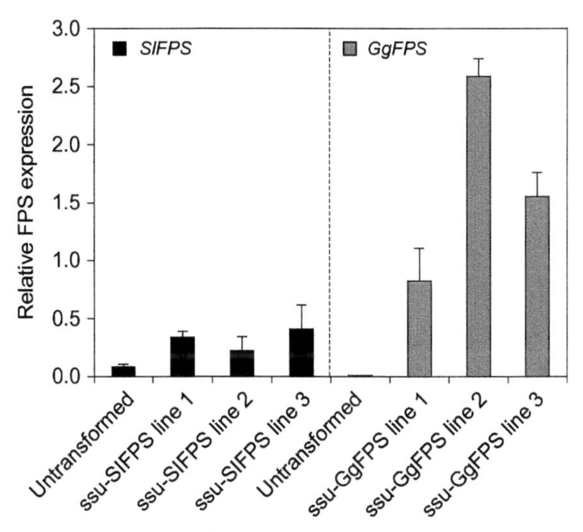

Fig. 5 Higher transcript levels of *SlIFPS* and *GgFPS* in transgenic tomato plants. Transcript levels of FPS in three independent transgenic T2 lines were analyzed by quantitative real-time PCR (qRT-PCR). Transcript levels of *ssu-SlFPS* lines were analyzed using *SlFPS*-specific primers (*black bars*) and transcript levels of *ssu-GgFPS* lines were analyzed using *GgFPS* specific primers (*gray bars*) with untransformed tomato plants as control. Transcript levels were normalized to *Rubisco conjugating enzyme 1* (*RCE1*). Bars represent means $\pm$ SE ($n=3$). Primers used: *SlFPS*; 5′-acagatgattctcgtaaatggg-3′ (f) and 5′-agatagtcctcggttcagcttc-3′ (r). *GgFPS*; 5′-ggagaggtacaaagccatcg-3′ (f) and 5′-ccttactgtcgatcccaacc-3′ (r). *RCE1*; 5′-gattctctctcatcaatcattcg-3′ (f) and 5′-tttggggacatcttcggatgaa-3′ (r).

in *SlFPS* transcript levels. Only driving *FPS* by the trichome-specific and constitutive *ShMKS1* promoter seems not suitable to get much higher expression levels. To boost transcript levels, using a nonplant ortholog of *FPS*, like *GgFPS*, can be a way to circumvent the regulatory mechanism.

3.2 Functional Validation of Trichome Engineering

3.2.1 Redirecting FPS to the Plastid Alters Terpenoid Production in Glandular Trichomes

Successful metabolic engineering largely depends on available precursor molecules to be utilized. Limited availability of precursor molecules makes it hard to produce an altered metabolic trait in transgenic plants. Also, metabolic fluxes between cell compartments in plants are only poorly understood (Rodríguez-Concepción & Boronat, 2015). To see if the enhanced expression levels of FPS also result in an increase of the FPP pool, we measured FPP levels from trichomes of 4-week-old *ssu*-*Gg*FPS (line 3) plants.

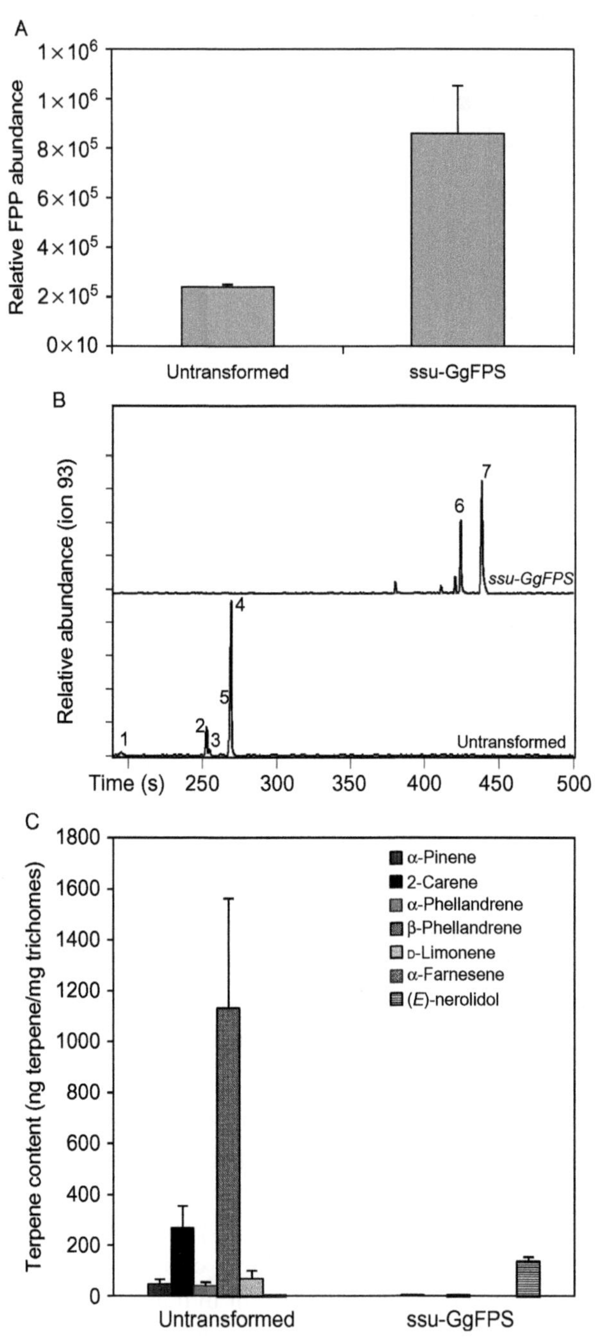

Fig. 6 Modified terpenoid metabolism in the trichomes of *ssu-GgFPS* transformed plants. (A) Measurements of sesquiterpene precursor farnesyl diphosphate (FPP). Bars represent mean values ± SE ($n = 3$). (B) GC-MS chromatograms of trichome extracts of untransformed (lower chromatogram) and *ssu-GgFPS* plants (upper chromatogram). Peaks: (1) α-pinene, (2) 2-carene, (3) α-phellandrene, (4) β-phellandrene, (5) D-limonene, (6) α-farnesene, and (7) (*E*)-nerolidol. (C) Quantified amounts of terpenes per mg trichomes. Bars represent mean values ± SE ($n = 3$).

The trichome content was extracted using methanol:water (1:3, v/v). Next, the extracts were purified on a CHROMABOND HR-XA column and quantified on LC–MS/MS. For a full description of the method, we refer to Nagel, Gershenzon, and Schmidt (2012).

Fig. 6A shows the relative amounts of FPP detected in the trichomes of untransformed plants and *ssu-GgFPS* transgenics ($n=3$). In the transgenic lines, the relative amount of FPP increased by ~3.5-fold compared to untransformed. So, targeting *Gg*FPS to the plastid did not only lead to elevated transcripts but also resulted in increased production of FPP.

We further analyzed the trichomes content of *ssu-GgFPS* transgenic plants by gas chromatography–mass spectrometry (GC–MS). In short, we isolated trichomes from stems of 4 weeks old tomato plants by shortly vortexing stem pieces frozen in liquid nitrogen in a 50-mL Greiner tube. Volatiles were extracted from frozen isolated trichomes with hexane spiked with 5 ng mL^{-1} benzyl acetate. Compounds were separated on a capillary DB-5 column (10 m × 180 μm, film thickness 0.18 μm; Hewlett-Packard) and mass spectra of eluting compounds were collected on a time-of-flight mass spectrometer (Leco Pegasus III). For a detailed description of the method, see Bleeker et al. (2009). We quantified the peak areas using the available terpene standards and normalized this to the fresh trichome weight.

Strikingly, we observed that terpene production shifted from mainly high levels of monoterpenes in untransformed trichomes, to only low levels of sesquiterpenes in trichomes of *ssu-GgFPS* transgenic lines (Fig. 6B). Compounds detected in untransformed plant samples were identified as α-pinene (1), 2-carene (2), α-phellandrene (3), β-phellandrene (4), and D-limonene (5). In the transgenic lines, two additional terpenes were detected in the trichome extract: α-farnesene (6) and (*E*)-nerolidol (7).

Quantification showed that the total terpene levels dropped dramatically in *ssu-GgFPS* lines. As shown in Fig. 6C, only trace amounts of monoterpenes were detected in the extracts of transgenic plants. Among these, the most abundant monoterpene was β-phellandrene (2.48 ng mg^{-1} trichomes, SE ± 0.27, $n=3$) which is a massive decrease compared to trichomes of untransformed plants (1131.69 ng mg^{-1} trichomes, SE ± 431.93, $n=3$). While we could not detect any sesquiterpenes in the extracts of untransformed trichomes, *ssu-GgFPS* lines produced α-farnesene (2.17 ng mg^{-1} trichomes, SE ± 0.61) and (*E*)-nerolidol (137.42 ng mg^{-1} trichomes, SE ± 16.46) as main products.

Although we did not introduce a sesquiterpene synthase in the plastid, additional sesquiterpenes were observed in the transgenic lines. We therefore assume there is a plastidial terpene synthase that can utilize FPP. In tomato, the plastidial MTS1 (TPS5) normally synthesizes the monoterpene (*R*)-linalool from GPP. However, in vitro experiments showed that MTS1 can utilize FPP resulting in the production of (*E*)-nerolidol (van Schie et al., 2007), providing an explanation for the production of (*E*)-nerolidol in the transgenic plants with FPS targeted to the plastid. The drastic depletion of monoterpene synthesis in the transgenic plants remains unexplained.

3.2.2 Biological Relevance of Changing Specialized Metabolites in Trichomes

Plant volatiles can function as a cue in insects' host preference. The whitefly *Bemisia tabaci* is an invasive generalist that constitutes a global problem in vegetable agriculture, as it serves as a vector for the geminiviruses causing crop damages recorded up to 95% (Polston & Anderson, 1997). Changing the volatile blend of commercial tomato, by the introduction of a novel sesquiterpene synthase from a wild species, has shown to improve repellence to *B. tabaci* (Bleeker et al., 2009, 2012). Here it was investigated if the observed change in terpenoid production of the *ssu-GgFPS* lines would also lead to improved repellence. The preference response of *B. tabaci* towards wild-type tomatoes and the transgenic line with an altered volatile blend was assessed in a free-choice bioassay. In this assay, we placed three untransformed, and one transgenic plant of the same age and size in the corners of a $1 \times 1\ m^2$ (Fig. 7A). A total of 150 whiteflies were released in the middle of the setup and recaptured from the plants after 20 min. In repetitions ($n = 4$) plants were rotated to avoid positional effects. The experiment was performed a total of four times with different *ssu-GgFPS* plants.

B. tabaci displayed significantly less preference for *ssu-GgFPS* plants compared to untransformed plants (Fig. 7B). There was no difference in the percentage of recaptured flies between untransformed lines (26%, 29%, and 32%, $p > 0.05$). In contrast, significantly fewer whiteflies were recaptured from *ssu-Gg*FPS transgenic plants (14%, $p < 0.01$). Thus, altering the production of terpenes in the trichomes can affect the plant's attractiveness to whiteflies. Whiteflies and other insects use, among other cues, semiochemical signals to detect their host plant (Birkett et al., 2004; Bleeker et al., 2009). The major loss of monoterpene production observed in *ssu-Gg*FPS plants therefore might have resulted in a loss of cues for *B. tabaci* to select its host plant. On the other hand, *ssu-GgFPS* lines produce (a low amount of) (*E*)-nerolidol that has been previously identified to play a role in the indirect defence of multiple plant species against herbivores

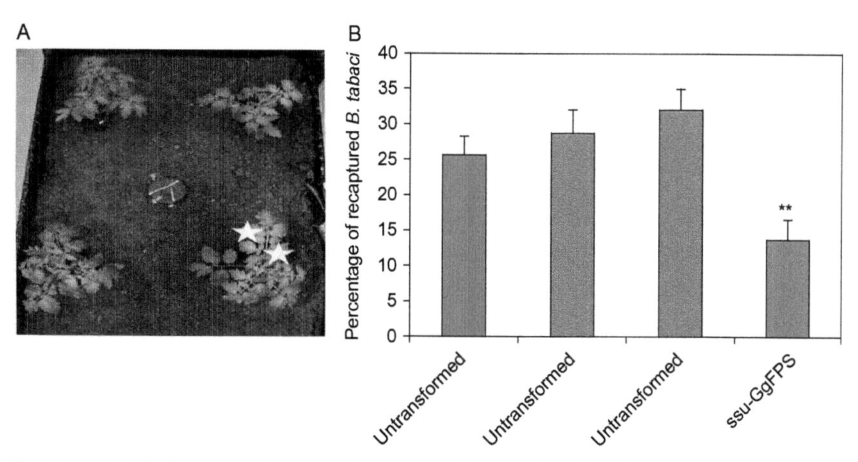

Fig. 7 ssu-GgFPS tomato plants are less attractive to whiteflies. (A) Experimental setup: 150 whiteflies (*Bemisia tabaci*) were released in the middle of a 1×1 m^2 with in three corners untransformed plants and in the fourth corner a *ssu-GgFPS* transgenic plant (indicated with *stars*). After 20 min, the whiteflies were recaptured from the plants and counted. After each assay, plants were rotated to exclude any positional effect. (B) Percentage of whiteflies that were recaptured on each plant. Significantly less whiteflies were recaptured on *ssu-GgFPS* transgenic plants (ANOVA: $p < 0.01$). Bars represent mean percentage of recaptured whiteflies ± SE ($n = 12$). (See the color plate.)

(Bouwmeester, Verstappen, Posthumus, & Dicke, 1999; Degenhardt & Gershenzon, 2000; Kant, Ament, Sabelis, Haring, & Schuurink, 2004; Kappers et al., 2005; Tholl, Sohrabi, Huh, & Lee, 2011). Also, the emission of various sesquiterpenes including 7-epizingiberene, (*R*)-curcumene, and (*E*,*E*)-α-farnesene have been shown to act as a repellent to insects and small herbivores (Bhatia, Maisnam, Jain, Sharma, & Bhattacharya, 2015; Bleeker, Diergaarde, et al., 2011; Bleeker, Spyropoulou, et al., 2011; Lopez, Quero, Iturrondobeitia, Guerrero, & Goldarazena, 2013). Despite the fact that nerolidol levels in the transgenic lines are low, we cannot exclude that emission of this novel compound played a role in the decreased attractiveness of plants expressing FPS in the plastid.

4. SUMMARY

In this chapter, we described a generic outline for metabolic engineering of glandular trichomes of plants by presenting a case study of tomato glandular trichomes. What is first and foremost needed for the plant of choice are one or more glandular trichome-specific promoters as any specific engineering of the glandular trichome will, in most cases, not affect normal plant development. Table 1 presents trichome-specific promoters that are readily available but screening of promoters can be rapidly done by particle bombardment of the trichomes of choice. Trichome databases (eg, http://www.planttrichome.org/trichomedb) are accessible and since the costs for RNAseq are rapidly coming down, transcriptomes of the trichomes of any plant are now within reach. Upon selecting putative trichome-specific genes, using for instance qRT-PCR on different tissues and organs, the specificity of the corresponding promoters still needs to be tested. Particle bombardment is a very fast and effective tool for this since expression of the reporter genes can be observed within days. In particular, as depicted in Fig. 3, GUS activity can be visualized with X-Gluc as substrate in trichomes. Since the gold particles not only hit the trichomes but also the epidermal cells and the mesophyll cells below, any activity of the promoters in these cells would directly be visible. In addition, one can also use other tissues and organs for particle bombardment to determine if expression is absent using a constitutive promoter as a positive control. Still transgenic plants with the promoter driving the reporter gene (Fig. 4) are necessary to know if the promoter is also expressed in other cells, if any, beside the glandular trichomes.

It is not necessary to mention that knowledge of the metabolism to be engineered is paramount. Trichomes are amenable to engineering, but even with extensive knowledge of the terpenoid biosynthesis pathways, the outcome may be unpredictable. By simply locating FPS in the plastid, we almost completely eliminated monoterpene biosynthesis (Fig. 6B and C). The mechanisms involved remain to be resolved. We did measure an increase in FPP levels (Fig. 6A) in trichomes expressing FPS in their plastids, but we do not know how this influenced, for instance, NDP levels, the substrate for the most abundant monoterpene, β-phellandrene. Our results suggest that high FPP levels inhibited NDPS, similarly as NPP inhibits GGPPS (Gutensohn et al., 2014). In the cytosol, FPP was thought to serve as a regulator of carbon flux in the cytosolic MVA pathway via feedback inhibition of HMGR (Closa et al., 2010). In addition, the plastidial linalool synthase that, in vitro, also accepts FPP as a substrate now produced the sesquiterpene nerolidol. This indicates that additional targeting of a sesquiterpene synthase to the plastid would result in production of higher levels of the desired sesquiterpenes. Modifiers of these sesquiterpenes, such as P450s, could then be added as well, under control of trichome-specific promoters.

ACKNOWLEDGMENTS

Ludek Tikovsky and Harold Lemereis are acknowledged for taking care of the tomato plants.

REFERENCES

Aharoni, A., Giri, A. P., Deuerlein, S., Griepink, F., de Kogel, W.-J., Verstappen, F. W., et al. (2003). Terpenoid metabolism in wild-type and transgenic Arabidopsis plants. *Plant Cell, 15*, 2866–2884.

Akhtar, T. A., Matsuba, Y., Schauvinhold, I., Yu, G., Lees, H. A., Klein, S. E., et al. (2013). The tomato cis-prenyltransferase gene family. *Plant Journal, 73*, 640–652. http://dx.doi.org/10.1111/tpj.12063.

Anandan, A., Gatehouse, L. N., Marshall, R. K., Murray, C., & Christeller, J. T. (2009). Two highly homologous promoters of a squash aspartic protease inhibitor (SQAPI) multigene family exhibit differential expression in transgenic tobacco phloem and trichome cells. *Plant Molecular Biology Reporter, 27*, 355–364. http://dx.doi.org/10.1007/s11105-009-0096-1.

Banerjee, A., Wu, Y., Banerjee, R., Li, Y., Yan, H., & Sharkey, T. D. (2013). Feedback inhibition of deoxy-D-xylulose-5-phosphate synthase regulates the methylerythritol 4-phosphate pathway. *Journal of Biological Chemistry, 288*, 16926–16936.

Bhatia, V., Maisnam, J., Jain, A., Sharma, K. K., & Bhattacharya, R. (2015). Aphid-repellent pheromone E-β-farnesene is generated in transgenic Arabidopsis thaliana over-expressing farnesyl diphosphate synthase2. *Annals of Botany, 115*(4), 581–591.

Birkett, M., Agelopoulos, N., Jensen, K., Jespersen, J. B., Pickett, J., Prijs, H., et al. (2004). The role of volatile semiochemicals in mediating host location and selection by nuisance and disease-transmitting cattle flies. *Medical and Veterinary Entomology, 18*, 313–322.

Bleeker, P. M., Diergaarde, P. J., Ament, K., Guerra, J., Weidner, M., Schütz, S., et al. (2009). The role of specific tomato volatiles in tomato-whitefly interaction. *Plant Physiology, 151*, 925–935.

Bleeker, P. M., Diergaarde, P. J., Ament, K., Schütz, S., Johne, B., Dijkink, J., et al. (2011). Tomato-produced 7-epizingiberene and R-curcumene act as repellents to whiteflies. *Phytochemistry, 72*, 68–73.

Bleeker, P. M., Mirabella, R., Diergaarde, P. J., VanDoorn, A., Tissier, A., Kant, M. R., et al. (2012). Improved herbivore resistance in cultivated tomato with the sesquiterpene biosynthetic pathway from a wild relative. *Proceedings of the National Academy of Sciences of the United States of America, 109*, 20124–20129.

Bleeker, P. M., Spyropoulou, E. A., Diergaarde, P. J., Volpin, H., De Both, M. T., Zerbe, P., et al. (2011). RNA-seq discovery, functional characterization, and comparison of sesquiterpene synthases from Solanum lycopersicum and Solanum habrochaites trichomes. *Plant Molecular Biology*, 77, 323–336. http://dx.doi.org/10.1007/s11103-011-9813-x.

Bouwmeester, H. J., Verstappen, F. W., Posthumus, M. A., & Dicke, M. (1999). Spider mite-induced (3S)-(*E*)-nerolidol synthase activity in cucumber and lima bean. The first dedicated step in acyclic C11-homoterpene biosynthesis. *Plant Physiology, 121*, 173–180.

Choi, Y. E., Lim, S., Kim, H. J., Han, J. Y., Lee, M. H., Yang, Y., et al. (2012). Tobacco NtLTP1, a glandular-specific lipid transfer protein, is required for lipid secretion from glandular trichomes. *Plant Journal, 70*, 480–491. http://dx.doi.org/10.1111/j.1365-313X.2011.04886.x.

Closa, M., Vranová, E., Bortolotti, C., Bigler, L., Arró, M., Ferrer, A., et al. (2010). The Arabidopsis thaliana FPP synthase isozymes have overlapping and specific functions in isoprenoid biosynthesis, and complete loss of FPP synthase activity causes early developmental arrest. *The Plant Journal, 63*, 512–525.

Cortina, C., & Culianez-Macia, F. A. (2004). Tomato transformation and transgenic plant production. *Plant Cell, Tissue and Organ Culture, 76*, 269–275. http://dx.doi.org/10.1023/B:Ticu.0000009249.14051.77.

Cunillera, N., Boronat, A., & Ferrer, A. (1997). The Arabidopsis thaliana FPS1 gene generates a novel mRNA that encodes a mitochondrial farnesyl-diphosphate synthase isoform. *Journal of Biological Chemistry, 272*, 15381–15388.

Degenhardt, J., & Gershenzon, J. (2000). Demonstration and characterization of (*E*)-nerolidol synthase from maize: A herbivore-inducible terpene synthase participating in (3*E*)-4, 8-dimethyl-1,3,7-nonatriene biosynthesis. *Planta, 210*, 815–822.

Dell, B., & McComb, A. (1978). Biosynthesis of resin terpenes in leaves and glandular hairs of Newcastelia viscida. *Journal of Experimental Botany, 29*, 89–95.

Dudareva, N., Andersson, S., Orlova, I., Gatto, N., Reichelt, M., Rhodes, D., et al. (2005). The nonmevalonate pathway supports both monoterpene and sesquiterpene formation in snapdragon flowers. *Proceedings of the National Academy of Sciences of the United States of America, 102*, 933–938.

Engene, N., Rottacker, E. C., Kaštovský, J., Byrum, T., Choi, H., Ellisman, M. H., et al. (2012). Moorea producens gen. nov., sp. nov. and Moorea bouillonii comb. nov., tropical marine cyanobacteria rich in bioactive secondary metabolites. *International Journal of Systematic and Evolutionary Microbiology, 62*, 1171–1178.

Ennajdaoui, H., Vachon, G., Giacalone, C., Besse, I., Sallaud, C., Herzog, M., et al. (2010). Trichome specific expression of the tobacco (Nicotiana sylvestris) cembratrien-ol synthase genes is controlled by both activating and repressing cis-regions. *Plant Molecular Biology, 73*, 673–685.

Falara, V., Akhtar, T. A., Nguyen, T. T., Spyropoulou, E. A., Bleeker, P. M., Schauvinhold, I., et al. (2011). The tomato terpene synthase gene family. *Plant Physiology, 157*, 770–789. http://dx.doi.org/10.1104/pp.111.179648.

Glas, J. J., Schimmel, B. C., Alba, J. M., Escobar-Bravo, R., Schuurink, R. C., & Kant, M. R. (2012). Plant glandular trichomes as targets for breeding or engineering of resistance to herbivores. *International Journal of Molecular Sciences, 13*, 17077–17103. http://dx.doi.org/10.3390/ijms131217077.

Gutensohn, M., Nguyen, T. T., McMahon, R. D., Kaplan, I., Pichersky, E., & Dudareva, N. (2014). Metabolic engineering of monoterpene biosynthesis in tomato fruits via introduction of the non-canonical substrate neryl diphosphate. *Metabolic Engineering, 24*, 107–116.

Huang, M., Abel, C., Sohrabi, R., Petri, J., Haupt, I., Cosimano, J., et al. (2010). Variation of herbivore-induced volatile terpenes among Arabidopsis ecotypes depends on allelic differences and subcellular targeting of two terpene synthases, TPS02 and TPS03. *Plant Physiology, 153*, 1293–1310.

Jiang, W., Lu, X., Qiu, B., Zhang, F., Shen, Q., Lv, Z., et al. (2013). Molecular cloning and characterization of a trichome-specific promoter of artemisinic aldehyde Δ11(13) reductase (DBR2) in Artemisia annua. *Plant Molecular Biology Reporter, 32*, 82–91. http://dx.doi.org/10.1007/s11105-013-0603-2.

Jindal, S., Longchar, B., Singh, A., & Gupta, V. (2015). Promoters of AaGL2 and AaMIXTA-like1 gene of Artemisia annua direct reporter gene expression in glandular and non-glandular trichomes. *Plant Signaling & Behavior, 10*(12), e1087629. http://dx.doi.org/10.1080/15592324.2015.1087629.

John, M. E., & Keller, G. (1996). Metabolic pathway engineering in cotton: Biosynthesis of polyhydroxybutyrate in fiber cells. *Proceedings of the National Academy of Sciences of the United States of America, 93*, 12768–12773.

Kant, M. R., Ament, K., Sabelis, M. W., Haring, M. A., & Schuurink, R. C. (2004). Differential timing of spider mite-induced direct and indirect defenses in tomato plants. *Plant Physiology, 135*, 483–495.

Kapila, J., De Rycke, R., Van Montagu, M., & Angenon, G. (1997). An Agrobacterium-mediated transient gene expression system for intact leaves. *Plant Science, 122*, 101–108.

Kappers, I. F., Aharoni, A., Van Herpen, T. W., Luckerhoff, L. L., Dicke, M., & Bouwmeester, H. J. (2005). Genetic engineering of terpenoid metabolism attracts bodyguards to Arabidopsis. *Science, 309*, 2070–2072.

Kremers, G.-J., Goedhart, J., van Munster, E. B., & Gadella, T. W. (2006). Cyan and yellow super fluorescent proteins with improved brightness, protein folding, and FRET Förster radius. *Biochemistry, 45*, 6570–6580.

Lange, B. M., Mahmoud, S. S., Wildung, M. R., Turner, G. W., Davis, E. M., Lange, I., et al. (2011). Improving peppermint essential oil yield and composition by metabolic engineering. *Proceedings of the National Academy of Sciences of the United States of America, 108*, 16944–16949.

Lopez, S., Quero, C., Iturrondobeitia, J. C., Guerrero, Á., & Goldarazena, A. (2013). Electrophysiological and behavioural responses of Pityophthorus pubescens (Coleoptera: Scolytinae) to (*E*, *E*)-α-farnesene,(*R*)-(+)-limonene and (*S*)-(−)-verbenone in Pinus radiata (Pinaceae) stands in northern Spain. *Pest Management Science, 69*, 40–47.

Mascarenhas, J. P., & Hamilton, D. A. (1992). Artifacts in the localization of GUS activity in anothers of petunia transformed with a CaMV 35S—GUS construct. *The Plant Journal, 2*, 405–408.

Nagel, R., Gershenzon, J., & Schmidt, A. (2012). Nonradioactive assay for detecting isoprenyl diphosphate synthase activity in crude plant extracts using liquid chromatography coupled with tandem mass spectrometry. *Analytical Biochemistry, 422*, 33–38.

Ning, J., Moghe, G. D., Leong, B., Kim, J., Ofner, I., Wang, Z., et al. (2015). A feedback-insensitive isopropylmalate synthase affects acylsugar composition in cultivated and wild tomato. *Plant Physiology, 169*, 1821–1835. http://dx.doi.org/10.1104/pp.15.00474.

Phillips, M. A., D'Auria, J. C., Gershenzon, J., & Pichersky, E. (2008). The Arabidopsis thaliana type I isopentenyl diphosphate isomerases are targeted to multiple subcellular compartments and have overlapping functions in isoprenoid biosynthesis. *Plant Cell, 20*, 677–696. http://dx.doi.org/10.1105/tpc.107.053926.

Polston, J. E., & Anderson, P. K. (1997). The emergence of whitefly-transmitted geminiviruses in tomato in the western hemisphere. *Plant Disease, 81*, 1358–1369.

Rodríguez-Concepción, M., & Boronat, A. (2015). Breaking new ground in the regulation of the early steps of plant isoprenoid biosynthesis. *Current Opinion in Plant Biology, 25*, 17–22.

Rontein, D., Onillon, S., Herbette, G., Lesot, A., Werck-Reichhart, D., Sallaud, C., et al. (2008). CYP725A4 from yew catalyzes complex structural rearrangement of taxa-4 (5), 11 (12)-diene into the cyclic ether 5 (12)-oxa-3 (11)-cyclotaxane. *Journal of Biological Chemistry, 283*, 6067–6075.

Sallaud, C., Giacalone, C., Topfer, R., Goepfert, S., Bakaher, N., Rosti, S., et al. (2012). Characterization of two genes for the biosynthesis of the labdane diterpene Z-abienol in tobacco (Nicotiana tabacum) glandular trichomes. *Plant Journal, 72*, 1–17. http://dx.doi.org/10.1111/j.1365-313X.2012.05068.x.

Sallaud, C., Rontein, D., Onillon, S., Jabes, F., Duffe, P., Giacalone, C., et al. (2009). A novel pathway for sesquiterpene biosynthesis from Z,Z-farnesyl pyrophosphate in the wild tomato Solanum habrochaites. *Plant Cell, 21*, 301–317. http://dx.doi.org/10.1105/tpc.107.057885.

Schilmiller, A. L., Charbonneau, A. L., & Last, R. L. (2012). Identification of a BAHD acetyltransferase that produces protective acyl sugars in tomato trichomes. *Proceedings of the National Academy of Sciences of the United States of America, 109*, 16377–16382. http://dx.doi.org/10.1073/pnas.1207906109.

Schilmiller, A. L., Last, R. L., & Pichersky, E. (2008). Harnessing plant trichome biochemistry for the production of useful compounds. *The Plant Journal, 54*, 702–711.

Schilmiller, A. L., Moghe, G. D., Fan, P., Ghosh, B., Ning, J., Jones, A. D., et al. (2015). Functionally divergent alleles and duplicated Loci encoding an acyltransferase contribute to acylsugar metabolite diversity in Solanum trichomes. *Plant Cell, 27*, 1002–1017. http://dx.doi.org/10.1105/tpc.15.00087.

Schilmiller, A. L., Schauvinhold, I., Larson, M., Xu, R., Charbonneau, A. L., Schmidt, A., et al. (2009). Monoterpenes in the glandular trichomes of tomato are synthesized from a neryl diphosphate precursor rather than geranyl diphosphate. *Proceedings of the National Academy of Sciences of the United States of America*, *106*, 10865–10870.

Shepherd, R. W., Bass, W. T., Houtz, R. L., & Wagner, G. J. (2005). Phylloplanins of tobacco are defensive proteins deployed on aerial surfaces by short glandular trichomes. *Plant Cell*, *17*, 1851–1861. http://dx.doi.org/10.1105/tpc.105.031559.

Simmons, A. T., & Gurr, G. M. (2005). Trichomes of Lycopersicon species and their hybrids: Effects on pests and natural enemies. *Agricultural and Forest Entomology*, *7*, 265–276.

Spyropoulou, E. A., Haring, M. A., & Schuurink, R. C. (2014). Expression of terpenoids 1, a glandular trichome-specific transcription factor from tomato that activates the terpene synthase 5 promoter. *Plant Molecular Biology*, *84*, 345–357. http://dx.doi.org/10.1007/s11103-013-0142-0.

Tholl, D., & Lee, S. (2011). Terpene specialized metabolism in Arabidopsis thaliana. *The Arabidopsis book: Vol. 9*. Rockville, MD: American Society of Plant Biologists.

Tholl, D., Sohrabi, R., Huh, J.-H., & Lee, S. (2011). The biochemistry of homoterpenes—Common constituents of floral and herbivore-induced plant volatile bouquets. *Phytochemistry*, *72*, 1635–1646.

Tissier, A. (2012). Glandular trichomes: What comes after expressed sequence tags? *The Plant Journal*, *70*, 51–68.

van Engelen, F. A., Molthoff, J. W., Conner, A. J., Nap, J.-P., Pereira, A., & Stiekema, W. J. (1995). pBINPLUS: An improved plant transformation vector based on pBIN19. *Transgenic Research*, *4*, 288–290.

van Schie, C. C., Haring, M. A., & Schuurink, R. C. (2007). Tomato linalool synthase is induced in trichomes by jasmonic acid. *Plant Molecular Biology*, *64*, 251–263. http://dx.doi.org/10.1007/s11103-007-9149-8.

Verwoerd, T. C., Dekker, B., & Hoekema, A. (1989). A small-scale procedure for the rapid isolation of plant RNAs. *Nucleic Acids Research*, *17*, 2362.

Wagner, G. J. (1991). Secreting glandular trichomes: More than just hairs. *Plant Physiology*, *96*, 675–679.

Wang, E. (2002). Isolation and characterization of the CYP71D16 trichome-specific promoter from Nicotiana tabacum L. *Journal of Experimental Botany*, *53*, 1891–1897. http://dx.doi.org/10.1093/jxb/erf054.

Wang, H., Han, J., Kanagarajan, S., Lundgren, A., & Brodelius, P. E. (2013). Trichome-specific expression of the amorpha-4,11-diene 12-hydroxylase (cyp71av1) gene, encoding a key enzyme of artemisinin biosynthesis in Artemisia annua, as reported by a promoter-GUS fusion. *Plant Molecular Biology*, *81*, 119–138. http://dx.doi.org/10.1007/s11103-012-9986-y.

Wang, Y., Yang, K., Jing, F., Li, M., Deng, T., Huang, R., et al. (2011). Cloning and characterization of trichome-specific promoter of cpr71av1 gene involved in artemisinin biosynthesis in Artemisia annua L. *Molecular Biology*, *45*, 751–758.

Weinhold, A., & Baldwin, I. T. (2011). Trichome-derived O-acyl sugars are a first meal for caterpillars that tags them for predation. *Proceedings of the National Academy of Sciences of the United States of America*, *108*, 7855–7859.

Werker, E. (2000). Trichome diversity and development. *Advances in Botanical Research*, *31*, 1–35.

Wroblewski, T., Tomczak, A., & Michelmore, R. (2005). Optimization of Agrobacterium-mediated transient assays of gene expression in lettuce, tomato and Arabidopsis. *Plant Biotechnology Journal*, *3*, 259–273.

Wu, S., Jiang, Z., Kempinski, C., Eric Nybo, S., Husodo, S., Williams, R., et al. (2012). Engineering triterpene metabolism in tobacco. *Planta*, *236*, 867–877. http://dx.doi.org/10.1007/s00425-012-1680-4.

Wu, S., Schalk, M., Clark, A., Miles, R. B., Coates, R., & Chappell, J. (2006). Redirection of cytosolic or plastidic isoprenoid precursors elevates terpene production in plants. *Nature Biotechnology*, *24*, 1441–1447.
Yu, G., & Pichersky, E. (2014). Heterologous expression of methylketone synthase1 and methylketone synthase2 leads to production of methylketones and myristic acid in transgenic plants. *Plant Physiology*, *164*, 612–622.
Zhang, H., Niu, D., Wang, J., Zhang, S., Yang, Y., Jia, H., et al. (2015). Engineering a platform for photosynthetic pigment, hormone and cembrane-related diterpenoid production in Nicotiana tabacum. *Plant & Cell Physiology*, *56*, 2125–2138. http://dx.doi.org/10.1093/pcp/pcv131.

CHAPTER THIRTEEN

Tomato Fruits—A Platform for Metabolic Engineering of Terpenes

M. Gutensohn*, N. Dudareva†,[1]
*Davis College of Agriculture, Natural Resources and Design, West Virginia University, Morgantown, WV, United States
†Purdue University, West Lafayette, IN, United States
[1]Corresponding author: e-mail address: dudareva@purdue.edu

Contents

1. Introduction 334
2. Terpenoid Formation in Tomato Fruits 338
3. Transgene Expression in Ripening Tomato Fruits 340
   3.1 Stable Transformation Under Fruit Ripening-Specific Promoters 340
   3.2 Transient Transformation via *Agrobacterium* Injection into Fruits 342
4. Overexpression of Terpene Biosynthetic Genes in Tomato Fruits 343
   4.1 MEP and MVA Pathway Genes 344
   4.2 Prenyltransferases 344
   4.3 Terpene Synthases 346
   4.4 Pyramiding Multiple Terpene Biosynthetic Genes 348
5. Analysis of Terpenes in Tomato Fruits 349
   5.1 Analysis of Emitted Terpenes 350
   5.2 Analysis of Internal Terpene Pools 350
6. Conclusions 350
Acknowledgments 352
References 352

Abstract

Terpenoids are a large and diverse class of plant metabolites including mono-, sesqui-, and diterpenes. They have numerous functions in basic physiological processes as well as the interaction of plants with their biotic and abiotic environment. Due to the tight regulation of biosynthetic pathways and the resulting limited natural availability of terpenes, there is a strong interest in increasing their production in plants by metabolic engineering for agricultural, pharmaceutical, and industrial applications. The tomato fruit system was developed as a platform for metabolic engineering of terpenes to overcome detrimental effects on overall plant growth and photosynthesis traits, which are affected when terpenoid engineering is performed in vegetative tissues. Here we describe how the use of fruit-specific promoters for transgene expression can avoid

Methods in Enzymology, Volume 576
ISSN 0076-6879
http://dx.doi.org/10.1016/bs.mie.2016.03.012

these unwanted effects. In addition, targeting the expression of the introduced terpene biosynthetic gene to fruit tissue can take advantage of the large precursor pool provided by the methylerythritol-phosphate (MEP) pathway, which is highly active during tomato fruit ripening to facilitate the accumulation of carotenoids. We also discuss how the production of high levels of target terpene compounds can be achieved in fruits by the expression of individual or a combination of (i) the MEP or mevalonic acid pathway enzymes, (ii) prenyltransferases, and/or (iii) terpene synthases. Finally, we provide a brief outline of how the emitted as well as internal pools of terpenes can be analyzed in transgenic tomato fruits.

1. INTRODUCTION

Terpenoids are a large and diverse class of plant metabolites that are involved in numerous physiological and ecological processes. These compounds are vital for basic plant processes like photosynthesis and respiration (the phytyl side chain of chlorophylls and quinones), regulation of growth and development (the hormones gibberellins, abscisic acid, strigolactones, and brassinosteroids), modulation of membrane fluidity (sterols), and photoprotection and energy transfer (carotenoids). In addition, this metabolite class includes monoterpenes (C_{10}), sesquiterpenes (C_{15}), and diterpenes (C_{20}), which play important roles in the interactions of plants with their biotic and abiotic environment. Due to their volatile nature, many mono-, sesqui-, and some diterpenes contribute to plant reproduction by attracting pollinators and seed dispersers, as well as to the direct and indirect plant defense against pests and pathogens (Gershenzon & Dudareva, 2007; Pichersky & Gershenzon, 2002; Unsicker, Kunert, & Gershenzon, 2009).

All terpenoids originate from the universal five-carbon building blocks, isopentenyl diphosphate (IPP) and its allylic isomer dimethylallyl diphosphate (DMAPP), both of which are produced by two independent, compartmentally separated pathways in plants (Ashour, Wink, & Gershenzon, 2010; Hemmerlin, Harwood, & Bach, 2012; and references within). The mevalonic acid (MVA) pathway is localized in the cytosol and partially in peroxisomes, while the methylerythritol-phosphate (MEP) pathway operates exclusively in plastids (Fig. 1). Although these two pathways are located in different subcellular compartments, there is ample evidence for metabolic cross talk between them, with IPP- and to a lesser extent DMAPP-exchange occurring in both directions (Dudareva et al., 2005; Furumoto et al., 2011; Hemmerlin et al., 2003; Kasahara

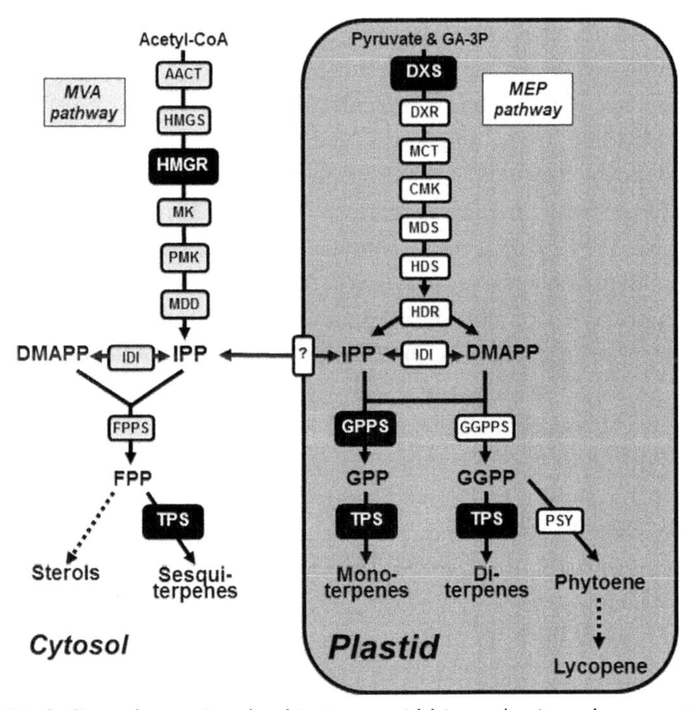

Fig. 1 Metabolic pathways involved in terpenoid biosynthesis and enzymatic steps introduced for terpene metabolic engineering in tomato fruits. Enzymes of the MEP pathway and lycopene biosynthesis are localized in plastids and are highlighted in *white*. Enzymes of the MVA and sterol biosynthesis are primarily localized in the cytosol and are highlighted in *green/gray*. Enzymes engineered into transgenic tomato fruits are highlighted in *black*. The enzymatic steps are indicated by *arrows* and *dashed lines* indicate multiple enzymatic steps. Individual enzymes are depicted as *boxes* with the abbreviation of their names. Abbreviations: *AACT*, acetoacetyl-CoA thiolase; *CMK*, 4-(cytidine 5′-diphospho)-2-*C*-methyl-D-erythritol kinase; *DMAPP*, dimethylallyl diphosphate; *DXR*, 1-deoxy-D-xylulose 5-phosphate reductoisomerase; *DXS*, 1-deoxy-D-xylulose 5-phosphate synthase; *FPP*, farnesyl diphosphate; *FPPS*, farnesyl diphosphate synthase; *GA-3P*, D-glyceraldehyde 3-phosphate; *GGPP*, geranylgeranyl diphosphate; *GGPPS*, geranylgeranyl diphosphate synthase; *GPP*, geranyl diphosphate; *GPPS*, geranyl diphosphate synthase; *HDR*, (*E*)-4-hydroxy-3-methylbut-2-enyl diphosphate reductase; *HDS*, (*E*)-4-hydroxy-3-methylbut-2-enyl diphosphate synthase; *HMG-CoA*, 3-hydroxy-3-methylglutaryl-CoA; *HMGR*, 3-hydroxy-3-methylglutaryl-CoA reductase; *HMGS*, 3-hydroxy-3-methylglutaryl-CoA synthase; *IDI*, isopentenyl diphosphate isomerase; *IPP*, isopentenyl diphosphate; *MCT*, 2-*C*-methyl-D-erythritol 4-phosphate cytidylyltransferase; *MDS*, 2-*C*-methyl-D-erythritol 2,4-cyclodiphosphate synthase; *MEP*, 2-*C*-methyl-D-erythritol 4-phosphate; *MK*, mevalonate kinase; *MVA*, mevalonate; *MDD*, mevalonate diphosphate decarboxylase; *PMK*, phosphomevalonate kinase; *PSY*, phytoene synthase; *TPS*, terpene synthases (including mono-, sesqui-, and diterpene synthases). (See the color plate.)

et al., 2002; Laule et al., 2003; Nagata, Suzuki, Yoshida, & Muranaka, 2002; Schuhr et al., 2003) via a yet unidentified transporter (Bick & Lange, 2003; Flügge & Gao, 2005; Soler, Clastre, Bantignies, Marigo, & Ambid, 1993). IPP and DMAPP are subsequently used in multiple compartments as substrates by short-chain *trans*-prenyltransferases to form larger prenyl diphosphate intermediates, including geranyl diphosphate (GPP, C10), farnesyl diphosphate (FPP, C15), and geranylgeranyl diphosphate (GGPP, C20), that ultimately serve as substrates for terpene synthases (TPSs) (Fig. 1). GPP, the precursor of all monoterpenes, is formed in the *trans* (*E*) configuration by the head-to-tail condensation of one DMAPP molecule with one IPP molecule in a reaction catalyzed by GPP synthase (GPPS), which is localized exclusively in plastids. FPP is the result of the condensation of one DMAPP with two IPP molecules and FPP synthase (FPPS) is responsible for its formation. FPPSs are localized in the cytosol and mitochondria, thus providing FPP for sesquiterpene, homoterpene, triterpene, sterol, brassinosteroid, and polyprenol biosynthesis (Fig. 1). GGPP synthase (GGPPS) catalyzes the condensation of one DMAPP with three IPP molecules to form GGPP. GGPPSs reside in plastids, mitochondria, and the endoplasmic reticulum, producing precursors for diterpenes, gibberellins, homoterpenes, carotenoids, phytyl side chains for chlorophyll/tocopherols/quinones, polyprenols, oligoprenols, abscisic acid, and strigolactones. Recently, in addition to *trans*-prenyltransferases, *cis*-prenyltransferases that synthesize *cis*-FPP and neryl diphosphate (NPP), the *cis*-isomer of GPP, from IPP and DMAPP, have been isolated from trichomes of wild and cultivated tomato along with TPSs, which utilize these cisoid substrates for sesquiterpene and monoterpene formation, respectively (Akhtar et al., 2013; Gonzales-Vigil, Hufnagel, Kim, Last, & Barry, 2012; Sallaud et al., 2009; Schilmiller et al., 2009).

Due to their involvement in specialized biological processes the biosynthesis of monoterpenes, sesquiterpenes, and diterpenes in plants is highly regulated. Their formation is often: (i) restricted to specific tissues or cell types such as flowers (Aros et al., 2012; Dudareva, Cseke, Blanc, & Pichersky, 1996; Dudareva et al., 2003; Lee & Chappell, 2008; Nieuwenhuizen et al., 2009) or glandular trichomes found at the surface of plant organs (Iijima et al., 2004; Schilmiller, Last, & Pichersky, 2008), (ii) limited to distinct developmental stages and/or times of the day (eg, formation of terpenes in flowers takes place after anthesis and is frequently diurnally or nocturnally regulated) (Dudareva et al., 1996, 2003; Guterman et al., 2002), and (iii) induced by an environmental stimulus such as a herbivore attack (Arimura et al., 2008; Fäldt, Arimura, Gershenzon, Takabayashi, & Bohlmann, 2003; Martin, Gershenzon, & Bohlmann, 2003; Schnee, Köllner, Gershenzon, & Degenhardt, 2002; Unsicker et al., 2009) and pathogen infection (Huang et al., 2003; Leitner et al., 2008). Due to this tight regulation of terpene formation in plants, natural sources often do not produce sufficient amounts of compounds required for agricultural or industrial uses. Moreover, terpene production in many crop plants has been severely compromised over the last decades by conventional breeding efforts that have focused on other attributes such as overall yield, shelf life, or resistance to environmental stresses, thus often resulting in the concomitant loss of unselected traits such as aroma and flavor. In addition to terpenoid production being a valuable agronomic trait, plant-derived terpenes are routinely used as flavors, fragrances, cosmetics, preservatives, insect repellents, and pharmaceutical products. However, many of these high-value terpene compounds are not produced in crop or model plants and found only in wild plants, which, as a consequence, are at risk of being overexploited and even going extinct.

There are two potential approaches to overcome limitations in terpene production in plants for agricultural, pharmaceutical, and industrial applications. When traits contributing to sufficient and/or required formation of target terpene compounds are available in the germplasm of a selected plant species, breeding and back-crossing can be used to introduce, combine, or improve the desired traits. While this approach is potentially useful and has been utilized in some plants, including crops (eg, corn [Degenhardt et al., 2009; Schnee et al., 2006]) and medicinal plants (eg, *Artemisia annua* [Graham et al., 2010]), it is a laborious and lengthy process. Furthermore, a breeding approach is not a viable option for many wild plant species, which are rich in terpenes but lack genetic resources, or in crop plants that do not produce a target terpene compound. In these cases, metabolic engineering represents an alternative and often highly efficient approach to either increase or modify terpene formation in a naturally producing plant species amenable to genetic modification, or to introduce a terpene biosynthetic pathway into a crop or model plant naturally devoid of the target compound. Numerous attempts have been made over the last decade to engineer the formation of selected terpene products in different plant species (summarized in Lange & Ahkami, 2013; Vickers, Bongers, Liu, Delatte, & Bouwmeester, 2014). While some of these attempts indeed achieved the formation of significant amounts of the terpene compound(s) of interest, others were less successful or resulted in unexpected outcomes including undesired product modifications or detrimental effects on plant growth and performance. The expression of a prenyl transferase or of several TPSs in vegetative tissues of *Arabidopsis thaliana*, potato, or tobacco led to severe growth retardation and in some cases to chlorotic phenotypes (Aharoni et al., 2003, 2006; Besumbes et al., 2004; Orlova et al., 2009). Further analyses of these transgenic potato and tobacco plants demonstrated that chlorophyll, carotenoid, and/or gibberellin levels were reduced, which could be due to (i) depletion of the isoprenoid precursor pool by the introduced biosynthetic pathway (Aharoni et al., 2003), (ii) potential toxic effects of the newly introduced terpene products in plant cells (Aharoni et al., 2003), or (iii) inhibitory effects of the introduced products on endogenous terpenoid biosynthetic pathways (Gutensohn et al., 2014). To overcome these detrimental effects on overall plant growth and photosynthesis, which ultimately result in reduced biomass and productivity, subsequent engineering strategies were designed and successfully carried out to restrict terpene production to specific plant tissues or organs such as fruits or glandular trichomes (Bleeker et al., 2012; Zhang et al., 2015).

2. TERPENOID FORMATION IN TOMATO FRUITS

The fruits of cultivated tomato (*Solanum lycopersicum*) contain various distinct terpenoid compounds that accumulate at different stages of fruit development and ripening. The early stages of fruit development are highly active in the production of sterols, which are essential for rapid cell division and subsequent cell expansion (Gillaspy, Ben-David, & Gruissem, 1993). At these early stages of tomato fruit development hydroxymethylglutaryl-CoA reductase (HMGR1), a key regulatory enzyme of the MVA pathway, and FPPS are highly expressed, but not during later stages of fruit ripening (Gaffe et al., 2000; Jelesko, Jenkins, Rodríguez-Concepción, & Gruissem, 1999; Narita & Gruissem, 1989).

The genome of cultivated tomato contains 44 TPS genes with 29 being functional or potentially functional (Falara et al., 2011). Expression of only six TPS genes was detected in red fruits (Falara et al., 2011). Sixteen TPS genes are expressed in green immature fruits, which possess glandular trichomes (Li et al., 2004) that dry up and fall off during fruit maturation. A large number of TPS genes are expressed in glandular trichomes of young leaves and stem (14 and 16 genes, respectively). These trichomes contain high quantities of terpenes, which likely act as defense compounds (Kang et al., 2010; Schilmiller et al., 2009; van Schie, Haring, & Schuurink, 2007).

The ripening process of tomato fruits is characterized by the massive accumulation of carotenoids, primarily lycopene, which are all tetraterpene pigments (Fraser, Truesdale, Bird, Schuch, & Bramley, 1994). Accumulation of lycopene begins at the "breaker" stage when the first flush of color becomes visible, which occurs only after the tomato fruit has reached its final size at the "mature green" stage and cell expansion ceases. The first step of the plastid localized lycopene biosynthetic pathway (Cunningham & Gantt, 1998), the head-to-head condensation of two GGPP molecules, is catalyzed by phytoene synthase (PSY). Subsequently two enzymes, phytoene desaturase (PDS) and ζ-carotene desaturase (ZDS), introduce double bonds to ultimately convert phytoene to lycopene. Transcript levels of the *Psy* and *Pds* genes were shown to increase significantly in tomato fruits at the "breaker" stage and to remain high until fruits are fully ripe (Giuliano, Bartley, & Scolnik, 1993; Pecker, Chamovitz, Linden, Sandmann, & Hirschberg, 1992; Ronen, Cohen, Zamir, & Hirschberg, 1999). In line with this, the expression of two genes encoding key enzymes of the MEP pathway, 1-deoxy-D-xylulose 5-phosphate synthase (DXS) and hydroxymethylbutenyl diphosphate reductase (HDR), is strongly upregulated during ripening of tomato fruits to provide GGPP substrate for PSY (Botella-Pavia et al., 2004; Lois, Rodríguez-Concepción, Gallego, Campos, & Boronat, 2000; Paetzold et al., 2010). Despite the fact that the MVA and MEP pathways are highly active during fruit growth and ripening, respectively, tomato fruits produce only very small amounts of monoterpenes and no detectable sesquiterpenes (Baldwin, Scott, Shewmaker, & Schuch, 2000; Buttery, Teranishi, Ling, & Turnbaugh, 1990; Davidovich-Rikanati et al., 2007, 2008; Lewinsohn et al., 2001). Such limited terpene production is consistent with the fact that only a small number of endogenous monoterpene synthases and no sesquiterpene synthases are expressed in tomato fruits (Bleeker et al., 2011; Falara et al., 2011). Moreover, the few terpenes found in ripe tomato fruits are primarily formed by degradation of carotenoids (Lewinsohn, Sitrit, Bar, Azulay, Ibdah, et al., 2005; Lewinsohn, Sitrit, Bar, Azulay, Meir, et al., 2005) rather than by de novo synthesis from prenyl diphosphate intermediates.

Ripening tomato fruits thus represent an ideal platform for the metabolic engineering of terpene production because: (i) the production of terpenes can be restricted to the ripening fruits through the use of fruit-specific promoters so that the growth, development, and performance of the rest of the tomato plant will not be affected; (ii) the MEP pathway, which provides precursors for newly introduced terpene compounds, is highly active during fruit ripening; (iii) many fruits can be obtained simultaneously; (iv) changes in the carbon flux toward terpene production often can be visually assessed based on the level of carotenoids; and (v) low levels of endogenous terpenes in ripening fruits make the metabolic profiling of transgenic plants easy. In this chapter we outline the potential approaches that can be utilized to achieve high-level terpene production in tomato fruits and summarize the outcome of respective metabolic engineering studies performed in our and other labs over the recent years.

3. TRANSGENE EXPRESSION IN RIPENING TOMATO FRUITS

3.1 Stable Transformation Under Fruit Ripening-Specific Promoters

In general, the first step in metabolic engineering is the introduction of transgene(s) encoding the biosynthetic enzymes involved in the formation of the target compound into a respective host plant. Tomato plants, like many other *Solanaceae* plants, are highly amenable for the integration of transgenes, and protocols for *Agrobacterium tumefaciens*-mediated stable transformation are well established. A critical aspect of terpene metabolic engineering in tomato fruits is the selection of suitable promoters that regulate the introduced transgene and restrict its expression to ripening fruits (Fig. 2A), thereby avoiding potential negative effects on vegetative tissues and plant development, and synchronizing the engineered terpene biosynthesis with high MEP pathway activity and precursor availability. To date three different promoters have been successfully used to express terpene metabolic genes in ripening tomato fruits.

The first fruit-specific promoter selected for terpene metabolic engineering in tomato was the promoter region of the ethylene-responsive tomato *E8* gene (Deikman & Fischer, 1988). Although the function of the *E8* gene is not completely understood, its promoter activity and expression have been extensively characterized. The *E8* promoter is activated at the onset of tomato fruit ripening (Deikman & Fischer, 1988; Deikman, Kline, & Fischer, 1992; Lincoln & Fischer, 1988) and gives rise to uniform transgene expression throughout the fruit pericarp (Kneissl & Deikman, 1996). The *E8* promoter has already been successfully used for genetic engineering in tomato fruits (Giovannoni, DellaPenna, Bennett, & Fischer, 1989; Good et al., 1994; Sandhu et al., 2000) including production of folates, carotenoids, and aromatic amino acids (Díaz de la Garza, Gregory, & Hanson,

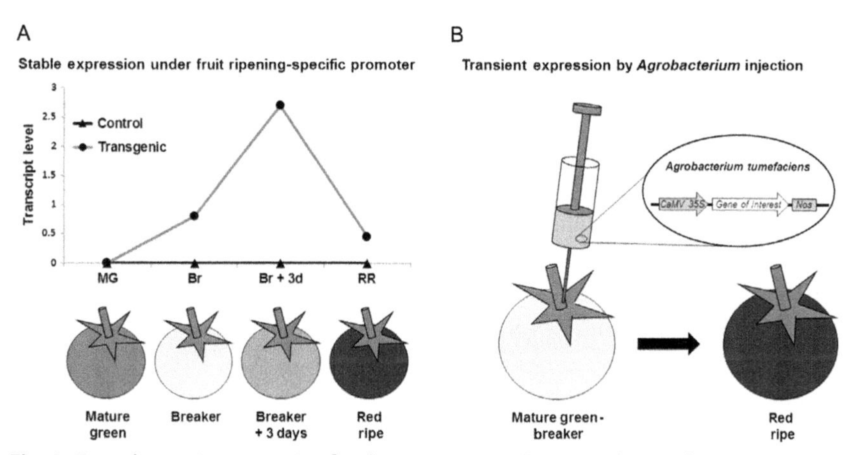

Fig. 2 Two alternative strategies for the expression of terpene biosynthetic transgenes in ripening tomato fruits. (A) Expression of transgene in stable transgenic tomato plants under control of a fruit ripening-specific promoter that restricts expression to the ripening period between the breaker and red ripe stages when carotenoids accumulate. (B) Transient expression of transgene in ripening tomato fruits upon *Agrobacterium* injection of fruits prior to or at the breaker stage. (See the color plate.)

2007; Sun et al., 2012; Tzin et al., 2013). When the *E8* promoter was used for terpene metabolic engineering, it has been shown that fruits of transgenic tomato plants carrying a linalool synthase gene (*LIS*), responsible for formation of the monoterpene linalool, contained no monoterpene products at the "mature green" stage, while monoterpenes began to accumulate at the "breaker" stages reaching maximum in fully ripe fruits (Lewinsohn et al., 2001).

A second promoter utilized for metabolic engineering in tomato fruits originated from the pepper *fibrillin* (*fib*) gene, which encodes a major chromoplast protein involved in the deposition of carotenoids into lipoprotein fibrils in ripening pepper fruits (Deruère, Bouvier, et al., 1994; Deruère, Römer, et al., 1994). When this promoter is used to drive transgene expression in tomato fruits, practically no expression is observed at early stages of fruit development. However, transgene expression levels increase between the "mature green" and "breaker" stages, and remain high throughout ripening of tomato fruits (Kuntz et al., 1998; Simkin et al., 2007). When the key MEP pathway enzyme DXS was expressed in transgenic tomato plants under the control of the *fibrillin* promoter, DXS activity was significantly increased in ripening tomato fruits 4–6 days after the "breaker" stage (Enfissi et al., 2005).

The tomato *polygalacturonase* (*PG*) promoter is the most frequently used promoter for metabolic engineering of terpenes in ripening tomato fruits. Polygalacturonase is responsible for the degradation of polyuronides in the fruit cell wall, thus contributing to the fruit softening. The expression of the *PG* gene is tightly regulated with no expression in vegetative tissues including roots, stems, and leaves (Biggs & Handa, 1989). Furthermore, no transcripts are detected at the earlier stages of fruit development up to the "mature green" stage. *PG* mRNA begins to accumulate at the "breaker" stage, reaches the highest level a few days after the "breaker" stage, and then drops once the fruit becomes fully ripe (Biggs & Handa, 1989; DellaPenna, Lincoln, Fischer, & Bennett, 1989). In contrast, the amounts of polygalacturonase protein and enzyme activity continue to increase consistently after their first appearance at the "breaker" stage and reach the highest levels in red ripe fruits (Biggs & Handa, 1989). Essentially identical fruit ripening-specific expression patterns were observed when two reporter genes were expressed in transgenic tomato plants under the control of either a short (1.4 kb) or a long (4.8 kb) version of the *PG* promoter (Bird et al., 1988; Fernandez et al., 2009; Montgomery, Pollard, Deikman, & Fischer, 1993; Nicholass, Smith, Schuch, Bird, & Grierson, 1995). While both promoter fragments were capable of driving transgene expression in the pericarp of ripening fruits, higher expression levels were achieved with the longer one. This longer version of the *PG* promoter has been successfully used to express two prenyltransferases as well as various TPSs in transgenic tomato plants (Davidovich-Rikanati et al., 2008, 2007; Gutensohn et al., 2014, 2013; Kovacs et al., 2007). While no transcripts of the respective transgenes were detected in fruits at the "mature green" stage, they began to accumulate at the "breaker" stage as expected, reached the highest level 3 days after the "breaker" stage, and declined in ripe fruits (Davidovich-Rikanati et al., 2008; Gutensohn et al., 2014, 2013).

3.2 Transient Transformation via *Agrobacterium* Injection into Fruits

Although stable transformation of tomato plants is now routine, it is still a time-consuming procedure involving tissue culture and plant regeneration, and it takes several months before fruit-bearing plants are available. Considering the often uncertain outcome of new metabolic engineering strategies, one would prefer to have a rapid and efficient way of testing gene expression constructs and the effect of a given transgene on host plant metabolism prior to the more labor-intensive generation of stable transgenic lines. Transient transformation via *Agrobacterium* infiltration into the intercellular spaces of leaves was successfully established in various plant species including *Nicotiana benthamiana*, *A. thaliana*, and tomato and is now commonly used for the assessment of transgene function. A comparable strategy for the transient transformation of fleshy fruits was first reported by Spolaore, Trainotti, and Casadoro (2001), in which an *Agrobacterium* suspension was injected into fruits using a syringe with hypodermic needle. This agroinjection procedure was subsequently further optimized for the transient expression and analysis of transgenes in ripening tomato fruits (Orzaez, Mirabel, Wieland, & Granell, 2006). Injection of *Agrobacterium* carrying β-glucuronidase and yellow fluorescent protein reporter gene constructs under control of the *Cauliflower Mosaic Virus* (*CaMV*) *35S* promoter into tomato fruits at the "mature green" stage through the fruit stylar apex resulted in high levels of reporter gene expression around the placenta tissue and moderate levels in the fruit pericarp. When this agroinjection technique was subsequently combined with the *Tobacco Rattle Virus*-based "virus induced gene silencing" (VIGS) to downregulate the carotenoid biosynthetic gene PDS in tomato fruits, these fruits accumulated lower levels of lycopene and lycopene-derived volatiles (Orzaez et al., 2009, 2006). Transient expression also offers the advantage of being able to test gene stacks through introduction of additional constructs in stably transformed plants. Indeed, we have recently used the agroinjection protocol as described by Orzaez et al. (2006) to transiently overexpress MEP and MVA pathway genes under the *CaMV 35S* promoter in the background of stable transgenic tomato lines expressing TPSs under the fruit ripening-specific *PG* promoter (Gutensohn & Dudareva, unpublished data).

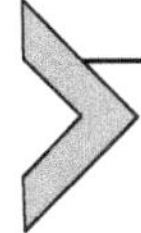

4. OVEREXPRESSION OF TERPENE BIOSYNTHETIC GENES IN TOMATO FRUITS

In the past decade numerous attempts have been made to modify the terpenoid profile of tomato fruits. In general, three types of genes encoding enzymes involved in the biosynthesis of terpenes were used to achieve formation of the desired product. They include: (i) MVA and MEP pathway genes to increase the pool of available IPP and DMAPP precursors, (ii) prenyltransferase genes to increase the pool of a required prenyl diphosphate intermediate, and (iii) TPS genes to direct the metabolic flux toward the formation of a target compound. Below we outline examples of terpene metabolic engineering in ripening tomato fruits that have utilized genes from one of these classes or combinations of genes from several classes and discuss the respective outcomes.

4.1 MEP and MVA Pathway Genes

Despite the fact that the plastidic MEP pathway is highly active in ripening tomato fruits to provide precursors for carotenoid biosynthesis, attempts were made to even further increase flux through this pathway. Expression of *Escherichia coli* DXS targeted to plastids under the control of the *fibrillin* promoter resulted in twofold increases in both DXS enzyme activity and 1-deoxy-D-xylulose 5-phosphate internal pool in ripening transgenic tomato fruits relative to controls (Enfissi et al., 2005). In addition, the carotenoid levels were increased 1.6-fold, indicating that *DXS* overexpression had indeed increased the metabolic flux through the plastidic MEP pathway toward the biosynthesis of downstream carotenoids.

In contrast to the MEP pathway, the cytosolic MVA pathway is highly active only during early stages of fruit development and not during later stages of fruit ripening. Thus, the key MVA pathway enzyme HMGR from *Arabidopsis* was overexpressed in transgenic tomato plants under the control of the constitutive *CaMV 35S* promoter (Enfissi et al., 2005). Despite the fact that no significant increases in HMGR enzyme activity were detected, the phytosterol content in ripe fruits of these transgenic lines was increased up to 2.4-fold compared to controls. This indicates that overexpression of *HMGR* increases metabolic flux through the cytosolic MVA pathway toward the biosynthesis of downstream phytosterols, similar to the effect of *DXS* overexpression on carotenoid biosynthesis. Despite these encouraging results up to date the stable overexpression of *DXS* or *HMGR* has not been further utilized for the metabolic engineering of terpene formation in tomato fruits.

4.2 Prenyltransferases

Besides the importance of having high metabolic flux through the upstream MEP and MVA pathways, the availability of sufficient pools of the required prenyl diphosphate intermediates is critical for efficient engineering of terpene formation. Of the two GGPP synthases present in tomato, *GGPPS2* is highly expressed in ripening tomato fruits (Ament, Van Schie, Bouwmeester, Haring, & Schuurink, 2006) and provides GGPP for the massive accumulation of lycopene and other carotenoids. In contrast, FPPS is highly expressed during tomato fruit development (Gaffe et al., 2000), while the putative small subunit of a heterodimeric GPPS is only weakly expressed in ripening tomato fruits (Gutensohn et al., 2013; Wang & Dixon, 2009). The *FPPS* and *GPPS* expression profiles suggest that only small pools of the FPP and GPP precursors are likely available in ripening fruits for sesqui- and monoterpene biosynthesis, respectively.

To increase the amount of available GPP and divert the metabolic flux from GGPP and carotenoid formation toward monoterpene biosynthesis we expressed a noncatalytic small subunit of snapdragon (*Antirrhinum majus*) GPPS (*GPPS-SSU*) under control of the *PG* promoter in tomato (Gutensohn et al., 2013). Heterodimeric GPPSs, like that found in snapdragon, contain a catalytically inactive GPPS-SSU that interacts with a *bona fide* GGPPS and changes its product specificity, thus forming a heterodimer, which catalyzes the formation of GPP (Tholl et al., 2004; Wang & Dixon, 2009). Overexpression of snapdragon *GPPS-SSU* significantly increased the total GPPS activity in ripening tomato fruits (Gutensohn et al., 2013). In contrast to control fruits, which contain only minute amounts of monoterpenes due to expression of limited number of endogenous TPSs (Bleeker et al., 2011; Falara et al., 2011), transgenic *GPPS-SSU* tomato fruits produced an array of monoterpenes including geraniol, the dephosphorylated form of GPP, geranial, neral, citronellol, citronellal, and geranic acid. The amounts of monoterpenes positively correlated with the *GPPS-SSU* transcript and protein levels in various transgenic lines. However, the monoterpenes produced were not a result of the action of endogenous monoterpene synthases, but of endogenous phosphatases (Ganjewala & Luthra, 2009; Izumi, Ashida, Yamamitsu, & Hirata, 1996; Perez, Taucher, & Cori, 1980), reductase(s) and alcohol dehydrogenase(s) (Bicsak, Kann, Reiter, & Chase, 1982; Davidovich-Rikanati et al., 2007), activity of which were previously shown to be present in tomato fruits (Davidovich-Rikanati et al., 2007). While *GPPS-SSU* expression had no deleterious effects on vegetative tissues and development of transgenic tomato plants, the lycopene levels of ripe fruits were strongly reduced. This severe reduction of lycopene could not be exclusively attributed to a redirection of metabolic flux from GGPP to GPP formation, but was the result of a posttranscriptional control likely via feed-forward regulation of lycopene biosynthetic enzymes by one (or several) product(s) formed in transgenic lines (Gutensohn et al., 2013). Thus, the observed monoterpene formation and reduction of lycopene levels in the transgenic fruits indicate that the expression of *GPPS-SSU* indeed leads to the redirection of metabolic flux and an increase in the GPP pool, which can be utilized for further metabolic engineering of monoterpenes in ripening tomato fruits.

Although most monoterpene synthases use GPP as substrate, it was recently found that some monoterpene synthases accept NPP, the *cis*-isomer of GPP, as substrate (Falara et al., 2011; Gonzales-Vigil et al., 2012; Matsuba et al., 2013; Schilmiller et al., 2009; Zhang, Liu, Li, & Yu, 2013). However, in tomato plants *NDPS1*, encoding the *cis*-prenyltransferase that catalyzes the NPP formation, is primarily expressed in trichomes of leaves and stems but not in fruits (Akhtar et al., 2013; Gutensohn et al., 2014; Schilmiller et al., 2009). Thus, to achieve formation of NPP-derived monoterpenes in tomato fruits, we have expressed the tomato *NDPS1* under the control of the *PG* promoter (Gutensohn et al., 2014). *NDPS1* transgenic fruits produced large amounts of nerol, the dephosphorylated form of NPP, as well as minor amounts of neral, geranial, and geraniol. Similar to the conversion of GPP into an array of monoterpenes in transgenic *GPPS-SSU* fruits, produced NPP was subjected to the action of endogenous phosphatases and alcohol dehydrogenases. As in the case of the *GPPS-SSU* expressing fruits, *NDPS1* expression led to a severe reduction in lycopene levels, which can be only partially explained by the redirection of metabolic flux from carotenoids toward NPP formation. Indeed, subsequent analyses demonstrated that the produced NPP serves as an inhibitor of GGPPS activity in tomato fruits. This finding was further corroborated by the fact that transcript levels of *GGPPS2* were 2.5-fold increased in *NDPS1* fruits (Gutensohn et al., 2014).

In contrast to our successful engineering of GPP and NPP formation in ripening tomato fruits by expression of *GPPS-SSU* and *NDPS1*, respectively (Gutensohn et al., 2014, 2013), to the best of our knowledge no comparable attempt has been made so far to increase the available FPP pool by expression of an FPPS in ripening tomato fruits.

4.3 Terpene Synthases

While the overexpression of MEP and MVA pathway genes increases the metabolic flux through the respective pathways, and overexpression of prenyltransferases, such as GPPS and NPPS, redirects the metabolic flux from carotenoids toward increased formation of corresponding prenyl diphosphate intermediates, only the expression of TPSs ultimately determines which terpene products, eg, mono-, sesqui-, or diterpenes, are formed in transgenic tomato fruits. The expression of *Clarkia breweri S-linalool synthase* (*LIS*) under the control of the *E8* promoter represents a first example of monoterpene engineering in tomato fruits (Lewinsohn et al., 2001). The resulting transgenic fruits indeed produced high levels of *S*-linalool as well as substantial levels of 8-hydroxylinalool and a number of unknown products. The appearance of 8-hydroxylinalool suggests the presence of an endogenous hydroxylating activity in ripening tomato fruits that accepts linalool as substrate. Despite the accumulation of monoterpenes in the *LIS* transgenic fruits the levels of other terpenoids such as carotenoids and tocopherols were unaltered. Expression of the *Ocimum basilicum geraniol synthase* (*GES*), a second monoterpene synthase used for tomato fruit metabolic engineering, under the control of the *PG* promoter, resulted in the formation of geraniol in transgenic fruits (Davidovich-Rikanati et al., 2007). In addition, these *GES* transgenic fruits accumulated other monoterpenes, including nerol, citronellol, geranial, neral, citronellal, geranic acid, neric acid, and citronellic acid, which all are derived from geraniol through the action of endogenous enzymes such as alcohol dehydrogenases and reductases. The amount of geraniol and its derivatives accumulated in the *GES* fruits (about 3500 ng/g FW) was significantly higher than the amount of linalool and its derivatives detected in the *LIS* fruits (400–800 ng/g FW) (Lewinsohn et al., 2001). In contrast to *LIS* fruits, lycopene levels in *GES* fruits were reduced by 50% indicating a substantial redirection of the metabolic flux from carotenoids toward monoterpenes. Recently, *phellandrene synthase 1* (*PHS1*), which is naturally only expressed in tomato trichomes and accepts both NPP and GPP substrates to form different monoterpenes, respectively (Schilmiller et al., 2009), was expressed in tomato fruits under control of the fruit ripening-specific *PG* promoter (Gutensohn et al., 2014). The resulting transgenic tomato fruits produced small amounts of myrcene and ocimene, the GPP-derived monoterpene products of PHS1 (Schilmiller et al., 2009). The low level of total monoterpene production in *PHS1* fruits (~50 ng/g FW) relative to that in the *LIS* and *GES* fruits (Davidovich-Rikanati et al., 2007; Lewinsohn et al., 2001) is likely due to the low affinity of PHS1 for GPP (Schilmiller et al., 2009). Overexpression of *PHS1* also uncovered the absence of an endogenous NPP pool in the fruits. Taken together, the above-described metabolic engineering efforts with three monoterpene synthases clearly demonstrated that despite the low expression level of a *GPPS* small subunit and the low GPPS activity in ripening tomato fruits (Gutensohn et al., 2013; Wang & Dixon, 2009) a sufficient GPP pool is available for the introduced monoterpene synthases.

The positive outcomes of the described monoterpene engineering approaches in tomato fruits fulfilled expectations because the high activity of the MEP pathway during ripening provides precursors for the overexpressed monoterpene synthases in plastids. In contrast, it was questionable if sesquiterpene formation could also be successfully achieved in ripening tomato fruits. Sesquiterpene synthases are localized in the cytosol where the MVA pathway genes, like *HMGR1*, as well as *FPPS* are expressed at low levels during fruit ripening (Gaffe et al., 2000; Jelesko et al., 1999; Narita & Gruissem, 1989). However, when *α-zingiberene synthase* (*ZIS*), a sesquiterpene synthase from *Ocimum basilicum*, was expressed in tomato fruits under the control of the *PG* promoter, the resulting transgenic fruits accumulated high levels of α-zingiberene and other sesquiterpenes such as α-bergamotene, 7-*epi*-sesquithujene, β-bisabolene, and β-curcumene (Davidovich-Rikanati et al., 2008). No significant differences between *ZIS* transgenic and control fruits were found in sterol levels, which are also synthesized from MVA pathway-derived precursors. In contrast, a slight decrease in carotenoids was observed in *ZIS* transgenic fruits. The significant accumulation of sesquiterpenes (up to 1000 ng α-zingiberene/g FW) in *ZIS* transgenic fruits indicates that despite the low expression level of *HMGR1* and *FPPS* there is a sufficient supply of FPP in ripening tomato fruits to drive the production of engineered sesquiterpenes.

Besides mono- and sesquiterpenes, the formation of diterpenes has been successfully engineered in tomato fruits as well. Taxans, such as taxadiene, are polycyclic diterpenes that are naturally produced in small amounts in various species of yew trees (*Taxus* spec.) and used as anticancer drugs. The taxadiene synthase from *Taxus baccata* was expressed under the control of both, the constitutive *CaMV 35S* promoter and the fruit ripening-specific *PG* promoter, in the background of the *yellow flesh* tomato mutant (Kovacs et al., 2007). This tomato mutant is deficient in the fruit-specific PSY and thus does not consume GGPP for carotenoid biosynthesis (Fray & Grierson, 1993). Resulting transgenic fruits accumulate taxadiene, with amounts being higher upon fruit-specific promoter-driven expression (400–600 μg/g DW) relative to constitutive promoter-driven expression (140–400 μg/g DW). Independent of the promoter used, seed formation in transgenic fruits was severely affected, which could, however, be fully restored by treatment of flowers with giberellic acid.

4.4 Pyramiding Multiple Terpene Biosynthetic Genes

After the successful engineering of individual metabolic steps in the terpene biosynthetic pathway including MEP/MVA pathway genes, prenyltransferases, and TPSs, the next step to achieve high levels of terpene production in ripening tomato fruits is to combine several transgenes in the same host plant. To obtain transgenic lines coexpressing a prenyltransferase and a TPS in ripening tomato fruits we have crossed parental lines that express respective transgenes under the control of the *PG* promoter. The resulting F1 plants were initially tested for the presence of both transgenes by PCR on genomic DNA using gene-specific primers. Subsequently, the expression levels of both transgenes were analyzed by quantitative real-time PCR in ripening fruits of these F1 plants and compared to the expression levels in the respective parental lines. Coexpression of the snapdragon *GPPS-SSU* with the *Ocimum basilicum GES* led to a more than threefold increase in the overall formation of monoterpenes in tomato fruits relative to the parental *GES* line (Gutensohn et al., 2013). Remarkably, geraniol emission from *GPPS-SSU* × *GES* fruits was sevenfold higher than from parental *GES* fruits, while no significant differences in internal monoterpene pools were detected. These results suggest that the metabolic flux through the engineered pathway was increased, while the internal monoterpene pools were saturated. A similar increase in monoterpene formation was achieved upon coexpression of snapdragon *GPPS-SSU* and tomato *PHS1* in tomato fruits, with a more than twofold increase in myrcene and ocimene emission, as well as a 3.8- and 8-fold increase in respective internal pools in *GPPS-SSU* × *PHS1* fruits compared to *PHS1* fruits (Gutensohn et al., 2014). In order to attain metabolic engineering of NPP-derived monoterpenes in tomato fruits, we generated transgenic lines coexpressing both tomato *NDPS1* and *PHS1* (Gutensohn et al., 2014). The monoterpenes found in the fruits of *NDPS1* and *PHS1* parental lines, nerol and myrcene/ocimene, respectively, were absent in *NDPS1* × *PHS1* fruits and instead they produced a blend of monoterpenes including β-phellandrene. These results indicate that the NPP pool produced by the overexpressed *NDPS1* is utilized by the introduced PHS1 for monoterpene formation in tomato fruits.

5. ANALYSIS OF TERPENES IN TOMATO FRUITS

For the development of metabolic engineering strategies it is not only important to express genes required for the production of a target compound in the host plant, but also to have sensitive methods for its detection. Here we outline some technical aspects of the characterization of volatile terpene production, which are specific to tomato fruits and different from the common analysis of leaf or flower volatiles. A comprehensive overview of the terpenoids produced in transgenic tomato fruits can be obtained by analyzing the terpenes emitted from fruits as well as the internal terpene pools accumulated in fruits.

5.1 Analysis of Emitted Terpenes

Volatiles emitted from ripe tomato fruits are collected using a closed-loop stripping method (Dudareva et al., 2005) at 21°C under growth chamber conditions. Ripe tomato fruits are freshly harvested, washed with cold tab water, dried with paper towels, and then depending on the size of the fruit vertically cut into six to eight triangular slices. Ten milliliters of a 5% (w/v) sucrose solution is pipetted into the volatile collection chamber. The tomato fruit slices (in total 30–50 g fresh weight) are then inserted into the collection chamber upright in such a way that the side of the fruit that had been attached to the plant has some contact with the sucrose solution, thus allowing the fruit to take up sucrose during the period of the collection. Volatile collections are performed for 24 h using Porapak Q traps and collected volatiles are subsequently eluted from the traps with 200 μL dichloromethane.

5.2 Analysis of Internal Terpene Pools

For the analysis of internal volatile pools approximately 30 g of pericarp from fresh ripe tomato fruits are cut into small pieces, transferred into 250 mL screwcup bottles, and extracted with 100 mL methyl *tert*-butyl ether. After vigorous shaking overnight, the extract is filtered to remove the fruit pieces. Since the resulting extract contains substantial amounts of water from the fruit tissue in addition to the organic solvent, it is subsequently allowed to separate into two phases. The ethereal phase is separated off, additionally dried with anhydrous $CaSO_4$ to remove residual water, and concentrated to 200 μL under a gentle N_2 stream.

The collected emitted as well as the extracted tomato fruit volatiles are supplemented with 3.33 μg of naphthalene as internal standard, and a 2 μL aliquot is subsequently analyzed by gas chromatography–mass spectrometry (GC–MS) (Dudareva et al., 2005). Representative terpene standards are used to determine response factors, which then allow the quantification of analyzed compounds.

6. CONCLUSIONS

Progress in the discovery and characterization of TPSs, prenyltransferases, and the MEP and MVA pathway biosynthetic genes made it possible to metabolically engineer terpenoid formation in plants. However, when genetic manipulations were performed in vegetative tissue they often resulted in undesired effects on the overall performance of the host plant. Therefore, the tomato fruit system was developed as an alternative platform for the efficient production of high levels of terpenes using fruit-specific promoters. This platform is beneficial not only because it avoids the deleterious effects of engineering on normal plant development, but also because it provides a sufficient pool of precursors for terpene biosynthesis, as the MEP pathway is highly active during the accumulation of carotenoids, which accompanies tomato fruit ripening. Thus, the tomato fruits can be used not only as an experimental system for studying metabolic pathways and engineering steps but also as biofactories for a large-scale production of desired target terpene compounds, since it is easy to grow this horticultural crop and scale up its production.

Another benefit of the tomato fruit system is the availability of transient transgene expression via the developed agroinjection method (Orzaez et al., 2006), which allows rapid testing of the effects of introduced transgenes on production of the target compounds prior to the more labor-intensive and time-consuming creation of stable transgenic lines. This transient expression system also permits assessment of the effectiveness of gene stacks via introduction of additional transgenes in stably transformed plants. Earlier terpene engineering studies have demonstrated that the expression of individual biosynthetic transgenes in host plants often resulted in a lower terpene production than transgenic plants coexpressing multiple terpene biosynthetic enzymes, for example, prenyltransferases and TPSs (Wu et al., 2006). Previously, stable gene stacks have been generated in the tomato fruit system by crossing of individual parental lines containing single transgenes (Gutensohn et al., 2014, 2013). However, this approach is impractical for staking a larger number of transgenes in one plant. Thus, the next step in the metabolic engineering of terpene formation in the tomato fruits will be utilization of newly developed techniques that allow the introduction of multiple transgenes under the control of one promoter via one transformation event. In one such approach, multiple transgenes are fused in a single transcriptional unit such that the coding sequences of the individual proteins are separated by the recognition sequence of the tobacco vein mottling virus (TVMV) NIa proteinase (Dafny-Yelin & Tzfira, 2007). The resulting polyprotein is then processed and cleaved into individual proteins by the coexpressed TVMV NIa proteinase. An alternative approach for multicistronic expression of transgenes in plants is based on the use of either viral internal ribosomal entry sites (Ngoi, Chien, & Lee, 2004) or viral 2A sequences which result in ribosomal skipping (de Felipe et al., 2006; Donnelly et al., 2001), to separate translation of individual transgenes in such transcriptional units for synchronized production of multiple proteins. The latter two techniques have recently been successfully used to engineer the formation of carotenoids in rice endosperm (Ha et al., 2010).

Metabolic engineering often leads to unpredicted results, highlighting our lack of a comprehensive understanding of plant metabolism including silent metabolism (Lewinsohn & Gijzen, 2009). In general, the production and metabolic fate of a newly synthesized compound depend on the entire biochemical repertoire within a given plant tissue; thus it is difficult to predict how much of the target compound will remain in the desired form. The newly synthesized compound may be affected by enzymes that are endogenously present in the cell and have broad substrate specificity, such as dehydrogenases, phosphatases, and others. A comprehensive understanding of the entire terpenoid metabolic network in plants is required for future rational and precise metabolic engineering, as was recently exemplified by the unexpected discovery of isopentenyl phosphate kinase and its role in terpenoid biosynthesis in plants (Henry, Gutensohn, Thomas, Noel, & Dudareva, 2015). Undoubtedly the application of omics approaches, such as transcriptomics and metabolomics, in tomato fruits will improve our understanding of the entire metabolic network in this platform, including silent metabolism. The identification and subsequent manipulation of enzymatic steps, which convert the desired terpene target compound into undesired downstream products, will result in more predictable outcomes of terpene metabolic engineering in the tomato fruit system.

ACKNOWLEDGMENTS

The authors work on metabolic engineering of terpenes in tomato fruits was supported by Agricultural and Food Research Initiative competitive Grant Number 2008-35318-04541 to N.D. and by United States/Israel Bi-national Agriculture Research and Development Grant Number IS-4125-08C to N.D. The work of M.G. is supported by the Ray Marsh and Arthur Pingree Dye Professorship.

REFERENCES

Aharoni, A., Giri, A. P., Deuerlein, S., Griepink, F., de-Kogel, W.-J., Verstappen, F. W. A., et al. (2003). Terpenoid metabolism in wild-type and transgenic Arabidopsis plants. *Plant Cell*, *15*, 2866–2884.

Aharoni, A., Jongsma, M. A., Kim, T.-K., Ri, M.-B., Giri, A. P., Verstappen, F. W. A., et al. (2006). Metabolic engineering of terpenoid biosynthesis in plants. *Phytochemistry Reviews*, *5*, 49–58.

Akhtar, T. A., Matsuba, Y., Schauvinhold, I., Yu, G., Lees, H. A., Klein, S. E., et al. (2013). The tomato *cis*-prenyltransferase gene family. *The Plant Journal*, *73*, 640–652.

Ament, K., Van Schie, C. C., Bouwmeester, H. J., Haring, M. A., & Schuurink, R. C. (2006). Induction of a leaf specific geranylgeranyl pyrophosphate synthase and emission of (E,E)-4,8,12-trimethyltrideca-1,3,7,11-tetraene in tomato are dependent on both jasmonic acid and salicylic acid signaling pathways. *Planta, 224,* 1197–1208.

Arimura, G., Garms, S., Maffei, M., Bossi, S., Schulze, B., Leitner, M., et al. (2008). Herbivore-induced terpenoid emission in *Medicago truncatula*: Concerted action of jasmonate, ethylene and calcium signaling. *Planta, 227,* 453–464.

Aros, D., Gonzalez, V., Allemann, R. K., Müller, C. T., Rosati, C., & Rogers, H. J. (2012). Volatile emissions of scented *Alstroemeria* genotypes are dominated by terpenes, and a myrcene synthase gene is highly expressed in scented *Alstroemeria* flowers. *Journal of Experimental Botany, 63,* 2739–2752.

Ashour, M., Wink, M., & Gershenzon, J. (2010). Biochemistry of terpenoids: Monoterpenes, sesquiterpenes and diterpenes. *Annual Plant Reviews, 40,* 258–303.

Baldwin, E. A., Scott, J. W., Shewmaker, C. K., & Schuch, W. (2000). Flavor trivia and tomato aroma: Biochemistry and possible mechanisms for control of important aroma components. *Horticultural Science, 35,* 1013–1022.

Besumbes, O., Sauret-Güeto, S., Phillips, M. A., Imperial, S., Rodríguez-Concepción, M., & Boronat, A. (2004). Metabolic engineering of isoprenoid biosynthesis in *Arabidopsis* for the production of taxadiene, the first committed precursor of Taxol. *Biotechnology and Bioengineering, 88,* 168–175.

Bick, J. A., & Lange, B. M. (2003). Metabolic cross talk between cytosolic and plastidial pathways of isoprenoid biosynthesis: Unidirectional transport of intermediates across the chloroplast envelope membrane. *Archives of Biochemistry and Biophysics, 415,* 146–154.

Bicsak, T. A., Kann, L. R., Reiter, A., & Chase, T. (1982). Tomato alcohol dehydrogenase: Purification and substrate specificity. *Archives of Biochemistry and Biophysics, 216,* 605–615.

Biggs, M. S., & Handa, A. K. (1989). Temporal regulation of polygalacturonase gene expression in fruits of normal, mutant, and heterozygous tomato genotypes. *Plant Physiology, 89,* 117–125.

Bird, C. R., Smith, C. J. S., Ray, J. A., Moureau, P., Bevan, M. W., Bird, A. S., et al. (1988). The tomato polygalacturonase gene and ripening-specific expression in transgenic plants. *Plant Molecular Biology, 11,* 651–662.

Bleeker, P. M., Mirabella, R., Diergaarde, P. J., VanDoorn, A., Tissier, A., Kant, M. R., et al. (2012). Improved herbivore resistance in cultivated tomato with the sesquiterpene biosynthetic pathway from a wild relative. *Proceedings of the National Academy of Sciences of the United States of America, 109,* 20124–20129.

Bleeker, P. M., Spyropoulous, E. A., Diergaarde, P. J., Volpin, H., De Both, M. T. J., Zerbe, P., et al. (2011). RNA-seq discovery, functional characterization, and comparison of sesquiterpene synthases from Solanum lycopersicum and Solanum habrochaites trichomes. *Plant Molecular Biology,* 77, 323–336.

Botella-Pavia, P., Besumbes, O., Phillips, M. A., Carretero-Paulet, L., Boronat, A., & Rodríguez-Concepción, M. (2004). Regulation of carotenoid biosynthesis in plants: Evidence for a key role of hydroxymethylbutenyl diphosphate reductase in controlling the supply of plastidial isoprenoid precursors. *The Plant Journal, 40,* 188–199.

Buttery, R. G., Teranishi, R., Ling, L. C., & Turnbaugh, J. G. (1990). Quantitative and sensory studies on tomato paste volatiles. *Journal of Agricultural and Food Chemistry, 38,* 336–340.

Cunningham, F. X., & Gantt, E. (1998). Genes and enzymes of carotenoid biosynthesis in plants. *Annual Review of Plant Physiology and Plant Molecular Biology, 49,* 557–583.

Dafny-Yelin, M., & Tzfira, T. (2007). Delivery of multiple transgenes to plant cells. *Plant Physiology, 145,* 1118–1128.

Davidovich-Rikanati, R., Lewinsohn, E., Bar, E., Iijima, Y., Pichersky, E., & Sitrit, Y. (2008). Overexpression of the lemon basil α-zingiberene synthase gene increases both mono- and sesquiterpene contents in tomato fruit. *The Plant Journal, 56*, 228–238.

Davidovich-Rikanati, R., Sitrit, Y., Tadmor, Y., Iijima, Y., Bilenko, N., Bar, E., et al. (2007). Enrichment of tomato flavor by diversion of the early plastidial terpenoid pathway. *Nature Biotechnology, 25*, 899–901.

de Felipe, P., Luke, G. A., Hughes, L. E., Gani, D., Halpin, C., & Ryan, M. D. (2006). E unum pluribus: Multiple proteins from a self-processing polyprotein. *Trends in Biotechnology, 24*, 68–75.

Degenhardt, J., Hiltpold, I., Kollner, T. G., Frey, M., Gierl, A., Gershenzon, J., et al. (2009). Restoring a maize root signal that attracts insect-killing nematodes to control a major pest. *Proceedings of the National Academy of Sciences of the United States of America, 106*, 13213–13218.

Deikman, J., & Fischer, R. L. (1988). Interaction of a DNA binding factor with the 5′-flanking region of an ethylene-responsive fruit ripening gene from tomato. *The EMBO Journal, 7*, 3315–3320.

Deikman, J., Kline, R., & Fischer, R. L. (1992). Organization of ripening and ethylene regulatory regions in a fruit-specific promoter from tomato (*Lycopersicon esculentum*). *Plant Physiology, 100*, 2013–2017.

DellaPenna, D., Lincoln, J. E., Fischer, R. L., & Bennett, A. B. (1989). Transcriptional analysis of polygalacturonase and other ripening associated genes in Rutgers, rin, nor, and Nr tomato fruit. *Plant Physiology, 90*, 1372–1377.

Deruère, J., Bouvier, F., Steppuhn, J., Klein, A., Camara, B., & Kuntz, M. (1994). Structure and expression of two plant genes encoding chromoplast-specific proteins: Occurrence of partially spliced transcripts. *Biochemical and Biophysical Research Communications, 199*, 1144–1150.

Deruère, J., Römer, S., d'Harlingue, A., Backhaus, R. A., Kuntz, M., & Camara, B. (1994). Fibril assembly and carotenoid overaccumulation in chromoplasts: A model for supramolecular lipoprotein structures. *Plant Cell, 6*, 119–133.

Díaz de la Garza, R. I., Gregory, J. F., & Hanson, A. D. (2007). Folate biofortification of tomato fruit. *Proceedings of the National Academy of Sciences of the United States of America, 104*, 4218–4222.

Donnelly, M. L., Luke, G., Mehrotra, A., Li, X., Hughes, L. E., Gani, D., et al. (2001). Analysis of the aphthovirus 2A/2B polyprotein 'cleavage' mechanism indicates not a proteolytic reaction, but a novel translational effect: A putative ribosomal 'skip'. *Journal of General Virology, 82*, 1013–1025.

Dudareva, N., Andersson, S., Orlova, I., Gatto, N., Reichelt, M., Rhodes, D., et al. (2005). The nonmevalonate pathway supports both monoterpene and sesquiterpene formation in snapdragon flowers. *Proceedings of the National Academy of Sciences of the United States of America, 102*, 933–938.

Dudareva, N., Cseke, L., Blanc, V. M., & Pichersky, E. (1996). Evolution of floral scent in *Clarkia*: Novel patterns of S-linalool synthase gene expression in the *C. breweri* flower. *The Plant Cell, 8*, 1137–1148.

Dudareva, N., Martin, D., Kish, C. M., Kolosova, N., Gorenstein, N., Fäldt, J., et al. (2003). (E)-ß-Ocimene and myrcene synthase genes of floral scent biosynthesis in snapdragon: Function and expression of three terpene synthase genes of a new terpene synthase subfamily. *The Plant Cell, 15*, 1227–1241.

Enfissi, E. M., Fraser, P. D., Lois, L. M., Boronat, A., Schuch, W., & Bramley, P. M. (2005). Metabolic engineering of the mevalonate and non-mevalonate isopentenyl diphosphate-forming pathways for the production of health-promoting isoprenoids in tomato. *Plant Biotechnology Journal, 3*, 17–27.

Falara, V., Akhtar, T. A., Nguyen, T. T., Spyropoulou, E. A., Bleeker, P. M., Schauvinhold, I., et al. (2011). The tomato terpene synthase gene family. *Plant Physiology, 157*, 770–789.

Fäldt, J., Arimura, G., Gershenzon, J., Takabayashi, J., & Bohlmann, J. (2003). Functional identification of AtTPS03 as (E)-β-ocimene synthase: A monoterpene synthase catalyzing jasmonate- and wound-induced volatile formation in *Arabidopsis thaliana*. *Planta, 216*, 745–751.

Fernandez, A. I., Viron, N., Alhagdow, M., Karimi, M., Jones, M., Amsellem, Z., et al. (2009). Flexible tools for gene expression and silencing in tomato. *Plant Physiology, 151*, 1729–1740.

Flügge, U. I., & Gao, W. (2005). Transport of isoprenoid intermediates across chloroplast envelope membranes. *Plant Biology*, 7, 91–97.

Fraser, P. D., Truesdale, M. R., Bird, C. R., Schuch, W., & Bramley, P. M. (1994). Carotenoid biosynthesis during tomato fruit development. *Plant Physiology, 105*, 405–413.

Fray, R. G., & Grierson, D. (1993). Identification and genetic analysis of normal and mutant phytoene synthase genes of tomato by sequencing, complementation and co-suppression. *Plant Molecular Biology, 22*, 589–602.

Furumoto, T., Yamaguchi, T., Ohshima-Ichie, Y., Nakamura, M., Tsuchida-Iwata, Y., Shimamura, M., et al. (2011). A plastidial sodium-dependent pyruvate transporter. *Nature, 476*, 472–475.

Gaffe, J., Bru, J.-P., Causse, M., Vidal, A., Stamitti-Bert, L., Carde, J.-P., et al. (2000). *LEFPS1*, a tomato farnesyl pyrophosphate gene highly expressed during early fruit development. *Plant Physiology, 123*, 1351–1362.

Ganjewala, D., & Luthra, R. (2009). Geranyl acetate esterase controls and regulates the level of geraniol in lemongrass (*Cymbopogon flexuosus* Nees ex Steud.) mutant cv. GRL-1 leaves. *Zeitschrift FüR Naturforschung C, Journal of Biosciences, 64*, 251–259.

Gershenzon, J., & Dudareva, N. (2007). The function of terpene natural products in the natural world. *Nature Chemical Biology, 3*, 408–414.

Gillaspy, G., Ben-David, H., & Gruissem, W. (1993). Fruits: A developmental perspective. *The Plant Cell, 5*, 1439–1451.

Giovannoni, J. J., DellaPenna, D., Bennett, A. B., & Fischer, R. L. (1989). Expression of a chimeric polygalacturonase gene in transgenic rin (ripening inhibitor) tomato fruit results in polyuronide degradation but not fruit softening. *The Plant Cell, 1*, 53–63.

Giuliano, G., Bartley, G. E., & Scolnik, P. A. (1993). Regulation of carotenoid biosynthesis during tomato development. *The Plant Cell, 5*, 379–387.

Gonzales-Vigil, E., Hufnagel, D. E., Kim, J., Last, R. L., & Barry, C. S. (2012). Evolution of TPS20-related terpene synthases influences chemical diversity in the glandular trichomes of the wild tomato relative *Solanum habrochaites*. *The Plant Journal, 71*, 921–935.

Good, X., Kellogg, J. A., Wagoner, W., Langhoff, D., Matsumura, W., & Bestwick, R. K. (1994). Reduced ethylene synthesis by transgenic tomatoes expressing S-adenosylmethionine hydrolase. *Plant Molecular Biology, 26*, 781–790.

Graham, I. A., Besser, K., Blumer, S., Branigan, C. A., Czechowski, T., Elias, L., et al. (2010). The genetic map of *Artemisia annua* L. identifies loci affecting yield of the anti-malarial drug artemisinin. *Science, 327*, 328–331.

Gutensohn, M., Nguyen, T. T., McMahon, R. D., Kaplan, I., Pichersky, E., & Dudareva, N. (2014). Metabolic engineering of monoterpene biosynthesis in tomato fruits via introduction of the non-canonical substrate neryl diphosphate. *Metabolic Engineering, 24*, 107–116.

Gutensohn, M., Orlova, I., Nguyen, T. T., Davidovich-Rikanati, R., Ferruzzi, M. G., Sitrit, Y., et al. (2013). Cytosolic monoterpene biosynthesis is supported by plastid-generated geranyl diphosphate substrate in transgenic tomato fruits. *The Plant Journal, 75*, 351–363.

Guterman, I., Shalit, M., Menda, N., Piestun, D., Dafny-Yelin, M., Shalev, G., et al. (2002). Rose scent: Genomics approach to discovering novel floral fragrance-related genes. *The Plant Cell, 14*, 2325–2338.

Ha, S. H., Liang, Y. S., Jung, H., Ahn, M. J., Suh, S. C., Kweon, S. J., et al. (2010). Application of two bicistronic systems involving 2A and IRES sequences to the biosynthesis of carotenoids in rice endosperm. *Plant Biotechnology Journal, 8*, 928–938.

Hemmerlin, A., Harwood, J. L., & Bach, T. J. (2012). A raison d'être for two distinct pathways in the early steps of plant isoprenoid biosynthesis? *Progress in Lipid Research, 51*, 95–148.

Hemmerlin, A., Hoeffler, J. F., Meyer, O., Tritsch, D., Kagan, I. A., Grosdemange-Billiard, C., et al. (2003). Cross-talk between the cytosolic mevalonate and the plastidial methylerythritol phosphate pathways in tobacco Bright Yellow-2 cells. *Journal of Biological Chemistry, 278*, 26666–26676.

Henry, L. K., Gutensohn, M., Thomas, S. T., Noel, J. P., & Dudareva, N. (2015). Orthologs of the archaeal isopentenyl phosphate kinase regulate terpenoid production in plants. *Proceedings of the National Academy of Sciences of the United States of America, 112*, 10050–10055.

Huang, J., Cardoza, Y. J., Schmelz, E. A., Raina, R., Engelberth, J., & Tumlinson, J. H. (2003). Differential volatile emissions and salicylic acid levels from tobacco plants in response to different strains of *Pseudomonas syringae*. *Planta, 217*, 767–775.

Iijima, Y., Davidovich-Rikanati, R., Fridman, E., Gang, D. R., Bar, E., Lewinsohn, E., et al. (2004). The biochemical and molecular basis for the divergent patterns in the biosynthesis of terpenes and polypropenes in the peltate glands of three cultivars of basil. *Plant Physiology, 136*, 3724–3736.

Izumi, S., Ashida, Y., Yamamitsu, T., & Hirata, T. (1996). Hydrolysis of isoprenyl diphosphates with the acid phosphatase from *Cinnamomum camphora*. *Cellular and Molecular Life Sciences, 52*, 81–84.

Jelesko, J. G., Jenkins, S. M., Rodríguez-Concepción, M., & Gruissem, W. (1999). Regulation of tomato *HMG1* during cell proliferation and growth. *Planta, 208*, 310–318.

Kang, J. H., Liu, G., Shi, F., Jones, A. D., Beaudry, R. M., & Howe, G. A. (2010). The tomato *odorless-2* mutant is defective in trichome based production of diverse specialized metabolites and broad-spectrum resistance to insect herbivores. *Plant Physiology, 154*, 262–272.

Kasahara, H., Hanada, A., Kuzuyama, T., Takagi, M., Kamiya, Y., & Yamaguchi, S. (2002). Contribution of the mevalonate and methylerythritol phosphate pathways to the biosynthesis of gibberellins in Arabidopsis. *Journal of Biological Chemistry, 277*, 45188–45194.

Kneissl, M. L., & Deikman, J. (1996). The tomato E8 gene influences ethylene biosynthesis in fruit but not in flowers. *Plant Physiology, 112*, 537–547.

Kovacs, K., Zhang, L., Linforth, R. S., Whittaker, B., Hayes, C. J., & Fray, R. G. (2007). Redirection of carotenoid metabolism for the efficient production of taxadiene [taxa-4(5),11(12)-diene] in transgenic tomato fruit. *Transgenic Research, 16*, 121–126.

Kuntz, M., Chen, H. C., Simkin, A. J., Römer, S., Shipton, C. A., Drake, R., et al. (1998). Upregulation of two ripening-related genes from a non-climacteric plant (pepper) in a transgenic climacteric plant (tomato). *The Plant Journal, 13*, 351–361.

Lange, B. M., & Ahkami, A. (2013). Metabolic engineering of plant monoterpenes, sesquiterpenes and diterpenes—Current status and future opportunities. *Plant Biotechnology Journal, 11*, 169–196.

Laule, O., Fürholz, A., Chang, H. S., Zhu, T., Wang, X., Heifetz, P. B., et al. (2003). Crosstalk between cytosolic and plastidial pathways of isoprenoid biosynthesis in *Arabidopsis thaliana*. *Proceedings of the National Academy of Sciences of the United States of America, 100*, 6866–6871.

Lee, S., & Chappell, J. (2008). Biochemical and genomic characterization of terpene synthases in *Magnolia grandiflora*. *Plant Physiology*, *147*, 1017–1033.

Leitner, M., Kaiser, R., Rasmussen, M. O., Driguez, H., Boland, W., & Mithöfer, A. (2008). Microbial oligosaccharides differentially induce volatiles and signaling components in *Medicago truncatula*. *Phytochemistry*, *69*, 2029–2040.

Lewinsohn, E., & Gijzen, M. (2009). Phytochemical diversity: The sounds of silent metabolism. *Plant Science*, *176*, 161–169.

Lewinsohn, E., Schalechet, F., Wilkinson, J., Matsui, K., Tadmor, Y., Nam, K. H., et al. (2001). Enhanced levels of the aroma and flavor compound S-linalool by metabolic engineering of the terpenoid pathway in tomato fruits. *Plant Physiology*, *127*, 1256–1265.

Lewinsohn, E., Sitrit, Y., Bar, E., Azulay, Y., Ibdah, M., Meir, A., et al. (2005a). Not just colors—Carotenoid degradation as a link between pigmentation and aroma in tomato and watermelon fruit. *Trends in Food Science & Technology*, *16*, 407–415.

Lewinsohn, E., Sitrit, Y., Bar, E., Azulay, Y., Meir, A., Zamir, D., et al. (2005b). Carotenoid pigmentation affects the volatile composition of tomato and watermelon fruits, as revealed by comparative genetic analyses. *Journal of Agricultural and Food Chemistry*, *53*, 3142–3148.

Li, L., Zhao, Y., McCaig, B. C., Wingerd, B. A., Wang, J., Whalon, M. E., et al. (2004). The tomato homolog of CORONATINE-INSENSITIVE1 is required for the maternal control of seed maturation, jasmonate-signaled defense responses, and glandular trichome development. *The Plant Cell*, *16*, 126–143.

Lincoln, J. E., & Fischer, R. L. (1988). Diverse mechanisms for the regulation of ethylene-inducible gene expression. *Molecular & General Genetics*, *212*, 71–75.

Lois, L. M., Rodríguez-Concepción, M., Gallego, F., Campos, N., & Boronat, A. (2000). Carotenoid biosynthesis during tomato fruit development: Regulatory role of 1-deoxy-D-xylulose 5-phosphate synthase. *The Plant Journal*, *22*, 503–513.

Martin, D. M., Gershenzon, J., & Bohlmann, J. (2003). Induction of volatile terpene biosynthesis and diurnal emission by methyl jasmonate in foliage of Norway spruce. *Plant Physiology*, *132*, 1586–1599.

Matsuba, Y., Nguyen, T. T., Wiegert, K., Falara, V., Gonzales-Vigil, E., Leong, B., et al. (2013). Evolution of a complex locus for terpene biosynthesis in *Solanum*. *The Plant Cell*, *25*, 2022–2036.

Montgomery, J., Pollard, V., Deikman, J., & Fischer, R. L. (1993). Positive and negative regulatory regions control the spatial distribution of polygalacturonase transcription in tomato fruit pericarp. *The Plant Cell*, *5*, 1049–1062.

Nagata, N., Suzuki, M., Yoshida, S., & Muranaka, T. (2002). Mevalonic acid partially restores chloroplast and etioplast development in Arabidopsis lacking the non-mevalonate pathway. *Planta*, *216*, 345–350.

Narita, J. O., & Gruissem, W. (1989). Tomato hydroxymethylglutaryl-CoA reductase is required early in fruit development but not during ripening. *The Plant Cell*, *1*, 181–190.

Ngoi, S. M., Chien, A. C., & Lee, C. G. (2004). Exploiting internal ribosome entry sites in gene therapy vector design. *Current Gene Therapy*, *4*, 15–31.

Nicholass, F. J., Smith, C. J., Schuch, W., Bird, C. R., & Grierson, D. (1995). High levels of ripening-specific reporter gene expression directed by tomato fruit polygalacturonase gene-flanking regions. *Plant Molecular Biology*, *28*, 423–435.

Nieuwenhuizen, N. J., Wang, M. Y., Matich, A. J., Green, S. A., Chen, X., Yauk, Y. K., et al. (2009). Two terpene synthases are responsible for the major sesquiterpenes emitted from the flowers of kiwifruit (*Actinidia deliciosa*). *Journal of Experimental Botany*, *60*, 3203–3219.

Orlova, I., Nagegowda, D. A., Kish, C. M., Gutensohn, M., Maeda, H., Varbanova, M., et al. (2009). The small subunit of snapdragon geranyl diphosphate synthase modifies

the chain length specificity of tobacco geranylgeranyl diphosphate synthase in planta. *The Plant Cell, 21*, 4002–4017.
Orzaez, D., Medina, A., Torre, S., Fernández-Moreno, J. P., Rambla, J. L., Fernández-Del-Carmen, A., et al. (2009). A visual reporter system for virus-induced gene silencing in tomato fruit based on anthocyanin accumulation. *Plant Physiology, 150*, 1122–1134.
Orzaez, D., Mirabel, S., Wieland, W. H., & Granell, A. (2006). Agroinjection of tomato fruits. A tool for rapid functional analysis of transgenes directly in fruit. *Plant Physiology, 140*, 3–11.
Paetzold, H., Garms, S., Bartram, S., Wieczorek, J., Urós-Gracia, E. M., Rodríguez-Concepción, M., et al. (2010). The isogene 1-deoxy-D-xylulose 5-phosphate synthase 2 controls isoprenoid profiles, precursor pathway allocation, and density of tomato trichomes. *Molecular Plant, 3*, 904–916.
Pecker, I., Chamovitz, D., Linden, H., Sandmann, G., & Hirschberg, J. (1992). A single polypeptide catalyzing the conversion of phytoene to zeta-carotene is transcriptionally regulated during tomato fruit ripening. *Proceedings of the National Academy of Sciences of the United States of America, 89*, 4962–4966.
Perez, L. M., Taucher, G., & Cori, O. (1980). Hydrolysis of allylic phosphates by enzymes from the flavedo of *Citrus sinensis*. *Phytochemistry, 19*, 183–187.
Pichersky, E., & Gershenzon, J. (2002). The formation and function of plant volatiles: Perfumes for pollinator attraction and defense. *Current Opinion in Plant Biology, 5*, 237–243.
Ronen, G., Cohen, M., Zamir, D., & Hirschberg, J. (1999). Regulation of carotenoid biosynthesis during tomato fruit development: Expression of the gene for lycopene epsilon-cyclase is down-regulated during ripening and is elevated in the mutant *Delta*. *The Plant Journal, 17*, 341–351.
Sallaud, C., Rontein, D., Onillon, S., Jabès, F., Duffé, P., Giacalone, C., et al. (2009). A novel pathway for sesquiterpene biosynthesis from Z,Z-farnesyl pyrophosphate in the wild tomato Solanum habrochaites. *The Plant Cell, 21*, 301–317.
Sandhu, J. S., Krasnyanski, S. F., Domier, L. L., Korban, S. S., Osadjan, M. D., & Buetow, D. E. (2000). Oral immunization of mice with transgenic tomato fruit expressing respiratory syncytial virus-F protein induces a systemic immune response. *Transgenic Research, 9*, 127–135.
Schilmiller, A. L., Last, R. L., & Pichersky, E. (2008). Harnessing plant trichome biochemistry for the production of useful compounds. *The Plant Journal, 54*, 702–711.
Schilmiller, A. L., Schauvinhold, I., Larson, M., Xu, R., Charbonneau, A. L., Schmidt, A., et al. (2009). Monoterpenes in the glandular trichomes of tomato are synthesized from a neryl diphosphate precursor rather than geranyl diphosphate. *Proceedings of the National Academy of Sciences of the United States of America, 106*, 10865–10870.
Schnee, C., Köllner, T. G., Gershenzon, J., & Degenhardt, J. (2002). The maize gene terpene synthase 1 encodes a sesquiterpene synthase catalyzing the formation of (E)-β-farnesene, (E)-nerolidol, and (E,E)-farnesol after herbivore damage. *Plant Physiology, 130*, 2049–2060.
Schnee, C., Kollner, T. G., Held, M., Turlings, T. C., Gershenzon, J., & Degenhardt, J. (2006). The products of a single maize sesquiterpene synthase form a volatile defense signal that attracts natural enemies of maize herbivores. *Proceedings of the National Academy of Sciences of the United States of America, 103*, 1129–1134.
Schuhr, C. A., Radykewicz, T., Sagner, S., Latzel, C., Zenk, M. H., Arigoni, D., et al. (2003). Quantitative assessment of crosstalk between the two isoprenoid biosynthesis pathways in plants by NMR spectroscopy. *Phytochemistry Reviews, 2*, 3–16.

Simkin, A. J., Gaffé, J., Alcaraz, J. P., Carde, J. P., Bramley, P. M., Fraser, P. D., et al. (2007). Fibrillin influence on plastid ultrastructure and pigment content in tomato fruit. *Phytochemistry, 68*, 1545–1556.

Soler, E., Clastre, M., Bantignies, B., Marigo, G., & Ambid, C. (1993). Uptake of isopentenyl diphosphate by plastids isolated from *Vitis vinifera* L. cell suspensions. *Planta, 191*, 324–329.

Spolaore, S., Trainotti, L., & Casadoro, G. (2001). A simple protocol for transient gene expression in ripe fleshy fruit mediated by *Agrobacterium*. *Journal of Experimental Botany, 52*, 845–850.

Sun, L., Yuan, B., Zhang, M., Wang, L., Cui, M., Wang, Q., et al. (2012). Fruit-specific RNAi-mediated suppression of SlNCED1 increases both lycopene and β-carotene contents in tomato fruit. *Journal of Experimental Botany, 63*, 3097–3108.

Tholl, D., Kish, C. M., Orlova, I., Sherman, D., Gershenzon, J., Pichersky, E., et al. (2004). Formation of monoterpenes in *Antirrhinum majus* and *Clarkia breweri* flowers involves heterodimeric geranyl diphosphate synthases. *The Plant Cell, 16*, 977–992.

Tzin, V., Rogachev, I., Meir, S., Moyal Ben Zvi, M., Masci, T., Vainstein, A., et al. (2013). Tomato fruits expressing a bacterial feedback-insensitive 3-deoxy-D-arabino-heptulosonate 7-phosphate synthase of the shikimate pathway possess enhanced levels of multiple specialized metabolites and upgraded aroma. *Journal of Experimental Botany, 64*, 4441–4452.

Unsicker, S. B., Kunert, G., & Gershenzon, J. (2009). Protective perfumes: The role of vegetative volatiles in plant defense against herbivores. *Current Opinion in Plant Biology, 12*, 479–485.

van Schie, C. C., Haring, M. A., & Schuurink, R. C. (2007). Tomato linalool synthase is induced in trichomes by jasmonic acid. *Plant Molecular Biology, 64*, 251–263.

Vickers, C. E., Bongers, M., Liu, Q., Delatte, T., & Bouwmeester, H. (2014). Metabolic engineering of volatile isoprenoids in plants and microbes. *Plant, Cell & Environment, 37*, 1753–1775.

Wang, G., & Dixon, R. A. (2009). Heterodimeric geranyl(geranyl)diphosphate synthase from hop (*Humulus lupulus*) and the evolution of monoterpene biosynthesis. *Proceedings of the National Academy of Sciences of the United States of America, 106*, 9914–9919.

Wu, S., Schalk, M., Clark, A., Miles, R. B., Coates, R., & Chappell, J. (2006). Redirection of cytosolic or plastidic isoprenoid precursors elevates terpene production in plants. *Nature Biotechnology, 24*, 1441–1447.

Zhang, M., Liu, J., Li, K., & Yu, D. (2013). Identification and characterization of a novel monoterpene synthase from soybean restricted to neryl diphosphate precursor. *PLoS One, 8*, e75972.

Zhang, H., Niu, D., Wang, J., Zhang, S., Yang, Y., Jia, H., et al. (2015). Engineering a platform for photosynthetic pigment, hormone and cembrane-related diterpenoid production in *Nicotiana tabacum*. *Plant & Cell Physiology, 56*, 2125–2138.

CHAPTER FOURTEEN

Libraries of Synthetic TALE-Activated Promoters: Methods and Applications

T. Schreiber, A. Tissier[1]
Leibniz Institute of Plant Biochemistry, Halle (Saale), Germany
[1]Corresponding author: e-mail address: alain.tissier@ipb-halle.de

Contents

1. Introduction 361
2. Construction of Libraries of Synthetic Promoters Using Golden Gate Cloning 366
3. Analyzing Promoter Activity in Transient Assays 371
4. Conclusion 373
References 375

Abstract

The discovery of proteins with programmable DNA-binding specificities triggered a whole array of applications in synthetic biology, including genome editing, regulation of transcription, and epigenetic modifications. Among those, transcription activator-like effectors (TALEs) due to their natural function as transcription regulators, are especially well-suited for the development of orthogonal systems for the control of gene expression. We describe here the construction and testing of libraries of synthetic TALE-activated promoters which are under the control of a single TALE with a given DNA-binding specificity. These libraries consist of a fixed DNA-binding element for the TALE, a TATA box, and variable sequences of 19 bases upstream and 43 bases downstream of the DNA-binding element. These libraries were cloned using a Golden Gate cloning strategy making them usable as standard parts in a modular cloning system. The broad range of promoter activities detected and the versatility of these promoter libraries make them valuable tools for applications in the fine-tuning of expression in metabolic engineering projects or in the design and implementation of regulatory circuits.

1. INTRODUCTION

Transcription represents the first level of gene expression control and therefore constitutes a key entry point to regulate gene expression. Additionally, our understanding of the mechanisms of posttranscriptional

Methods in Enzymology, Volume 576
ISSN 0076-6879
http://dx.doi.org/10.1016/bs.mie.2016.03.004

regulation of gene expression, apart from gene silencing, is still partial, making their use for engineering purposes difficult, at best on a case by case basis. Thus, modulating gene expression at the level of transcription (initiation and strength) is an accessible and universal approach for genetic engineering purposes. Genetic engineering projects can be classified in two main categories. The first are those that deal with regulatory circuits, that is, how one or several inputs can be translated into a response. Pathway engineering constitutes the second major type, where, for example, genes for a metabolic pathway need to be coexpressed for the synthesis of the product of interest. Both approaches can be combined, eg, when the engineered metabolic pathway is the output in a circuit engineering approach. In both cases, fine-tuning of gene expression is required to achieve optimal signal integration in the regulatory circuit or maximal metabolic flux for pathway engineering. Fine-tuning means that specific steps in a regulatory circuit or in a metabolic pathway should be expressed at a specific level to optimally fulfill their task. Such expression levels can be predicted by modeling approaches, for example, by metabolic control analysis for metabolic pathways (Jensen & Hammer, 1998a). Experimentally, this requires the availability of a set of transcriptional promoters which cover a range of expression levels sufficiently broad to satisfy the modeling criteria. To build such libraries of promoters with predictable and verifiable strength different strategies have been adopted (Dehli, Solem, & Jensen, 2012). The first consists in amplifying a given promoter or parts thereof by error-prone PCR, thereby introducing random mutations which will result in modified expression strength (Qin et al., 2011). However, since single mutations are introduced few promoters will differ significantly from the original promoter, thus requiring extensive screening to identify promoters covering a broad range of activity. An alternative to error-prone PCR are synthetic promoter libraries (SPL). Here, specific motifs are conserved, such as a TATA box or DNA-binding elements, and in between these motifs, random sequences of a fixed length are introduced. This can be easily achieved by using synthetic oligonucleotides that contain these randomized fragments. Using this approach, a broad range of promoter activity can be identified within a relatively small set of sequences. This was first shown in 1998 for bacterial promoters (Jensen & Hammer, 1998b), and later with Eukaryotic promoters (Tornoe, Kusk, Johansen, & Jensen, 2002). One issue with these promoters is that they are based on an endogenous chassis and therefore are not orthogonal. Inducibility can be introduced via a TET operator and the use of the inhibition of the Tet repressor by anhydrotetracycline and orthogonality can be implemented on such promoters by the introduction of an effector-binding element of so-called transcription activator-like orthogonal repressors (TALORs) (Blount, Weenink, Vasylechko, & Ellis, 2012). However, with this design only repression of a constitutive promoter can be achieved. The use of transcription factors with a customizable DNA-binding recognition domain provides the opportunity to design such orthogonal promoters. This was first done with zinc finger domain TF (ZFs), followed by transcription activator-like effectors (TALEs) and more recently with the CRISPR/dCas9 system (Dominguez, Lim, & Qi, 2016; Khalil et al., 2012; Perez-Pinera, Kocak, et al., 2013). In all these cases it was shown that increasing the number of DNA-binding (or recognition) elements increases gene expression or that more than one activator for one promoter, thus providing a way to tune expression (Perez-Pinera, Kocak, et al., 2013; Perez-Pinera, Ousterout, et al., 2013). Nonetheless, these approaches do not give rise to libraries of promoters with a wide range of expression levels. Such sets of promoters are highly desirable, for example, for metabolic engineering, where multiple genes with distinct expression levels should be expressed under the control of a single TF. Designing SPL using TF factors with customizable DNA-binding domain would be the solution. Among those, TALEs present a number of advantages over the other systems. ZFs are customizable but their binding specificity is difficult to predict requiring extensive testing. The CRISPR/Cas system is a natural bacterial defense system against viruses which specifically cleaves viral DNA using a guide RNA (gRNA). The most commonly used CRISPR/Cas system is that of *Streptococcus pyogenes* (Jinek et al., 2012). Specific mutants of the Cas protein (dCas) inactivate its nuclease activity without compromising DNA recognition specificity (Qi et al., 2013). The CRISPR/dCas can be used to activate or repress transcription depending on the position of the binding element in the promoter or by fusing a transcription activation or repressor domain to it (Farzadfard, Perli, & Lu, 2013; Gilbert et al., 2013, 2014; Perez-Pinera, Kocak, et al., 2013). Transcriptional activation by Cas9-based TFs in plants was accomplished by fusing the activation domain (AD) of TALEs; however, comparison to the performance of a natural TALE was not shown (Piatek et al., 2015). One issue with the CRISPR/Cas system is that in Eukaryotes the gRNA which confers sequence specificity has to be transcribed from RNA Pol III promoters, so that the gRNA can stay in the nucleus. This restricts possibilities to express the gRNA in a tissue specific or developmentally regulated fashion, since most RNA Pol III promoters are constitutive. This could be circumvented by the use of a sequence-specific RNAse, Csy4, which could then release the gRNA from a transcript generated by a RNA Pol II promoter (Nissim, Perli, Fridkin, Perez-Pinera, & Lu, 2014). Another potential issue with CRISPR/dCas is the fact there is a single protein which recognizes all the gRNA. This is problematic for regulatory circuits where distinct proteins with or without different function (eg, activation or repression) would be required. Recently however, other CRISPR/Cas systems (eg, from *Streptococcus thermophilus* or *Staphylococcus aureus*) could also be used for the same purposes (Kleinstiver, Prew, Tsai, Nguyen, et al., 2015; Kleinstiver, Prew, Tsai, Topkar, et al., 2015).

Concerning transcriptional induction in plants naturally evolved designer TALEs (dTALEs) seem to be the TFs of choice. TALEs, such as AvrBs3, are proteins from plant-pathogenic *Xanthomonas* species, which are translocated into the host cell and induce transcription of plant genes to promote disease (Boch, Bonas, & Lahaye, 2014). TALEs are composed of three domains. The N-terminal region (NTR) with the type III secretion (T3S) signal, the C-terminal region (CTR) containing nuclear localization signals (NLS) as well as an acidic AD and the central DNA-binding domain compromised of a varying number of tandem 33–35 amino acid repeats (Boch & Bonas, 2010) (Fig. 1). The amino acid sequences of the repeats vary predominantly at position 12 and 13, which are designated as repeat-variable diresidue (RVD; Fig. 1). TALEs bind to DNA by a one repeat one base pair recognition mode, in which the RVD defines the specificity of a given repeat (Boch et al., 2009; Moscou & Bogdanove, 2009). Although all possible amino acid combinations were analyzed in the context of the TALE repeat, the most common natural RVDs HD, NI, NG, and NN confer efficient and modular specificity to cytosine, adenine, thymine, and guanine/adenine, respectively (Cong, Zhou, Kuo, Cunniff, & Zhang, 2012; Juillerat et al., 2015; Miller et al., 2015; Streubel, Blucher, Landgraf, & Boch, 2012; Yang et al., 2014) (Fig. 1). Structural data reveal that only amino acid 13 makes specific contacts to the corresponding base by formation of hydrogen bonds or Van der Waals interactions, whereas amino acid 12 has structural functions that help to expose residue 13 toward the major groove of the DNA double strand (Deng, Yan, Wu, Pan, & Yan, 2014; Deng et al., 2012; Mak, Bradley, Cernadas, Bogdanove, & Stoddard, 2012). The DNA sequence which is bound by a TALE is defined by the number of repeats and the order of RVDs (Fig. 1). Therefore, it is possible to generate dTALEs with any desired DNA-binding specificity by rearrangement of repeats (Cermak et al., 2011; Geissler et al., 2011). Replacement of the AD by other executor domains possessing different

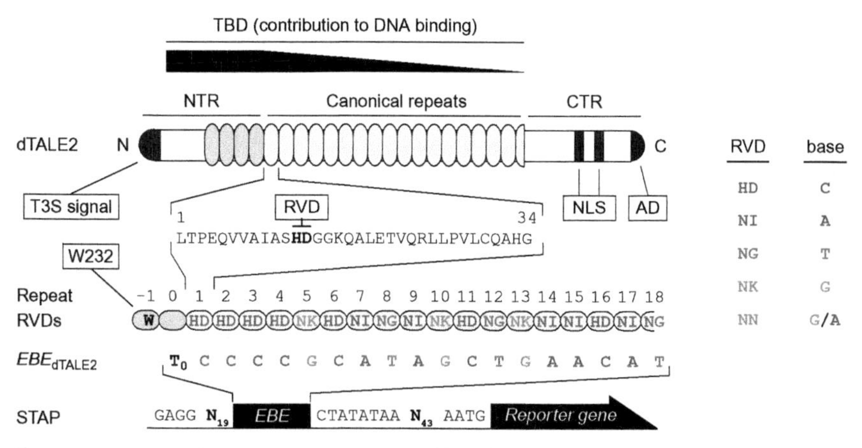

Fig. 1 Properties of transcription activator-like effectors (TALEs) using the example of dTALE2. TALEs are composed of three domains: the N-terminal region (NTR), the repeat region, and the C-terminal region (CTR). The NTR contains a type III secretion and translocation (T3S) signal and a crucial part of the TALE DNA-binding domain (TBD). It includes at least four degenerated repeats (−3 to 0) and serves as nucleation site for the TALE–DNA interaction. The repeat region consists of 17.5 tandem arrayed 34-amino acid motifs (repeats), conferring the binding specificity. The repeats are highly conserved except for amino acid 12 and 13 designated as repeat-variable diresidue (RVD). One repeat mediates binding to one base pair of the target sequence in which the RVD defines the specificity of a given repeat. The most common RVDs are HD, NI, NG, NK, and NN which are specific for cytosine (C), adenine (A), thymine (T), guanine (G), and guanine/adenine (G/A), respectively. The order of the RVDs defines the DNA target sequence that is bound by a TALE. The RVD-defined target sequence is preceded by a 5′-thymine (T_0) and is part of the effector-binding element (EBE). T_0 is coordinated by a tryptophan residue (W232) of the degenerated repeat −1 that facilitates the RVD–base interaction of the following canonical repeats. The CTR contains two nuclear localization signals and an acidic activation domain (AD). A reporter gene fused to a synthetic TALE-activated promoter (STAP) can be transcriptionally induced by a dTALE. (See the color plate.)

function converts the transcriptional activator to sequence-specific artificial TALE-based repressors (KRAB, TALORs), nucleases (*Fok*I, TALENs), recombinases (Gin, TALERs), or epigenetic modifiers (LSD1, epiTALEs) (Cong et al., 2012; de Lange, Binder, & Lahaye, 2014; Mendenhall et al., 2013; Mercer, Gaj, Fuller, & Barbas, 2012; Miller et al., 2011).

However, the application of the TALE DNA-binding domain (TBD) is limited by some constraints. In nature, almost all RVD-defined target sequences are preceded by a 5′-thymine (T_0), which was shown to be beneficial for the DNA-binding affinity and activity of a TALE (Doyle et al., 2013; Gao, Wu, Chai, & Han, 2012; Römer, Recht, & Lahaye, 2009;

Schreiber & Bonas, 2014) (Fig. 1). Structural analysis identified at least four degenerated repeats in the TALE NTR termed repeat −3 to 0, whereas only repeat −1 contains a RVD loop-like structure with a tryptophan residue (W232) on it, coordinating T_0 by nonpolar Van der Waals forces (Gao et al., 2012; Mak et al., 2012) (Fig. 1). Although T_0-independent NTRs were generated by directed evolution, there is evidence that their functionality is RVD-context dependent and that the W232 and T_0 combination lead to the highest state of modularity (Doyle et al., 2013; Lamb, Mercer, & Barbas, 2013; Schreiber & Bonas, 2014; Tsuji, Futaki, & Imanishi, 2013). In addition, functional analysis of AvrBs3 revealed that the TBD is extended into the NTR beyond the four degenerated repeats (Schreiber et al., 2015) (Fig. 1). Beside the T_0 constraint, targeting stretches of identical nucleotides seem to be unfavorable when using TALEs and depend on the strength of the given RVD–base interaction (Streubel et al., 2012). Strong RVD–base interactions like HD-cytosine and NN-guanine allow polynucleotide binding, whereas weak RVD–base interactions like NI-adenine and NG-thymine do not (Streubel et al., 2012). Thus, the RVD composition directly affects the DNA-binding affinity and thus the activity (transcriptional induction) of a given TALE (Meckler et al., 2013). Therefore, the design of the repeat region, precisely the RVD order, allows the fine-tuning of the TALE-mediated transcriptional induction even if the same DNA sequence is targeted (eg, guanine-targeting RVDs, NN, and NK) (Streubel et al., 2012). This added flexibility cannot be achieved by the CRISPR/Cas system.

Here, we describe methods for the construction and assay of libraries of synthetic promoters activated by dTALEs (synthetic TALE-activated promoters, STAPs). The construction of the libraries is designed to be integrated into a Golden Gate-based cloning strategy, where the promoters are assigned a fixed position in a transcription unit. Possible variations on the design of these promoters are also presented.

2. CONSTRUCTION OF LIBRARIES OF SYNTHETIC PROMOTERS USING GOLDEN GATE CLONING

The TALE promoter libraries were designed based on the observations that a TATA box should be present and that DNA-binding site of the TALE should be between 50 and 100 bp from the transcription start site (Brückner et al., 2015; Hummel, Doyle, & Bogdanove, 2012; Kay,

Hahn, Marois, Wieduwild, & Bonas, 2009). The TALE promoter libraries were generated by PCR using degenerated oligonucleotides containing a *Bsa*I restriction site in the 3′ part and the effector-binding element of dTALE2 (EBE_{dTALE2}) (Weber, Gruetzner, Werner, Engler, & Marillonnet, 2011) in the 5′ part (Talpro7 (fwd) and Talpro8 (rev1) or Talpro9 (rev2); Table 1; Fig. 2). Oligonucleotide combination Talpro7 (fwd) and Talpro8 (rev1) results in STAP modules for a 3′-fusion directly to the coding sequence of reporter genes (GGAG-*STAP*-AATG; STAP library 1 [L1]) (Brückner et al., 2015). In contrast, combination of Talpro7 (fwd) with Talpro9 (rev2) allows the cloning of a 5′-UTR between the STAP and the coding sequence of reporter genes (GGAG-*STAP*-TACT; STAP library 2 [L2]; Fig. 2). Primer extension was performed with KOD Polymerase (Merk Millipore) in a 50 μL reaction according to customer instructions with a concentration of 0.2 μ*M* of each oligonucleotide (fwd and rev). The reaction was performed with denaturing for 5 min at 94°C, annealing for 30 s at 60°C and elongation for 30 s at 72°C. The PCR product was purified using Qiaquick columns (Qiagen) according to the manufacturer's instructions and cloned into pAGM1311 vector (level −1) by standard Golden Gate cloning procedure with *Bsa*I (NEB) and T4 DNA ligase (Promega) (Werner, Engler, Weber, Gruetzner, & Marillonnet, 2012). The reaction was transformed into *E. coli* DH10B (Thermo Fisher), inoculated in liquid culture (LB media) with selective antibiotics and incubated over night at 37°C. The complete level −1 promoter library (STAP-L1 [pAGT559L-n] or STAP-L2 [pAGT1597L-n])

Table 1 Used Oligonucleotides

Name	Sequence (5′→3′)
Talpro7	TTTGGTCTCAACATGGAGNNNNNNNNNNNN NNNNNNNTCCCCGCATAGCTGAACATC
Talpro8	TTTGGTCTCAACAACATTNNNNNNNNNNNN NNNNNNNNNNNNNNNNNNNNNNNNNNNNNN NTTATATAGATGTTCAGCTATGCGGGG
Talpro9	TTTGGTCTCAACAAAGTANNNNNNNNNNNN NNNNNNNNNNNNNNNNNNNNNNNNNNNNNN NTTATATAGATGTTCAGCTATGCGGGG
pICH41308_for	AGCGAGGAAGCGGAAGAGCG

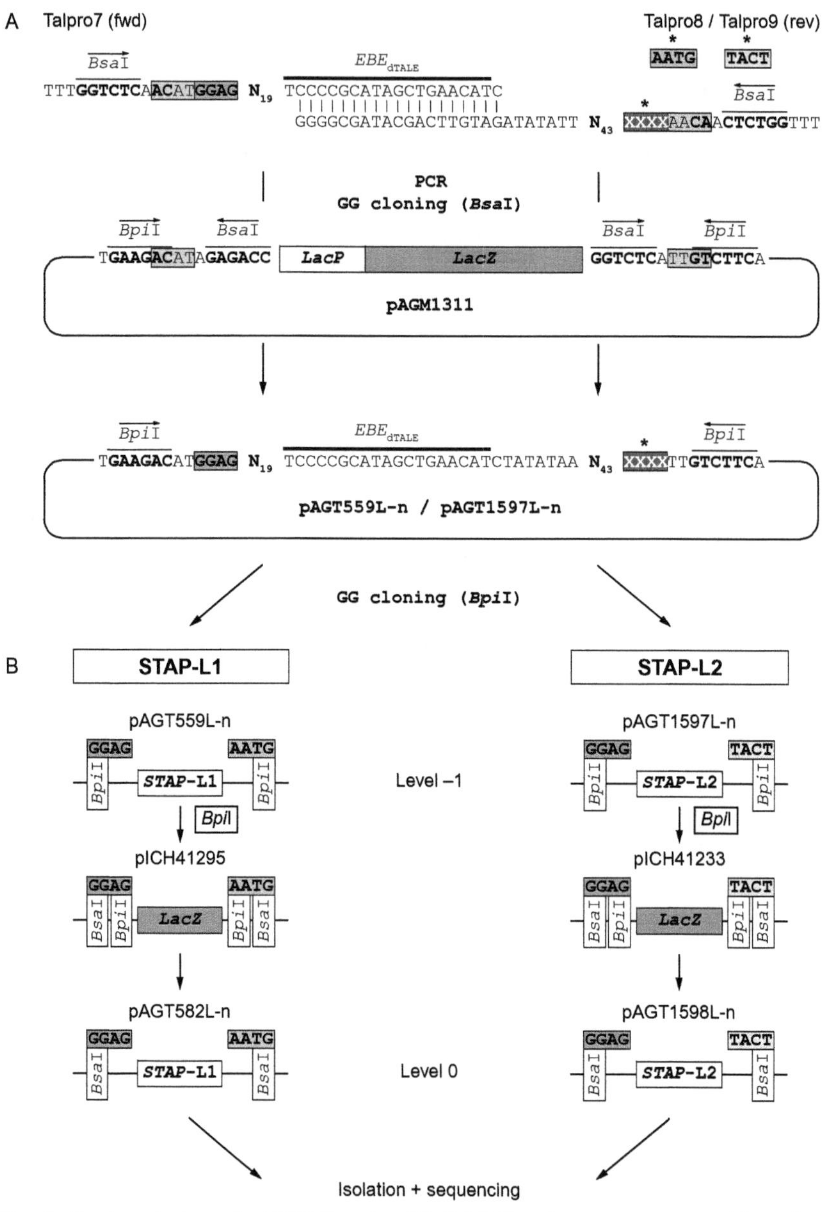

Fig. 2 Cloning strategy for STAP libraries. (A) STAP libraries were generated by primer extension using degenerated oligonucleotides (Talpro7 and Talpro8 (L1) or Talpro7 and Talpro9 (L2)). The first amplification step leads to a *Bsa*I site-flanked PCR product which allows directed cloning into pAGM1311 (level −1) by restriction ligation using *Bsa*I and T4 DNA ligase (Golden Gate [GG] cloning). The *Bsa*I overhangs are depicted in *purple*. The resulting level −1 STAP library is designated as pAGT559L-n and pAGT1597L-n

was isolated by plasmid preparation using NucleoSpin® Plasmid EasyPure (Macherey-Nagel, Düren, Germany) and used for further cloning into the level 0 vectors pICH41295 (GGAG-*LacZ*-AATG; STAP-L1) or pICH41233 (GGAG-*LacZ*-TACT; STAP-L2) by standard Golden Gate cloning procedure with *Bpi*I (Fermentas) and T4 DNA ligase (Promega) (Engler et al., 2014) (Fig. 2B). The reaction was transformed into *E. coli* DH10B and spread on agar with selective antibiotics and X-gal [$c = 20\ \mu M$]. For isolation of individual STAPs single white clones were picked, inoculated into liquid media with selective antibiotics and incubated overnight at 37°C. Individual level 0 STAPs were isolated by plasmid preparation using NucleoSpin® Plasmid EasyPure (Macherey-Nagel), analyzed by digestion and sequenced with *Bsa*I and vector-specific oligonucleotide pICH41308_for, respectively (Table 1). Individual clones of STAP-L1 are designated as pAGT582L-n and individual clones of STAP-L2 are designated as pAGT1598L-n (Fig. 2B). The reason for going through two cloning steps (*Bsa*I and *Bpi*I cloning) before sequencing, is to eliminate all STAPs that contain *Bsa*I or *Bpi*I sites, which would cause trouble in the subsequent cloning using the MoClo-system (Engler et al., 2014) and to eliminate dimer or oligomers of the promoters which may occur during cloning.

To analyze the activity of individual STAP-L1 promoters, the level 0 modules (pAGT582L-n) were cloned in front of the β-glucuronidase reporter gene (*GUS*; pICH75111) (Engler et al., 2014) or the green fluorescent protein (*GFP*; pICH4153) (Engler et al., 2014) followed by the octopine synthase gene terminator (*tOCS*; pICH41432) (Engler et al., 2014) into the level 1 vector pICH75044 (Engler et al., 2014), resulting in the vectors pAGT615L-n and pATG917L-n, respectively (Fig. 3A). To analyze the activity of individual STAP-L2 promoters a module with the Ω enhancer sequence of tobacco mosaic virus (pICH46501) (Engler et al., 2014) was cloned in between STAP-L2 and *GUS*, resulting in the pAGT2082L-n vectors (Fig. 3B).

for STAP-L1 and STAP-L2, respectively. Cloning of the PCR products reconstitutes *Bpi*I sites with corresponding overhangs that were already introduced by the degenerated oligonucleotides (5′-overhang in *blue*; 3′-overhang in *red* [L1—AATG; L2—TACT]). (B) Level −1 libraries STAP-L1 (pAGT559L-n) and STAP-L2 (pAGT1597L-n) were further cloned by GG cloning using *Bpi*I into the level 0 vectors pICH41295 and pICH41233, respectively. (See the color plate.)

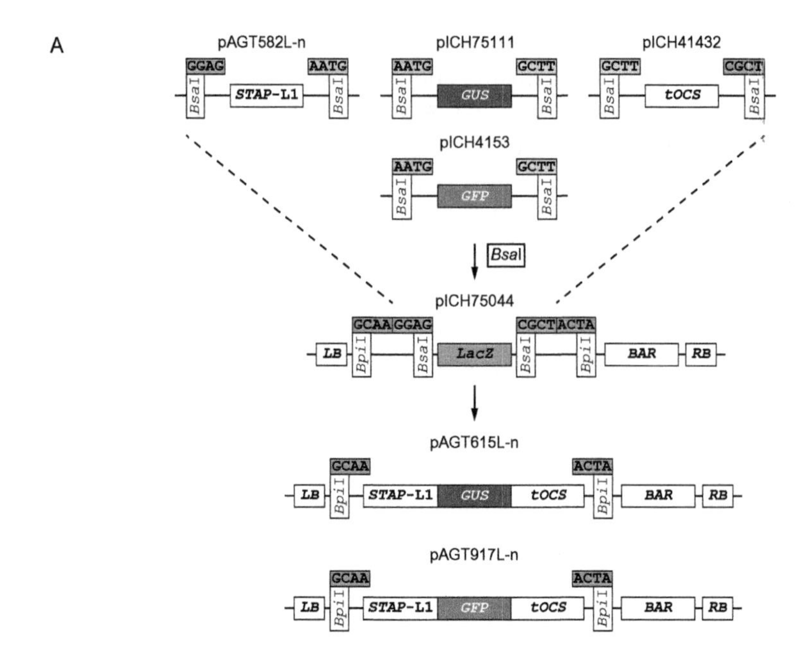

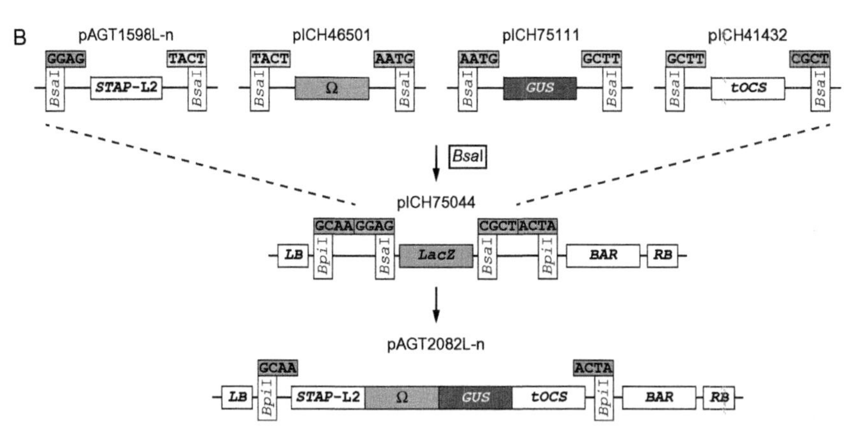

Fig. 3 Cloning strategy for STAP reporters. (A) To analyze the dTALE2-mediated transcriptional induction of single STAPs several T-DNA reporter construct were generated. Level 0 modules were cloned by GG cloning with *Bsa*I into the level 1 vector pICH75044. Promoter modules of the STAP library L1 (pAGT582L-n) were combined with the ORF modules of *GUS* (pICH75111) and *GFP* (pICH4153) together with the terminator module tOCS (pICH41432) resulting in the T-DNA vectors pAGT615L-n and pAGT917L-n, respectively. (B) Promoter modules of the STAP library L2 were combined with the Ω enhancer module (pICH46501), the ORF module of *GUS* (pICH75111), and the terminator module *tOCS* (pICH41432) resulting in the T-DNA vector pAGT2082L-n. (See the color plate.)

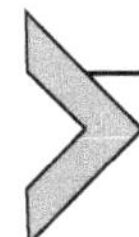

3. ANALYZING PROMOTER ACTIVITY IN TRANSIENT ASSAYS

T-DNA containing vectors were transformed into *Agrobacterium* strain GV3101 (Koncz & Schell, 1986). Transient assays were performed in *Nicotiana benthamiana* plants grown in a greenhouse with constant temperature and humidity of 22°C and 82%, respectively. For transient GUS assays, strains with the T-DNA containing the STAP fused to β-glucuronidase (GUS; pAGT615-L; *STAP-L1:GUS:tOCS*) (Brückner et al., 2015) were coinoculated with two other strains harboring a T-DNA for constitutive expression of the dTALE2 (pICH74043; *act2:dTALE2:tOCS*) (Weber et al., 2011) and GFP (pAGM4731; *35S:GFP:tOCS*) (Engler et al., 2014), respectively (Fig. 4). The constitutive expressed dTALE2, specific for the *EBE2*, was used to induce the GUS reporter construct while the constitutively expressed GFP served as normalization of transformation efficiency. Coinoculation without the *Agrobacterium* strain expressing the dTALE2 serves as negative control to monitor the basal activity of the STAPs in *N. benthamiana*. An *Agrobacterium* strain with a T-DNA containing a constitutively expressed GUS (pICH75181; *35S:GUS:tOCS*) (Engler et al., 2014) served as positive control and was also coinoculated with the strain holding a T-DNA for constitutive expression of GFP (pAGM4731) for normalization. Inoculation was done into different sections of the leaf to eliminate position effects (Fig. 4). *Agrobacterium* strains were cultivated in Luria–Bertani medium (LB) and resuspended into *Agrobacterium* infiltration media (AIM; 10 m*M* 2-(*N*-morpholine)-ethanesulfonic acid (MES); 10 m*M* $MgSO_4$) to an OD_{600} of 0.3. Leaf material (leaf discs diameter 0.9 cm) was harvested 5 days postinoculation (5 dpi), transferred into 2 mL tubes together with 0.2 mm steel beads and directly frozen in liquid nitrogen. Frozen leaf material was ground twice in a bead miller (Mixer Mill MM 400; Retsch) at 30 Hz for 1 min, resuspended in 1 mL GUS extraction buffer (50 m*M* NaH_2PO_4, pH 7.0; 10 m*M* EDTA; 0.1% Triton X-100; 0.1% sodium lauroyl sarcosine; 10 m*M* β-mercaptoethanol) (Jefferson, Kavanagh, & Bevan, 1987) and vortexed. The cell lysate was then filtered on a 0.2 μ*M* polytetrafluoroethylene filter (Chromafil, Macherey-Nagel) and transferred into a 96-well plate. 200 μL were used for normalization by measurement of GFP fluorescence (extinction: 395 nm, emission: 509 nm; Fig. 4). The remaining 800 μL were incubated with 1 m*M* 4-methylumbelliferyl-β-D-glucuronide (MUG) at 37°C, aliquots of 200 μL were removed at four time points (0, 10, 30, and

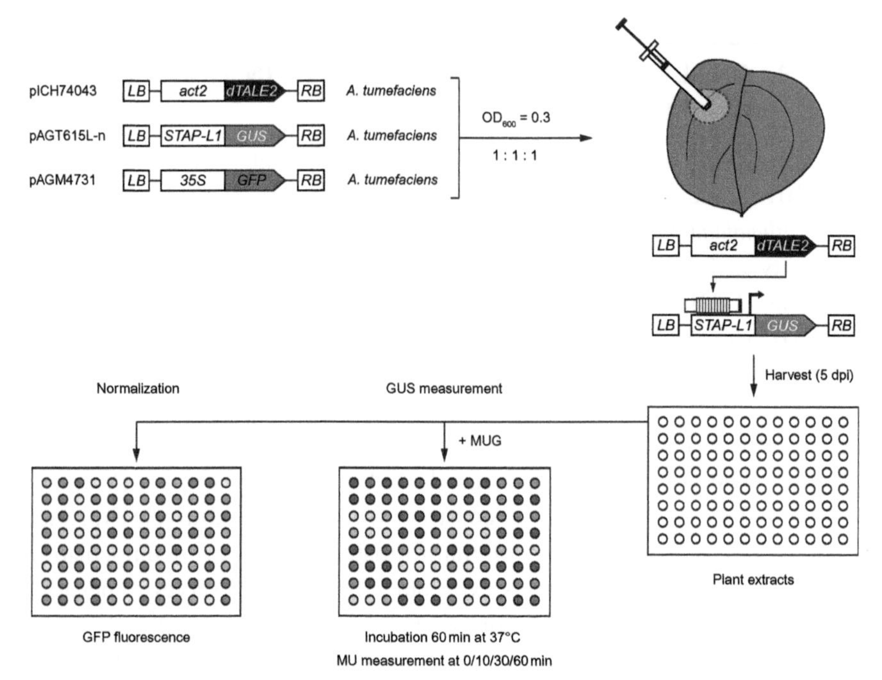

Fig. 4 Workflow of the GUS reporter assay. The *Agrobacterium* strains carrying the different T-DNA constructs (pICH74043, *act2:dTALE2*; pAGT615L-n, *STAP-L1:GUS*; pAGM4731, *35S:GFP*) were resuspended in *Agrobacterium infiltration media* (AIM) and diluted to an OD_{600} of 0.3. The strains were mixed to equal amounts and inoculated in *N. benthamiana*. Five days postinoculation (5 dpi) leaf material was harvested, processed to plant extracts, and samples transferred into a 96-well plate by filtration. Plant extracts were separated for normalization by GFP measurement and measurement of GUS activity. For the measurement of GUS activity, 1 m*M* 4-methylumbelliferyl-β-D-glucuronide (MUG) was added and incubated for 60 min at 37°C. Reaction was stopped and measured at subsequent time points (0, 10, 30, and 60 min). (See the color plate.)

60 min) and the reaction stopped by adding 800 μL of 0.2 *M* Na_2CO_3 (Jefferson et al., 1987) (Fig. 4). The fluorimeter was calibrated with standards of 0.1 μ*M* and 1 μ*M* 4-methylumbelliferone in 0.2 *M* Na_2CO_3, respectively. The activity of the individual *STAP-L1:GUS:tOCS* reporters was expressed as a percentage of the activity of *35S:GUS:tOCS*, based on the control from the same leaf. Results with a STAP-L1 library of 43 promoters are already published and gave a broad spectrum of promoter activity ranging from a few percent to 100% equivalent to that of the 35S promoter (Brückner et al., 2015). Importantly no or only very low activity was seen in the absence of the dTALE. Results with the STAP-L2 library for some

members indicate a promoter activity that is even stronger than that of the 35S promoter (unpublished results).

4. CONCLUSION

With the STAPs at hand, it is now possible to implement circuits in plants involving fine-tuning of gene expression of multiple genes in a pathway by using a single transcription activator. This can be a metabolic or a signaling pathway where in both cases individual genes have to be expressed at defined levels to optimally fulfill their role (Fig. 5). One can imagine more complex set-ups, for example, where a metabolic pathway is divided in two blocks, each expressed in different tissues, with transport from one tissue to the other of the intermediate between the two parts. With the STAPs one would need simply two dTALEs and two series of STAPS for each block. Although side effects cannot be excluded, it is possible to exchange the EBE in already existing STAPs to obtain promoters which can be activated by different TALEs but maintain a similar strength of promoter activity. The promoters controlling the expression of the dTALE can be either tissue specific or induced under certain conditions, for example, upon pathogen attack or various abiotic stresses (Fig. 5). Only a single promoter with the required expression profile is needed. With such a tool, one can, for example, rapidly test which tissues will provide the best performance for the production of a metabolite of interest without affecting plant growth. Generalist promoters such as 35S may not have the required specificity to achieve the best production levels. Another application of STAPs is for regulatory networks. Activation cascades with positive or negative feedback loops can be implemented (Fig. 5) also as highlighted in de Lange et al. (2014). In all these design configurations, more than one dTALE is required. The cloning of the dTALEs now with the modular cloning tools available is an easy and rapid procedure (Geissler et al., 2011; Weber et al., 2011), which cannot be considered as an obstacle to the implementation of such strategies. An alternative to the TALEs is the CRISPR/Cas system. In plants, artificial CRISPR/Cas9-based transcriptional activators need to be optimized and probably never reach the level of performance compared to a naturally evolved TALEs. Concerning the usage of synthetic promoters for fine-tuning, it remains to be shown whether libraries of promoters with a broad spectrum of expression levels can be generated in combination with artificial CRISPR/Cas-based activators in plants. Furthermore, the tissue-specific expression of the gRNAs is still problematic, since they are typically

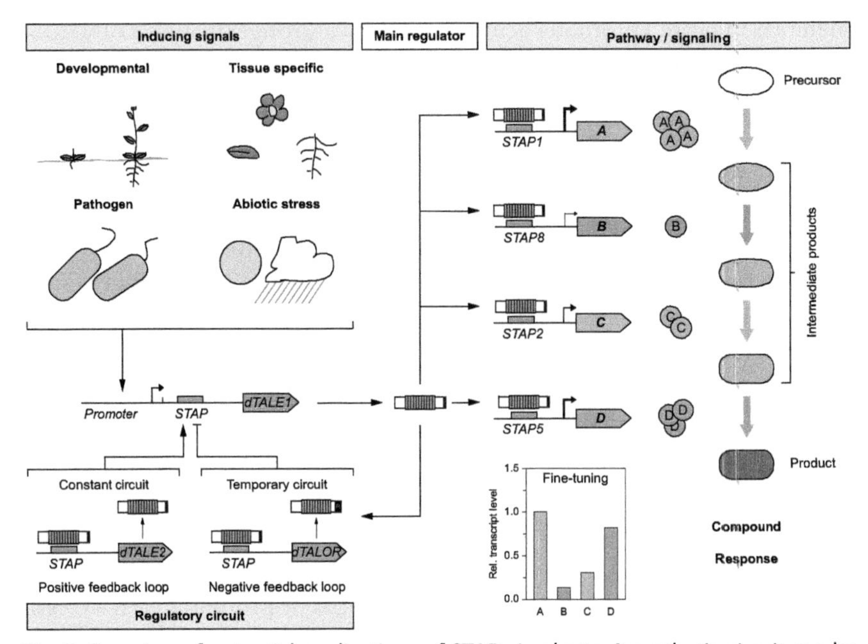

Fig. 5 Overview of potential applications of STAPs *in planta*. A synthetic circuit can be initiated by different external or internal signals (developmental signals, abiotic stresses, etc.). A simple way to generate an output from these signals is to put a dTALE under the control of a signal-responsive promoter, in which the dTALE gets expressed under certain conditions or developmental stages to induce a signal-adapted reaction. Besides an induced reaction this system can also be used to induce a pathway resulting in production of a high-value compound. A STAP library can be used to induce enzymes of a pathway with different expression levels by a single activator (dTALE1). This enables the generation of efficient biosynthetic pathways by fine-tuning of intermediate enzymatic steps (eg, no accumulation of intermediate products). In some cases, inducing signals are only transient. To ensure a constant activity of a pathway one can introduce positive feedback loops. In this scenario, dTALE1 may also induce another dTALE with different DNA-binding specificity (dTALE2), which then in turn induces the dTALE1 if a dTALE2-responsible STAP is fused in front of dTALE1. Responses however, sometimes only need to occur transiently. To accomplish a transient expression one can introduce a negative feedback loop. Therefore, dTALE1 may induce a dTALE2-based transcriptional repressor, which then in turn repress the expression of dTALE1 and hence the subsequent dTALE1-induced response. (See the color plate.)

transcribed by RNA Pol III promoters which are typically constitutive. Although a solution to this problem could be shown in mammalian cells (Nissim et al., 2014), it requires the expression of additional proteins and its effectiveness as well as functionality in plants still needs to be proven. The creation of regulatory circuits requires sequence-specific activators and repressors. The sequence-specificity of CRISPR/dCas9 is defined by the single guide RNA (sgRNA). To circumvent binding of activators and repressors to the same loci two different strategies have been followed up. The first strategy used orthogonal dCas9 variants from different bacterial species (*Neisseria meningitides*, *S. thermophiles*, *S. pyogenes*), that recognize different sgRNAs (Ma et al., 2015). The second strategy consists in adding protein recruitment sites to the sgRNAs which are bound by either an activator or a repressor (Zalatan et al., 2015). However those strategies require further optimization or the expression of additional proteins. This increases the complexity, which should be reduced to a minimum for circuit applications. Therefore, dTALEs and their associated STAPs represent promising tools for various aspects of engineering in plants.

REFERENCES

Blount, B. A., Weenink, T., Vasylechko, S., & Ellis, T. (2012). Rational diversification of a promoter providing fine-tuned expression and orthogonal regulation for synthetic biology. *PLoS One*, 7(3), e33279.

Boch, J., & Bonas, U. (2010). Xanthomonas AvrBs3 family-type III effectors: Discovery and function. *Annual Review of Phytopathology*, *48*, 419–436.

Boch, J., Bonas, U., & Lahaye, T. (2014). TAL effectors-pathogen strategies and plant resistance engineering. *New Phytologist*, *204*(4), 823–832.

Boch, J., Scholze, H., Schornack, S., Landgraf, A., Hahn, S., Kay, S., ... Bonas, U. (2009). Breaking the code of DNA binding specificity of TAL-type III effectors. *Science*, *326*(5959), 1509–1512.

Brückner, K., Schafer, P., Weber, E., Grutzner, R., Marillonnet, S., & Tissier, A. (2015). A library of synthetic transcription activator-like effector-activated promoters for coordinated orthogonal gene expression in plants. *Plant Journal*, *82*(4), 707–716.

Cermak, T., Doyle, E. L., Christian, M., Wang, L., Zhang, Y., Schmidt, C., ... Voytas, D. F. (2011). Efficient design and assembly of custom TALEN and other TAL effector-based constructs for DNA targeting. *Nucleic Acids Research*, *39*(12), 11.

Cong, L., Zhou, R. H., Kuo, Y. C., Cunniff, M., & Zhang, F. (2012). Comprehensive interrogation of natural TALE DNA-binding modules and transcriptional repressor domains. *Nature Communications*, *3*, 6.

de Lange, O., Binder, A., & Lahaye, T. (2014). From dead leaf, to new life: TAL effectors as tools for synthetic biology. *The Plant Journal*, *78*(5), 753–771.

Dehli, T., Solem, C., & Jensen, P. R. (2012). Tunable promoters in synthetic and systems biology. *Subcellular Biochemistry*, *64*, 181–201.

Deng, D., Yan, C. Y., Pan, X. J., Mahfouz, M., Wang, J. W., Zhu, J. K., ... Yan, N. E. (2012). Structural basis for sequence-specific recognition of DNA by TAL effectors. *Science*, *335*(6069), 720–723.

Deng, D., Yan, C. Y., Wu, J. P., Pan, X. J., & Yan, N. (2014). Revisiting the TALE repeat. *Protein & Cell*, *5*(4), 297–306.

Dominguez, A. A., Lim, W. A., & Qi, L. S. (2016). Beyond editing: Repurposing CRISPR-Cas9 for precision genome regulation and interrogation. *Nature Reviews. Molecular Cell Biology*, *17*(1), 5–15.

Doyle, E. L., Hummel, A. W., Demorest, Z. L., Starker, C. G., Voytas, D. F., Bradley, P., & Bogdanove, A. J. (2013). TAL effector specificity for base 0 of the DNA target is altered

in a complex, effector- and assay-dependent manner by substitutions for the tryptophan in cryptic repeat-1. *PLoS One, 8*(12), e82120.
Engler, C., Youles, M., Gruetzner, R., Ehnert, T. M., Werner, S., Jones, J. D., ... Marillonnet, S. (2014). A golden gate modular cloning toolbox for plants. *ACS Synthetic Biology, 3*(11), 839–843.
Farzadfard, F., Perli, S. D., & Lu, T. K. (2013). Tunable and multifunctional eukaryotic transcription factors based on CRISPR/Cas. *ACS Synthetic Biology, 2*(10), 604–613.
Gao, H., Wu, X., Chai, J., & Han, Z. (2012). Crystal structure of a TALE protein reveals an extended N-terminal DNA binding region. *Cell Research, 22*(12), 1716–1720.
Geissler, R., Scholze, H., Hahn, S., Streubel, J., Bonas, U., Behrens, S. E., & Boch, J. (2011). Transcriptional activators of human genes with programmable DNA-specificity. *PLoS One, 6*(5), 7.
Gilbert, L. A., Horlbeck, M. A., Adamson, B., Villalta, J. E., Chen, Y., Whitehead, E. H., ... Weissman, J. S. (2014). Genome-scale CRISPR-mediated control of gene repression and activation. *Cell, 159*(3), 647–661.
Gilbert, L. A., Larson, M. H., Morsut, L., Liu, Z. R., Brar, G. A., Torres, S. E., ... Qi, L. S. (2013). CRISPR-mediated modular RNA-guided regulation of transcription in eukaryotes. *Cell, 154*(2), 442–451.
Hummel, A. W., Doyle, E. L., & Bogdanove, A. J. (2012). Addition of transcription activator-like effector binding sites to a pathogen strain-specific rice bacterial blight resistance gene makes it effective against additional strains and against bacterial leaf streak. *New Phytologist, 195*(4), 883–893.
Jefferson, R. A., Kavanagh, T. A., & Bevan, M. W. (1987). GUS fusions: Beta-glucuronidase as a sensitive and versatile gene fusion marker in higher plants. *EMBO Journal, 6*(13), 3901–3907.
Jensen, P. R., & Hammer, K. (1998a). Artificial promoters for metabolic optimization. *Biotechnology and Bioengineering, 58*(2–3), 191–195.
Jensen, P. R., & Hammer, K. (1998b). The sequence of spacers between the consensus sequences modulates the strength of prokaryotic promoters. *Applied and Environmental Microbiology, 64*(1), 82–87.
Jinek, M., Chylinski, K., Fonfara, I., Hauer, M., Doudna, J. A., & Charpentier, E. (2012). A programmable dual-RNA-guided DNA endonuclease in adaptive bacterial immunity. *Science, 337*(6096), 816–821.
Juillerat, A., Pessereau, C., Dubois, G., Guyot, V., Marechal, A., Valton, J., ... Duchateau, P. (2015). Optimized tuning of TALEN specificity using non-conventional RVDs. *Scientific Reports, 5*, 8150.
Kay, S., Hahn, S., Marois, E., Wieduwild, R., & Bonas, U. (2009). Detailed analysis of the DNA recognition motifs of the Xanthomonas type III effectors AvrBs3 and AvrBs3Δrep16. *The Plant Journal, 59*(6), 859–871.
Khalil, A. S., Lu, T. K., Bashor, C. J., Ramirez, C. L., Pyenson, N. C., Joung, J. K., & Collins, J. J. (2012). A synthetic biology framework for programming eukaryotic transcription functions. *Cell, 150*(3), 647–658.
Kleinstiver, B. P., Prew, M. S., Tsai, S. Q., Nguyen, N. T., Topkar, V. V., Zheng, Z. L., & Joung, J. K. (2015). Broadening the targeting range of Staphylococcus aureus CRISPR-Cas9 by modifying PAM recognition. *Nature Biotechnology, 33*(12), 1293–1298.
Kleinstiver, B. P., Prew, M. S., Tsai, S. Q., Topkar, V. V., Nguyen, N. T., Zheng, Z. L., ... Joung, J. K. (2015). Engineered CRISPR-Cas9 nucleases with altered PAM specificities. *Nature, 523*(7561), 481–485.
Koncz, C., & Schell, J. (1986). The promoter of Tl-DNA gene 5 controls the tissue-specific expression of chimeric genes carried by a novel type of Agrobacterium binary vector. *Molecular & General Genetics, 204*(3), 383–396.

Lamb, B. M., Mercer, A. C., & Barbas, C. F. (2013). Directed evolution of the TALE N-terminal domain for recognition of all 5′ bases. *Nucleic Acids Research*, *41*(21), 9779–9785.

Ma, H., Naseri, A., Reyes-Gutierrez, P., Wolfe, S. A., Zhang, S., & Pederson, T. (2015). Multicolor CRISPR labeling of chromosomal loci in human cells. *Proceedings of the National Academy of Sciences of the United States of America*, *112*(10), 3002–3007.

Mak, A. N. S., Bradley, P., Cernadas, R. A., Bogdanove, A. J., & Stoddard, B. L. (2012). The crystal structure of TAL effector PthXo1 bound to its DNA target. *Science*, *335*(6069), 716–719.

Meckler, J. F., Bhakta, M. S., Kim, M. S., Ovadia, R., Habrian, C. H., Zykovich, A., ... Baldwin, E. P. (2013). Quantitative analysis of TALE-DNA interactions suggests polarity effects. *Nucleic Acids Research*, *41*(7), 4118–4128.

Mendenhall, E. M., Williamson, K. E., Reyon, D., Zou, J. Y., Ram, O., Joung, J. K., & Bernstein, B. E. (2013). Locus-specific editing of histone modifications at endogenous enhancers. *Nature Biotechnology*, *31*(12), 1133–1136.

Mercer, A. C., Gaj, T., Fuller, R. P., & Barbas, C. F., 3rd. (2012). Chimeric TALE recombinases with programmable DNA sequence specificity. *Nucleic Acids Research*, *40*(21), 11163–11172.

Miller, J. C., Tan, S., Qiao, G., Barlow, K. A., Wang, J., Xia, D. F., ... Rebar, E. J. (2011). A TALE nuclease architecture for efficient genome editing. *Nature Biotechnology*, *29*(2), 143–148.

Miller, J. C., Zhang, L., Xia, D. F., Campo, J. J., Ankoudinova, I. V., Guschin, D. Y., ... Rebar, E. J. (2015). Improved specificity of TALE-based genome editing using an expanded RVD repertoire. *Nature Methods*, *12*(5), 465–471.

Moscou, M. J., & Bogdanove, A. J. (2009). A simple cipher governs DNA recognition by TAL effectors. *Science*, *326*(5959), 1501.

Nissim, L., Perli, S. D., Fridkin, A., Perez-Pinera, P., & Lu, T. K. (2014). Multiplexed and programmable regulation of gene networks with an integrated RNA and CRISPR/Cas toolkit in human cells. *Molecular Cell*, *54*(4), 698–710.

Perez-Pinera, P., Kocak, D. D., Vockley, C. M., Adler, A. F., Kabadi, A. M., Polstein, L. R., ... Gersbach, C. A. (2013). RNA-guided gene activation by CRISPR-Cas9-based transcription factors. *Nature Methods*, *10*(10), 973–976.

Perez-Pinera, P., Ousterout, D. G., Brunger, J. M., Farin, A. M., Glass, K. A., Guilak, F., ... Gersbach, C. A. (2013). Synergistic and tunable human gene activation by combinations of synthetic transcription factors. *Nature Methods*, *10*(3), 239–242.

Piatek, A., Ali, Z., Baazim, H., Li, L., Abulfaraj, A., Al-Shareef, S., ... Mahfouz, M. M. (2015). RNA-guided transcriptional regulation in planta via synthetic dCas9-based transcription factors. *Plant Biotechnology Journal*, *13*, 578–589.

Qi, L. S., Larson, M. H., Gilbert, L. A., Doudna, J. A., Weissman, J. S., Arkin, A. P., & Lim, W. A. (2013). Repurposing CRISPR as an RNA-guided platform for sequence-specific control of gene expression. *Cell*, *152*(5), 1173–1183.

Qin, X., Qian, J., Yao, G., Zhuang, Y., Zhang, S., & Chu, J. (2011). GAP promoter library for fine-tuning of gene expression in Pichia pastoris. *Applied and Environmental Microbiology*, 77(11), 3600–3608.

Römer, P., Recht, S., & Lahaye, T. (2009). A single plant resistance gene promoter engineered to recognize multiple TAL effectors from disparate pathogens. *Proceedings of the National Academy of Sciences of the United States of America*, *106*(48), 20526–20531.

Schreiber, T., & Bonas, U. (2014). Repeat 1 of TAL effectors affects target specificity for the base at position zero. *Nucleic Acids Research*, *42*(11), 7160–7169.

Schreiber, T., Sorgatz, A., List, F., Blueher, D., Thieme, S., Wilmanns, M., & Bonas, U. (2015). Refined requirements for protein regions important for activity of the TALE AvrBs3. *PLoS One*, *10*(3), e0120214.

Streubel, J., Blucher, C., Landgraf, A., & Boch, J. (2012). TAL effector RVD specificities and efficiencies. *Nature Biotechnology*, *30*(7), 593–595.

Tornoe, J., Kusk, P., Johansen, T. E., & Jensen, P. R. (2002). Generation of a synthetic mammalian promoter library by modification of sequences spacing transcription factor binding sites. *Gene*, *297*(1–2), 21–32.

Tsuji, S., Futaki, S., & Imanishi, M. (2013). Creating a TALE protein with unbiased 5′-T binding. *Biochemical and Biophysical Research Communications*, *441*(1), 262–265.

Weber, E., Gruetzner, R., Werner, S., Engler, C., & Marillonnet, S. (2011). Assembly of designer TAL effectors by golden gate cloning. *PLoS One*, *6*(5), e19722.

Werner, S., Engler, C., Weber, E., Gruetzner, R., & Marillonnet, S. (2012). Fast track assembly of multigene constructs using Golden Gate cloning and the MoClo system. *Bioengineered Bugs*, *3*(1), 38–43.

Yang, J. J., Zhang, Y., Yuan, P. F., Zhou, Y. X., Cai, C. Z., Ren, Q. P., … Wei, W. S. (2014). Complete decoding of TAL effectors for DNA recognition. *Cell Research*, *24*(5), 628–631.

Zalatan, J. G., Lee, M. E., Almeida, R., Gilbert, L. A., Whitehead, E. H., La Russa, M., … Lim, W. A. (2015). Engineering complex synthetic transcriptional programs with CRISPR RNA scaffolds. *Cell*, *160*(1–2), 339–350.

AUTHOR INDEX

Note: Page numbers followed by "*f*" indicate figures, and "*t*" indicate tables.

A

Abbott, E., 53–54
Abbruscato, P., 264*t*
Abe, I., 103–105
Abe, N., 270–271
Abel, C., 308–309
Abel, E.W., 168–169
Abouheif, E., 254
Abulfaraj, A., 361–364
Abuqamar, S., 266
Ackerman, J.J.H., 234
Adams, K.L., 259–260
Adams, P.D., 2, 86–87
Adams, R.P., 162
Adamson, B., 361–364
Adebesin, F., 48–49
Adler, A.F., 361–364
Adsuara, J.E.E., 273–274
Aebi, M., 128, 142–143
Afonine, P.V., 86–87
Agarkova, I., 107–109
Agelopoulos, N., 324–325
Aggerbeck, L., 21
Agier, N., 21
Aharoni, A., 306–309, 324–325, 337–338
Ahkami, A., 148, 337–338
Ahn, J.H., 88–89
Ahn, M.J., 351–352
Airoldi, C.A., 268
Aitken, E.A., 266
Ajikumar, P.K., 88–89, 227–229
Akhtar, T.A., 310–313, 311*t*, 320–321, 334–336, 338–339, 345–346
Alba, J.M., 306–307
Albà, M.M., 273–274
Alcaraz, J.P., 341
Alejandro, S., 208–209, 210*t*
Alexeev, D., 109
Alexeeva, M., 109
Alhagdow, M., 342
Ali, Z., 361–364
Allam, B., 109, 109*f*
Allemann, R.K., 336–337
Allen, E., 107–110, 108*f*
Allen, E.E., 109, 109*f*
Allis, C.D., 266–267
Almeida, R., 373–375
Alonso-Gutierrez, J., 2
Alper, H., 208
Alpert, A.J., 237, 239
Alseekh, S., 264*t*
Al-Shareef, S., 361–364
Al-Suwailem, A.M., 102–103
Altaf-Ul-Amin, M., 37
Altman, W.E., 30–31
Alvarez, J.M., 263–266
Amann-Zalcenstein, D., 267–268
Amaral-Zettler, L.A., 109, 109*f*
Ambid, C., 334–336
Ament, K., 318, 323–325, 344–345
Amit, I., 32–34, 73–75, 267–268
Amsellem, Z., 342
Amslinger, S., 244
Amstislavskiy, V., 29–30
Anand, S., 103–107
Anandan, A., 311*t*
Anders, S., 35–36, 58, 181
Andersen, K.R., 83
Andersen, T.G., 209–211
Andersen-Ranberg, J., 49
Anderson, P.K., 324
Andersson, D., 26–27
Andersson, S., 307, 334–336, 350
Andes Hess, B., 86–87
Angenent, G.C., 267–268
Angenon, G., 310–313
Ankoudinova, I.V., 364–365
Annala, M., 273–274
Ansari, M.Z., 103–107
Anthony, L.C., 227–229
Antonescu, C.M., 173
Antonescu, V., 172
Antontseva, E.V., 259–260
Ao, W., 277–278

Aoyama, T., 272–273
Apweiler, R., 37
Archibald, J.M., 110
Arda, H.E., 277–279
Arendt, P., 48–50
Argiriou, A., 59–61
Arigoni, D., 226–227, 334–336
Arimura, G., 336–337
Ariyurek, Y., 256
Arkin, A.P., 361–364
Armbruster, D.A., 243
Armstrong, B.L., 237
Aros, D., 336–337
Arrivault, S., 234
Arró, M., 308–309, 326
Arroyo, A., 229
Arsovski, A.A., 258–259
Asakawa, N., 239
Asakawa, Y., 239
Asano, T., 38–39
Asawatreratanakul, K., 125
Asayama, M., 105–106
Asencio, M., 169–170
Ashburner, M., 37
Ashida, Y., 345
Ashour, M., 334–336
Assmann, S.M., 210*t*
Atkinson, N.J., 70
Atsumi, S.M., 187–189
Attiya, S., 30–31
Attwood, T.K., 37
Au, K.F., 74, 76–77
Auclair, K., 83–84
Auger, D.L., 254
Austin, M.B., 86–87
Autran, D., 20, 266
Auxillos, J., 89
Avagyan, V., 36
Avci-Adali, M., 270
Avila, C., 281
Ayre, B.G., 258–259
Azevedo, L., 210*t*
Azulay, Y., 339
Azzarello, E., 210*t*

B

Baart, G.J., 37
Baazim, H., 361–364
Babon, J.A., 277–279
Bach, T.J., 334–336
Bacher, A., 226–227, 244
Bachmann, B.O., 106–107, 110
Bachmann, H., 83–84
Back, K., 122–123
Backhaus, R.A., 341
Backstrom, S., 266
Baden, D., 111, 111*f*
Bader, J.S., 30–31
Bader, R., 258–259
Badger, J.H., 107–110, 108*f*
Bai, L., 256
Baidoo, E.E.K., 227, 232–234, 244
Baijot, A., 209
Bailey, P., 264*t*
Bailly, A., 209–211, 210*t*
Bakaher, N., 309–310, 311*t*
Baker, J.M., 227
Bakhtiar, R., 243
Balasubramanian, S., 30–31
Balcells, L., 71–72
Balcke, G.U., 244
Baldwin, E.A., 339
Baldwin, E.P., 365–366
Baldwin, I.T., 2–3, 306–307
Balhoff, J.P., 254
Balk, J., 210*t*
Ball, C.A., 37
Ballard, C.C., 86–87
Ban, Z., 83–84
Banaru, M., 29–30
Banba, M., 23
Bandyopadhyay, A., 209–211, 210*t*
Banerjee, A., 308–309
Banerjee, R., 308–309
Bantignies, B., 334–336
Bar, E., 336–337, 339, 342, 345–348
Barbas, C.F., 26, 364–366
Barbosa-Morais, N.L., 277–278
Barbour, E.L., 50–51
Barco, B., 85
Bargmann, B.O.R., 272–273
Bari, R.P., 21–22
Bar-Joseph, Z., 268
Barlow, K.A., 364–365
Barnes, H.J., 83–84
Barnett-Itzhaki, Z., 267–268

Barrasa, M.I., 277–279
Barrera, L.O., 267
Barreto, V., 236–237
Barry, C.S., 4, 334–336, 346
Barthelson, R.A., 26–27
Bartley, G.E., 339
Bartram, S., 339
Bashor, C.J., 361–364
Baskar, R., 258–259
Bass, W.T., 2–3, 311*t*
Bassard, J.E., 76
Bateman, A., 37
Battenhouse, A., 259–260
Batth, T.S., 2
Batut, P., 254–256
Batzoglou, S., 268
Baulcombe, D., 209–211, 214–215
Baumann, S., 154, 159
Baxter, C., 254
Baxter, R.L., 109
Bazin, I., 209–211, 210*t*
Beale, M.H., 227
Bealer, K., 36
Bearss, D., 262–263
Beatty, P.H., 264*t*
Beauchesne, K., 100
Beaudoin, G.A.W., 38–39
Beaudry, R.M., 338
Beemster, G.T., 20, 266
Behrens, S.E., 364–365, 373–375
Beijnen, J.H., 229–230
Beis, K., 209
Belcram, K., 20, 266
Belele, C., 258–259
Belew, Z.M., 208–221
Bemben, L.A., 30–31
Ben-David, H., 338
Bender, S., 254
Benes, V., 24–26
Bengtsson, M., 26–27
Benhamed, M., 267
Benito, M.J., 269–270
Benjamin, K., 89, 148, 169–170
Bennett, A.B., 340–342
Bennetzen, J., 261
Bennetzen, J.L., 258–259
Bennewitz, S., 244
Benning, C., 21
Ben-Nissan, G., 85
Bentley, D.R., 30–31
Berger, A., 82
Berger, M.F., 271–274
Bernatavichute, Y.V., 267
Bernstein, B.E., 268, 364–365
Berri, S., 264*t*
Berry, C.C., 23
Berthiller, F., 83–84
Bertsch, A., 78–80
Besse, I., 309, 311*t*
Besseau, S., 168–203
Besser, K., 337–338
Bestwick, R.K., 340–341
Besumbes, O., 337–339
Beszteri, S., 113
Betel, D., 35–36
Bethe, B., 114–115
Beutler, J.A., 70–71
Bevan, M.W., 342, 371–373
Bhakta, M.S., 365–366
Bhatia, V., 324–325
Bhattacharya, D., 113
Bhattacharya, R., 324–325
Bi, Y.M., 264*t*
Bi, Z., 237–238
Bick, J.A., 334–336
Bicsak, T.A., 345
Biggs, M.S., 342
Bigler, F., 263
Bigler, L., 308–309, 326
Bilenko, N., 339, 342, 345–347
Binder, A., 364–365, 373–375
Birchler, J.A., 254, 260
Bird, A.S., 342
Bird, C.R., 339, 342
Bird, I.M., 24–26
Birkemeyer, C., 243
Birkett, M., 324–325
Birnbaum, K.D., 26–27, 272–273
Birney, E., 32–34, 56–57, 172, 258–259
Birol, I., 32–34
Bitton, F., 267
Bjorklund, S., 266
Blake, J.A., 37, 83–84
Blakeslee, J.J., 209–211, 210*t*
Blanc, G., 107–109
Blanc, N., 191–192, 201–202

Blanc, V.M., 336–337
Blatchley, M.R., 266
Blecher-Gonen, R., 267–268
Bleeker, P.M., 306–326, 337–339, 345–346
Blin, K., 106, 110–111
Blood, P.D., 75, 172–173
Blount, B.A., 361–364
Blucher, C., 364–366
Blueher, D., 365–366
Blumer, S., 337–338
Boch, J., 364–366, 373–375
Bode, H.B., 114–115
Boersema, P.J., 239
Bogdanove, A.J., 364–369
Bohlmann, J., 48–64, 82, 148, 336–337
Bohn, L.M., 168–169
Bohner, S., 272–273
Boisson, A.M., 244
Boland, W., 336–337
Bolanos, A., 258–259
Bolduc, N., 268
Bolger, A.M., 32, 56–57
Bolivar, F., 88–89
Bolouri, H., 252
Bomblies, K., 272–273
Bonaldi, K., 278–279
Bonas, U., 364–369, 373–375
Bonawitz, N.D., 266
Bongers, M., 337–338
Borg, M., 114–115
Bork, P., 101*f*
Bornscheuer, U.T., 87
Borodina, T., 29–30
Boronat, A., 71–72, 226–227, 308–309, 321–323, 337–339, 341, 344
Bortolotti, C., 308–309, 326
Bossi, S., 336–337
Botella-Pavia, P., 339
Botstein, D., 37
Botterman, J., 84
Bottner, S., 281
Boubakir, Z., 74–75
Bouchard, R., 209–211, 210*t*
Boudet, J., 125, 134–135, 142–143
Boutry, M., 209
Bouvier, F., 226, 341
Bouwmeester, H.J., 59–60, 306–309, 324–325, 337–338, 344–345
Bovet, L., 210*t*
Bowden, J., 75, 172–173
Bowen, B.A., 254, 259–260, 269–270
Bowman, M.E., 86–87
Bowman, S.K., 258–259
Bowser, M.T., 270
Boyce, J.M., 266
Bradley, P., 364–366
Bradnam, K., 56–57
Bramley, P.M., 284*t*, 339, 341, 344
Brandt, A.J., 272
Branigan, C.A., 337–338
Brar, G.A., 361–364
Brasher, A.K., 168–169
Brasher, M.I., 124–125, 134–136, 142
Brasileiro, A.C., 264*t*
Brehm, A., 262–263
Brem, R.B., 259
Brenner, S., 22
Breton, G., 278–279, 281
Brezinski, M.L., 271–272
Brichta, J.L., 124
Bridgham, J., 22
Briggs, S.D., 266
Brill, Y., 284*t*
BringerMeyer, S., 226–227
Britovsek, G., 168–169
Brizuela, E., 281
Brodelius, P.E., 311*t*
Brown, C.G., 30–31
Brown, J.B., 255
Brown, M., 267
Brown, M.J., 227–229
Brown, P.O., 20–21
Brown, S., 169–170
Bru, J.-P., 338, 344–345, 347–348
Bruccoleri, R., 106, 111
Brückner, K., 366–369, 371–373
Brudno, M., 269
Brueggeman, A., 107–109
Brunelle, J.L., 152, 160
Brunger, A.T., 86–87
Brunger, J.M., 361–364
Brunk, B., 273–274
Brunner, S., 254
Brutnell, T.P., 29, 185
Bryant, D., 76–77
Bryzgalov, L.O., 259–260

Bubb, K.L., 258–259
Buchanan, B., 226
Bucher, P., 270–272
Buchner, K., 270
Buck, M.J., 267
Buell, C.R., 113, 172
Buetow, D.E., 340–341
Buisine, N., 267
Bulger, M., 258–259
Bülow, L., 284*t*
Bulyk, M.L., 271–274
Bunkoczi, G., 86–87
Bunsupa, S., 35–37
Burchell, B., 83–84
Burdo, B., 275*f*, 278–279, 282*t*, 286
Burgess, D., 76–77
Burki, F., 109–110, 109*f*
Burla, B., 208–209, 210*t*
Burlat, V., 74–75, 170–172, 176, 191–195, 197–198
Burley, S.K., 260–261
Burow, M., 209–211
Burraco, P., 24–26
Burton, D.R., 26
Busch, M.A., 272–273
Bushman, B.S., 124
Busing, R.T., 70–71
Busso, S., 270
Bustillo, J., 31
Buszewski, B., 237, 239
Butcher, R.A., 106–107
Butelli, E., 264*t*
Butler, H., 37
Buttery, R.G., 339
Byrne, P.F., 254
Byrum, T., 306

C

Caelles, C., 71–72
Cai, C.Z., 364–365
Cai, X., 194
Cairney, J., 168–169
Cairns, J., 268
Caissard, J.-C., 48–49
Cakir, M., 284*t*, 286
Calenge, F., 23
Callewaert, N., 48–50
Camacho, C., 36
Camacho-Sanchez, M., 24–26
Camara, B., 226, 341
Camargo, A.A., 270
Cameron, M.D., 168–169
Campbell, M.M., 281
Campbell, M.S., 76
Campo, J.J., 364–365
Campopiano, D.J., 109
Campos, N., 339
Canas, R., 281
Cane, D.E., 122–123
Caniard, A., 48–49
Cankar, K., 59–60
Canovas, F.M., 281
Canton, F.R., 32
Cao, J., 254
Cao, X., 259–260, 266–267
Caparros-Ruiz, D., 263–266
Caputi, L., 87
Carabelli, M., 272–273
Carabias-Martinez, R., 237, 239
Carbonero, P., 282*t*
Carda, A., 254–257
Carde, J.-P., 338, 341, 344–345, 347–348
Cardoza, Y.J., 336–337
Carninci, P., 254–258
Carpenter, E.J., 23, 81
Carqueijeiro, I., 168–203
Carrasco, J.L.L., 272
Carrasquilla-Garcia, N., 32–34
Carretero-Paulet, L., 339
Carroll, J.S., 267
Cartayrade, A., 226–227
Carter, C.W., 86–87
Carver, J., 113–114
Casadoro, G., 342–343
Casaretto, J.A., 281
Casas, M.I., 262
Cassels, K.B., 169–170
Castillejo, P., 281
Castillo, S., 78–80
Castoldi, M., 24–26
Castrillo, G., 282*t*
Cattolico, R.A., 111–112, 112*f*
Causse, M., 338, 344–345, 347–348
Celedon, J.M., 48–64
Cella, M.A., 48–49
Century, K., 263–266

Cerdan, P.D., 266
Cermak, T., 364–365
Cernadas, R.A., 364–366
Cerutti, H., 115–116
Cervantes, B., 89–90
Cha, J.Y., 237
Cha, M.N., 88–89
Chai, C., 269–270
Chai, J., 365–366
Chakrabarty, R., 124–125, 134–136, 142
Chakrobarty, M., 281
Challabathula, D., 76–77
Challis, G.L., 71–72
Chamovitz, D., 339
Chan, G., 83–84
Chan, R., 2
Chan, Y.A., 103–105
Chandler, V.L., 258–260, 262, 269–270
Chang, H.S., 334–336
Chang, K., 70–71
Chang, M.C., 169–170
Chang, T.C., 173
Chang, Z., 106
Chappell, J., 122–123, 172, 308–310, 336–337, 351–352
Chapple, C., 78–80, 83–85
Charbonneau, A.L., 2–3, 10, 307, 311*t*, 334–336, 338, 346–347
Charpentier, E., 361–364
Chase, T., 345
Chater, K.F., 71–72
Chatterjee, A., 142–143
Chau, N.H., 272–273
Chechetka, S.A., 23
Chefson, A., 83–84
Chehab, W.E., 227, 232–234
Chemler, J.A., 70–71
Chen, C.C., 259–260
Chen, D.-F., 59–60
Chen, F., 49, 57–58, 148
Chen, G., 264*t*
Chen, H., 264*t*, 270
Chen, H.C., 341
Chen, J., 37, 88–89
Chen, J.C., 187–188
Chen, K., 263–266
Chen, L., 26, 264*t*
Chen, M., 264*t*
Chen, Q.K., 273–274
Chen, S., 209–211, 210*t*
Chen, V.B., 86–87
Chen, W., 264*t*
Chen, X., 264*t*, 266, 273–274, 284*t*, 336–337
Chen, Y., 227–229, 361–364
Chen, Z., 81–82, 125, 134–135, 142–143
Chen, Z.C., 210*t*
Cheng, L., 270–271
Cheng, M., 256
Cheng, Q., 63
Chénieux, J.C., 194
Cheremushkin, E., 273–274
Cherry, J.M., 37
Chetverina, D., 258–259
Chiang, A., 49–51, 58–61, 64, 148
Chiba, M., 38–39
Chien, A.C., 351–352
Childs, K.L., 36, 76–77, 113
Chiu, R., 32–34, 56–57
Cho, B.K., 87
Choi, H., 208–209, 306
Choi, Y., 210*t*, 266–267
Choi, Y.E., 311*t*
Chomczynski, P., 24–26
Chory, J., 266
Choumane, W., 259–260
Christeller, J.T., 311*t*
Christian, M., 364–365
Christianson, D.W., 122–123
Christianson, L.M., 73–74
Chu, C., 264*t*
Chu, J., 361–364
Chua, N.H., 272–273
Chudalayandi, S., 260
Church, G.M., 271–274
Chye, M.L., 281
Chylinski, K., 361–364
Cicek, A.E., 268
Cierlik, I., 272–273
Clardy, J., 70–71, 106–107, 114–115
Clark, A., 308–310, 351–352
Clark, R.M., 258–259
Clark, S.M., 74–75
Claros, M.G., 32
Clastre, M., 168–203, 334–336
Clavijo, B.J., 36, 76–77

Clay, N.K., 85
Clemente, T.E., 82
Clinton, R., 259
Closa, M., 308–309, 326
Clynes, D., 252–253
Coates, R., 308–310, 351–352
Coates, R.M., 122–123, 148–150
Cock, P.J.A., 31–32
Coego, A., 281
Coen, E., 254
Cohade, A., 48–49
Cohen, M., 339
Cokus, S., 267
Collins, B.D., 270
Collins, J.J., 361–364
Collins-Silva, J., 124
Comelli, R.N., 281
Compagnon, V., 76
Compeau, P.E.C., 32–34
Conaway, J.W., 266
Conaway, R.C., 266
Conboy, C.M., 277–278
Conesa, A., 35–37, 57–58
Cong, L., 364–365
Conget, P., 169–170
Conner, A.J., 317–320
Contin, A., 198
Contrino, S., 286
Cook, D., 82
Cook, D.R., 32–34
Corbett, R., 32–34
Corcoran, D.L., 254–257
Corcoran, K., 22
Cordoba, E., 226–227
Cori, O., 345
Cornish, K., 124
Cornish-Bowden, A., 84–85
Corrêa, L.G.G., 284*t*
Cortes, M.E., 229
Cortina, C., 317–318
Coruzzi, G.M., 272–273
Cosimano, J., 308–309
Costa, J.A.V., 102
Costa, M.M., 254
Costello, M., 76–77
Cote, A., 273–274
Coulouris, G., 36
Coulson, A.R., 30
Coupland, G., 282*t*
Courdavault, V., 74–75, 168–203
Covello, P.S., 49–50, 56
Cowtan, K.D., 86–87
Cox, K.D., 258–259
Cragg, G.M., 70–72, 168–169
Crasta, O., 259–260
Crawford, D.L., 21–22
Crespi, M., 23
Croteau, R.B., 48–49, 122–123, 148–150, 154, 159, 226
Crothers, D.M., 279–281
Crowley, J.J., 259
Crüsemann, M., 113–114
Cseke, L., 336–337
Cueto, M., 114–115
Cui, G., 38–39
Cui, J., 142–143
Cui, M., 340–341
Culianez-Macia, F.A., 317–318
Cunillera, N., 308–309
Cunniff, M., 364–365
Cunningham, F.X., 339
Curtis, B.A., 110
Curtis, M.D., 258–259, 281
Czechowski, T., 21–22, 337–338

D

d'Avignon, D.A., 234
Dafny-Yelin, M., 336–337, 351–352
Dai, M., 264*t*
Dai, X., 264*t*
Dalca, A.V., 269
Dan, K., 209–211, 210*t*
Dancis, A., 209–211, 210*t*
Daniel Jones, A., 2–3
Datta, H.S., 70–72
D'Auria, J.C., 263–266, 308–309
Davey, M., 31
Davidovich-Rikanati, R., 336–337, 339, 342, 344–349, 351–352
Davidson, E.H., 252, 254
Davis, E.M., 148–149, 309–310
Davis, I.W., 86–87
Davis, R.W., 20–21, 272–273
Davison, B.H., 168–169
Davisson, V.J., 152
Dawson, J.A., 74–75

Dayan, F.E., 78–80, 82
De Both, M.T.J., 306–326, 339, 345
de Carvalho, L.P., 74–75, 85–86
De Clercq, R., 267
de Felipe, P., 351–352
de Kogel, W.-J., 308–309
de Laat, W., 258–259
de Lange, O., 364–365, 373–375
De Luca, V., 85, 187–189
de Lucas, M., 278–279
de Morais, E.G., 102
de Morais, M.G., 102
De Rycke, R., 310–313
Debayle, D., 76
Dedow, L.K., 29
Deforce, D.L., 38–39
Degenhardt, J., 70, 324–325, 336–338
Deguchi, Y., 23
Dehesh, K., 244
Dehli, T., 361–364
Deikman, J., 340–342
Dekker, B., 320
de-Kogel, W.-J., 337–338
del Pozo, J.C., 281
del Viso, F., 281
Delacroix, H., 21
Delatte, T., 337–338
Dell, B., 310–313
DellaPenna, D., 172, 340–342
Dellas, N., 86–87
DeLoache, W.C., 89–90
Demissie, Z.A., 48–49
Demorest, Z.L., 365–366
Deng, D., 364–365
Deng, L., 264*t*
Deng, T., 311*t*
Deodato, C.R., 111–112, 112*f*
Depicker, A., 281
Deplancke, B., 274–278, 281
Deruère, J., 341
Dervan, P.B., 271–272
Descrimes, M., 29
Determan, M.K., 148
DeTure, M., 111, 111*f*
Deuerlein, S., 308–309, 337–338
Dev, S., 70–71
Dewey, C.N., 34–35, 74–75, 175–176
Dey, B., 281
Deyholos, M., 266
d'Harlingue, A., 341
Dhawan, R., 266
Diao, J.-Z., 105–106
Díaz de la Garza, R.I., 340–341
Diaz-Chavez, M.L., 49–51, 58–61, 64
Dicke, M., 306–309, 324–325
Diergaarde, P.J., 309–310, 318, 323–325, 337–339, 345
Diergaarden, P., 306–326
Dijkink, J., 318, 324–325
Ding, H., 81–82, 84–85
Ditta, G.S., 272–273
Dittrich, H., 71–72
Dixon, R.A., 168–169, 344–347
Dobin, A., 254–256
Dodson, E.J., 86–87
Doebley, J.F., 254, 258–259, 263–266
Doerge, R.W., 259
Doherty, C.J., 278–279
Doireau, P., 194
Domier, L.L., 340–341
Dominguez, A.A., 361–364
Donath, A., 81
Dong, C.H., 272–273
Dong, J., 89
Dong, L., 270
Dong, T., 264*t*
Donnelly, M.L., 351–352
Dopazo, J., 35–36
Dorrestein, P.C., 106, 113–114
Dorweiler, J.E., 258–259
Doseff, A.I., 252–287
Douce, R., 244
Doucette-Stamm, L., 277–279
Doudna, J.A., 361–364
Douglas, C.J., 21, 55
Douglas, D.J., 100
Doyle, E.L., 364–369
Drake, R., 341
Drewes, S.E., 168–169
Driguez, H., 336–337
Droge-Laser, W., 281
Drott, D., 83
Drummond, B.J., 269–270
Du, B., 105–106
Du, G., 88–89
Du, H.N., 264*t*, 266

Du, Y.L., 282*t*
Duan, H., 60–61
Duan, X., 268
Duarte, P., 194–195
Dubois, G., 364–365
Duchateau, P., 364–365
Duchtig, P., 210*t*
Dudareva, N., 48–49, 81–82, 122, 226, 307, 326, 334–352
Duddela, S., 106, 111
Dudek, P., 252–253
Dueber, J.E., 89–90
Duffé, P., 307, 309–310, 334–336
Dugé de Bernonville, T., 168–203
Duncan, K.R., 113–114
Dunigan, D.D., 107–109
Dupuy, D., 274–278, 281
Durbin, R., 175–176
Durst, F., 83–84
Dusi, R.G., 102, 106–107, 110, 114–115
Dutta, A., 258–259

E

Eachus, R.A., 169–170
Earley, K.W., 281
Earnest, T.N., 86–87
Eaton, R.M., 270
Ebenstein, Y., 76–77
Echols, N., 86–87
Ecker, J.R., 23
Eckert, C.A., 168–169
Edgar, C.I., 266
Edger, P.P., 76–77
Edwards, R., 32
Eeckhoute, J., 268
Effenberger, F., 87
Efroni, I., 26–27, 272–273
Egener, T., 271–272
Egholm, M., 30–31
Ehlert, A., 281
Ehlting, J., 21
Ehnert, T.M., 366–369, 371–373
Eichhorn, S.E., 23–24
Eicker, A., 258–259
Eisen, M.B., 271–272
Eisenreich, W., 226–227, 244
Ekins, A., 38–39
Eklund, D.M., 272–273
Elewa, A.M., 277–278
Elfving, N., 266
Elias, L., 337–338
El-Kereamy, A., 264*t*
Ellington, A.D., 269–270
Ellis, B.E., 55
Ellis, T., 361–364
Ellisman, M.H., 306
Elmore, C.C., 149–150
Emsley, P., 86–87
ENCODE Project Consortium, 252, 258–259
Enfissi, E.M., 341, 344
Eng, D., 227–229
Enge, M., 270–271
Engelberth, J., 336–337
Engene, N., 306
Engler, C., 84, 366–369, 371–375
English, M.A., 262–263
Ennajdaoui, H., 309, 311*t*
Epping, J., 124–125, 134–136, 142–143
Erban, A., 243
Eric Nybo, S., 309–310
Eriksson, C., 272–273
Eriksson, R., 48–49
Erland, L.A.E., 48–49
Erokhin, M., 258–259
Escalante, A., 88–89
Escobar-Bravo, R., 306–307
Estep, P.W., 271–272
Evans, P.R., 86–87
Evans, R.M., 272–273
Evans, S.K., 266
Eveland, A.L., 268
Evert, R.F., 23–24

F

Fabris, M., 37
Facchini, P.J., 38–39, 49–50, 56, 82, 85, 102–103
Faivre-Rampant, O., 264*t*
Falara, V., 82, 310–313, 338–339, 345–346
Fäldt, J., 48–49, 336–337
Falgueras, J., 32
Fallon, T.R., 70–91
Fan, B., 81–82
Fan, P., 2–16, 311*t*
Fan, X., 209–211

Fan, Z.F., 26
Fang, L., 37
Fang, Y., 264*t*
Farin, A.M., 361–364
Farinelli, L., 267–268
Farley, E.K., 272
Farré, D., 273–274
Farrona, S., 267
Farzadfard, F., 361–364
Fedewa, G., 113, 172
Feil, R., 234, 236–237
Feldmann, K.A., 71–72
Feller, A., 262
Feng, H., 175–176, 209–211
Feng, J., 78–80
Fengler, K.A., 254, 259–260
Fenical, W., 114–115
Ferlanti, E.S., 286
Fernandez, A.I., 342
Fernández-Del-Carmen, A., 342–343
Fernández-Moreno, J.P., 342–343
Fernandez-Pozo, N., 32
Fernie, A.R., 56–57
Ferrer, A., 35–36, 71–72, 308–309, 326
Ferrer, J.L., 86–87
Ferreyra, L.F., 262
Ferruzzi, M.G., 342, 344–349, 351–352
Fett-Neto, E., 194–195
Feyereisen, R., 83–84
Field, B., 76, 112–113
Field, M., 32–34
Field, R.A., 100–116
Fields, C.J., 31–32
Fields, S., 274–276
Figueroa, P., 267
Figueroa, R.I., 109–110
Filipski, A., 82
Finnegan, P.M., 49–51, 58–61, 64
Fischbach, M.A., 106–107
Fischer, M., 244
Fischer, R.L., 340–342
Fischer, U., 281
Fisher, C.L., 15, 153
Fisher, K.J., 89, 148, 169–170
Fitz, J., 254
Fiume, M., 269
Flatt, P.M., 106
Fleming, T.C., 113–114
Flett, F., 114–115
Flores, N., 88–89
Flores-Perez, U., 229
Flügge, U.I., 334–336
Flynn, C.M., 102–103
Foerster, A.M., 266
Foister, S., 271–272
Foley, R., 263–266
Fonfara, I., 361–364
Forster, S., 87
Fossati, E., 38–39
Foureau, E., 168–203
Fowler, Z.L., 88–89
Fox, S., 254
Frame, K., 262
Francis, W.R., 73–74
Francoeur, N.J., 272–273
Francois, N., 21
Franco-Zorrilla, J.M.M., 272
Fraser, C.M., 76
Fraser, P.D., 339, 341, 344
Fray, R.G., 342, 348
Fredslund, J., 284*t*
Freitas, L.G., 243
Frelet-Barrand, A., 209–211, 210*t*
Frey, M., 337–338
Fricke, E., 273–274
Fricke, J., 124–125, 134–136, 142–143
Frickenhaus, S., 113
Fridkin, A., 361–364, 373–375
Fridman, E., 336–337
Fried, M.G., 279–281
Friedman, K., 236–237
Friml, J., 209–211, 210*t*
Frith, M.C., 255
Frugier, F., 23
Fu, L., 34, 172
Fuertes, A., 269–270
Fujihashi, M., 123–124
Fujisaki, S., 125–126
Fujita, Y., 264*t*
Fukao, Y., 210*t*
Fukuda, S., 209–211, 210*t*
Fukui, M., 100, 103–105
Fukushima, A., 38–39
Fuller, R.P., 364–365
Fumasoni, I., 264*t*
Fürholz, A., 334–336

Furumichi, M., 37
Furumoto, T., 334–336
Fushimi, H., 21
Fuss, E., 258–259
Futaki, S., 365–366

G

Gadella, T.W., 313
Gaffé, J., 338, 341, 344–345, 347–348
Gagliano, W.B., 258–259
Gagne, S.J., 74–75
Gaiki, S., 239
Gaj, T., 364–365
Galbraith, D.W., 26–27, 266
Galeano, N.F., 255–258
Gallego, F., 339
Gamborg, O.L., 194
Gammerman, A.J., 284*t*
Gang, D.R., 70–71, 336–337
Gani, D., 351–352
Ganjewala, D., 345
Gantt, E., 339
Gao, G., 284*t*
Gao, H., 365–366
Gao, S.Q., 264*t*
Gao, W., 38–39, 334–336
Gao, Y., 172, 254–256, 267
Garcia-Alcalde, F., 35–36
Garcia-Fuentes, B., 284*t*, 286
Garcia-Gomez, D., 237, 239
García-Gómez, J.M., 37, 57–58
Garcia-Mas, J., 237–238
Gardiner, J.M., 259–260
Garg, N., 113–114
Garms, S., 336–337, 339
Garner, M.M., 279–280
Garrard, W.T., 258–259
Garvey, G.S., 13, 14*f*
Gas, E., 229
Gassmann, M., 26
Gatehouse, L.N., 311*t*
Gatto, N., 307, 334–336, 350
Gatz, C., 272–273
Gaudinier, A., 278–279
Gautheret, D., 29
Ge, X., 234
Geerlings, A., 198
Geffers, R., 273–274
Geigl, J.B., 26–27
Geisler, M., 209–211, 210*t*
Geissler, R., 364–365, 373–375
Gentalen, E., 271–272
Georgantea, P., 2
George, J., 168–169
Georgiev, P., 258–259
Germano, A., 194–195
Gersbach, C.A., 361–364
Gershenzon, J., 2–3, 48–49, 70, 122, 149–150, 226–244, 263–266, 308–309, 321–325, 334–338, 345
Gerstein, M., 23
Gertz, J., 20
Gerwick, W.H., 106
Geu-Flores, F., 74–75, 87, 211
Gheyselinck, J., 258–259
Ghirardo, A., 227–229, 232–234, 237–238
Ghisalberti, E.L., 48–51
Ghosh, B., 2–4, 7, 10, 311*t*
Giacalone, C., 307, 309–310, 311*t*, 334–336
Gibon, Y., 236–237
Gibson, D.G., 211
Gibson, G., 254
Gierl, A., 337–338
Giglioli-Guivarc'h, N., 168–203
Gijzen, M., 352
Gilad, Y., 24–26
Gilbert, L.A., 361–364, 373–375
Gilbert, W., 30
Gilham, D.E., 83–84
Gillaspy, G., 81–82, 338
Giner, J.L., 78–80
Gingeras, T.R., 254–256, 258–259
Giovannoni, J.J., 21, 340–341
Giri, A.P., 308–309, 337–338
Giritch, A., 84
Girke, T., 21
Giuliano, G., 339
Glancey, M.J., 83–84
Glas, J.J., 306–307
Glass, C.K., 262–263
Glass, K.A., 361–364
Glatigny, A., 21
Glenn, W.S., 74–75
Glévarec, G., 168–203
Glieder, A., 84
Glöckner, G., 113

Godoy, M., 272
Goedhart, J., 313
Goepfert, S., 309–310, 311*t*
Goetting-Minesky, M.P., 275*f*, 278–279, 282*t*, 286
Goff, L., 35–36, 173
Goff, S.A., 262
Gokhale, R.S., 106–107
Gold, L., 269–270
Gold, S.E., 272–273
Golda, G., 22
Goldarazena, A., 324–325
Golden, M.C., 270
Goldschmidt, A., 268
Goldsmith, M., 87
Gomez-Maldonado, J., 281
Gomez-Mestre, I., 24–26
Gong, W., 282*t*
Góngora-Castillo, E., 113, 172
Gonzales-Vigil, E., 4, 82, 334–336, 346
Gonzalez, D.H., 281
Gonzalez, V., 336–337
Gonzalez-Cabanelas, D., 226–244
Good, A.G., 264*t*
Good, X., 340–341
Goossens, A., 37–39, 48–50, 168–169
Gopal, G.J., 83
Gordon, L., 114–115
Gorenstein, N., 48–49, 336–337
Gosset, G., 88–89
Gossling, E., 273–274
Goto, N., 31–32
Goto, S., 37
Gottardo, R., 267
Gotz, C.W., 243
Götz, S., 37, 57–58
Gout, E., 244
Grabherr, M.G., 32–34, 73–75, 172–173
Grabińska, K.A., 124–125, 135–136, 140–143
Graham, I.A., 337–338
Grandjean, O., 20, 266
Grando, S., 259–260
Granell, A., 342–343, 351–352
Gray, J., 255–258, 261, 263–266, 275*f*, 278–279, 282*t*, 284*t*, 286
Green, M.R., 266
Green, P., 209–211, 210*t*
Green, R., 152, 160
Green, S.A., 336–337
Greer, J., 71–72
Gregory, B.D., 23
Gregory, J.F., 340–341
Griepink, F., 308–309, 337–338
Griep-Raming, J., 236–237
Grierson, D., 342, 348
Grimwood, J., 107–109
Grishin, N.V., 81–82
Gritti, F., 239
Grob, H., 209–211, 210*t*
Groenke, K., 236–237
Groer, C.E., 168–169
Groisman, E.A., 23
Gropl, C., 78–80
Grosdemange-Billiard, C., 334–336
Gross, D.S., 258–259
Grossel, M., 262–263
Grossniklaus, U., 258–259, 281
Grotewold, E., 252–287, 284*t*
Groth, M., 109–110
Groudine, M., 258–259
Grove, C.A., 277–278
Gruber, A., 110
Gruetzner, R., 366–369, 371–375
Gruissem, W., 226, 338, 347–348
Grutzner, R., 366–369, 371–373
Gu, G., 270
Gu, L., 4–5
Gu, T., 281, 282*t*
Gu, X., 284*t*
Guan, X.Y., 30–31
Guan, Z., 124–125, 135–136, 140–143
Guengerich, F.P., 63
Guenther, M., 234
Guerra, J., 323–325
Guerrero, Á., 324–325
Guevara-Garcia, A.A., 229
Guigo, R., 258–259
Guihur, A., 191–192, 197–198, 201–202
Guilak, F., 361–364
Guillaume, E., 267
Guillot, A., 194
Guiochon, G., 239
Guirimand, G., 191–195, 197–198, 201–202
Gunther, T., 254

Guo, A.-Y.Y., 284*t*
Guo, J.R., 26
Guo, M., 254, 259–260
Guo, Y., 89, 239
Gupta, S., 281
Gupta, V., 311*t*
Gurr, G.M., 2–3, 306–307
Guschin, D.Y., 364–365
Gustincich, S., 255
Gutensohn, M., 326, 334–352
Guterman, I., 336–337
Gutierrez, R.A., 263–266
Gutierrez-Nava, M.L., 229
Guyot, V., 364–365

H

Ha, S.H., 351–352
Haag, J.R., 281
Haas, B.J., 20, 32–34, 56–57, 73–75, 172–173
Habben, J., 254, 259–260
Haberle, V., 257–258
Habrian, C.H., 365–366
Haddock, S.H., 73–74
Hadvary, P., 71–72
Hadzhiev, Y., 257–258
Hagel, J.M., 38–39
Hahlbrock, K., 281
Hahn, M.W., 254
Hahn, S., 364–369, 373–375
Hahnen, J., 256
Hajduk, P.J., 71–72
Hake, S., 261
Halkier, B.A., 208–221
Hall, A.W., 259–260
Hall, D., 53–54
Hallen, L., 29–30
Hallikas, O., 272
Halpern, C.B., 70–71
Halpin, C., 351–352
Hamamoto, M., 209–211, 210*t*
Hamann, M.T., 105–106
Hamann, T., 60–61
Hamberger, B., 49, 53–54, 57–58, 82, 148
Hamburger, M., 113–114
Hamilton, C.J., 102, 106–107, 110, 114–115
Hamilton, D.A., 315–316
Hamilton, J.P., 36, 76–77, 113
Hammell, M., 277–278
Hammer, K., 361–364
Hammerbacher, A., 226–244
Han, B., 78–80
Han, J.Y., 311*t*
Han, R., 20–39
Han, S., 264*t*
Han, Z., 365–366
Hanada, A., 334–336
Hancock, J.M., 284*t*
Handa, A.K., 342
Hangauer, M.J., 29–30
Hanlon, M.R., 286
Hanna, A.I., 254
Hanna-Rose, W., 262–263
Hannick, L.I., 20
Hansen, B.G., 213, 216
Hansen, E.H., 59–60
Hansen, U., 262–263
Hanson, A.D., 340–341
Hanson, J.R., 168–169
Hao, P., 56–57
Hare, E.E., 77–78
Harich, K., 81–82
Haring, M.A., 258–259, 310–313, 311*t*, 323–325, 338, 344–345
Hartemink, A.J., 273–274
Hartmann, C., 23
Hartmann, L., 281
Hartmannová, H., 124–125, 135–136, 140–143
Haruki, M., 105–106
Harvey, B., 114–115
Harwood, D.T., 109–110
Harwood, J.L., 334–336
Hase, K.-I., 105–106
Hasegawa, M., 105–106
Hashimoto, S., 254–256
Haskins, W.E., 280–281
Hasnain, G., 37
Hata, S., 23
Hathwaik, U., 124
Hauer, M., 361–364
Haun, W.J., 259–260
Haupt, I., 308–309
Hauschild, K.E., 271–272
Hawkins, K.M., 169–170

Hawkins, N.D., 227
Hawthorne, D.J., 2–3
Hayashizaki, Y., 255
Hayden, D.M., 227, 232–234
Hayes, C.J., 342, 348
He, F.S., 271–272
He, H., 264*t*
He, K., 267
He, X., 268
Head, S.R., 29–30, 38–39
Hecht, S., 244
Hecht, S.M., 168–169
Heck, A.J., 239
Hegardt, F.G., 71–72
Hegarty, R., 76–77
Hehl, R., 273–274, 284*t*
Heifetz, P.B., 334–336
Heintzman, N.D., 258–259
Held, M., 337–338
Hemberg, M., 26–27
Hemmerlin, A., 334–336
Hendrickson, D.G., 35–36
Hendriks, J.H., 236–237
Henrissat, B., 115–116
Henry, L.K., 352
Herbette, G., 309–310
Hermanson, P.J., 259–260
Hernandez, J.M., 262
Herold, M., 272–273
Herrera-Estrella, L., 71–72
Herrmann, J., 87
Hershey, D.M., 148
Hert, J., 70–71
Hertweck, C., 103–109, 111–115
Hertz, G.Z., 273–274
Herwig, C., 84
Herz, S., 244
Herzog, M., 309, 311*t*
Hestand, M.S., 256
Heuer, M.L., 31–32
Hey, J., 259
Heyndrickx, K.S., 272
Hickman, R., 56–57
Hicks, L.M., 244
Hidalgo, C.A., 277–279
Higo, K., 284*t*
Higuchi, Y., 123–124
Hildebrandt, A., 78–80
Hill, L., 102, 106–107, 110, 114–115, 264*t*
Hillebrand, A., 124–125
Hillwig, M.L., 38–39
Hiltpold, I., 70, 337–338
Hinz, W., 31
Hirata, A., 125–126
Hirata, T., 345
Hiratsu, K., 262–263, 272–273
Hirschberg, J., 339
Hirzmann, J., 256
Hobom, G., 256
Hochuli, E., 71–72
Hodaňová, K., 124–125, 135–136, 140–143
Hodge, D., 244
Hodo, Y., 256
Hoeffler, J.F., 334–336
Hoekema, A., 320
Hoffman, M., 270
Hofs, R., 114–115
Hojati, Z., 114–115
Hollinger, A., 76–77
Holmes, D., 2–3
Holt, C., 76
Honda, M., 256
Honeycutt, E., 254
Hong, K.K., 89
Hong, S., 208
Hoof, I., 110
Hope, I.A., 277–278
Hopwood, D.A., 70–72, 106–107, 110, 187–189
Horbach, S., 226–227
Horie, K., 237, 240–241
Horlbeck, M.A., 361–364
Horning, T., 227–229
Horwitz, S.B., 70–72
Hoskins, R.A., 255
Houlès, T., 268
Houtz, R.L., 2–3, 311*t*
Hovde, B.T., 111–112, 112*f*
Hovel, I., 258–259
Howat, S., 208
Howe, G.A., 338
Howell, J., 78–80
Hrmova, M., 263–266
Hsu, M.H., 83–84
Hu, H., 264*t*, 270
Hu, S., 278

Hu, Z., 264*t*
Huang, C.F., 209–211, 210*t*
Huang, J., 336–337
Huang, L., 38–39
Huang, M., 284*t*, 308–309
Huang, R., 311*t*
Huang, S., 259
Huang, W., 32–34
Huang, X., 172
Huang, Y., 172, 263
Huang, Z., 152
Hubbard, L., 254
Huber, C., 124–125, 134–136, 142–143
Huber, W., 35–36, 58, 181
Huber-Allanach, K.L., 82
Hueller, T., 88–89
Huerta, M., 273–274
Hufnagel, D.E., 334–336, 346
Hughes, L.E., 351–352
Huh, J.-H., 324–325
Hummel, A.W., 365–369
Hung, L.H., 23, 81
Hunsperger, H.M., 111–112, 112*f*
Hunstad, D.A., 244
Hunt, M., 275*f*, 278–279, 282*t*, 286
Hunter, A., 284*t*, 286
Hunter, S., 37
Hunter, W.N., 227
Hurlock, A.K., 266
Husodo, S., 309–310
Huson, D.H., 110
Hussong, R., 78–80
Hutchison, C.A., 211
Huttner, E., 258–259
Hwang, H.J., 32–34
Hwang, J.U., 210*t*
Hyatt, D.C., 148–149

I

Ibberson, D., 24–26
Ibdah, M., 339
Ichikawa, J., 30–31
Ignea, C., 2, 59–61
Iijima, Y., 336–337, 339, 342, 345–348
Ikeda, M., 208, 262–263, 282*t*
Ikegami, T., 237, 240–241
Ikeura, E., 35–36
Illarionova, V., 244
Imanaka, T., 105–106
Imanishi, M., 365–366
Imperial, S., 337–338
Inouye, C., 262–263
Inze, D., 281
Ioannidis, P., 73–74
Ioannou, E., 2
Ip, P.L., 26–27
Irimia, M., 110
Irisawa, Y., 270
Irudayaraj, J., 29
Irwin, J.J., 70–71
Ishiba, M., 239
Ishibashi, Y., 100, 103–105
Ito, R., 37
Ito, Y., 264*t*
Itoh, M., 254–256
Iturrondobeitia, J.C., 324–325
Ivakov, A., 234
Iven, T., 281
Iwamoto, M., 284*t*
Iyer, V.R., 259–260
Izumi, S., 345

J

Jabès, F., 307, 309–310, 334–336
Jackman, S.D., 32–34
Jackson, A., 38–39
Jacobsen, S.E., 266–267
Jaeckisch, N., 113
Jain, A., 324–325
Jaini, R., 48–49
Jaitin, D., 267–268
Jakobsen, K.S., 102
Jang, M.O., 208
Janies, D., 284*t*, 286
Jarosch, F., 270
Jarrett, H.W., 280–281
Jarvis, P., 229
Jauregui, R., 268
Jefferson, R.A., 258–259, 371–373
Jelesko, J.G., 338, 347–348
Jenkins, G.I., 70
Jenkins, S.M., 338, 347–348
Jensen, J.K., 213, 216
Jensen, K., 324–325
Jensen, M.K., 272

Jensen, P.R., 107–110, 108*f*, 113–115, 361–364
Jeon, B., 210*t*
Jespersen, J.B., 324–325
Jez, J.M., 87
Jha, R.K., 111–112, 112*f*
Jia, H., 309, 337–338
Jia, J., 106
Jia, Y., 280–281
Jiang, D., 280–281
Jiang, J., 258–259, 272
Jiang, L., 278
Jiang, W., 311*t*
Jiang, Z., 309–310
Jin, J., 284*t*
Jin, Y., 122–123
Jin, Y.W., 208
Jindal, S., 311*t*
Jinek, M., 361–364
Jing, F., 311*t*
Jirschitzka, J., 263–266
Johansen, T.E., 361–364
John, M.E., 309–310
John, U., 109–110, 113
Johne, B., 318, 324–325
Johnson, D.S., 22, 267–268
Johnson, M., 22, 106–107
Johnson, P.L., 273–274
Johnson, S.M., 30–31
Johnson, W.E., 267
Johnston, J.S., 77–78
Jolma, A., 270–271
Jona, G., 271–274
Jonak, C., 20, 266
Jones, A.D., 2–4, 7, 10, 13, 311*t*, 338
Jones, C.G., 48–51
Jones, D., 4–5
Jones, D.R., 27
Jones, D.T., 208
Jones, J.D., 366–369, 371–373
Jones, M., 342
Jones, P., 37
Jones, R., 226
Jongedijk, E., 59–60
Jongsma, M.A., 337–338
Jorgensen, M.E., 209–211, 216–217
Joshi, T., 284*t*, 286
Joung, J.K., 361–365
Juehne, T., 281
Juillerat, A., 364–365
Jung, H., 351–352
Juven-Gershon, T., 257

K

Kabadi, A.M., 361–364
Kadonaga, J.T., 257
Kagan, I.A., 334–336
Kai, C., 255
Kaiser, J., 244
Kaiser, R., 336–337
Kamileen, M.O., 87
Kamimoto, Y., 209–211, 210*t*
Kaminaga, Y., 85
Kamiya, A., 209–211, 210*t*, 219–220
Kamiya, Y., 267, 334–336
Kampranis, S.C., 59–61
Kanagarajan, S., 311*t*
Kanaya, S., 37, 105–106
Kanehisa, M., 37
Kanellis, A.K., 2, 59–61
Kang, H., 267
Kang, J.H., 208–209, 210*t*, 338
Kang, K., 4
Kang, S.E., 278–279
Kang, T.J., 87
Kann, L.R., 345
Kant, M.R., 306–307, 309–310, 324–325, 337–338
Kapila, J., 310–313
Kaplan, I., 326, 337–338, 342, 346–349, 351–352
Kapoor, S., 266
Kapp, J.A., 70–71
Kapp, U., 87
Kappers, I.F., 306–309, 324–325
Karamycheva, S., 172, 286
Karimi, M., 281, 342
Karin, M., 260–261
Kasahara, H., 334–336
Kasahara, Y., 270
Kasai, Y., 254–256
Kashina, E.V., 259–260
Kaspera, R., 48–49
Kaštovský, J., 306
Kasuga, M., 254
Kasukawa, T., 255

Katahira, E.J., 154, 159
Katayama, K., 35–36
Katayama, S., 255, 257–258
Katayama, T., 88–89
Kato, J., 125–126
Kato, S., 255–256
Kauffman, C., 114–115
Kaufmann, K., 267–268
Kavanagh, T.A., 371–373
Kawai, J., 255
Kawaji, H., 255
Kawashima, M., 37
Kay, S.A., 281, 364–369
Kaysser, L., 106
Kazakevich, Y.V., 239
Kazan, K., 262–263
Kazlauskas, R.J., 87
Kazmaier, M., 83–84
Ke, Y., 70–71
Keasling, J.D., 2, 169–170, 244
Keatinge-Clay, A.T., 103–105, 104*f*
Kebarle, P., 242
Keeling, C.I., 48–51
Keeling, P.J., 109, 109*f*
Keiser, M.J., 70–71
Kel, A.E., 273–274
Keller, G., 309–310
Keller, U., 105–106
Kelley, D.R., 173
Kelley, L.A., 13
Kellmann, R., 102
Kellner, F., 36, 76–77, 191
Kellogg, J.A., 340–341
Kel-Margoulis, O.V., 273–274
Kemp, L.M., 244
Kempinski, C., 309–310
Kera, K., 131–132, 132*f*
Kerr, I.D., 209
Kersten, B., 284*t*
Kersten, R.D., 102–103
Keshava, N., 70–71
Ketchum, R.E.B., 48–49
Kevany, B.M., 103–105
Khalil, A.S., 361–364
Khan, F., 168–169
Khanin, R., 35–36
Kheradpour, P., 268
Kidd, B.N., 266
Kidokoro, S., 264*t*
Kigawa, K., 209–211, 210*t*
Kiko, R., 73–74
Kim, B.G., 88–89, 264*t*
Kim, B.H., 81–82
Kim, D., 173
Kim, H.J., 88–89, 311*t*
Kim, H.U., 106, 111
Kim, J., 2–4, 36, 76–77, 87, 311*t*, 334–336, 346
Kim, J.S., 88–89
Kim, K., 259
Kim, M.S., 286, 365–366
Kim, S.C., 32–34
Kim, T.H., 266–267
Kim, T.K., 258–259
Kim, T.-K., 337–338
Kim, Y.J., 259, 266
Kim, Y.Y., 210*t*
Kin, K., 34–35
King, A.J., 83–84
Kingsford, C., 34–35, 58
Kinoshita, T., 266–267
Kinoshita, Y., 266–267
Kinzler, K.W., 22, 259
Kirby, J., 244
Kirsh, O., 268
Kis, K., 244
Kish, C.M., 48–49, 85, 336–338, 345
Kishitani, S., 264*t*
Kitamura, A., 256
Kivioja, T., 270–272
Klein, A., 341
Klein, A.P., 78–80
Klein, C.A., 26–27
Klein, M., 209–211, 210*t*
Klein, S.E., 311*t*, 320–321, 334–336, 346
Klein-Marcuschamer, D., 227–229
Kleinstiver, B.P., 361–364
Klempien, A., 48–49
Kliebenstein, D.J., 259
Klimyuk, V., 84
Kline, R., 340–341
Klingenhoff, A., 256
Klinkenberg, A., 237–238
Klompmaker, M., 59–60
Klussmann, S., 270
Knani, M., 226–227

Kneissl, M.L., 340–341
Knierim, E., 29
Knight, H., 266
Knudsen, J.T., 48–49
Kobayashi, J., 103–105
Kobayashi, K., 229
Kobler, C., 87
Kocak, D.D., 361–364
Koch, T.H., 270
Kock, G.R., 59–60
Koffas, M.A.G., 70–71, 88–89
Kohlbacher, O., 110
Kohli, G.S., 109–110
Koksal, M., 122–123
Köllner, T.G., 70, 336–338
Kolosova, N., 48–49, 55, 336–337
Kolter, R., 106–107
Kolukisaoglu, H.U., 210*t*
Koma, D., 88–89
Komatsu, K., 21
Komori, H.K., 29–30
Koncz, C., 281, 371–373
Kondo, S., 255
Kondou, Y., 282*t*
Kong, Y.-M., 56–57
Kongstad, K.T., 49
Kopka, J., 243
Korban, S.S., 340–341
Koren, S., 34, 77–78
Korenaga, T., 284*t*
Korf, I., 56–57, 270
Korfhage, U., 281
Kortbeek, R.W.J., 306–326
Kourtzelis, I., 59–61
Kouzarides, T., 262–263
Kovacs, K., 342, 348
Kowalski, N., 124
Koyama, T., 123–124, 131–132, 132*f*, 262–263
Koyanagi, T., 88–89
Krainer, F.W., 84
Krajewski, P., 267–268
Krasnyanski, S.F., 340–341
Kratochvil, N.C., 271–272
Krek, A., 35–36
Kremers, G.-J., 313
Kretzschmar, T., 208–209
Kries, H., 87
Krishnakumar, V., 286
Krishnan, S., 281
Kriventseva, E.V., 73–74
Krizek, B.A., 258–259
Krobitsch, S., 29–30
Krogh, A., 255
Kronbach, T., 83–84
Kronenberger, J., 20, 266
Krug, D., 106, 111
Kruglyak, L., 259
Kruse, I., 210*t*
Krzyzanek, V., 124
Kubista, M., 26–27
Kubota, T., 103–105
Kuhlmann, U., 70
Kumar, A., 83
Kumar, K.K., 266
Kumar, N., 103–107
Kumar, S., 13, 82
Kundaje, A., 268
Kunert, G., 334, 336–337
Kuntz, M., 341
Kuo, A., 107–109
Kuo, M.H., 266–267
Kuo, Y.C., 364–365
Kupfer, E., 71–72
Kuromori, T., 209–211, 210*t*, 219–220
Kurz, K., 284*t*, 286
Kusk, P., 361–364
Kutchan, T.M., 71–72, 226
Kutrzeba, L., 78–80
Kuwahara, M., 270
Kuzuyama, T., 334–336
Kvam, C., 256
Kwan, G., 175–176
Kweon, S.J., 351–352
Kwon, E.-J.G., 122–143
Kwon, M., 122–143

L

La Claire, J.W., 103–105
La Russa, M., 373–375
Lacroute, P., 269
Lafontaine, F., 168–203
Laggner, C., 70–71
Lahaye, T., 364–366, 373–375
Lahr, G., 26–27
Lam, T.W., 32–34

Lamb, B.M., 365–366
Lambert, G.M., 26–27
LaMere, S.A., 29–30
Lammerhofer, M., 239
LaMonica, M., 30–31
Lan, X., 70–71
Lanckmans, K., 243
Landgraf, A., 364–366
Landolin, J.M., 255
Landt, S.G., 268
Lane, H.E., 100
Lange, B.M., 53–54, 148–164, 309–310, 334–338
Lange, E., 78–80
Lange, I., 148–164, 309–310
Langhoff, D., 340–341
Langmead, B., 175–176, 269
Langowski, L., 209–211, 210*t*
Laniel, M.A., 252–253
Lanoue, A., 168–203
Lanzotti, V., 71–72
Lao, K.Q., 26–27
Lara, A.J., 32
Lara-Astiaso, D., 267–268
Larrainzar, E., 32–34
Larrouy-Maumus, G., 74–75, 85–86
Larson, M., 307, 334–336, 338, 346–347
Larson, M.H., 361–364
Lassmann, T., 254–256
Last, R.L., 2–16, 306, 309, 311*t*, 334–337, 346
Latchman, D.S., 260–261
Latzel, C., 334–336
Lau, W., 2, 90
Lauble, H., 87
Laule, O., 334–336
Lawson, J., 270
Lazarevic, D., 255
Lazear, M., 84–85
Le Cam, L., 268
Le Goff, G., 84
Le, H.-S.S., 268
Lecharny, A., 282*t*
Lechner, A., 113–114
Lee, B.S., 87
Lee, C.G., 351–352
Lee, D., 267
Lee, E.A., 254
Lee, E.K., 208
Lee, H., 267
Lee, J., 267
Lee, M., 168–169, 210*t*
Lee, M.E., 89–90, 373–375
Lee, M.H., 311*t*
Lee, O.R., 209–211, 210*t*
Lee, S., 308–309, 324–325, 336–337
Lee, S.K., 264*t*
Lee, S.M., 208
Lee, T.I., 260–261, 266
Lee, Y., 208–209, 210*t*
Lees, H.A., 311*t*, 320–321, 334–336, 346
Lefevre, F., 209
Legrand, A.M., 100, 103–105
Legrand, S., 48–49
Lehtonen, R., 272
Leiber, M., 26
Leitner, M., 336–337
Leksa, N.C., 83
Lelandais-Briere, C., 23
Lemiere, F., 239
Lempe, J., 258–259
Lenders, M., 124–125
Leng, N., 74–75
Lengsfeld, H., 71–72
Lenhard, B., 255, 257–258
Lenihan, J.R., 227–229
Lenk, I., 272–273
Lennon, N.J., 76–77
Leon, P., 71–72, 226–227, 229
Leonard, E., 88–89, 227–229
Leonard, J.A., 24–26
Leong, B., 82, 124–125, 134–136, 142, 311*t*, 346
Lesburg, C.A., 122–123
Lesot, A., 309–310
Letunic, I., 101*f*
Levac, D., 187–189
Leveugle, M., 282*t*
Levin, J.Z., 32–34, 73–75
Levine, M.S., 272
Levy-Sakin, M., 76–77
Lewinsohn, E., 70, 336–337, 339–342, 346–348, 352
Lewis, K., 168–169
Lewis, M.W., 268
Lewis, N.G., 226

Leyva, A., 269–270
Li, B., 34–35, 74–75, 175–176
Li, C., 254, 262–263, 268
Li, H., 83–84, 175–176, 256
Li, J., 81–82, 84–85, 113–114, 263–266
Li, K., 346
Li, L., 338, 361–364
Li, L.C., 264*t*
Li, L.P., 168–169
Li, M., 311*t*
Li, N., 257–258
Li, P., 185
Li, S., 254–257
Li, T., 89, 275*f*, 278–279, 282*t*, 286
Li, W., 34, 172, 252–287
Li, X., 56–57, 84–85, 264*t*, 351–352
Li, X.-M., 274–276
Li, X.Y., 123–124
Li, Y., 78–80, 272–273, 308–309
Li, Z., 32–34, 234, 237, 244
Liang, F., 172
Liang, P.-H., 123–124
Liang, X., 284*t*, 286
Liang, Y., 35–36
Liang, Y.S., 351–352
Liang, Z., 263–266
Liao, D.J., 75–76
Liao, X.-J., 105–106
Libault, M., 26–27, 284*t*, 286
Licht, J.D., 262–263
Lichtenthaler, H.K., 226–227
Lieb, J.D., 267
Liebich, I., 273–274
Lim, C.G., 88–89
Lim, S., 311*t*
Lim, W.A., 361–364, 373–375
Lim, Y.W., 32
Lin, C.P., 187–188
Lin, J., 278
Lin, L., 113, 278–279
Lincoln, J.E., 340–342
Lindemose, S., 272
Linden, H., 339
Lindgreen, S., 32
Lindner, W., 239
Linforth, R.S., 342, 348
Ling, L.C., 339
Liscombe, D.K., 85–88, 113, 187–189
Liseron-Monfils, C., 268
Lissi, A.E., 169–170
List, F., 365–366
Lister, R., 23
Liu, B., 32–34, 258–259
Liu, C., 264*t*
Liu, C.J., 233
Liu, D., 270–271
Liu, E., 74–75
Liu, G., 338
Liu, H., 168–169
Liu, J., 105–106, 346
Liu, P., 29, 81–82, 185
Liu, Q., 254, 337–338
Liu, S.N., 281, 282*t*
Liu, S.Q., 75–76, 81
Liu, T., 268
Liu, W., 168–169
Liu, X., 2–4, 13, 70–71
Liu, X.-C.C., 284*t*
Liu, X.S., 267–268
Liu, Y., 35–36
Liu, Z.R., 258–259, 278–279, 361–364
Llewellyn, A.M., 227
Lloyd, A.M., 272–273
Lloyd, D.H., 22
Loake, G.J., 208
Lobkovsky, E., 114–115
Lockhart, D.J., 271–272
Lohr, M., 106–109, 111–113
Lohse, M., 32, 56–57
Lois, L.M., 244, 339, 341, 344
Lomonossoff, G.P., 90
Londono, M.A., 170
Long, R.M., 48–49
Longchar, B., 311*t*
Lopato, S., 263–266
Lopez, S., 324–325
López-Legentil, S., 111, 111*f*
López-Vidriero, I., 272
Lorenz, P., 168–169
Lorenzo, O., 272
Loupassaki, S., 2
Louwers, M., 258–259
Lovenberg, W., 85
Lowry, L., 148
Lu, C., 264*t*
Lu, L., 102–103

Lu, T.K., 361–364, 373–375
Lu, X., 311*t*
Luang, S., 263–266
Lucke, B., 29
Luckerhoff, L.L., 306–309, 324–325
Luckner, M., 70
Lucyshyn, D., 83–84
Luedemann, A., 243
Luke, G.A., 351–352
Lukens, L., 254, 259, 263–266
Lundgren, A., 311*t*
Lunn, J.E., 56–57, 236–237
Luo, B., 236–237
Luo, D., 56–57, 256
Luo, H., 266
Luo, J., 264*t*
Luo, R., 32–34
Luthra, R., 345
Lv, Z., 311*t*
Lynch, A.G., 268
Lynch, V.J., 34–35
Lyver, E.R., 209–211, 210*t*

M

Ma, H., 373–375
Ma, J.F., 209–211, 210*t*
Ma, L., 83–84
Ma, L.G., 282*t*
Ma, N., 36
Ma, X., 70–71, 115–116
Macgregor, A., 284*t*, 286
Madden, T.L., 36, 106–107
Madilao, L.L., 49–51, 58–61, 64, 148
Madsen, S.R., 209–211
Maeda, H., 337–338
Mael, L.E., 270
Maeshima, M., 210*t*
Maffei, M.E., 59–60, 336–337
Magallanes-Lundback, M., 113
Magnard, J.-L., 48–49
Mahfouz, M.M., 361–365
Mahmoud, S.S., 48–49, 309–310
Mahroug, S., 191–192, 201–202
Maisnam, J., 324–325
Maiti, R., 20
Majdic, T., 59–60
Majumdar, T.K., 243
Mak, A.N.S., 364–366
Makris, A.M., 59–60
Malone, A., 86–87
Malpica, J.M., 269–270
Man, T.K., 273–274
Mandel, M.A., 71–72
Manghani, C., 281
Manley, J.L., 262–263
Mann, R.S., 270–271
Manners, J.M., 266
Manning, S.R., 103–105
Mano, M., 168–169
Marahiel, M.A., 105–106
Marais, E., 187–188
Marechal, A., 364–365
Margueron, R., 266–267
Margulies, E.H., 258–259
Margulies, M., 30–31
Marian Walhout, A.J., 277–279
Marigo, G., 334–336
Marillonnet, S., 84, 366–369, 371–375
Marinov, G.K., 20, 268
Marois, E., 366–369
Marshall, R.K., 311*t*
Marshall-Colon, A., 272–273
Martell, R.E., 270
Martin, D.M., 48–49, 336–337
Martin, J.A., 56–57, 172
Martin, M.V., 63
Martin, V.J.J., 38–39
Martinez, J.A., 88–89
Martinez, N.J., 277–278
Martin-Magniette, M.-L.L., 267
Martinoia, E., 208–211, 210*t*
Maruyama, K., 254–256, 264*t*
Maruyama, S., 110
Mascarenhas, J.P., 315–316
Masci, T., 340–341
Masini, E., 187–188
Maston, G.A., 266
Matasci, N., 23, 81
Mateos-Vivas, M., 237, 239
Mathelier, A., 273–274
Mathur, S., 266
Matich, A.J., 336–337
Matsuba, Y., 82, 311*t*, 320–321, 334–336, 346
Matsuda, F., 36
Matsui, K., 262–263, 339–341, 346–347

Matsumura, W., 340–341
Matsuno, M., 76
Mattern, D.J., 263–266
Mattes, R., 154, 159
Matveeva, M.Y., 259–260
Maxam, A.M., 30
Mayer, C., 81
McCaig, B.C., 338
McCarthy, D.J., 35–36, 58, 181
McComb, A., 310–313
McCormick, S.P., 13, 14*f*
McCue, K., 20
McDonald, M.D., 21–22
McDowall, K.J., 279–280
McGarry, R.C., 258–259
McGarvey, D.J., 154, 159, 226
McGeady, P., 149–150
McGinnis, S., 106–107
McInerney, P., 27
McMahon, R.D., 326, 337–338, 342, 346–349, 351–352
McManus, M.T., 29–30
McMullen, M.D., 254
McPhee, D., 89, 148, 169–170
Mears, K.S., 270
Meckler, J.F., 365–366
Medema, M.H., 106, 110–111
Medina, A., 342–343
Mehrotra, A., 351–352
Mehta, P., 273–274
Meier, I., 281
Meinhard, J., 82
Meir, A., 339
Meir, S., 340–341
Meissle, M., 263
Mejia-Guerra, M.K., 253, 255–258, 262, 266–269, 277–278, 284*t*, 286
Melin, C., 168–203
Mello, A., 26–27
Mellor, J., 252–253
Memelink, J., 194–195, 198
Menda, N., 336–337
Mendell, J.T., 173
Mendenhall, E.M., 364–365
Meng, X., 264*t*
Mercer, A.C., 364–366
Merezhuk, Y., 106–107
Mérillon, J.M., 194
Mermod, N., 270–272
Merryman, C., 211
Mestre, P., 209–211, 214–215
Metyger, J.O., 236–237
Metzger, M., 270
Metzker, M.L., 31
Meusemann, K., 81
Meyer, A.J., 210*t*
Meyer, C.A., 267–268
Meyer, O., 334–336
Meyerowitz, E.M., 272–273
Mezulis, S., 13
Michelmore, R., 310–313
Michelmore, R.W., 259
Michotte, Y., 243
Micklefield, J., 114–115
Miehlich, B., 87
Mierendorf, R., 83
Miki, K., 123–124
Miles, R.B., 308–310, 351–352
Mileski, W., 31
Millar, A.H., 23
Miller, A.J., 209–211
Miller, A.M., 2–4, 13
Miller, B., 55
Miller, J.C., 364–365
Miller, J.R., 34
Miller, M.L., 168–169
Miller, R.A., 194
Milne, C., 114–115
Milos, P.M., 27
Milton, J., 30–31
Minami, H., 88–89
Minguet, E.G., 268, 273–274
Minkoff, L., 270
Minor, W., 86–87
Mintseris, J., 271–272
Mirabel, S., 342–343, 351–352
Mirabella, R., 309–310, 324, 337–338
Mirarab, S., 23, 81
Misof, B., 81
Mitani, N., 209–211, 210*t*
Mitchell, A., 37
Mithöfer, A., 336–337
Mitra, S.K., 70–72
Mitsuda, N., 282*t*
Mittag, M., 106–109, 111–113
Mitterbauer, R., 83–84

Miura, A., 266–267
Miura, S., 254
Miyaji, T., 209–211, 210*t*, 219–220
Mizoi, J., 263–266
Mo, B., 266
Mo, H., 78–80
Mo, Q., 268
Mochida, K., 38–39
Moczek, A.P., 87–88
Moeller, P.D., 100
Moghe, G.D., 2–16, 311*t*
Moghul, I., 75
Mohammed, S., 239
Mohanty, D., 103–107
Mohring, T., 236–237
Møller, B.L., 59–61
Molthoff, J.W., 317–320
Moniodis, J., 48–51, 59–61
Monniaux, M., 268, 273–274
Monroe, E.A., 109–110
Montgomery, J., 342
Mooibroek, H., 124
Moolhuijzen, P., 284*t*, 286
Moon, S.J., 264*t*
Moore, B., 76
Moore, B.S., 102–106
Moore, J.M., 258–259
Moore, S., 21
Mootz, H.D., 105–106
Moran, T.M., 272–273
Morcuende, R., 236–237
Mordvinov, V.A., 259–260
Morgan, J.A., 48–49
Morgante, M., 254, 259–260
Morikawa, M., 105–106
Morillon, A., 29
Morimoto, H., 149–150
Morioka, R., 37
Morita, H., 103–105
Moriyoshi, K., 88–89
Morohashi, K., 253, 262, 266–268, 272–273, 277–278
Morohoshi, K., 254–256
Morrone, D., 148
Morsut, L., 361–364
Mortazavi, A., 175–176, 267–268
Morton, T., 254–257
Moscou, M.J., 364–365
Moses, T., 168–169
Mosing, R.K., 270
Mount, S.M., 34–35, 58
Moureau, P., 342
Moyal Ben Zvi, M., 340–341
Moyano, T.C., 263–266
Moyroud, E., 268, 273–274
Msanne, J., 115–116
Mucchielli, M.H., 21
Muckenthaler, M.U., 24–26
Muehlbacher, M., 152
Mueller, O., 26, 55
Mueller-Roeber, B., 284*t*
Muhlemann, J.K., 48–49
Muiño, J.M., 267–268
Mukherjee, S., 271–274
Mukhopadhyay, A., 277–278
Müller, C.T., 336–337
Muller, D., 258–259
Muller, S.R., 243
Mumm, R., 59–60
Munagala, A., 75
Munt, O., 125
Muranaka, T., 334–336
Murata, M., 100, 103–105, 254–256
Murphy, A.S., 210*t*
Murray, C., 311*t*
Murray, S.A., 102, 109–110
Mutschler, M.A., 2–3
Muzzey, D., 271–274
Myers, R.M., 20, 267–268

N

Nagalakshmi, U., 23
Nagamura, Y., 209–211, 210*t*
Nagaoka, S., 254–255
Nagasaki, K., 113
Nagata, N., 229, 334–336
Nagegowda, D.A., 337–338
Nagel, A., 56–57
Nagel, D.H., 278–279
Nagel, R., 306–326
Nagy, R., 209–211, 210*t*
Naidong, W., 237
Nakabayashi, R., 36, 38–39
Nakagawa, A., 88–89
Nakamura, M., 20–39, 334–336
Nakano, A., 125–126

Nakano, T., 105–106
Nakashima, K., 263–266
Nakayama, T., 131–132, 132*f*
Nam, K.H., 339–341, 346–347
Nanao, M.H., 87
Nanayakkara, N.P., 82
Nap, J.-P., 317–320
Narita, J.O., 338, 347–348
Naseri, A., 373–375
Nawy, T., 26–27
Naya, L., 23
Ndungu, J.M., 169–170
Neal, T.R., 152
Nebenführ, A., 194
Nei, M., 13
Neilan, B.A., 102
Nelson, B.K., 194
Nelson, D., 57–58
Nepal, C., 257–258
Nevins, J.R., 270
Newman, D.J., 70–72, 168–169
Newman, K.L., 169–170
Newton, R.P., 239
Ngoi, S.M., 351–352
Nguyen, H.T., 284*t*, 286
Nguyen, N.T., 361–364
Nguyen, T.-D., 124–125, 134–136, 142
Nguyen, T.T., 82, 310–313, 326, 337–339, 342, 344–349, 351–352
Ni, T., 254–256
Ni, Y., 259–260
Nicholass, F.J., 342
Nicklen, S., 30
Nielsen, J., 89, 227–229
Nielsen, M.T., 49, 211
Niephaus, E., 124–125, 134–136, 142–143
Nieuwenhuizen, N.J., 336–337
Niggeweg, R., 264*t*
Niida, R., 36
Nikolayeva, O., 268
Nikolov, D.B., 260–261
Nilsson, R., 266
Ning, J., 2–4, 10, 311*t*
Nishikawa, S., 125–126
Nishimoto, M., 244
Nishimura, Y., 125–126
Nishiyama, M.Y., 284*t*, 286
Nishiyama, Y., 254–255
Nishizawa, A., 105–106
Nissim, L., 361–364, 373–375
Niu, B., 34, 172
Niu, D., 309, 337–338
Niu, X., 254, 259–260
Nobel, A.B., 267
Noel, J.P., 70, 80–81, 83, 85–88, 106, 122–123, 352
Noga, S., 237, 239
Noguchi, T., 105–106
Nordborg, M., 254
Norel, R., 273–274
Norholm, M.H., 213, 216
Nour-Eldin, H.H., 208–221
Novy, R., 83
Nural, A.T., 124
Nyren, P., 30

O

O'Connor, S.E., 169–170, 187–189
O'Neill, E.C., 100–116
Obata, K., 209–211, 210*t*
Obika, S., 270
Ochman, H., 23
Ochoa-Villarreal, M., 208
O'Connor, D., 268
O'Connor, S.E., 36, 71–72, 85–88
Offner, S., 26–27
Ofner, I., 2–4, 13, 311*t*
Ogasawara, N., 37
Oh, E., 267
Ohlrogge, J., 21
Ohme-Takagi, M., 262–263
Ohmoto, T., 88–89
Ohnishi, T., 53–54, 82
Ohshima-Ichie, Y., 334–336
Ohsumi, T., 254–256
Ohta, M., 262–263
Ohyama, K., 125, 134–135, 142–143, 229
Oikawa, A., 35–36
Ojima, I., 168–169
Ojima, K., 194
Okazaki, Y., 256
Okusako, Y., 23
Oldiges, M., 236–237
Oldre, A.G., 259–260
Oliveros, J.C., 272
Olsen, B.A., 237

Olsen, C.E., 208–221
Olson, K.M., 272
O'Maille, P.E., 86–87
O'Malley, R.C., 23
Onate-Sanchez, L., 263–266, 281
O'Neill, E.C., 102, 106–107, 110, 114–115
Onillon, S., 307, 309–310, 334–336
Onishi, A., 278
Onkokesung, N., 227, 232–234, 237–238
Onouchi, H., 272–273
Openshaw, M.R., 266
Opper, K., 281
Ordoukhanian, P., 29–30
Oresic, M., 78–80
Orlova, I., 307, 334–338, 342, 344–352
Orr, R.J.S., 102
Ortiz-Alcaide, M., 227–229, 232–234, 237–238
Orzaez, D., 342–343, 351–352
Osadjan, M.D., 340–341
Osbourn, A.E., 71–72, 76, 112–113
O'Shea, C., 272
Oshima, T., 37
Osorio, S., 208–209, 210*t*
Ossowski, S., 254
Østerås, M., 267–268
Ostrozhenkova, E., 244
Osuna, D., 236–237
Ott, F., 268, 273–274
Otten-Bruggeman, I., 306–326
Otwinowski, Z., 86–87
Ou, B., 281, 282*t*
Oudin, A., 168–203
Ouellet, M., 169–170
Ouma, W.Z., 252–287
Ousterout, D.G., 361–364
Ovadia, R., 365–366
Overton, G.C., 273–274
Ow, M.C., 277–278
Oyler, A.R., 237
Ozaki, H., 270
Ozawa, C., 239
Ozsolak, F., 27

P

Pabo, C.O., 261
Pachter, L., 34–36, 175–176
Paddon, C.J., 48–49, 89, 148, 169–170, 227–229
Padidam, M., 272–273
Paetz, C., 227, 232–234, 237–238
Paetzold, H., 339
Page, J.E., 74–75
Pai, A.A., 24–26
Pakatci, I.K., 259
Pan, S.-S., 105–106
Pan, X.J., 364–365
Pan, Y., 264*t*, 282*t*
Pandey, M., 273–274
Papadopoulos, J., 36
Papadopoulou, K.K., 76, 112–113
Papanicolaou, A., 56–57, 75, 172–173
Papon, N., 168–203
Para, A., 272–273, 281
Paradise, E.M., 169–170
Parage, C., 170
Pareja-Tobes, P., 268–269
Park, E.J., 124–125, 135–136, 140–143
Park, H.Y., 87
Park, J., 208–209, 210*t*, 267
Park, S.R., 264*t*
Parker, R., 24–26
Parkhomchuk, D., 29–30
Parr, A., 264*t*
Parra, G., 56–57
Parra, R.G., 273–274
Paszkowski, J., 266–267
Patro, R., 34–35, 58
Patwardhan, B., 70–72
Paul, A., 270
Pauli, F., 268
Pautler, M., 268
Payton, P., 21
Paz-Ares, J., 269–270, 282*t*
Pea, G., 254
Pecker, I., 339
Peckham, H., 30–31
Pederson, T., 373–375
Peel, G., 85
Pei, G.K., 15, 153
Pei, J., 81–82
Pekny, M., 26–27
Pellegrini, M., 267
Peng, J., 26
Peng, W., 256

Peng, Y., 113–114
Peng, Z.Y., 75–76
Penn, K., 107–110, 108*f*
Pereira, A., 317–320
Perez, L.M., 345
Perez-Castro, C., 273–274
Perez-Gil, J., 244
Pérez-Pérez, J., 272
Perez-Pinera, P., 361–364, 373–375
Pérez-Rodríguez, P., 284*t*
Perez-Trabado, G., 32
Perle, N., 270
Perli, S.D., 361–364, 373–375
Perola, E., 71–72
Pertea, G., 172–173, 175–176
Pertea, G.M., 173
Pertea, M., 173
Pessereau, C., 364–365
Petat-Dutter, K., 26–27
Peterbauer, C., 83–84
Peters, R.J., 148
Peters, R.S., 81
Petersen, M., 82
Peterson, C.L., 252–253
Peterson, D., 13, 82
Peterson, N., 13
Peterson, T., 269–270
Petri, J., 308–309
Petricka, J., 254–257
Petrovska, B.B., 23–24
Petzold, C.J., 2
Pevzner, P.A., 32–34
Peyret, H., 90
Pfennig, F., 105–106
Pfluger, J., 266–267
Phelan, V.V., 113–114
Philippakis, A.A., 271–272
Philippe, R.N., 70, 80–81
Phillippy, A.M., 77–78
Phillips, M.A., 197–198, 227, 237–238, 308–309, 337–339
Piatek, A., 361–364
Pichersky, E., 2–3, 49, 57–58, 70, 124–125, 134–136, 142, 148, 306, 308–310, 311*t*, 326, 334, 336–339, 342, 345–349, 351–352
Pickens, L.B., 103–105
Pickett, J., 324–325
Piestun, D., 336–337
Pieterse, C.M.J., 56–57
Pirun, M., 35–36
Pitcher, J., 124–125, 134–136, 142
Pitera, D.J., 48–49, 89, 148, 169–170
Pizer, M., 254
Platt, A.R., 27
Plaza, S., 210*t*
Plessy, C., 254–258
Pletzenauer, R., 84
Plummer, J.A., 48–51
Pluskal, T., 78–80
Podell, S., 107–110, 108*f*
Podevels, A.M., 103–105
Pohnert, G., 106–109, 111–113
Poirier, Y., 124
Pollard, K.S., 187
Pollard, V., 342
Pollier, J., 37, 48–50, 168–169
Polstein, L.R., 361–364
Polston, J.E., 324
Pomeranz, M., 253, 266–267, 277–278
Ponjavic, J., 255, 257–258
Pontes, O., 281
Pontini, M., 59–60
Pop, M., 175–176, 269
Poppenberger, B., 83–84
Posé, D., 268, 273–274
Post, J.J., 124
Posthumus, M.A., 324–325
Potterat, O., 113–114
Poulter, C.D., 152
Poulter, D.C., 152
Pourcel, L., 262
Powers, B.L., 266
Powers, M.L., 73–74
Prasad, M.V.R., 103–107
Preinerstorfer, B., 239
Prestridge, D.S., 273–274
Previti, C., 257–258
Prew, M.S., 361–364
Prijs, H., 324–325
Pritchard, M.P., 83–84
Priyam, A., 75
Prosser, G.A., 74–75, 85–86
Pruneda-Paz, J.L., 278–279, 281
Pry, T., 243
Pu, J.J., 26

Pu, L., 278–279
Pulido, P., 237–238
Pyenson, N.C., 361–364
Pyun, H.J., 149–150

Q

Qi, L.S., 361–364
Qi, W., 268
Qi, X., 76, 112–113
Qi, Z., 264*t*
Qian, J., 361–364
Qiao, G., 364–365
Qiao, Z.Z., 26–27, 268
Qin, H., 83–84
Qin, X., 361–364
Qiu, B., 311*t*
Qiu, F., 70–71
Qiu, Y., 84
Qu, Y., 124–125, 134–136, 142
Qualley, A., 81–82
Qualley, A.V., 48–49
Quatrano, R.S., 281
Quero, C., 324–325
Quijada, P., 258–259
Quin, M.B., 102–103
Qureshi, A.M., 271–272

R

Raatz, B., 258–259
Rabara, R.C., 263–266
Rabin, S., 236–237
Rach, E.A., 254–256
Radovic, S., 259–260
Radykewicz, T., 334–336
Raehal, K.M., 168–169
Raetz, C.R., 142–143
Rafalski, A., 254
Ragauskas, A.J., 168–169
Ragg, T., 26
Raghavan, R., 23
Raguschke, B., 226–244
Raha, D., 23
Rahier, A., 226
Rahier, N.J., 168–169
Rai, A., 20–39
Rai, V., 75
Raina, R., 336–337
Rajaonarivony, J.I., 149–150
Rajniak, J., 85
Rallapalli, G., 264*t*
Ralph, J., 76
Ralph, S., 55
Ram, O., 364–365
Ramakrishna, W., 258–259
Rambla, J.L., 342–343
Ramirez, A., 306–326
Ramirez, C.L., 361–364
Ramos, Y., 256
Ramsey, U.P., 100
Ranade, S., 30–31
Ranathunge, K., 264*t*
Rapaport, F., 35–36
Rapoport, H., 78–80
Rasmann, S., 70
Rasmussen, M.O., 336–337
Rastogi, C., 270–271
Ratcliffe, O.J., 261, 263–266
Rattanapittayaporn, A., 125
Rausch, C., 110
Ravel, J., 106–107, 110
Raven, P.H., 23–24
Rawat, R., 281
Ray, J.A., 342
Ray, L., 103–105
Rayment, I., 13, 14*f*
Raytselis, Y., 106–107
Rea, P.A., 209–211, 210*t*
Rearick, T.M., 31
Rebar, E.J., 364–365
Recht, S., 365–366
Redondo, F., 198
Reece-Hoyes, J.S., 275*f*, 277–279
Regalado, A., 30–31
Regev, A., 32–34
Regnier, S., 83–84
Reichelt, M., 307, 334–336, 350
Reichelt, R., 125
Reifenberger, J.G., 27
Reinberg, D., 266–267
Reiter, A., 345
Rejzek, M., 102, 106–107, 110, 114–115
Ren, B., 258–259, 266–267
Ren, Q.P., 364–365
Rensing, S.A., 284*t*
Repas, T., 115–116
Reuber, T.L., 263–266

Reuter, I., 273–274
Revzin, A., 279–280
Reyes-Gutierrez, P., 373–375
Reyon, D., 364–365
Rheault, M.R., 48–49
Rhoads, A., 74, 76–77
Rhodes, D., 307, 334–336, 350
Rhodes, L.L., 109–110
Ri, M.-B., 337–338
Riaño-Pachón, D.M., 284*t*
Rice, L.M., 86–87
Rice, P.M., 31–32
Richardson, T.H., 83–84
Riddle, N.C., 254
Rideau, M., 194
Riechmann, J.L., 261
Riely, B.K., 32–34
Rieping, M., 272–273
Riethmuller, G., 26–27
Riley, T.R., 270–271, 273–274
Rimando, A.M., 82
Ringler, P., 263–266
Rinn, J.L., 35–36
Rissman, A.I., 74–75
Ritland, K., 55
Rivas, S., 209–211, 214–215
Rivasseau, C., 244
Ro, D.K., 122–143, 169–170
Ro, M., 262–263
Robbins, P.W., 142–143
Roberts, A., 34–35, 173, 175–176
Roberts, M.A., 106
Roberts, S.G., 261–262
Robertson, G., 32–34, 56–57
Robinson, M.D., 35–36, 58, 181, 268
Robinson, R.I., 237
Robles, M., 37, 57–58
Roccia, A., 48–49
Rocha-Sosa, M., 71–72
Rockman, M.V., 254
Rodier, G., 268
Rodriguez, A., 88–89
Rodríguez-Concepción, M., 226–229, 232–234, 237–238, 321–323, 337–339, 347–348
Rodriguez-Gonzalo, E., 237, 239
Roeder, R.G., 260–261
Rogachev, I., 340–341
Rogers, H.J., 336–337
Rohmer, M., 226–227
Rohr, C.O., 273–274
Rohwer, F., 32
Rohwer, J.M., 227–229, 232–234, 237–238
Rokhsar, D.S., 272
Rombauts, S., 37
Romeis, J., 263
Römer, P., 365–366
Römer, S., 341
Romero, I.G., 24–26, 269–270
Ronaghi, M., 30
Ronen, G., 339
Ronning, C.M., 20
Rontein, D., 307, 309–310, 334–336
Roongsawang, N., 105–106
Rorsman, P., 26–27
Rosati, C., 336–337
Rose, D.W., 262–263
Roselló, L., 273–274
Rosenfeld, M.G., 262–263
Rosenkranz, M., 237–238
Roset, R., 273–274
Rosing, H., 229–230
Ross, A.C., 102–103
Ross, G.M., 102
Ross, M.G., 76–77
Rossetti, L., 84
Rossi, J.V., 106
Rossmann, S., 258–259
Rosti, S., 309–310, 311*t*
Rothberg, J.M., 30–31
Rothstein, S.J., 264*t*
Rottacker, E.C., 306
Röttig, M., 110
Roudier, F., 267
Roulet, E., 270–272
Ruan, S., 273–274
Ruberti, I., 272–273
Rudd, B.A., 70–72
Ruff, M., 243
Ruffel, S., 272–273
Ruiz, S., 281
Rullkotter, J., 236–237
Rumble, S.M., 269
Ruotti, V., 74–75
Rupe, M.A., 254, 259–260
Rusconi, C., 270

Rush, J.S., 128, 142–143
Rushton, P.J., 263–266
Rusnak, F., 105–106
Russ, C., 76–77
Rutledge, P.J., 71–72
Rutters, H., 236–237
Ruuska, S., 21
Ruzicka, K., 209–211, 210*t*
Ryan, M.D., 351–352
Ryken, S.A., 111–112, 112*f*

S

Saalbach, G., 100–116
Sabelis, M.W., 324–325
Sablowski, R.W., 272–273
Sabo, P.J., 258–259
Sacchi, N., 24–26
Sagner, S., 334–336
Sahm, H., 226–227
Sainz, M.B., 262, 269–270
Saito, K., 20–39
Sakai, K., 88–89
Sakaitani, M., 105–106
Saladie, M., 237–238
Salas, F., 273–274
Salcedo, T., 113
Salim, V., 187–189, 191
Sallaud, C., 307, 309–310, 311*t*, 334–336
Sallet, E., 23
Salmi, M., 226–227
Salomon, D.R., 29–30
Salowsky, R., 26
Salvi, S., 254, 259–260
Salzberg, S.L., 173, 175–176, 269
Samach, A., 272–273
Sammons, R.D., 234
San Roman, C., 229
Sanchez, L., 113–114
Sanchez-Fernandez, R., 209–211, 210*t*
Sandelin, A., 255, 257–258
Sandhu, H.K., 148
Sandhu, J.S., 340–341
Sandler, J.E., 255
Sandmann, G., 339
Sang, T., 254
Sanger, F., 30
Santelia, D., 210*t*
Santhamma, B., 148–149
Sardet, C., 268
Sarkar, A., 113–114
Sarkar, T.R., 29
Sarker, L.S., 48–49
Sarre, S., 243
Sass, L.E., 27
Sasso, S., 106–109, 111–113
Sato, F., 88–89
Sato, K., 125–126
Sato, M., 125–126
Sato, Y., 37
Satoh, K., 264*t*
Sattely, E.S., 2, 78–80, 85, 90
Sauer, R.T., 261
Sauret-Güeto, S., 229, 244, 337–338
Sauvageau, M., 35–36
Savchenko, T., 227, 232–234
Sawyer, L., 109
Scaffidi, A., 48–51
Scardino, E., 270
Schaedler, T.A., 210*t*
Schafer, P., 366–369, 371–373
Schaffer, S., 88–89
Schalechet, F., 339–341, 346–347
Schalk, M., 308–310, 351–352
Schaller, H., 124
Schauer, S.E., 258–259
Schauvinhold, I., 307, 310–313, 311*t*, 320–321, 334–336, 338–339, 345–347
Schauwecker, F., 105–106
Scheible, W.R., 21–22
Schein, J., 32–34, 56–57
Schell, J., 371–373
Schena, M., 20–21, 272–273
Schenk, B., 128, 142–143
Schenk, P.M., 154, 159, 266
Scherf, M., 256
Scherlach, K., 114–115
Scheuer, P.J., 105–106
Schibeci, D., 284*t*, 286
Schiesel, S., 239
Schiff, P.B., 70–72
Schilmiller, A.L., 2–4, 10, 13, 306–307, 309, 311*t*, 334–338, 346–347
Schimmel, B.C., 306–307
Schluesener, H.J., 269–270
Schluter, P.M., 258–259
Schmelz, E.A., 336–337

Schmidt, A., 306–326, 334–336, 338, 346–347
Schmidt, C., 364–365
Schmidt, D., 277–278
Schmidt, T., 125
Schmidt-Dannert, C., 102–103
Schmidt-Kittler, O., 26–27
Schmieder, R., 32
Schmitt, M., 76
Schmitz, G., 258–259
Schnee, C., 336–338
Schneeberger, K., 254
Schneider, B., 191
Schnepp, J., 85
Schoch, G.A., 76
Schoenherr, J.A., 76
Scholte, A.A., 124
Scholze, H., 364–365, 373–375
Schorn, M., 106
Schornack, S., 364–365
Schrader, J., 48–50
Schreiber, T., 361–375
Schroder, J., 82, 86–87
Schröder, W., 105–106
Schroeder, A., 26, 55
Schroeder, F.C., 106–107
Schroth, G.P., 20
Schuch, W., 339, 341–342, 344
Schuelke, M., 29
Schuhr, C.A., 244, 334–336
Schuler, M.A., 60–61
Schultz, J., 31
Schulz, M.H., 56–57, 172, 268
Schulze, B., 336–337
Schütz, S., 318, 323–325
Schutze, K., 26–27
Schuurink, R.C., 306–326, 311*t*, 338, 344–345
Schwab, D.J., 273–274
Schwartz, T.U., 83
Schwarz, J.M., 29
Schwarz, M., 226–227
Schwarzlander, M., 210*t*
Schwarz-Sommer, Z., 272–273
Schwender, J., 226–227
Scolnik, P.A., 339
Scott, C.P., 21–22
Scott, D.J., 124
Scott, J.A., 102
Scott, J.K., 26
Scott, J.W., 339
Scoville, D.J., 270
Seals, L.A., 266
Seelow, D., 29
Seemann, M., 226–227, 244
Segal, D.J., 270
Séguin, A., 53–54, 82
Seidl, S., 26–27
Seki, H., 229
Sengupta, A.M., 273–274
Senhorinho, G.N.A., 102
Seo, J.H., 87
Sepúlveda, L.J., 187–188
Sequerra, R., 277–279
Servet, C., 267
Sese, J., 254–256
Sessa, G., 272–273
Seto, E., 262–263
Seyedsayamdost, M.R., 70–71
Seymour, G.B., 264*t*
Shahmuradov, I.A., 284*t*
Shalev, G., 336–337
Shalit, M., 336–337
Shallcross, J.A., 270
Shalon, D., 20–21
Shan, Q., 263–266
Shaner, N.C., 73–74
Shanks, J., 70–71
Shao, Y., 29
Shao, Z., 102–103
Shapiro, J.A., 2–3
Shapiro, L.J., 274–276
Sharkey, T.D., 234, 237, 244, 308–309
Sharma, K.K., 324–325
Sharrocks, A.D., 262–263
Sheehan, I., 86–87
Shehara, J., 103–107
Shen, B., 103–105
Shen, H., 168–169, 264*t*
Shen, Q.J., 263–266, 311*t*
Shen, Y.P., 256, 282*t*
Shepherd, R.W., 2–3, 311*t*
Sherden, N.H., 74–75, 87
Sheriha, G.M., 78–80
Sherlock, G., 37
Sherman, D., 345

Shewmaker, C.K., 339
Shi, F., 2–4, 338
Shi, M.J., 26
Shi, Y.G., 86
Shibata, D., 32–34
Shiekhattar, R., 258–259
Shilov, A.G., 259–260
Shimamura, M., 334–336
Shimizu, H., 209–211, 210*t*, 219–220
Shimoda, Y., 23
Shimokawa, K., 255, 257–258
Shin, D., 264*t*
Shin, H., 268
Shingles, J., 277–278
Shinohara, A., 105–106
Shinozaki, K., 209–211, 210*t*, 219–220, 254, 263–266, 264*t*
Shinshi, H., 262–263
Shipton, C.A., 341
Shirai, M., 105–106
Shiraki, T., 255
Shirley, A.M., 76, 83–84
Shiroto, Y., 264*t*
Shitan, N., 209–211, 210*t*
Shoichet, B.K., 70–71
Shou, C., 23
Si, Y.Q., 29, 185
Sidow, A., 269
Siewers, V., 227–229
Silverman, G.J., 26
Simao, F.A., 73–74
Simeon, F., 88–89, 227–229
Simkin, A.J., 197–198, 341
Simkovsky, R., 113–114
Simmons, A.T., 2–3, 306–307
Simone, R., 255
Simonin, P., 226–227
Simonite, T., 30–31
Simonsen, H.T., 59–60
Simpson, A.J., 270
Sinelnikov, I.V., 78–80
Singer, H.P., 243
Singer, S.D., 258–259
Singh, A., 311*t*
Singh, K., 263–266
Sinlapadech, T., 76
Sitachitta, N., 106
Sitrit, Y., 339, 342, 344–349, 351–352
Sjoback, R., 26–27
Skaggs, A., 124
Slattery, M., 270–271
Slegers, H., 239
Smaczniak, C., 268
Smale, S.T., 257
Smentek, L., 86–87
Smith, A.J., 30–31
Smith, C., 284*t*, 286
Smith, C.J.S., 342
Smith, G.P., 30–31
Smith, H.O., 211
Smith, M.L., 268
Smith, O.S., 254, 259–260
Smith, R.K., 20
Smits, B.M., 74–75
Smolders, I., 243
Smolke, C.D., 169–170
Smyth, G.K., 35–36, 58, 181
Snook, M.E., 254
Snyder, M., 23, 271–274
Soetaert, S.S., 38–39
Sohl, C.D., 63
Sohrabi, R., 308–309, 324–325
Solano, R., 272
Soldatov, A., 29–30
Solem, C., 361–364
Soler, E., 334–336
Solovyev, V.V., 284*t*
Soltau, W.L., 266
Somssich, I.E., 263–266, 281
Song, B., 111, 111*f*
Song, C.X., 259–260
Song, K., 281
Song, O., 274–276
Song, S., 254–256
Sorgatz, A., 365–366
Sornaraj, P., 263–266
Sottomayor, M., 194–195
Souza, G.M., 284*t*, 286
Spadiut, O., 84
Spana, E.P., 254–256
Speisky, H., 169–170
Spies, T.A., 70–71
Spiro, R.G., 123–124
Splinter, E., 258–259
Spolaore, S., 342–343
Sponza, G., 254, 259–260

Springer, N.M., 254, 259–260
Spyropoulou, E.A., 306–326, 311*t*, 338–339, 345–346
Spyropoulous, E.A., 339, 345
Spyrou, C., 268
Srividya, N., 148–164
Stachelhaus, T., 105–106
Ståhl, B., 48–49
Stahlberg, A., 26–27
Ståldal, V., 272–273
Stam, M., 258–259
Stamati, K., 259–260
Stamatoyannopoulos, J.A., 258–259
Stamitti-Bert, L., 338, 344–345, 347–348
Stansbury, M.S., 87–88
Stark, R., 268
Starker, C.G., 365–366
Starks, C.M., 122–123
Stead, J.A., 279–280
Stec, A., 254, 259
Stecher, G., 13, 82
Steinbiβ, H.H., 154, 159
Steinbüchel, A., 227–229
Stephanopoulos, G., 227–229, 232–234, 244
Steppuhn, J., 341
Stermitz, F.R., 168–169
Sternberg, M.J.E., 13
Stevenson, C.E., 87
Stiekema, W.J., 317–320
Stillian, J., 236–237
Stitt, M., 21–22, 56–57
Stocker, S., 26, 55
Stockley, P.G., 281
Stoddard, B.L., 364–366
Stoeckert, C.J., 273–274
Stokvis, E., 229–230
Stolc, V., 267
Stolze, A., 124–125, 134–136, 142–143
Stormo, G.D., 273–274
Stout, J.M., 74–75, 84–85
St-Pierre, B., 168–203
Strader, L.C., 209–211, 210*t*
Straight, P.D., 106–107
Stránecký, V., 124–125, 135–136, 140–143
Streb, P., 244
Stremler, K.E., 152
Streubel, J., 364–366, 373–375
Stuart, J., 30–31
Stüken, A., 102
Stupar, R.M., 254, 259–260
Sturm, M., 78–80
Sudre, D., 208–209, 210*t*
Sugano, S., 254–256
Sugimoto, E., 209–211, 210*t*, 219–220
Suh, S.C., 351–352
Suh, S.J., 210*t*
Sullenger, B.A., 270
Sullivan, A.M., 258–259
Sultana, R., 172
Sumner, J., 23–24
Sun, H.X., 38–39
Sun, L., 256, 340–341
Sun, W., 259
Sung, Y.C., 187–188
Surani, M.A., 26–27
Surmacz, L., 124–125, 134–136, 142
Sutoh, T., 131–132, 132*f*
Sutter, B., 226–227
Sutton, G., 34
Suzuki, H., 20–39
Suzuki, M., 229, 334–336
Suzuki, Y., 254–256
Suzuri, R., 23
Svec, D., 26–27
Sweet, R.M., 86–87
Swerdlow, H.P., 30–31
Swiezewska, E., 124–125, 134–136, 142
Szostak, J.W., 269–270

T

Taconnat, L., 267
Tadmor, Y., 339–342, 345–347
Tagaya, O., 239
Takabayashi, J., 336–337
Takada, S., 282*t*
Takagi, M., 334–336
Takahashi, H., 20–39, 254–256
Takahashi, S., 125, 131–132, 132*f*
Takanashi, K., 209–211, 210*t*
Takasaki, H., 263–266, 264*t*
Takayama, H., 36
Takiguchi, Y., 282*t*
Takors, R., 236–237
Takubo, H., 237, 240–241
Talón, M., 37, 57–58

Tamura, K., 13, 82
Tan, C., 70–71
Tan, Q., 270
Tan, S., 364–365
Tanabe, M., 37
Tanaka, N., 237, 240–241
Tang, F.C., 26–27
Tang, H., 76–77
Tang, J., 229
Tang, L., 242, 284*t*
Tang, N., 264*t*
Tang, T., 21
Tang, W., 284*t*
Tang, Y., 103–105
Tanifuji, G., 110
Tanksley, S., 21
Tarazona, S., 35–36
Tari, L.W., 82
Tariq, M., 266–267
Tarselli, M.A., 168–169
Taucher, G., 345
Tavaré, S., 268
Tawara, J.N., 168–169
Tawfik, D.S., 87
Taylor-Teeples, M., 278–279
Tekmal, R.R., 70–71
Teng, K.-H., 123–124
Ter, F., 75
Teranishi, R., 339
Terasaka, K., 209–211, 210*t*
Terol, J., 37, 57–58
Tesler, G., 32–34
Testa, C.A., 227–229
Teutsch, H.G., 83–84
Thamm, A.M.K., 191
Theodoulou, F.L., 209
Thermes, C., 29
Thieme, S., 365–366
Tholl, D., 49, 57–58, 122–123, 148, 308–309, 324–325, 345
Thomas, C.J., 168–169
Thomas, M.G., 103–105
Thomas, S.T., 352
Thompson, D.A., 32–34, 73–75
Thompson, M.G., 76
Thompson, T.J., 48–49
Thomson, J.A., 74–75
Thornton, J.D., 210*t*
Thukral, S., 281
Tillmann, U., 113
Tingey, S., 254
Tingey, W.M., 2–3
Tippmann, S., 227–229
Tissier, A., 2–3, 244, 309–310, 324, 337–338, 361–375
Tittiger, C., 84
To, K.Y., 21
Toal, T.W., 278–279
Todd, J., 21
Toepfer, S., 70
Tohge, T., 36, 208–209, 210*t*, 264*t*
Toivonen, J., 270–271
Tokida, N., 239
Toledo-Ortiz, G., 237–238
Tolstikov, V., 227, 232–234
Tomczak, A., 310–313
Tomes, D., 254, 259–260
Tomomatsu, K., 237, 240–241
Tonthat, T., 30–31
Tonti-Filippini, J., 23
Too, H.P., 232–234, 244
Toofan, M., 236–237
Topfer, R., 309–310, 311*t*
Topkar, V.V., 361–364
Toriyama, K., 264*t*
Tornoe, J., 361–364
Torre, F., 281
Torre, S., 342–343
Torrens-Spence, M.P., 70–91
Torres, S.E., 361–364
Town, C.D., 20
Toyooka, K., 35–36
Traas, J., 20
Trainotti, L., 342–343
Tran, H.T., 124–125, 134–136, 142
Trapnell, C., 35–36, 173, 175–176, 269
Trezzini, G.F., 281
Trial, R.M., 113–114
Trick, M., 102, 106–107, 110, 114–116
Triemer, R.E., 100
Trieu, W., 169–170
Triezenberg, S.J., 262–263
Trikka, F.A., 2, 59–61
Tripathi, P., 263–266
Tritsch, D., 334–336
Trojer, P., 266–267

Truesdale, M.R., 339
Tsai, S.Q., 361–364
Tsuchida-Iwata, Y., 334–336
Tsuda, M., 103–105
Tsuji, S., 365–366
Tsuruta, H., 227–229
Tuerk, C., 269–270
Tumlinson, J.H., 336–337
Tung, J., 24–26
Turck, F., 267, 282*t*
Turco, G., 278–279
Turlings, T.C., 337–338
Turnbaugh, J.G., 339
Turner, G.W., 309–310
Turunen, M., 272
Tuupanen, S., 272
Tuytten, R., 239
Tyagi, A.K., 266
Tybulewicz, V.L., 277–278
Tyo, K.E., 88–89, 227–229
Tzfira, T., 351–352
Tzin, V., 340–341

U

Udenfriend, S., 85
Udomson, N., 38–39
Udupa, S.M., 259–260
Udvardi, M.K., 21–22
Ueda, K., 209–211, 210*t*
Uehara, T., 78
Ugawa, Y., 284*t*
Uhlen, M., 30
Um, M., 262–263
Unsicker, S.B., 334, 336–337
Upadhyay, R.J., 113
Urban, P., 83–84
Uros, E.M., 244
Urós-Gracia, E.M., 339
Urwin, P.E., 70
Usadel, B., 32
Usera, A.R., 85–86

V

Vachon, G., 309, 311*t*
Vaillancourt, B., 36, 76–77
Vainstein, A., 340–341
Valen, E., 255
Valliyodan, B., 284*t*, 286
Valot, N., 48–49
Valouev, A., 30–31
Valsecchi, I., 272–273
Valton, J., 364–365
van Baren, M.J., 175–176
van Beilen, J.B., 124
Van de Velde, J., 272
van Deemter, J.J., 237–238
van Deenen, N., 124–125, 134–136, 142–143
van Der Heijden, R., 198
van der Laan, M.J., 187
Van Dolah, F.M., 109–110
van Dongen, J.T., 234
Van Dongen, W., 239
van Driel, R., 258–259
Van Eeckhaut, A., 243
van Eldik, G., 84
van Engelen, F.A., 317–320
Van Etten, J.L., 107–109
van Ham, R.C., 268
Van Herpen, T.W., 306–309, 324–325
van Leeuwen, H., 259
Van Moerkercke, A., 37
Van Montagu, M., 310–313
van Munster, E.B., 313
Van Neste, C.M., 38–39
Van Nieuwerburgh, F.C., 29–30, 38–39
Van Schie, C.C., 310–313, 323–324, 338, 344–345
van Veen, H.W., 210*t*
Van Verk, M.C., 56–57
van Workum, W., 256
VanBuren, R., 76–77
Vanderbeld, B., 263
Vandewoestyne, M.L., 38–39
VanDoorn, A., 309–310, 324, 337–338
Vanes, L., 277–278
Vanharanta, S., 272
VanWye, J., 21–22
Varala, K., 272–273
Varbanova, M., 337–338
Vassetzky, Y., 258–259
Vasylechko, S., 361–364
Vater, A., 270
Vaughn, I.W., 29–30
Vaz, B.d.S., 102
Vederas, J.C., 124

Veitia, R.A., 254
Velculescu, V.E., 22, 259
Venter, J.C., 211
Vera, P., 272
Veres, D., 208–221
Vermeirssen, V., 277–279
Verpoorte, R., 198
Verstappen, F.W.A., 308–309, 324–325, 337–338
Verwoerd, T.C., 320
Vickers, C.E., 337–338
Vickers, E., 262–263
Vidal, A., 338, 344–345, 347–348
Vidal, E.A., 263–266
Vidal, M., 255–258, 274–279, 275*f*, 281
Videla, A.L., 169–170
Villalta, J.E., 361–364
Villar-Briones, A., 78–80
Vincenzetti, V., 209–211, 210*t*
Vingron, M., 56–57, 172
Viola, I.L., 281
Viron, N., 342
Vockley, C.M., 361–364
Vogelstein, B., 22, 259
Voinnet, O., 209–211, 214–215
Volpin, H., 318, 324–325, 339, 345
von Guggenberg, R., 84–85
von Korff, M., 259–260
Vosloh, D., 234
Voytas, D.F., 364–366
Vranová, E., 308–309, 326
Vyas, S., 266

W

Waechter, C.J., 128, 142–143
Waern, K., 23
Wagler, T.N., 258–259
Wagner, C., 243
Wagner, D., 266–267
Wagner, G.J., 2–3, 306, 311*t*
Wagner, G.P., 34–35
Wagoner, W., 340–341
Wagschal, K.C., 149–150
Wahler, D., 124
Wajant, H., 87
Wajnberg, E., 84
Waki, K., 255
Walbot, V., 272–273
Walhout, A.J., 274–279, 275*f*, 281
Wall, M.E., 23–24
Walsh, C.T., 105–107
Walter, J.A., 100
Wan, J., 263
Wandrey, C., 236–237
Wang, C.K., 21
Wang, D.H., 282*t*
Wang, D.-Z., 113
Wang, E., 311*t*
Wang, G., 83–84, 344–347
Wang, H.M., 21, 311*t*
Wang, J.W., 37, 270, 309, 337–338, 364–365
Wang, L., 29, 340–341, 364–365
Wang, M., 113–114
Wang, M.Y., 336–337
Wang, Q., 256, 268, 340–341
Wang, R.L., 259
Wang, T., 259–260, 270
Wang, X., 334–336
Wang, X.S., 271–274
Wang, Y., 56–57, 88–89, 227–229, 256, 263–266, 264*t*, 268, 311*t*
Wang, Z., 23, 56–57, 172, 284*t*, 286, 311*t*
Wani, M.C., 23–24
Ward, J.L., 227
Warren, C.L., 271–272
Warren, G., 266
Wass, M.N., 13
Wasserman, W.W., 273–274
Wasson, T., 273–274
Waterhouse, R.M., 73–74
Waterman, M.R., 83–84
Watrous, J.D., 113–114
Watt, R.M., 109
Watts, K.M., 244
Weber, E., 366–369, 371–375
Weber, T., 106, 110–111
Webster, S.P., 109
Weder, B., 209–211, 210*t*
Weenink, T., 361–364
Wees, S.C.M. Van, 56–57
Wegel, E., 76, 112–113
Wehner, N., 281
Wei, G., 270–271
Wei, W.S., 364–365
Weibel, E.K., 71–72

Weidner, M., 323–325
Weigel, D., 254, 272–273
Weinhold, A., 2–3, 306–307
Weirauch, M.T., 258–259, 273–274
Weiss, D., 85
Weissbach, H., 85
Weissman, J.S., 361–364
Weiste, C., 281
Welch, L., 284*t*, 286
Wendel, H.P., 270
Wendel, J.F., 260
Weng, J.K., 70–91
Werck-Reichhart, D., 57–58, 83–84, 309–310
Werker, E., 310–313
Werner, S., 366–369, 371–375
Wery, M., 29
West, M.A., 259
Westbrook, T.C., 2–4, 7
Westfall, C.S., 87
Westfall, P.J., 48–49, 89, 148, 169–170
Westover, A., 254–255
Whalon, M.E., 338
Whisenant, T., 29–30
White, J., 21
White, K.P., 35–36
White, R., 270
White, S.E., 254
Whitehead, E.H., 361–364, 373–375
Whittaker, B., 342, 348
Wicker-Thomas, C., 84
Wickett, N.J., 23, 81
Widhalm, J.R., 48–49
Widstrom, N.W., 254
Wieczorek, J., 339
Wieduwild, R., 366–369
Wiegert, K., 82, 346
Wieland, W.H., 342–343, 351–352
Wilcox, H.M., 109, 109*f*
Wildung, M.R., 48–49, 309–310
Wilkinson, J., 339–341, 346–347
Williams, B.A., 20, 175–176
Williams, C.E., 269–270
Williams, C.K., 168–169
Williams, D.C., 149–150, 154, 159
Williams, P.G., 149–150
Williams, R., 309–310
Williamson, K.E., 364–365
Willis, M.C., 270
Wilmanns, M., 365–366
Wilson, M.D., 277–278
Wing Hui, J.M., 281, 282*t*
Wingender, E., 273–274
Wingerd, B.A., 338
Wingsle, G., 266
Wink, M., 168–169, 334–336
Winn, M.D., 86–87
Winter, C.M., 254–257
Wiseman, B.R., 254
Wishart, D.S., 78–80
Wititsuwannakul, D., 125
Wititsuwannakul, R., 125
Witters, E., 239
Wittler, B., 275*f*, 278–279, 282*t*, 286
Wohlleben, W., 110
Wolberg, A., 270
Wold, B., 267–268
Wold, B.J., 20
Wolf, C.R., 83–84
Wolfe, S.A., 373–375
Won, S.Y., 266
Wong, G.K., 23
Woodcroft, B.J., 75
Woods, D.R., 208
Woodside, A.B., 152
Woolsey, R., 124
Worden, A.Z., 109, 109*f*
Workman, C.T., 272
Wortman, J.R., 20
Wray, G.A., 254
Wright, J.L., 100
Wright, L.P., 226–244
Wright, M., 21
Wroblewski, T., 310–313
Wu, C., 74–75
Wu, J., 88–89
Wu, J.P., 364–365
Wu, S., 34, 308–310, 351–352
Wu, X., 264*t*, 365–366
Wu, Y.N., 267, 272–273, 308–309
Wungsintaweekul, J., 244
Wurbs, D., 124

X

Xia, D.F., 364–365
Xia, J., 78–80

Xia, L.Q., 264*t*
Xia, X., 209–211
Xia, Z.H., 26
Xiang, Y., 264*t*
Xiao, B., 264*t*
Xiao, H., 38–39
Xiao, S., 259–260
Xiao, W.H., 88–89, 227–229
Xiao, Y.M., 227, 232–234, 244
Xie, D., 259–260
Xie, Y., 32–34
Xie, Z., 266–267, 269–270, 278
Xiong, L., 264*t*
Xiu, H.J., 264*t*
Xu, B., 256
Xu, D., 208–221, 284*t*, 286
Xu, G., 209–211
Xu, J., 306–326
Xu, M., 148, 284*t*
Xu, N.Z., 75–76
Xu, R., 307, 334–336, 338, 346–347
Xu, S.-H., 105–106
Xu, W.-J., 105–106
Xu, X., 270
Xu, Y., 102–103
Xu, Z.F., 281
Xu, Z.S., 264*t*
Xue, G.-P.P., 269–270

Y

Yabuuchi, H., 209–211, 210*t*, 219–220
Yadav, G., 103–107
Yaeger, K., 83
Yamada, T., 254–256
Yamaguchi, S., 267, 334–336
Yamaguchi, T., 334–336
Yamaguchi-Shinozaki, K., 254, 263–266
Yamaji, N., 209–211, 210*t*
Yamamitsu, T., 345
Yamamoto, H., 37
Yamanaka, H., 88–89
Yamashita, T., 256
Yamazaki, M., 20–39
Yan, C.Y., 364–365
Yan, H., 259, 308–309
Yan, N.E., 364–365
Yan, Z., 23, 81
Yanagida, M., 78
Yandell, M., 76
Yang, A., 264*t*
Yang, D., 269–270
Yang, F., 252–287
Yang, H., 209–211, 210*t*
Yang, I., 113
Yang, J., 125, 134–135, 142–143
Yang, J.J., 364–365
Yang, J.Y., 113–114
Yang, K., 311*t*
Yang, Q., 237
Yang, S.H., 262–263
Yang, W.M., 258–259, 262–263
Yang, X., 259–260, 268
Yang, Y., 269–270, 281, 282*t*, 309, 311*t*, 337–338
Yankilevich, P., 273–274
Yano, M., 209–211, 210*t*
Yanofsky, M.F., 272–273
Yant, L., 268, 273–274
Yao, G., 361–364
Yao, H., 260
Yao, J., 264*t*
Yao, K.M., 281
Yassour, M., 32–34, 56–57, 73–75, 172–173
Yasumoto, T., 100, 103–105
Yates, C.M., 13
Yauk, Y.K., 336–337
Yazaki, K., 209–211, 210*t*
Ye, H., 264*t*
Ye, J., 37
Ye, K.Q., 70–71
Ye, W., 270
Ye, Z.H., 263–266
Yeo, Y., 172
Yilmaz, A., 262, 268–269, 284*t*, 286
Yin, K.Q., 281, 282*t*
Yin, W.C., 21
Yonekura-Sakakibara, K., 36
Yong, S.Y., 37
Yoshida, S., 334–336
Yoshimoto, N., 32–34, 37
Yoshizumi, T., 282*t*
Yost, W., 111–112, 112*f*
You, J., 264*t*
Youles, M., 366–369, 371–373
Youn, B., 148–149
Young, E., 208

Young, R.A., 260–261, 266
Young, S., 84
Yu, D., 346
Yu, F., 187–189
Yu, G., 309–310, 311*t*, 320–321, 334–336, 346
Yu, H.J., 32–34
Yu, X.H., 233, 278
Yu, Y., 266
Yuan, B., 340–341
Yuan, C.F., 75–76
Yuan, J., 32–34
Yuan, P.F., 364–365
Yuan, W., 259
Yuan, Y., 268
Yuen, M.M.S., 49–51, 57–61, 64, 148
Yuh, C.-H., 252
Yvert, G., 259

Z

Zabel, S., 244
Zalatan, J.G., 373–375
Zamir, D., 339
Zaretskaya, I., 106–107
Zdobnov, E.M., 73–74
Zeeck, A., 114–115
Zehnder, M., 271–272
Zellmer, L., 75–76
Zenewicz, L.A., 168–169
Zeng, L., 70–71
Zeng, Y., 262–263
Zenk, M.H., 226–227, 334–336
Zerbe, P., 48–50, 57–58, 148, 318, 324–325, 339, 345
Zerbino, D.R., 32–34, 56–57, 172
Zettler, J., 106
Zhabotynsky, V., 259
Zhai, G., 122–123
Zhan, X., 59–60
Zhang, B., 244
Zhang, C.Q., 26–27
Zhang, D.Z., 168–169
Zhang, F., 311*t*, 364–365
Zhang, H., 125, 134–135, 142–143, 284*t*, 309, 337–338
Zhang, J., 264*t*
Zhang, L., 22, 278–279, 342, 348, 364–365
Zhang, M., 125, 134–135, 142–143, 340–341, 346
Zhang, N., 256
Zhang, Q., 264*t*
Zhang, S.-F., 113, 309, 337–338, 361–364, 373–375
Zhang, T., 258–259
Zhang, W.H., 89, 258–259, 264*t*, 272
Zhang, X., 267
Zhang, Y.-H., 37, 59–60, 113, 263–266, 264*t*, 268, 364–365
Zhang, Y.W., 123–125
Zhang, Z., 37, 269–270
Zhao, J., 168–169
Zhao, Q.-Y., 56–57
Zhao, Y., 87–88, 148–149, 273–274, 284*t*, 338
Zheng, B., 266
Zheng, H., 37, 256
Zheng, M., 267
Zheng, W., 70–71, 284*t*
Zheng, Z.L., 81–82, 361–364
Zhong, R., 263–266
Zhong, Y.-F.F., 284*t*
Zhou, A., 254
Zhou, J., 35–36, 88–89, 264*t*
Zhou, K., 232–234, 244
Zhou, M.X., 237
Zhou, R.H., 364–365
Zhou, T., 88–89
Zhou, X.-L., 105–106, 264*t*
Zhou, Y.X., 364–365
Zhu, B., 258–259
Zhu, D., 70–71
Zhu, F., 168–169
Zhu, J.-K., 254, 364–365
Zhu, M., 264*t*
Zhu, Q.-H.H., 284*t*
Zhu, S., 21
Zhu, T., 334–336
Zhu, X., 264*t*
Zhu, Z., 34
Zhuang, Y., 259–260, 361–364
Ziegler, J., 102–103
Ziemer, G., 270
Ziemert, N., 107–110, 108*f*, 113–114
Zimba, P.V., 100

Zimmerman, E., 30–31
Zinselmeier, C., 254, 259–260
Zjawlony, J.K., 78–80
Zou, J.Y., 364–365
Zou, R., 232–234, 244
Zubieta, C., 87
Zuccolo, A., 254
Zuiderweg, F.J., 237–238
Zulak, K.G., 48–51
Zumbo, P., 35–36
Zuo, J., 272–273
Zykovich, A., 270, 365–366

SUBJECT INDEX

Note: Page numbers followed by "*f*" indicate figures, and "*t*" indicate tables.

A

Acyl-CoA substrate specificity, 13–15
Acylsugars, 2–3, 3*f*
  ASAT3 isoform, 10–12, 11*f*
  purifying from trichomes, 7–8
  in *S. habrochaites*, 10
  structure profiling by LC/MS, 5
Agrobacterium-mediated transient-transformation assay (ATTA), 310–313
Agrobacterium strains, 371–373, 372*f*
Alexandrium catenella, 113
Allele-specific expression (ASE), 259–260
Arabidopsis
  A. thaliana, 20–21, 308–309
  exon engineering in, 211–215, 212*f*
  experiment on, 26–27
  transcription factors, 21–22
Arabidopsis abscisic acid (ABA)
  AtABCG25, 219–220
  diffusion, 217–219
  import and export assay, 218*f*
  transport assay, 218*f*
Artemisia annua, 168–169
ASAT3 isoform, 10–12, 11*f*
  Acyl-CoA substrate specificity, 13–15
  in vitro functional variation, 13, 14*f*
ASATs, 4
ATP-binding cassette (ABC) transporters, 208–209
  ABA diffusion, 217–219, 218*f*
  AtABCG25, 219–220
  cDNA preparation, 211–215, 212*f*
  exon engineering, 209–215, 212*f*
  heterologous expression of, 209–211
  nonplant expression system, 210*t*
  subfamilies, 209
  in *Xenopus* oocytes, 215–217

B

BAHD acyltransferase, 13, 81–82
Basic Local Alignment Search Tool (BLAST), 36–37, 173
Bemisia tabaci, 324–325, 325*f*
Bigelowiella natans, 110
Biolistic-mediated VIGS, 187–188
  C. roseus, 190–191
  plant material and growth condition pretransformation, 188–189
  posttransformation treatments and analysis, 191
  silencing constructs, 189–190
Biosynthetic pathways, 23–24, 36
BLAST. *See* Basic Local Alignment Search Tool (BLAST)
Bowtie2, transcriptome alignment, 177–179
Bryopsis, 105–106

C

CAGE. *See* Cap analysis of gene expression (CAGE)
Camptotheca acuminata, 168–169
Candidate enzyme identification
  comparative biochemical analysis, 84–86
  crystallization, 86–87
  evolutionary approaches
    biosynthetic model, 81
    evolutionary history, 82
    exploitation, 80–81
    large-scale MSA, 81–82
    phylogenomics approach, 82
  natural enzymes, 86
  recombinant expression and purification, 83–84
  structural analyses, 87
Candidate genes
  identification, 171–172
    clustering procedures, 184–187
    correlation analysis, 183–184
    differential expression, 181–182
    transcript abundance estimation, 175–180
    transcriptome annotation, 173–175
    transcriptome assembly, 172–173

Candidate genes (*Continued*)
transcriptome postassembly analysis, 180
validation by VIGS, 187–188
C. roseus, biolistic-mediated transformation, 190–191
plant material and growth condition pretransformation, 188–189
posttransformation treatments and analysis, 191
silencing constructs, 189–190
Cap analysis of gene expression (CAGE), 255–256
Capillary electrophoresis SELEX (CE-SELEX), 270
Carboxyl-coated magnetic nanoparticles, 29
Catharanthus roseus, 168–169
candidate gene validation, 190–191
subcellular localization, 192
cell culture and plating, 194
expression, 192–194
imaging and microscopy, 195–197, 196*f*
targeting signals, 197–198
transient cell transformation by biolistic, 194–195
virus-induced gene silencing, 188, 189*f*
CDF97, abundance estimation on, 176–177
cDNA
library construction, 27–30
preparation by exon engineering, 211–215, 212*f*
RNA-Seq, 22–23
vs. TFome libraries, 277–278
Cell-type-specific analysis, 53–54
Chemical Abstracts Service (CAS) Registry, 70–71
Chemically regulated gene expression systems, 272–273
Chromatin immunoprecipitation (ChIP) approach, 266–269
Chrysochromulina
C. tobin, 111–112
PKS, 112*f*
Cinchona tree, 168–169
Cis-polyisoprenoids
biosynthesis of, 123–124
natural rubber, 124
Cis-prenyltransferase (CPT)
biochemical assay
bead beating, 136–137
in vitro CPT assays, 135–136
microsome preparation, 137–138
using yeast microsomes and ^{14}C-IPP, 138–139
yeast culture, 136
crystal structure, 123–124
dephosphorylation
chemical dephosphorylation, 139–140
enzymatic dephosphorylation, 139
enzymatic activity, 125
natural rubber, 124
protein complex, 124–125
results, 140–142
separation by thin layer chromatography, 141*f*
thin layer chromatography, 140
Cis-regulatory element (CRE)
cis-regulatory variation, 259–260
establishing TSSs, 254–256
promoters and enhancers, 257–259
TF with, 252
Clustering procedures
hierarchical clustering, 185–187
HOPACH, 187
partitioning, 184–185
Coccomyxa, 107–109, 112–113
C. subellipsoidea, 107–109
Coptis japonica MDR1 (CjMDR1), 209–211
CPT. *See Cis*-prenyltransferase (CPT)
CRE. *See Cis*-regulatory element (CRE)
CRISPR/Cas system, 361–364
Cryopreservation, 24–26

D

De Bruijn graphs, 32–34
Deep transcriptome, 22–24
cDNA library construction, 27–30
classical Sanger method, 30–31
data interpretation
de novo assembly, 32–34
example, 38
expression analysis of transcripts, 34–36
gene annotation, 37
sequence similarity search, 36–37

trimming/quality validation of raw reads, 31–32
DE analysis, 35–36
example of, 38
high-throughput sequencing, 20–22, 30–31
on medicinal plants, 38–39
plant materials, 24–27
RNA extraction, 24–27
for single-cell type, 26–27
workflow for, 22–23, 28*f*
DELTA-BLAST, 106–107
De novo assembly, RNA-Seq, 32–34
De novo transcriptome, specialized metabolic pathway
data uses, 74–76
genomic sequencing, 76–78
sequencing and assembly, 73–74
1-Deoxy-D-xylulose 5-phosphate (DXP), 227, 228*f*
DESeq, 58, 181
Designer TALEs (dTALEs), 364–365, 365*f*
Differential expression (DE) analysis, RNA-Seq, 35–36
Dimethylallyl diphosphate (DMADP/DMAPP), 49, 122–123, 226, 228*f*, 334–336
DNA
cDNA
library construction, 27–30
preparation by exon engineering, 211–215, 212*f*
RNA-Seq, 22–23
vs. TFome libraries, 277–278
gel loading dye, 151
microarray, 21
Dolichol
CPT reaction, 125–126
in eukaryote, 123–124
structures of, 123–124, 123*f*
Double bond isomer dimethylallyl diphosphate (DMAPP), 307

E

E. coli
cultivation and induction, 158
MEcDP isolation, 232–233
MEP pathway, 230–231, 231–232*f*
EdgeR, 181
Electrophoretic mobility shift assay (EMSA), 279–281
Enoyl-CoA hydratase domain, 111–112
Enzyme assays
buffer, 152
geranyl diphosphate stock solution, 152
Ephedra sinica, 168–169
Error-prone PCR, 361–364
Escherichia coli
cultivation and induction, 158
with labeled glucose, 230–231
specialized metabolic pathway, 88–89
transgenic, 232–233
Euglena gracilis, 102, 110, 114*f*
Euglenophycin, 100
Eukaryotic algae, 100
classes of, 102–103
genome mining for identification, 106–113
natural product discovery, 113–115, 114*f*
nonribosomal peptide synthetases, 105–106
polyketide synthases, 103–105
reaction catalyzed by, 104*f*
toxins produced by, 101*f*
Eutreptiella gymnastica, 109, 109*f*
Exon engineering, 209–211
in *Arabidopsis*, 211–215, 212*f*

F

False discovery rate (FDR), 181
Family-wise error rate, 181
FaNES, 308–309
Farnesyl diphosphate (FPP), 122–123
FASTA format, 34
FASTQ format, 31–32
Fluorescent protein (FP) fusions
C. roseus, 192
cell culture and plating, 194
expression, 192–194
imaging and microscopy, 195–197, 196*f*
targeting signals, 197–198
transient cell transformation by biolistic, 194–195
yeast cells, 198
competent cell preparation, 199–200
expression, 198–199

Fluorescent protein (FP) fusions (*Continued*)
protocols of transformation, 200
targeting signals, 201–202, 201–202*f*
Food and Drug Administration (FDA), 70–71

G

Galaxaura, 105–106
Gambierdiscus, 109–110
GC–FID chromatograms, 163*f*
Gene ontology (GO), 37
Gene regulatory grids (GRGs), 253
Gene regulatory networks (GRNs), 253, 253*f*
cis-regulatory apparatus
cis-regulatory variation, 259–260
establishing TSSs, 254–256
promoters and enhancers, 257–259
gene-centered approaches
EMSA based methods, 279–281
Y1H approaches, 274–279
resources
databases, 283–286, 283*f*, 284*t*
TFomes, 281–282, 282*t*
trans-acting factors
class of, 260–261
cofactors, 266
transcription factors, 261–266
transcription factor centered approaches
ChIP approaches, 266–269
ChIP-chip, 267
ChIPed, 267
ChIP-Seq, 268–269
in silico TF-location prediction approaches, 273–274
in vitro TF-DNA interaction approaches, 269–273
Genome mining
hybrid NRPS/PKSs, 111–112
NRPSs identification, 110
PKSs identification, 106–110
Geranyl diphosphate (GPP), 122–123
Geranyl geranyl diphosphate (GGPP), 122–123
Glandular trichomes
bombardment conditions, 315
focus on usage, 310
FPS
redirecting to plastid alters, 321–324
transcript levels, 320–321, 321*f*
GUS expression assay, 315
modified terpenoid metabolism, 322*f*
particle coating with DNA, 313–315
plant material and growing conditions, 315
on plants, 306
plasmids, 314*f*
pMKS1:ssu-SlFPS and pMKS1: ssu-GgFPS preparation, 318–320, 319*f*
promoter trichome specificity, 311*t*, 315–316, 316*f*
pSlTP5:GUS-sYFP1 construct preparation, 313
specialized metabolites
and application, 306–307
biological relevance of changing, 324–325, 325*f*
production, 307
stable transformation
pSlTPS9:GUS-sYFP1, 317
stable tomato transformation, 317–318
transgene expression, 318
stress resistance, 306
terpene production modification, 308–310
terpenoid biosynthesis in, 308*f*
transient-plant transformation, 310–316
Golden Gate cloning, 366–370
GRNs. *See* Gene regulatory networks (GRNs)

H

Harmful algal blooms, 100
Heterocapsa circularisquama, 113
Heterologous systems
biochemical and structural analyses, 87–88
E. coli, 88–89
N. benthamiana, 90
S. cerevisiae, 89–90
Hierarchical clustering, transcriptome, 185–187
High-throughput sequencing, RNA sequencing, 20–22, 30–31
HMMERscan, 173

HOPACH, transcriptome, 187
HpcRunningGrid collection, 173
Hydrophilic interaction liquid chromatography (HILIC), 240–241
Hydroxymethylglutaryl-CoA synthase, 111–112

I

In silico TF-location prediction approaches, 273–274
Ion exchange chromatography, 236–237
Isopentenyl diphosphate (IDP/IPP), 49, 122–123, 226, 334–336
Isoprenoids, 100, 122–123, 226–227
Isotope-labeled standards, 229–233, 241, 243
  E. coli cultures, 230–231, 231–232*f*
  MEcDP isolation, 232–233

K

Karenia brevis, 109–110, 111*f*

L

LC-MS/MS analysis, MEP pathway, 236–237
  HILIC, separation by, 237–240
  quantification, 242–243
  tandem mass spectrometry, detection by, 240–241
4S-Limonene synthase, 149–150
Lupinus angustifolius, 35–36

M

Medicinal plants
  RNA-Seq application on, 38–39
  sampling tissues of, 24–26
MEP pathway. *See* Methylerythritol phosphate (MEP) pathway
2-C-Methyl-D-erythitol 4-phosphate (MEP), 227
2-C-Methyl-D-erythritol-2,4-cyclodiphosphate (MEcDP), 227
  isolation, 232–233
Methylerythritol phosphate (MEP) pathway, 226–227, 228*f*, 307
  in animals, 227–229
  detection by tandem mass spectrometry, 240–241, 241*t*
  extraction
    A. thaliana rosette, 238*f*
    bacterial cultures, 235–236
    plant materials, 234–235
  HILIC stationary phase, 237–240, 240*t*
  ion exchange chromatography, 236–237
  limits of detection, 242
  quantification, 242–243
  regulation, 229
  stable isotope-labeled internal standards, 229–233
    E. coli cultures, 230–231, 231–232*f*
    MEcDP isolation, 232–233
Methylmalonyl-CoA, 103–105
Mevalonate (MVA) pathway, 226–227, 307
Mevalonic acid (MVA) pathway, 334–336
Michaelis–Menten saturation curve, 162–164
Microsome
  assays, 63
  expression cultures, 62
  isolation, 62
Molecular biology
  DNA gel loading dye, 151
  IPTG solution, 151
  TE stock solution, 150–151
Monoterpene synthases
  enzyme assays, 152
  equipment, 150
  expression constructs, generation of, 153
    overlap extension mutagenesis, 155–157
    quick change PCR, 154–155
  functional evaluation of recombinant
    analysis of kinetic data, 162–164
    enzymatic reaction, 161–162
    quantitation, 162
  molecular biology
    DNA gel loading dye, 151
    IPTG solution, 151
    TE stock solution, 150–151
  protein gel electrophoresis, 151–152
  recombinant enzyme purification, 151
  target enzyme
    E. coli cultivation and induction, 158
    isolation and assessment, 158–161
3-Morpholino-2-hydroxypropanesulfonic acid (MOPSO) buffer, 151

mRNAs
carboxyl-coated magnetic nanoparticles, 29
poly-A containing, 29
Multi-beads Shocker, 24–26
Multiplexed collision-induced dissociation (CID), 4–7
Mutagenesis, overlap extension
amplicons, 156–157
point mutations, 156

N

NaPDoS webserver, 107–110, 108*f*
Natural products
bacterial, 71–72
synthases
classes of, 102–103
discovery, 113–115, 114*f*
genome mining for identification, 106–113
nonribosomal peptide synthetases, 105–106
polyketide synthases, 103–105
reaction catalyzed by, 104*f*
therapeutic, 70–71
Natural rubber (NR), 124. *See also* *Cis*-prenyltransferase (CPT)
NCBI Conserved Domain Database, 106–107
Next-generation sequencing (NGS), 30–34
Nicotiana
N. benthamiana, 90
N. sylvestris, 309–310
N. tabacum, 308–309
Nonribosomal peptides (NRPs), 102–103
synthetases, 105–106
hybrid, 111–112
identification, 110
Nuclear magnetic resonance (NMR) analysis, 64
NucleoSpin® Plasmid EasyPure, 366–369

O

Oligo-capping, 254–255

P

Pacific yew tree *(Taxus brevifolia)*, 23–24
Paired-end analysis of transcription (PEAT), 256
Partitioning methods, transcriptome, 184–185
Pearson correlation coefficient (PCC), 183
Peptidyl carrier protein (PCP), 105–106
Phosphate buffer, 151
Polyacrylamide gel electrophoresis (PAGE), 279–280
Polygalacturonase (PG) promoter, 342
Polyketide synthase (PKS), 102–105
hybrid, 111–112
identification, 106–110
phylogeny of, 108*f*
Prenyltransferases, 344–346
Protein binding microarrays (PBMs), 271–272
Protein gel electrophoresis
destaining solution, 152
loading dye, 151
running buffer, 152
staining dye, 152
Protein subcellular localization
C. roseus, FP fusion, 192
cell culture and plating, 194
expression, 192–194
imaging and microscopy, 195–197, 196*f*
targeting signals, 197–198
transient cell transformation by biolistic, 194–195
specialized metabolites, 191–192
yeast cells, FP fusion, 198
competent cell preparation, 199–200
expression, 198–199
protocols of transformation, 200
targeting signals, 201–202, 201–202*f*
Prymnesium pavum, 113

Q

Quantitative real-time PCR (qRT-PCR), 21–22
Quick change PCR
mutated cDNA sequence amplification, 154
transformation, 154–155

R

Rauwolfia serpentina, 168–169
Reads per kilobase of transcript per million mapped reads (RPKM), 34–35
Regulatory circuits, 361–364

rer2Δ strain, 128–129
construct generation, 131–132
results, 134–135
selection in 5-FOA, 133–134
srt1 deletion
homologous DNA preparation, 129–131
primers, 130*t*
yeast transformation
with linear DNA, 129–131
with plasmids, 132–133
rer2 mutant, revertants observation, 125–128, 127*f*
RNA
purification, protocol in, 24–26
types, 27
RNA annotation and mapping of promoters for analysis of gene expression (RAMPAGE), 256
RNA sequencing (RNA-Seq), 22–24
cDNA library construction, 27–30
classical Sanger method, 30–31
data interpretation
de novo assembly, 32–34
example, 38
expression analysis of transcripts, 34–36
gene annotation, 37
sequence similarity search, 36–37
trimming/quality validation of raw reads, 31–32
DE analysis, 35–36
example of, 38
high-throughput sequencing, 20–22, 30–31
on medicinal plants, 38–39
plant materials, 24–27
RNA extraction, 24–27
for single-cell type, 26–27
workflow for, 22–23, 28*f*
RSEM, transcript abundance estimation, 179–180
Rubisco conjugating enzyme 1 (RCE1), 320

S

Saccharomyces cerevisiae, specialized metabolic pathway, 89–90
SAGE. *See* Serial analysis of gene expression (SAGE)
Saxitoxin gene cluster, 102
Secondary metabolites, 148
Secretingglandular trichomes, 2–3
SELEX-Seq, 270–271
Serial analysis of gene expression (SAGE), 22
Serial ChIP (sChIP), 266–267
Short oligonucleotide analysis package 2 (SOAPdenovo2), 32–34
Single molecule, real-time (SMRT) sequencing, 31
Solanum lycopersicum ASAT3 (Sl-ASAT3), 10
Specialized metabolism, 70
de novo transcriptome
data usesage, 74–76
genomic sequencing, 76–78
sequencing and assembly, 73–74
enzyme function divergence, 12–15
furanose ring acylation, 10–12
in heterologous systems
biochemical and structural analyses, 87–88
E. coli, 88–89
N. benthamiana, 90
S. cerevisiae, 89–90
LC/MS
acylsugar structure profiling, 5
plant trichome acylsugar extraction and, 5–7, 6*f*
metabolomics approaches, 78–80
NMR spectroscopy, 7–8
omics, 72–73
phylogeny-based screening, 8–9, 9*f*
physicochemical features, 70–71
staggering diversity of, 2
structure–function analysis
comparative biochemical analysis, 84–86
enzyme evolution, 86–87
evolutionary approaches, 80–82
recombinant expression and purification, 83–84
srt1Δ strain, 128–129
construct generation, 131–132
deletion
homologous DNA preparation, 129–131
primers, 130*t*
results, 134–135

srt1Δ strain (*Continued*)
selection in 5-FOA, 133–134
yeast transformation
with linear DNA, 129–131
with plasmids, 132–133
STAP. *See* Synthetic TALE-activated promoter (STAP)
Streptomyces coelicolor, 114–115
Subcellular localization
C. roseus, FP fusion, 192
cell culture and plating, 194
expression, 192–194
imaging and microscopy, 195–197, 196*f*
targeting signals, 197–198
transient cell transformation by biolistic, 194–195
specialized metabolites, 191–192
yeast cells, FP fusion, 198
competent cell preparation, 199–200
expression, 198–199
protocols of transformation, 200
targeting signals, 201–202, 201–202*f*
Synthetic biology, 169–171, 171*f*, 208
Synthetic promoter libraries (SPL), 361–364
Synthetic TALE-activated promoter (STAP)
applications, 374*f*
dTALE, 365*f*
libraries, cloning, 368*f*
oligonucleotide, 366–369
promoters, 369
reporters, cloning, 370*f*
Systematic evolution of ligands by exponential enrichment (SELEX), 269–271

T

Tabernaemontana ssp., 168–169
TALE DNA-binding domain (TBD), 365–366
TALEs. *See* Transcription activator-like effectors (TALEs)
Tandem mass spectrometry, MEP pathway, 240–241
Target enzyme
E. coli cultivation and induction, 158
isolation and assessment
batch purification, 159
cell disruption, 158–159
recombinant enzyme purity assessment, 160–161
total protein quantitation, 159–160
Taxus brevifolia, 168–169
Terpene fragrances
biosynthesis, 48–49
candidate genes
functional characterization of, 59–63
selection, 58–59
chemical diversity, 49
chemical identity
MS–MS, 64
NMR analysis, 64
differential expression, 58–59
high-quality RNA, isolation of, 55–56
industrial applications, 48–49
metabolite profiles, 53–55
P450s and development, 50–51
replication and statistical design, 51–52
sandalwood, 50–51
tissue sampling
logistics, 53
sandalwood tissues, 53
temporal and spatial variables, 52–53
transcriptome
mining and annotation, 57–58
sequence and de novo assembly, 56–57
Terpenes
biosynthetic genes, overexpression of
MEP and MVA pathway genes, 344
prenyltransferases, 344–346
pyramiding multiple terpene biosynthetic genes, 348–349
synthases, 346–348
breeding approach, 337–338
emitted analysis, 350
internal pools analysis, 350
metabolic engineering, 337–338
production, 337–338
synthases, 346–348
in tomato fruits
emitted analysis, 350
internal pools analysis, 350
metabolic engineering, 349
overexpression, of terpene biosynthetic genes, 343–349

terpenoid formation, 338–340
transgene expression, in ripening, 340–343
Terpenoids, 148, 306–307, 334
TFome
cDNA *vs.*, 277–278
collections, 281–282, 282*t*
Thioesterase (TE) domain, 103–105
Tobacco acid pyrophosphatase (TAP), 256
Tobacco vein mottling virus (TVMV), 351–352
Tomato fruits
overexpression, terpene biosynthetic gene
biosynthesis, 343–344
MEP and MVA pathway genes, 344
prenyltransferases, 344–346
pyramiding multiple terpene biosynthetic genes, 348–349
ripening process, 339
terpenes
emitted analysis, 350
internal pools analysis, 350
metabolic engineering, 349
overexpression, of terpene biosynthetic genes, 343–349
synthases, 346–348
terpenoid formation, 338–340
transgene expression, in ripening, 340–343
terpenoid formation in, 338–340
transgene expression
stable transformation, 340–342
transient transformation via *Agrobacterium* injection, 342–343
trichomes (*see* Glandular trichomes)
Transcription activator-like effectors (TALEs), 361–364
designer TALEs, 364–365, 365*f*
Golden Gate cloning, 366–370
GUS reporter assay, 371–373, 372*f*
oligonucleotides, 366–369, 367*t*
promoter activity, 371–373
Transcription activator-like orthogonal repressors (TALORs), 361–364
Transcriptional regulatory activity
functional assays, 261–262
TFs controlling agronomic traits, 262–263, 264*t*
TFs function as activators/repressors, 262–263
Transcription factors (TFs)
as activators/repressors, 262–263
Arabidopsis, 21–22
controlling agronomic traits, 262–263, 264*t*
with CRE, 252
with DNA, 252–253
gene expression control, 253
Transcription, gene expression control, 361–364
Transcription start sites (TSSs), 252
cis-regulatory apparatus, 254–256
Transcriptome
mining and annotation, 57–58
sequence and de novo assembly, 56–57
sequencing approach, 113
Transcriptome, candidate genes identification, 171–172
annotation, 173–175
assembly, 172–173
bowtie2, 177–179
postassembly analysis, 180
clustering procedures, 184–187
correlation analysis, 183–184
differential expression, 181–182
edgeR, 181
hierarchical clustering, 185–187
HOPACH, 187
partitioning, 184–185
PCC, 183
RSEM, 179–180
transcript abundance estimation, 175–180
Transgene expression, 318
Transient-plant transformation, 310–316
Transport engineering technology, 208, 208*f*
Trans-prenyltransferase (TPT), 122–125

U

Undecaprenol, 123–124, 123*f*
Uracil-*N*-glycosylase (UNG), 29–30

V

Virus-induced gene silencing (VIGS)
  candidate gene validation, 187–188
    C. roseus, 190–191
    plant material and growth condition pretransformation, 188–189
    posttransformation treatments and analysis, 191
    silencing constructs, 189–190
  in *Catharanthus roseus*, 188, 189*f*

W

Web gene ontology annotation plot (WEGO), 37

X

Xenopus oocytes, 209–211
  ABA diffusion, 217–219, 218*f*
  ABC transporter expression in, 215–217, 216*f*
  AtABCG25, 219–220

Y

Yeast bioengineering, FP fusion, 198
  competent cell preparation, 199–200
  expression, 198–199
  protocols of transformation, 200
  targeting signals, 201–202, 201–202*f*
Yeast in vivo assays
  expression cultures, 61
  extraction, of terpenoid metabolites, 61
  yeast transformation, 60
Yeast one-hybrid (Y1H) screening system
  cDNA *vs.* TFome libraries, 277–278
  limitations
    false negatives, 279
    false positives, 279
  principle and work flow, 274–277, 275*f*
  transformation *vs.* mating, 278–279

Z

Z-abienol production, 309–310
Zero-mode waveguide (ZMW), 31
Zinc-mediated RNA fragmentation method, 29

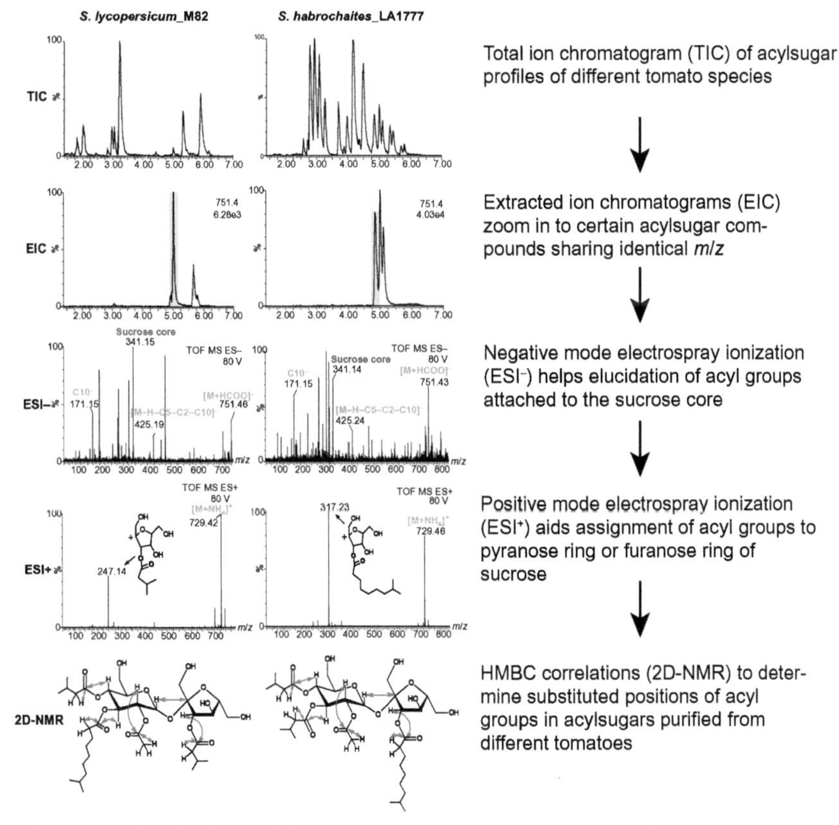

P. Fan *et al.*, Fig. 2 The workflow used to characterize structures of two acylsugar positional isomers extracted from *S. habrochaites* LA1777 and *S. lycopersicum* M82. The *top row* shows the total ion chromatograms of acylsugars extracted from the two species. The *second row* displays the extracted ion chromatograms of the acylsugar isomers with *m/z* of 751.4. The *yellow* highlighted peaks are selected for further ESI^- and ESI^+ analysis. The *third row* shows the mass spectra of the selected *yellow* highlighted peaks obtained using ESI^- with the CID Aperture 1 voltage of 80 V. $[M+HCOO]^-$ is the acylsugar formate adduct. $[M-H-C5-C2-C10]^-$ is the ion fragment that lost a proton and three acyl chains. The *fourth row* shows the mass spectra of the selected peaks obtained using ESI^+ with the CID Aperture 1 voltage at 80 V. The most abundant ion $[M+NH_4]^+$ is the acylsugar ammonium adduct. The ion fragments released from the acylsugars after breaking the glycosidic bond are also shown (*m/z* of 247.14 and 317.46). The *last row* shows the ^1H and ^{13}C correlations determined by 2D-NMR (HMBC) to elucidate the acyl groups substituted positions. *This figure is adapted from Ghosh, B., Westbrook, T.C., & Jones, A.D. (2014). Comparative structural profiling of trichome specialized metabolites in tomato (*Solanum lycopersicum*) and* S. habrochaites*: Acylsugar profiles revealed by UHPLC/MS and NMR.* Metabolomics, 10, *496–507.*

A

	Residue positions																			Enzyme activity S2:10 (5, 5) as acyl acceptor	
	5	6	20	35	41	107	109	114	116	131	163	205	207	210	256	261	296	327	329	iC5-CoA	nC12-CoA
Sl-ASAT3	T	I	S	F	Y	L	H	V	D	T	L	A	R	Q	T	S	A	E	R	Yes	No
LA1777-ASAT3-F	K	M	P	L	C	I	N	A	G	N	V	T	H	K	A	K	T	K	M	Yes	Yes
LA1731-ASAT3-F	K	M	P	L	C	I	N	A	G	N	V	T	H	K	A	K	T	K	M	Yes	Yes

B

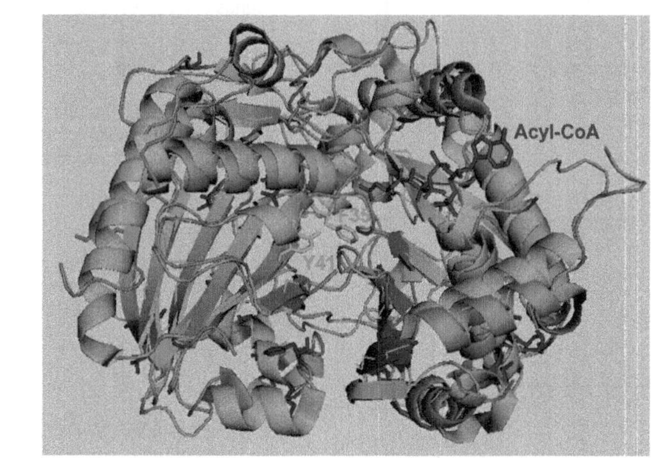

C

Sl-ASAT3_Y41C

S-CoA

Tyr41 → Cys41
TAT → TGT

P. Fan *et al.*, Fig. 5 Comparative sequence, in vitro function, and structure analysis lead to identification of a functionally important polymorphism in ASAT3. (A) The alignment of amino acid sequences to identify two positions (in *green*) that correlate with the enzyme function differences of three ASAT3 isoforms. (B) The homology-model-derived proposed structure of Sl-ASAT3 (*orange*) is superimposed on the crystallographic structure (*gray*) trichothecene 3-*O*-acetyltransferase–acyl-CoA complex (PDB ID: 3B2S) (Garvey, McCormick, & Rayment, 2008), which has the acyl-CoA (*red*) binding pocket resolved. The amino acid residues shown in (A) are labeled with *blue* and *green* in the Sl-ASAT3 modeled structure. The *green*-colored amino acids are those that correlate with activity and map close to the acyl-CoA binding pocket. (C) In vitro mutagenesis verified that a single nucleic acid substitution from A to G leads to the amino acid substitution from Tyr^{41} to Cys^{41} and enables ASAT3 to use the long-chain acyl-CoAs as substrates.

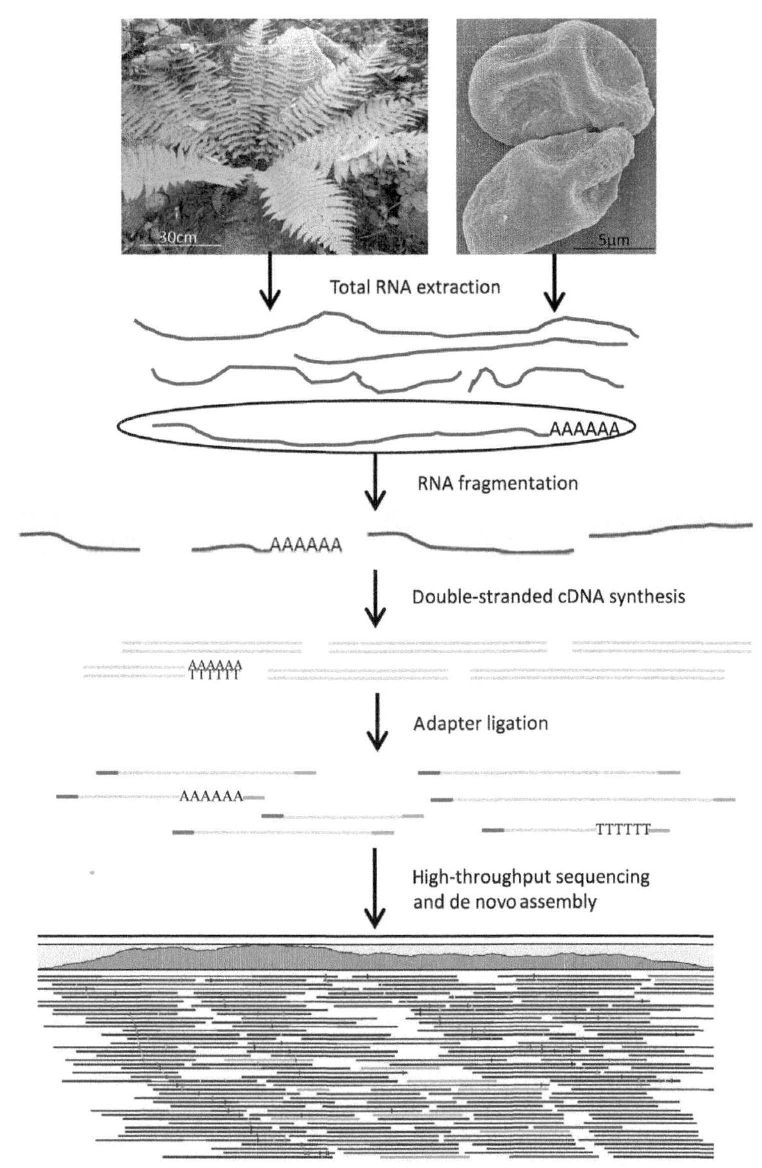

R. Han *et al.*, Fig. 1 A typical workflow for RNA-Seq. From plant tissues or single cells, total RNA is extracted. According to various purposes, specific types of RNA, eg, mRNA, are isolated and fragmented. After the synthesis of double-stranded cDNA, adapters are ligated for single or paired-end high-throughput sequencing. De novo assembly constructs transcripts from the obtained raw reads. *Plant photographs (*Dryopteris crassirhizoma *and its spores under scanning electron microscope) courtesy of Dr. Liang Xu, Liaoning University of TCM.*

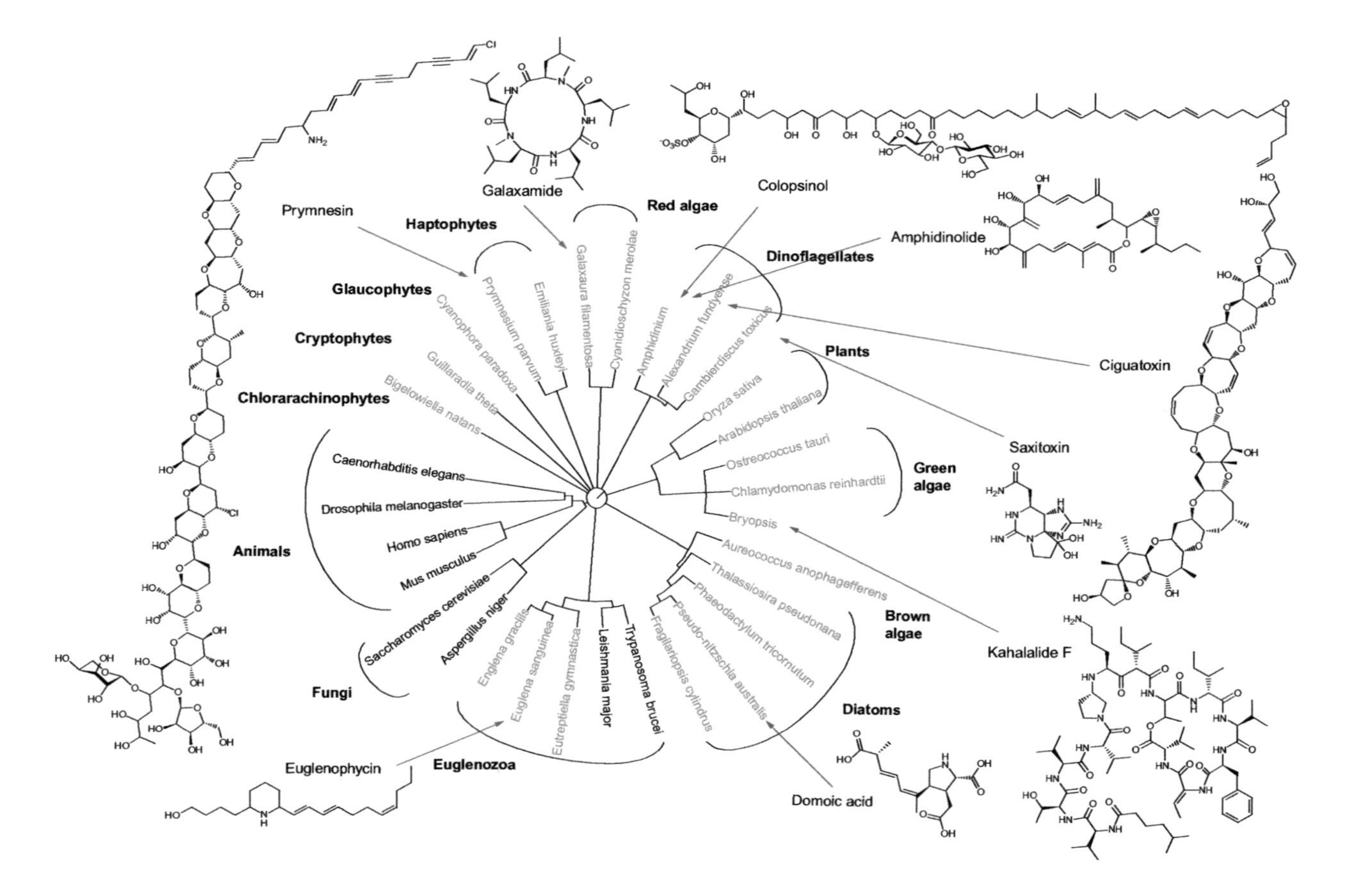

E.C. O'Neill *et al.*, Fig. 1 Toxins produced by eukaryotic algae. Phylogeny showing relationship between algae and model organisms (Letunic & Bork, 2011) and a selection of the toxins produced by alga. Species shown in *green* are photosynthetic.

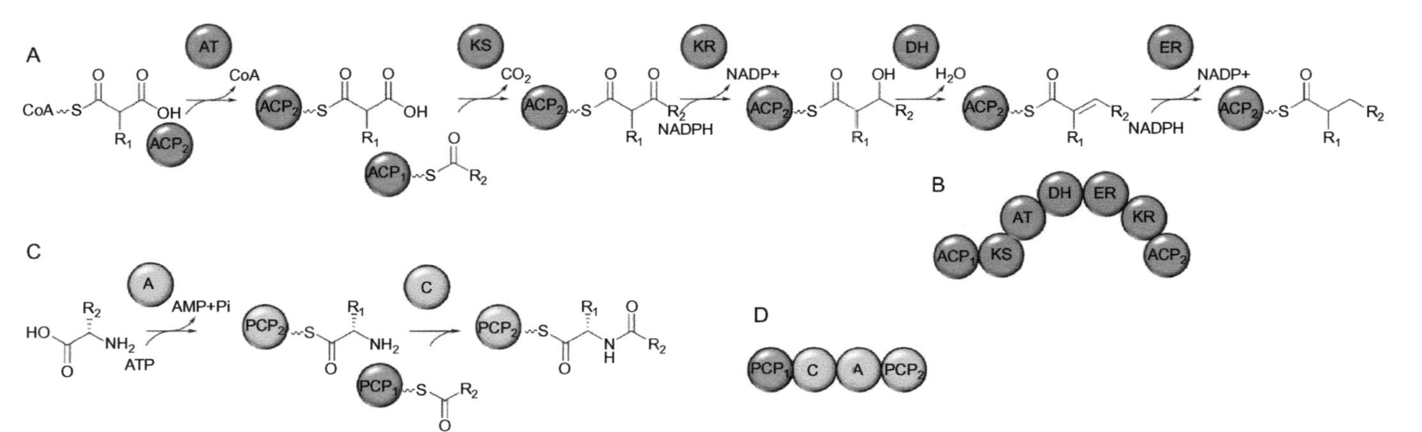

E.C. O'Neill *et al.*, Fig. 2 The reaction catalyzed by natural product synthase domains. (A) Reactions carried out by PKS domains. The acyltransferase (AT) domain selects an acyl CoA and transfers the dicarboxylic acid onto the acyl carrier protein (ACP). R_1 is typically H for malonyl-CoA but a range of other extender units can be used, selected by the AT. The ketosynthase (KS) catalyzes the decarboxylative condensation between the acyl group and the acyl held on the previous ACP. The ketoreductase (KR) reduces the ketone to a hydroxyl, the dehydratase (DH) domain removes water, leaving a double bond, and the enoyl reductase (ER) reduces this double bond (Keatinge-Clay, 2012). (B) The order of domains in a typical PKS—note this is not the order in which the reactions are carried out. Any of these domains may be missing or inactive. (C) Reactions carried out by NRPS domains. The adenylation (A) domain activates an amino acid and attaches it to the peptidyl carrier protein (PCP). The condensation (C) domain catalyzes the condensation between the new amino acid and the acid held on the previous PCP, forming a new amide bond. (D) The order of domains in a typical NRPS.

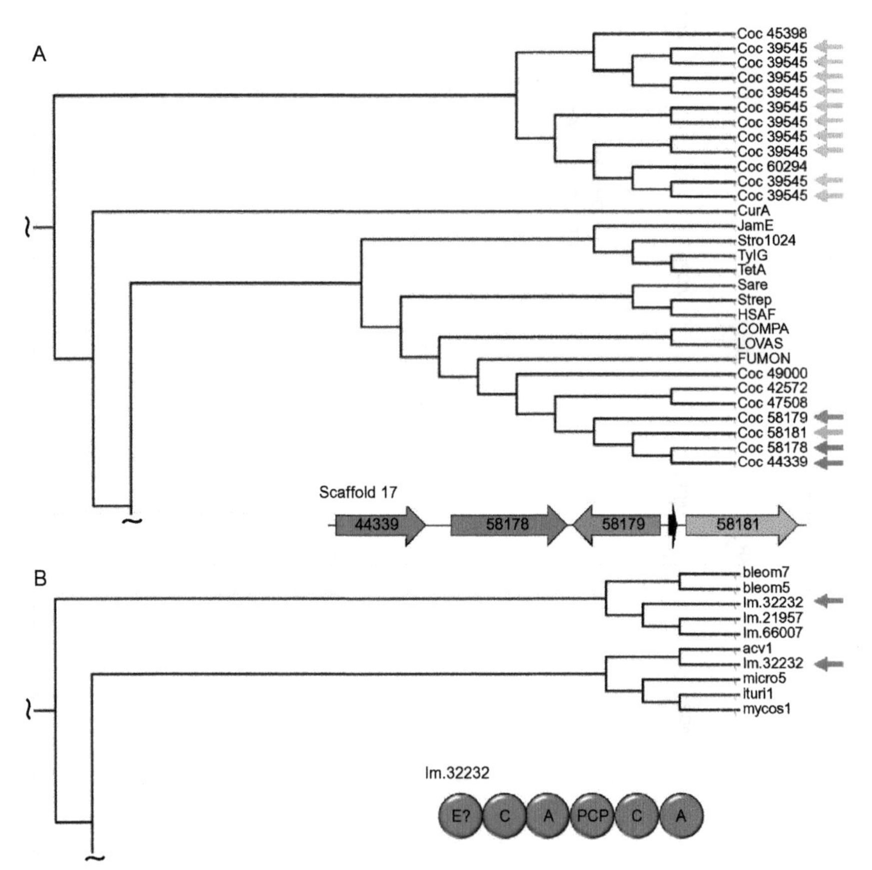

E.C. O'Neill *et al.*, Fig. 3 Phylogeny of selected algal KS and C domains. Sections of phylogenies constructed using NaPDoS (Ziemert et al., 2012). (A) KSs from *Coccomyxa* PKSs cluster in two clades among the standard KSs, with all the KSs from the 10 domain protein (Coc 39545, *gray arrows*) in one and all the KSs from the four PKS gene cluster (shown below as a 120-kb region of scaffold 17) in the other (*colored arrows*). (B) C domains from *E. gracilis* multidomain NRPSs are scattered among the standard C domains, with the two C domains from one protein (lm.32232, shown below, *colored arrows*) separated.

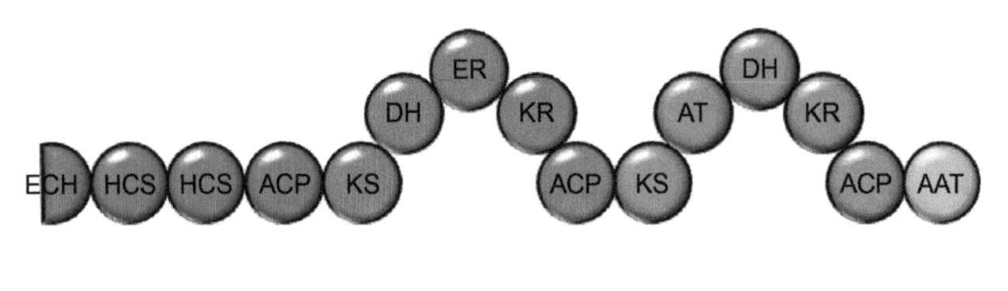

E.C. O'Neill *et al.*, Fig. 4 *Eutreptiella* PKS. The largest PKS in the *Eutreptiella* transcriptome does not appear to be complete (Keeling et al., 2014). At the N-terminus is a partial enoyl-CoA-hydratase (ECH) and two hydroxymethylglutaryl-CoA synthase domains (HCS). It is unclear what the compound on which these domains act (R_1). This is followed by two PKS domains, one fully reducing without an AT domain and one leaving a double bond. This is followed by a decarboxylative amino acid transferase (AAT) which can take an amino acid and add it to a small molecule, decarboxylating it and adding the side chain (R_2), and the terminal nitrogen, to give the possible product shown.

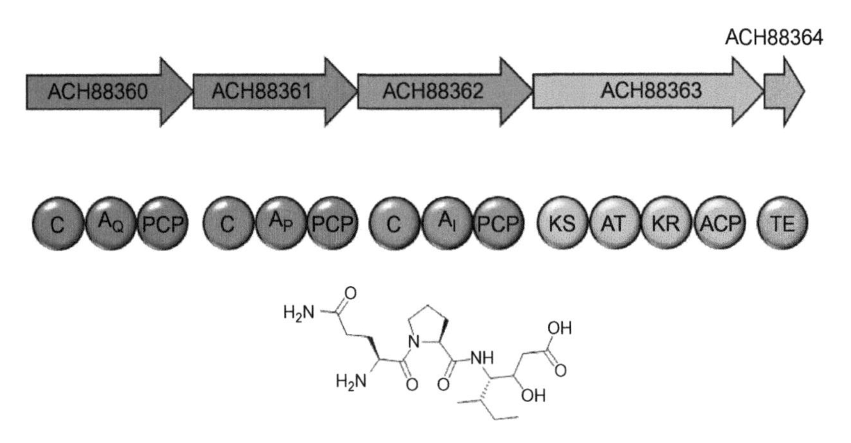

E.C. O'Neill *et al.*, Fig. 5 *Karenia brevis* hybrid NRPS/PKS gene cluster. The 16-kb gene cluster from *K. brevis* contains five genes, which encode three NRPS modules, one PKS module and a thioesterase (López-Legentil et al., 2010). This is tentatively predicted to produce the compound shown.

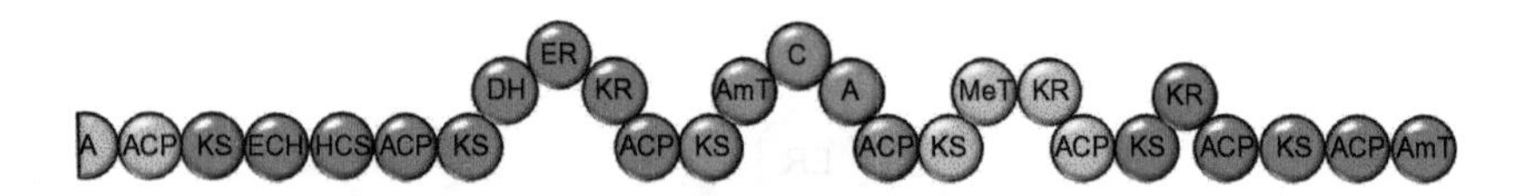

E.C. O'Neill *et al.*, Fig. 6 *Chrysochromulina* PKS. There is a hybrid PKS/NRPS encoded in the *Chrysochromulina* genome (Hovde et al., 2015), although it is apparently interrupted by two stop codons, within domains. Assuming these are sequencing errors and this is in reality one gene, the protein has the domain architecture shown with six ketosynthases, one NRPS domain and is initiated by an A domain activating an unknown substrate (R_1). There is also an enoyl-CoA-hydratase (ECH) and a hydroxymethylglutaryl-CoA synthase domain (HCS), two amino transferases (AmT), which likely replace a ketone with a nitrogen from glutamate, and a C-methyltransferase (MeT). Assuming each domain acts once and the product is released by hydrolysis (by no means certain) the product may resemble the structure shown, with R_2 representing the amino acid side chain which cannot be predicted.

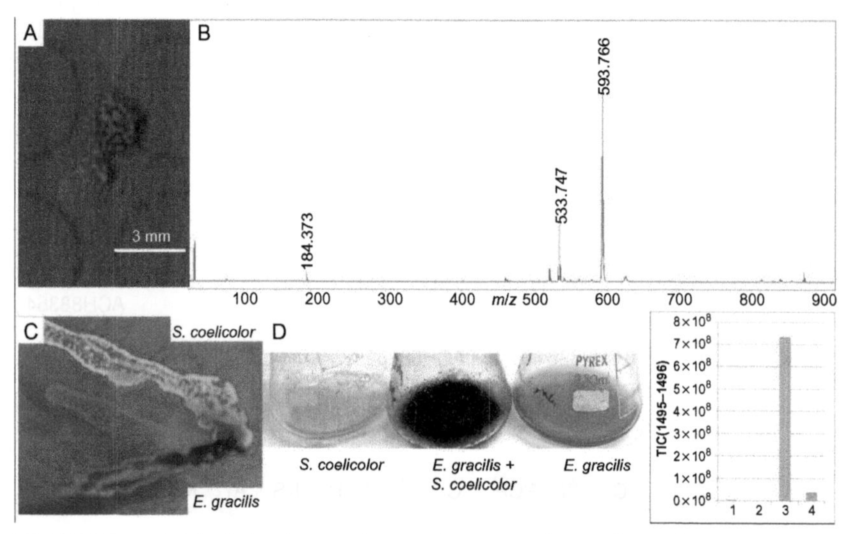

E.C. O'Neill *et al.*, Fig. 7 *Euglena* and natural products biosynthesis. (A) Imaging mass spectrometry of *Euglena gracilis* colonies grown on EG:JM + Glc agar, imaged using MALDI-ToF. The pixels containing ions corresponding to $m/z = 870–871$ Da are highlighted. (B) MS2 of the corresponding ion at $m/z = 870.553$. (C) *E. gracilis* and *Streptomyces coelicolor* streaked on EG:JM + Glc agar. (D) Cultures of *E. gracilis* and *S. coelicolor* grown in EG:JM + Glc media. After 7 days the coculture is darkly pigmented. LCMS analysis of the methanol extract of these cultures shows that the peak corresponding to CDA ($m/z = 1495.52$) is not detectable in either the *Euglena* (1) or *Streptomyces* (2) only extracts but is easily detectable in the coculture (3), corresponding to the extract from *S. coelicolor* grown on its preferred media (LB, 4).

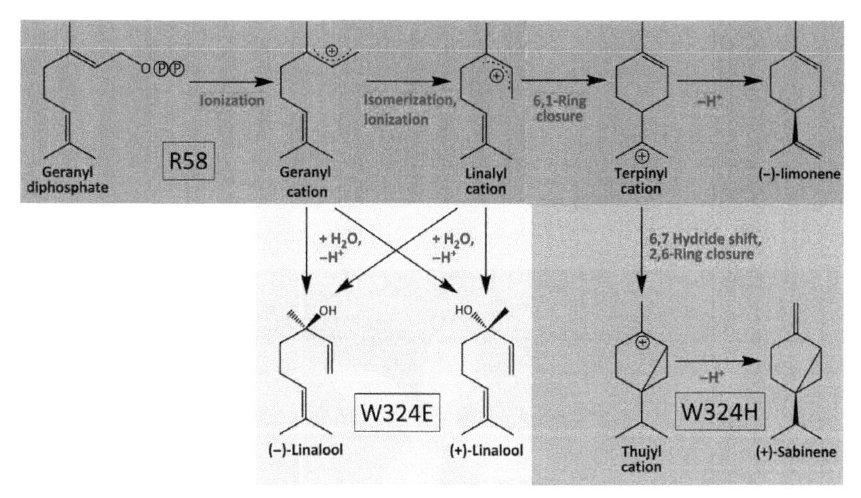

N. Srividya *et al.*, Fig. 1 Formation of novel monoterpenes by mutants of (4S)-limonene synthase. The primary reaction pathway toward (−)-limonene, as catalyzed by the pseudomature form of the wild-type enzyme (designated as R58), is highlighted in *dark gray*, while primary products formed by mutant enzymes W324H and W324E are highlighted with brighter shades of *gray*.

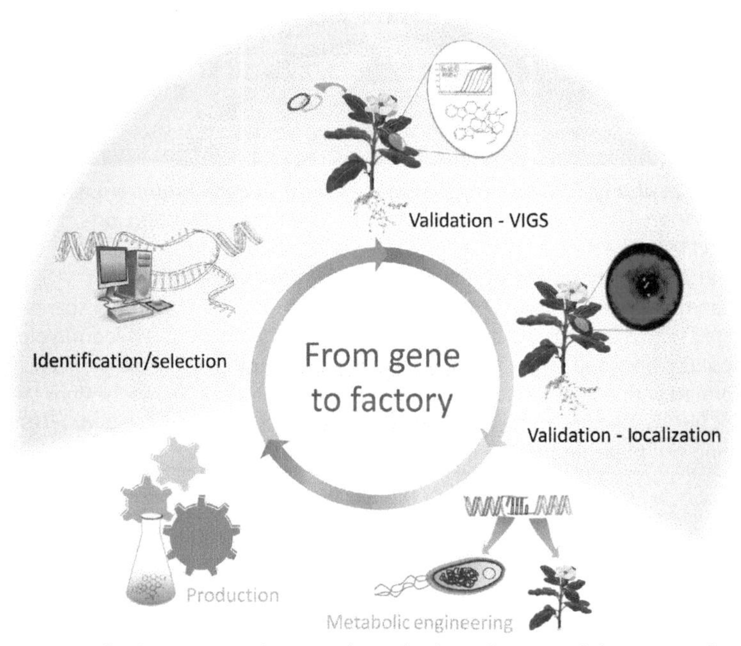

E. Foureau *et al.*, Fig. 1 Prequels to synthetic biology: from candidate gene identification and validation to their characterization and transfer into heterologous organisms.

E. Foureau *et al.*, Fig. 2 Virus-induced gene silencing in *Catharanthus roseus* by biolistic transformation (VIGS). 2 weeks old *C. roseus* plantlets presenting one pair of fully expanded leaves were used to perform the particle bombardment. (A) Time table representing the development of the leaves following transformation of VIGS vectors (B) prebombardment (0 dpb), 7 (7 dpb), and 20 (20 dpb) days after bombardment, as compared with a 20 days old nontransformed leaf (wild type). (C–F) Phenotypic aspect of *C. roseus* plants portraying different conditions including wild-type plants (C), plants transformed with empty vector pTRV2-EV (D), protoporphyrin IX magnesium chelatase pTRV2-ChlH depicting the characteristic yellow pigmentation (E), or pTRV2-PDS exhibiting the bleaching of the leaves (F), at 30 dpb.

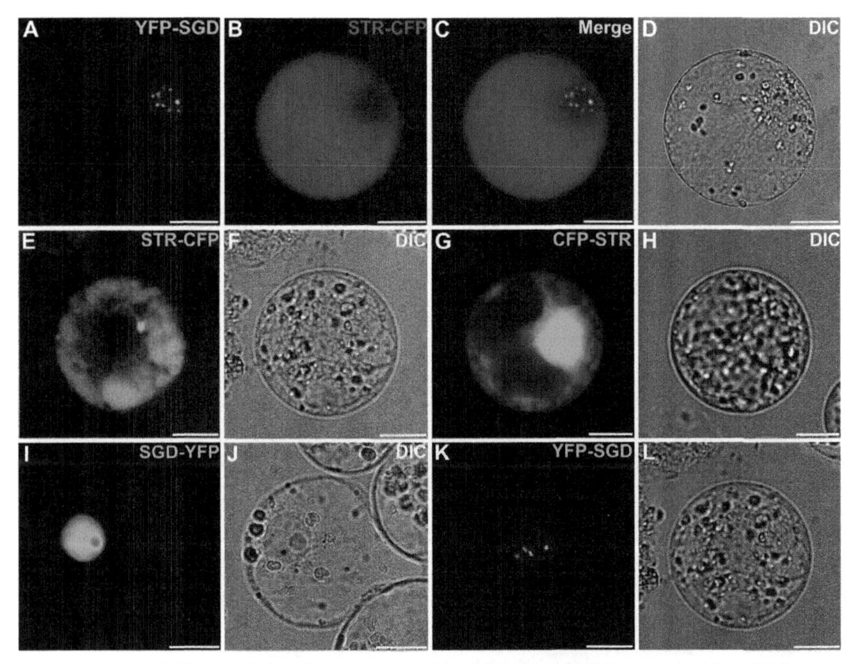

E. Foureau *et al.*, Fig. 3 Subcellular localization of STR and SGD expressed as FP fusions in plant cells. *C. roseus* cells were transiently cotransformed (A–D) or transformed (E–L) with constructs expressing YFP-SGD (A; K), STR-CFP (B; E), CFP-STR (G) of SGD-YFP (I). For cotransformation, superimposition of the two fluorescence signals appears on the merged image (C). Cell morphology (D; F; H; J; L) was observed with differential interference contrast (DIC). Bars = 10 μm.

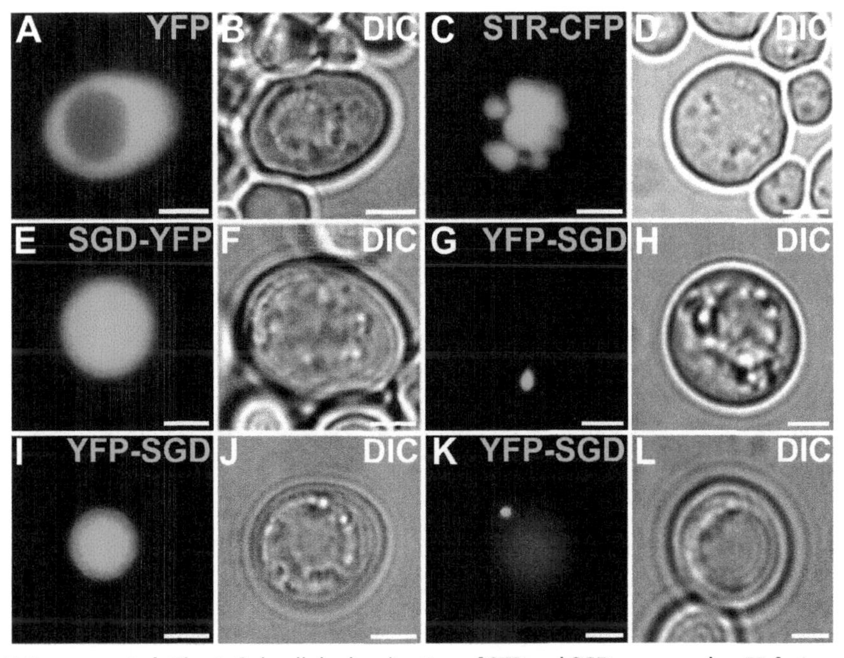

E. Foureau *et al.*, Fig. 4 Subcellular localization of STR and SGD expressed as FP fusions in yeast cells. Yeast cells were transformed with constructs expressing unfused YFP (A), STR-CFP (C), SGD-YFP (E), or YFP-SGD (G; I; K). Cell morphology (B; D; F; H; J; L) was observed with differential interference contrast (DIC). Bars = 2 μm.

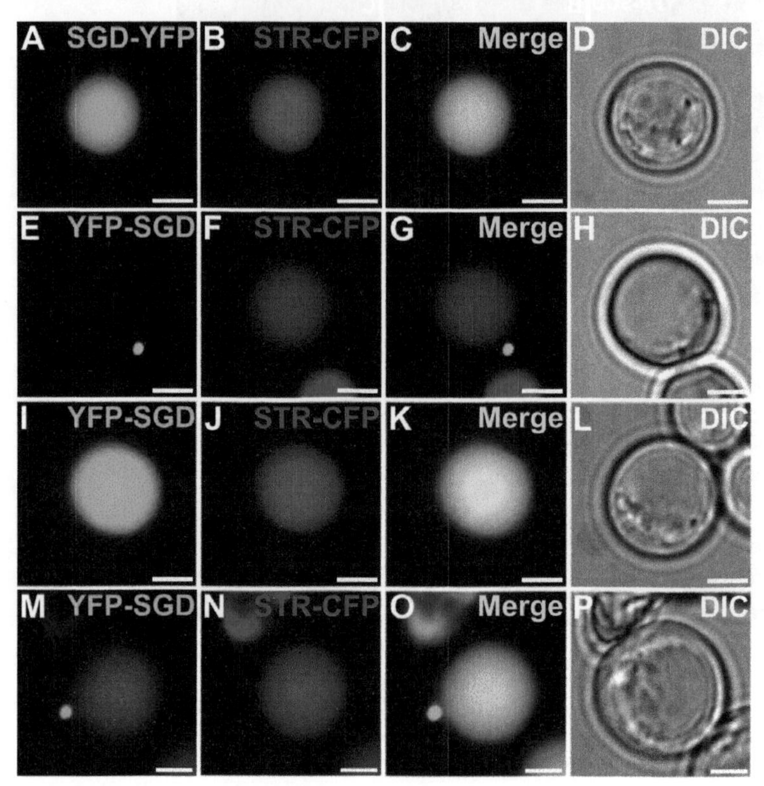

E. Foureau *et al.*, Fig. 5 Colocalization of STR and SGD expressed as FP fusions in yeast cells. Yeast cells were co transformed with constructs expressing SGD-YFP and STR-CFP (A–D) or YFP-SGD and STR-CFP (E–P). Colocalization of the two fluorescence signals appears on the merged images (C; G; K; O). Cell morphology (D; H; L; P) was observed with differential interference contrast (DIC). Bars = 2 μm.

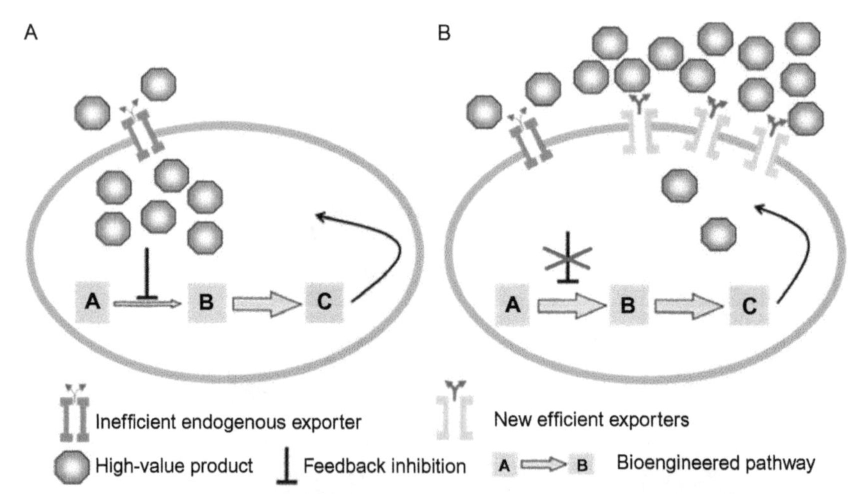

D. Xu *et al.*, Fig. 1 Schematic illustration of the potential impact of transport engineering. (A) Pathway engineered yeast strain with low yield due to end product feedback inhibition. Purification from cell lysates. (B) High-yielding pathway and transport engineered yeast strain with no feedback inhibition and purification from growth medium.

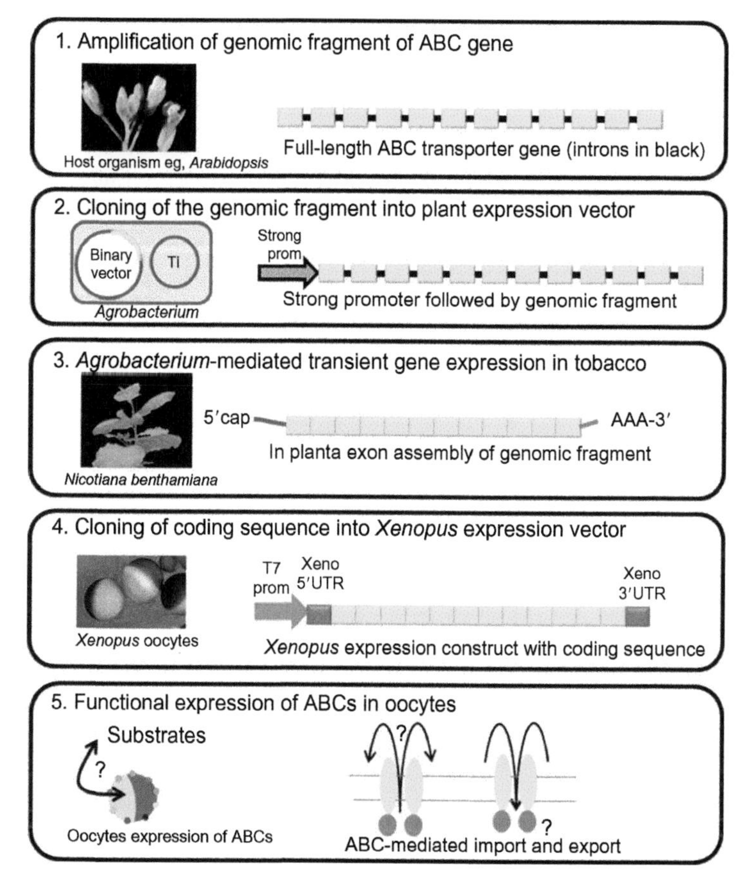

D. Xu *et al.*, Fig. 2 Work flow of cloning and functional characterization of an ABC transporter gene from a given host with "exon engineering" strategy. Diagram representing the scalable process for characterization an ABC protein of interest from its coding genomic DNA sequence in five steps with two heterologous systems: (1) transient overexpression of gDNA in *Nicotiana benthamiana* by *Agrobacterium* infiltration for cDNA preparation; (2) overexpression of cRNA in the *Xenopus laevis* oocyte system.

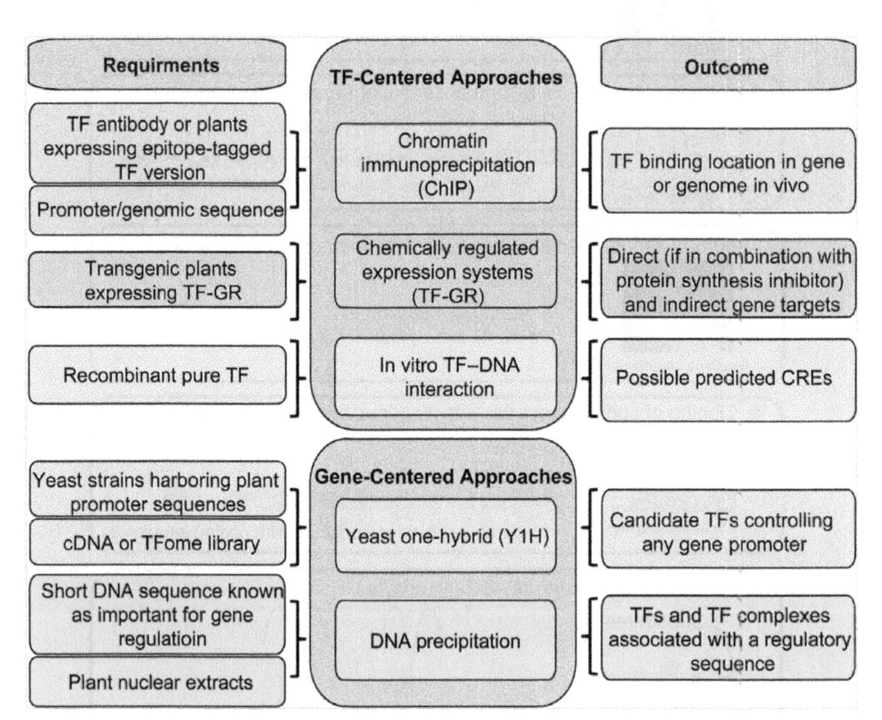

F. Yang *et al.*, Fig. 1 Approaches to study gene regulatory networks. Commonly used TF- and gene-centered approaches to establish gene regulatory networks with their requirements and expected outcomes.

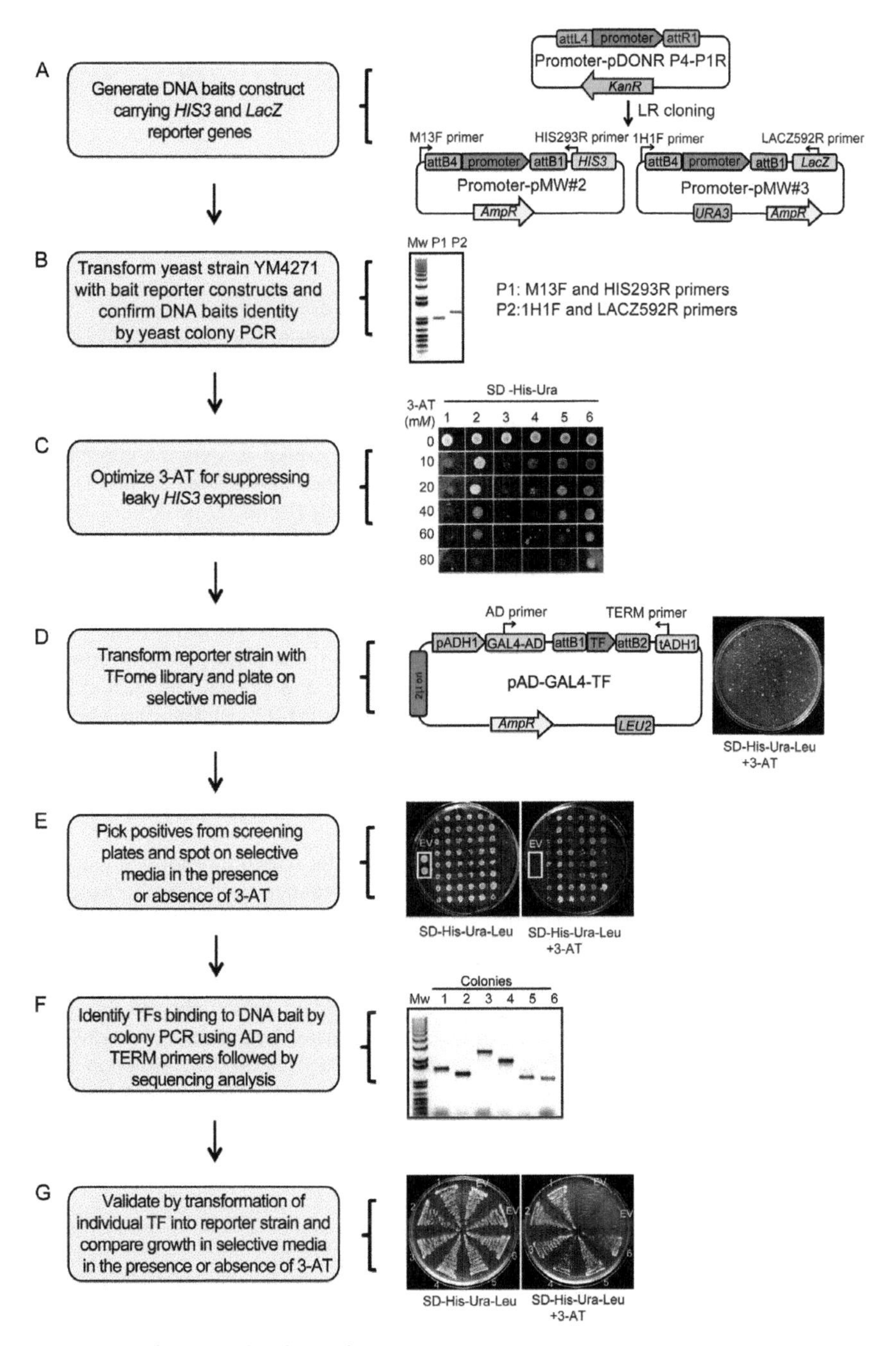

F. Yang *et al.*, Fig. 2 See legend on next page.

F. Yang *et al.*, Fig. 2 Y1H flowchart to identify maize PDIs. (A) DNA bait promoters cloned into the pDONR P4-P1R entry vector were subsequently subcloned into the Y1H destination vectors pMW#2 and pMW#3 containing the *HIS3* and *LacZ* reporter genes by LR Gateway cloning. (B) Y1H strains were generated by genome integration of the destination constructs into the YM4271 yeast strain (*MATa, ura3-52, his3-D 200, ade2-101, ade5, lys2-801, leu2-3,112, trp1-901, tyr1-501, gal4 D, gal80 D, ade5::hisG*). Genomic integration was tested by colony PCR using the primer pair P1 for the *His* construct (M13F: 5′-GTAAAACGACGGCCAGT-3′ and HIS293R: 5′-GGGACCACCCTTTAAAGAGA-3′) and primer pair P2 for the *LacZ* construct (1H1F: 5′-GTTCGGAGATTACCGAATCAA-3′ and LACZ592R, 5′-ATGCGCTCAGGTCAAATTCAGA-3′) (Reece-Hoyes & Walhout, 2012). (C) The generated DNA bait yeast strains were tested for autoactivation by growing colonies onto synthetic defined (SD) media lacking His and Ura with increasing 3-AT concentrations, as shown for six independent colonies (#1–6). (D) Transformation of a yeast reporter strain with a GAL4^{AD}-TFome library. The TFome library (Burdo et al., 2014) comprises TFs ORFs cloned into the pAD-GAL4-GW-C1 destination vector and PDIs were selected by growth on SD–His–Ura–Leu media with 3-AT. (E) Selected colonies were picked and spotted onto SD–His–Ura–Leu and SD–His–Ura–Leu with 3-AT plates. Colonies representing the yeast strain used for the screen transformed with the pAD-GAL4-GW-C1 empty vector (EV) were used as a negative control (white squares). (F) The identification of the TFs binding to the DNA baits was conducted by yeast colony PCR of DNA isolated from colonies selected in SD–His–Ura–Leu with 3-AT plates using primers AD and TERM primers (AD: 5′-CGCGTTTGGAATCACTACAGGG-3′ and TERM: 5′-GGAGACTTGACCAAACCTCTGGCG-3′) (Walhout & Vidal, 2001) followed by DNA sequencing. (G) Validation of Y1H interactions was conducted by transforming individual TFs into the yeast strain used for the original screening followed by selection onto SD–His–SD–His–Ura–Leu + 3-AT. TFs #1, 2, 3, 4, 5, and 6 represent confirmed Promoter–TF interactions. EV was used as a negative control.

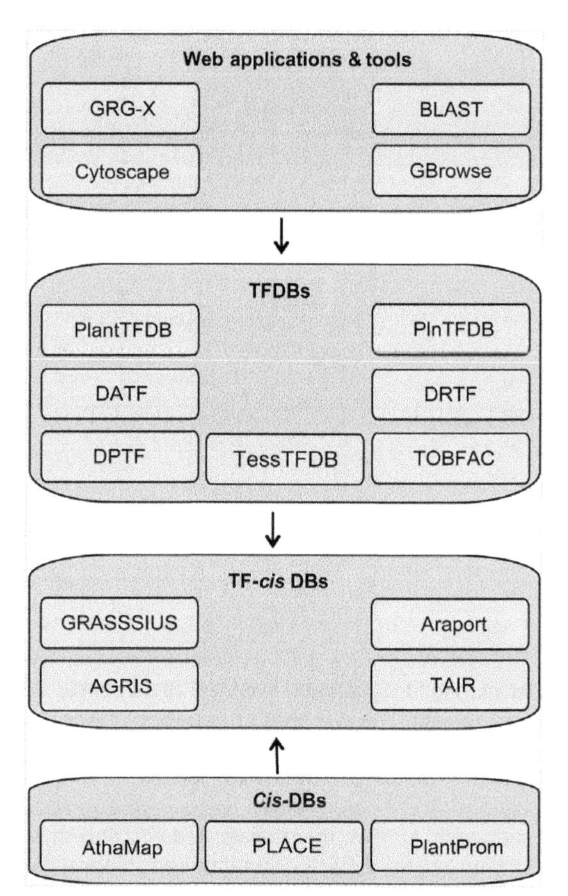

F. Yang *et al.*, Fig. 3 Databases and database applications for the study of GRNs and grids. TF-*cis*-DBs host information contained in both TFDBs (TF-related information) and *cis*-DBs (*cis*-regulatory element-related information).

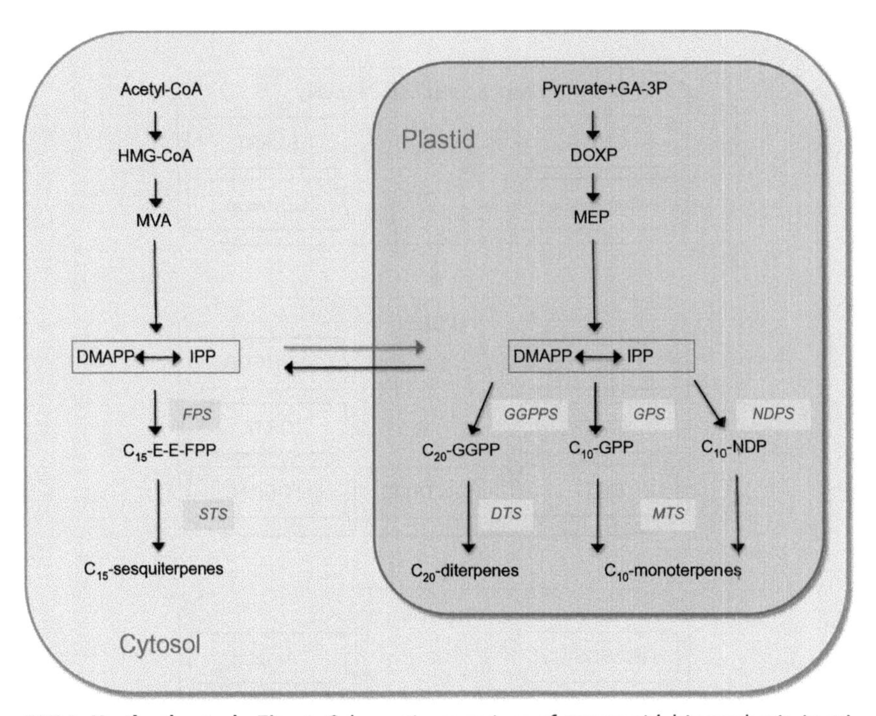

R.W.J. Kortbeek *et al.*, Fig. 1 Schematic overview of terpenoid biosynthesis in trichomes of cultivated tomato. The synthesis of terpenoids takes place in the cytosol through the mevalonate (MVA) pathway, in the plastid via the methyl-erythritol phosphate (MEP) pathway but also (partly) in the mitochondria, ER and/or peroxisomes (not depicted). Abbreviations: *Acetyl-CoA*, acetoacetyl-coenzyme A; *HMG-CoA*, 3-hydroxy-3-methylglutaryl-coenzyme A; *MVA*, mevalonate; *DMAPP*, dimethylallyl diphosphate; *IPP*, isopentenyl diphosphate; *E,E-FPP*, *trans*-farnesyl diphosphate; *FPS*, farnesyl diphosphate synthase; *STS*, sesquiterpene synthase; *GA-3P*, glyceraldehyde-3-phosphate; *DOXP*, 1-deoxy-D-xylulose 5-phosphate; *GGPP*, geranylgeranyl diphosphate; *GGPPS*, geranylgeranyl diphosphate synthase; *DPS*, diterpene synthase; *GPP*, geranyl diphosphate; *GPS*, geranyl diphosphate synthase; *NDP*, neryl diphosphate; *NDPS*, neryl diphosphate synthase; *MTS*, monoterpene synthase.

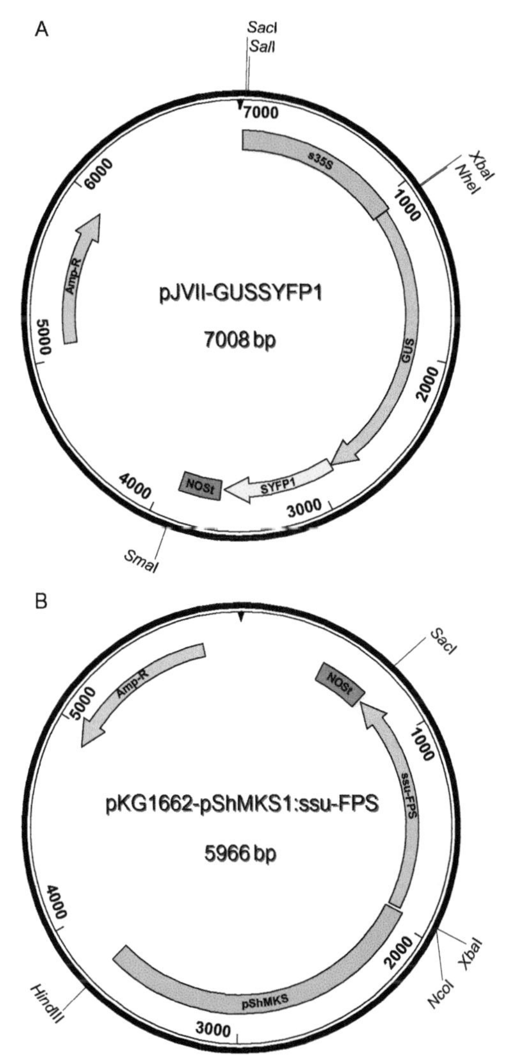

R.W.J. Kortbeek *et al.*, Fig. 2 Maps of plasmids used for engineering. (A) Map of the pJVII–GUS–sYFP plasmid used for transient and stable tomato transformation with trichome-specific promoters. For transient transformation by particle bombardment, the *S. lycopersicum* terpene synthase 5 (*SlTPS5*) promoter was used to drive GUS–sYFP. For stable transformation, the promoter of *S. lycopersicum* terpene synthase 9 (*SlTPS9*) was used. (B) pKG11662-pShMKS1:ssu-FPS construct used in the engineering of specialized metabolism in tomato trichomes. Abbreviations: *GUS*, β-glucuronidase; *sYFP1*, yellow fluorescent protein; *NOSt*, nopaline synthase terminator; *Amp-R*, β-lactamase ampicillin-resistance gene.

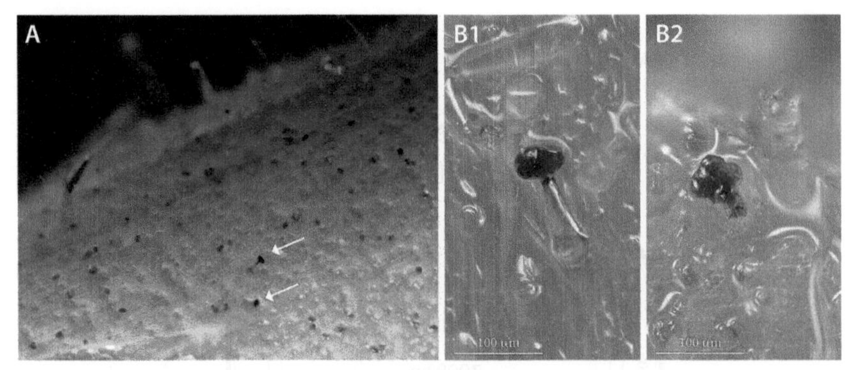

R.W.J. Kortbeek *et al.*, Fig. 3 GUS activity shows trichome-specific activity of the *SlTPS5* promoter. β-Glucuronidase (GUS) enzymatic activity in *S. lycopersicum* cv. Moneymaker stem trichomes transformed, by particle bombardment, with the pJVII-pSlTPS5:GUS construct containing GUS gene driven by the *S. lycopersicum* terpene synthase 5 (*SlTPS5*) promoter. Panoramic view (A) and close up (B1,2) of type VI glandular trichomes expressing SlTPS5:GUS. GUS activity was determined by incubation with X-Gluc (16 h). *Photograph by Jan van Arkel, IBED, University of Amsterdam.*

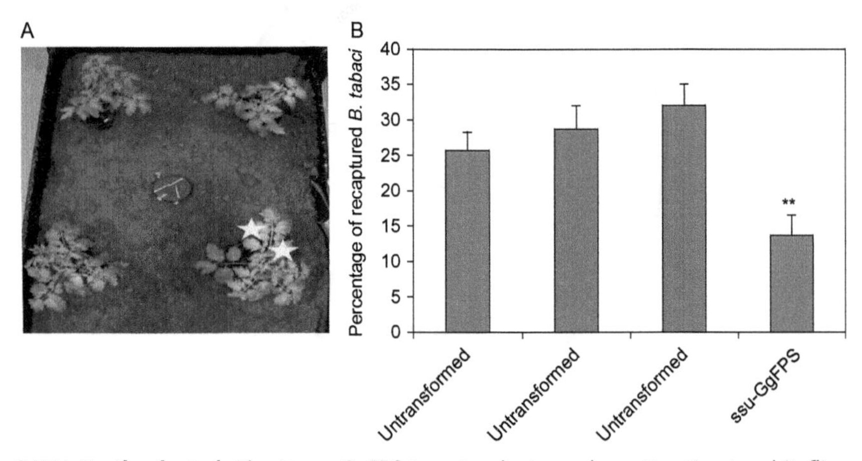

R.W.J. Kortbeek *et al.*, Fig. 7 ssu-GgFPS tomato plants are less attractive to whiteflies. (A) Experimental setup: 150 whiteflies (*Bemisia tabaci*) were released in the middle of a 1×1 m^2 with in three corners untransformed plants and in the fourth corner a *ssu-GgFPS* transgenic plant (indicated with *stars*). After 20 min, the whiteflies were recaptured from the plants and counted. After each assay, plants were rotated to exclude any positional effect. (B) Percentage of whiteflies that were recaptured on each plant. Significantly less whiteflies were recaptured on *ssu-GgFPS* transgenic plants (ANOVA: $p < 0.01$). Bars represent mean percentage of recaptured whiteflies $\pm$ SE ($n = 12$).

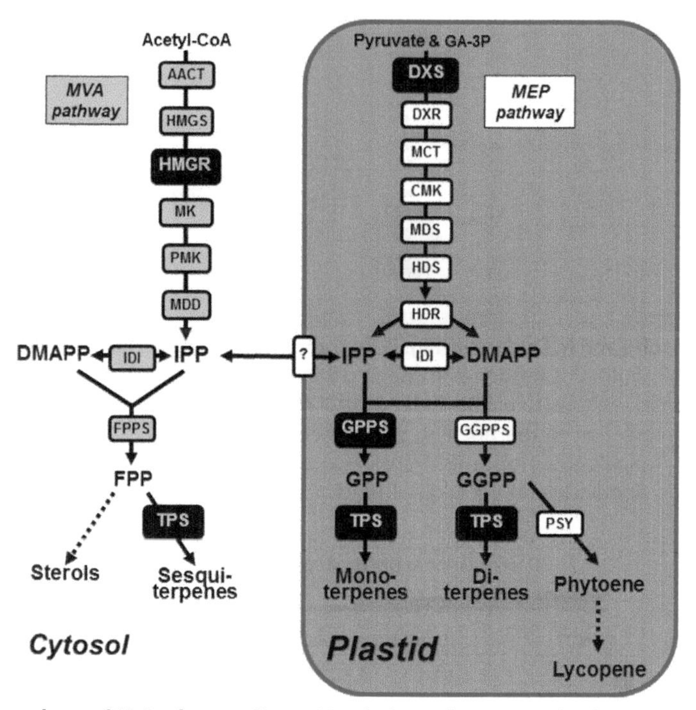

M. Gutensohn and N. Dudareva, Fig. 1 Metabolic pathways involved in terpenoid biosynthesis and enzymatic steps introduced for terpene metabolic engineering in tomato fruits. Enzymes of the MEP pathway and lycopene biosynthesis are localized in plastids and are highlighted in *white*. Enzymes of the MVA and sterol biosynthesis are primarily localized in the cytosol and are highlighted in *green/gray*. Enzymes engineered into transgenic tomato fruits are highlighted in *black*. The enzymatic steps are indicated by *arrows* and *dashed lines* indicate multiple enzymatic steps. Individual enzymes are depicted as *boxes* with the abbreviation of their names. Abbreviations: *AACT*, acetoacetyl-CoA thiolase; *CMK*, 4-(cytidine 5′-diphospho)-2-*C*-methyl-D-erythritol kinase; *DMAPP*, dimethylallyl diphosphate; *DXR*, 1-deoxy-D-xylulose 5-phosphate reductoisomerase; *DXS*, 1-deoxy-D-xylulose 5-phosphate synthase; *FPP*, farnesyl diphosphate; *FPPS*, farnesyl diphosphate synthase; *GA-3P*, D-glyceraldehyde 3-phosphate; *GGPP*, geranylgeranyl diphosphate; *GGPPS*, geranylgeranyl diphosphate synthase; *GPP*, geranyl diphosphate; *GPPS*, geranyl diphosphate synthase; *HDR*, (*E*)-4-hydroxy-3-methylbut-2-enyl diphosphate reductase; *HDS*, (*E*)-4-hydroxy-3-methylbut-2-enyl diphosphate synthase; *HMG-CoA*, 3-hydroxy-3-methylglutaryl-CoA; *HMGR*, 3-hydroxy-3-methylglutaryl-CoA reductase; *HMGS*, 3-hydroxy-3-methylglutaryl-CoA synthase; *IDI*, isopentenyl diphosphate isomerase; *IPP*, isopentenyl diphosphate; *MCT*, 2-*C*-methyl-D-erythritol 4-phosphate cytidylyltransferase; *MDS*, 2-*C*-methyl-D-erythritol 2,4-cyclodiphosphate synthase; *MEP*, 2-*C*-methyl-D-erythritol 4-phosphate; *MK*, mevalonate kinase; *MVA*, mevalonate; *MDD*, mevalonate diphosphate decarboxylase; *PMK*, phosphomevalonate kinase; *PSY*, phytoene synthase; *TPS*, terpene synthases (including mono-, sesqui-, and diterpene synthases).

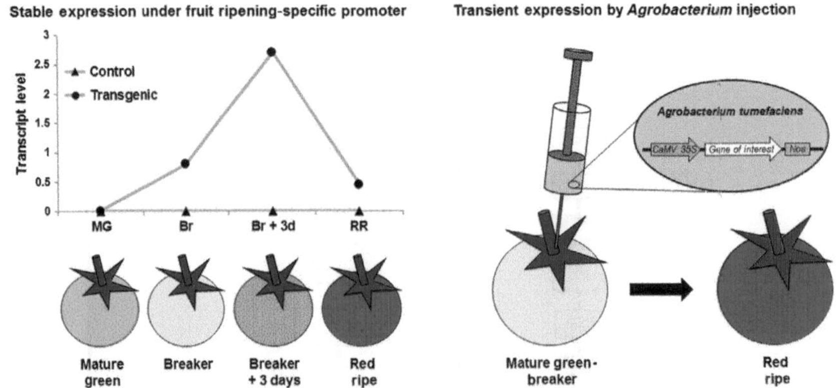

M. Gutensohn and N. Dudareva, Fig. 2 Two alternative strategies for the expression of terpene biosynthetic transgenes in ripening tomato fruits. (A) Expression of transgene in stable transgenic tomato plants under control of a fruit ripening-specific promoter that restricts expression to the ripening period between the breaker and red ripe stages when carotenoids accumulate. (B) Transient expression of transgene in ripening tomato fruits upon *Agrobacterium* injection of fruits prior to or at the breaker stage.

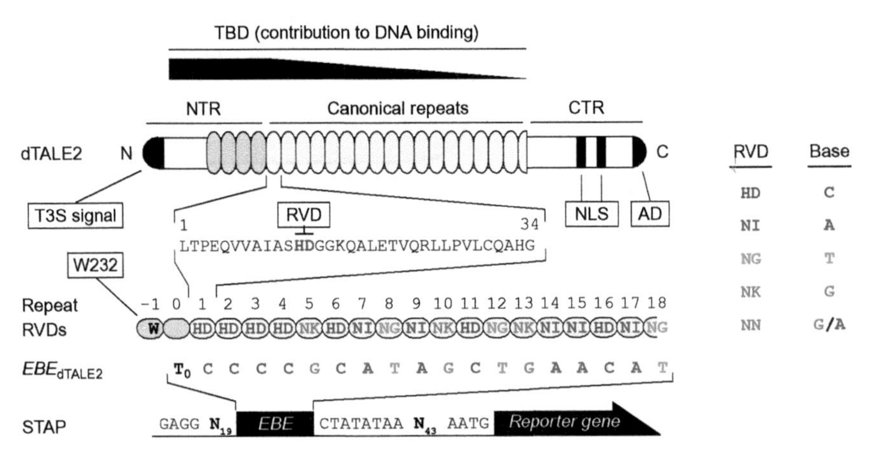

T. Schreiber and A. Tissier, Fig. 1 Properties of transcription activator-like effectors (TALEs) using the example of dTALE2. TALEs are composed of three domains: the N-terminal region (NTR), the repeat region, and the C-terminal region (CTR). The NTR contains a type III secretion and translocation (T3S) signal and a crucial part of the TALE DNA-binding domain (TBD). It includes at least four degenerated repeats (−3 to 0) and serves as nucleation site for the TALE–DNA interaction. The repeat region consists of 17.5 tandem arrayed 34-amino acid motifs (repeats), conferring the binding specificity. The repeats are highly conserved except for amino acid 12 and 13 designated as repeat-variable diresidue (RVD). One repeat mediates binding to one base pair of the target sequence in which the RVD defines the specificity of a given repeat. The most common RVDs are HD, NI, NG, NK, and NN which are specific for cytosine (C), adenine (A), thymine (T), guanine (G), and guanine/adenine (G/A), respectively. The order of the RVDs defines the DNA target sequence that is bound by a TALE. The RVD-defined target sequence is preceded by a 5′-thymine (T_0) and is part of the effector-binding element (EBE). T_0 is coordinated by a tryptophan residue (W232) of the degenerated repeat −1 that facilitates the RVD–base interaction of the following canonical repeats. The CTR contains two nuclear localization signals and an acidic activation domain (AD). A reporter gene fused to a synthetic TALE-activated promoter (STAP) can be transcriptionally induced by a dTALE.

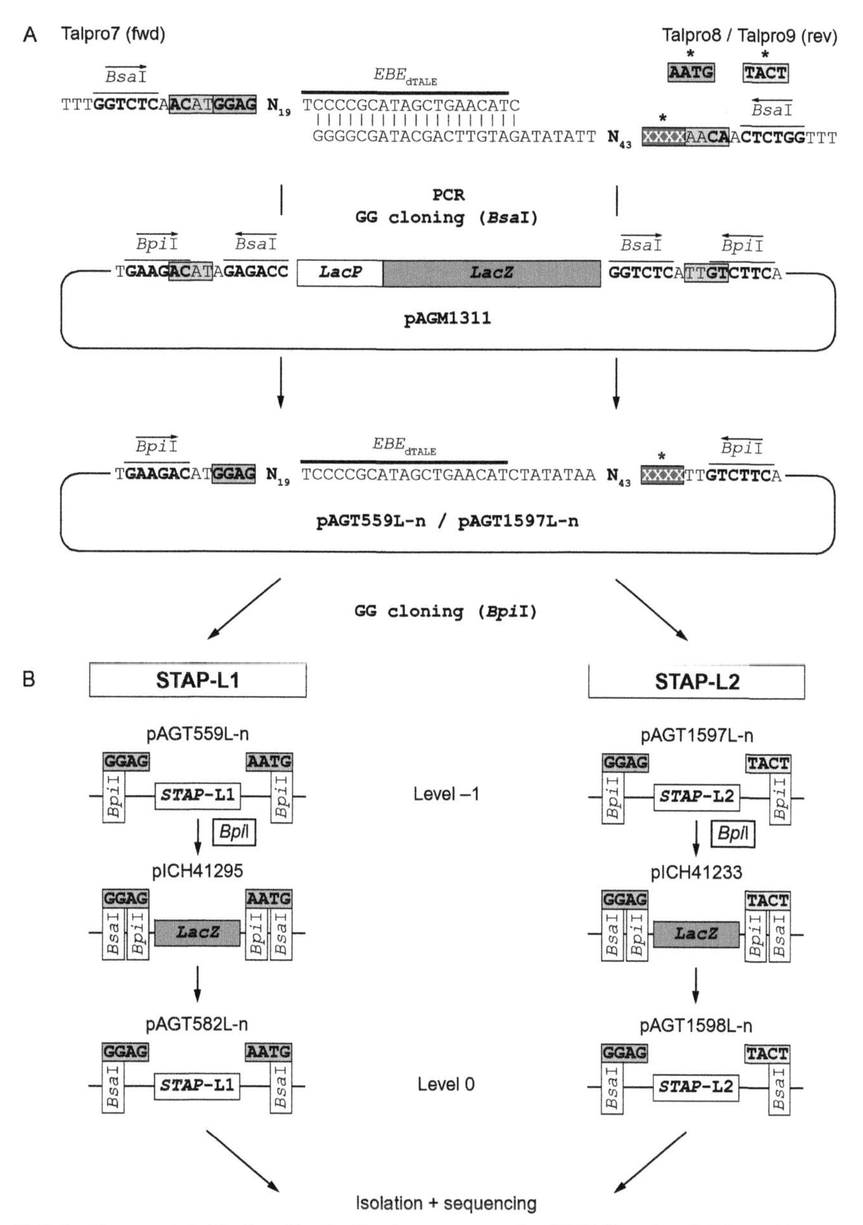

T. Schreiber and A. Tissier, Fig. 2 Cloning strategy for STAP libraries. (A) STAP libraries were generated by primer extension using degenerated oligonucleotides (Talpro7 and Talpro8 (L1) or Talpro7 and Talpro9 (L2)). The first amplification step leads to a *Bsa*I site-flanked PCR product which allows directed cloning into pAGM1311 (level −1) by restriction ligation using *Bsa*I and T4 DNA ligase (Golden Gate [GG] cloning). The *Bsa*I overhangs are depicted in *purple*. The resulting level −1 STAP library is designated as pAGT559L-n and pAGT1597L-n for STAP-L1 and STAP-L2, respectively. Cloning of the PCR products reconstitutes *Bpi*I sites with corresponding overhangs that were already introduced by the degenerated oligonucleotides (5′-overhang in *blue*; 3′-overhang in *red* [L1—AATG; L2—TACT]). (B) Level −1 libraries STAP-L1 (pAGT559L-n) and STAP-L2 (pAGT1597L-n) were further cloned by GG cloning using *Bpi*I into the level 0 vectors pICH41295 and pICH41233, respectively.

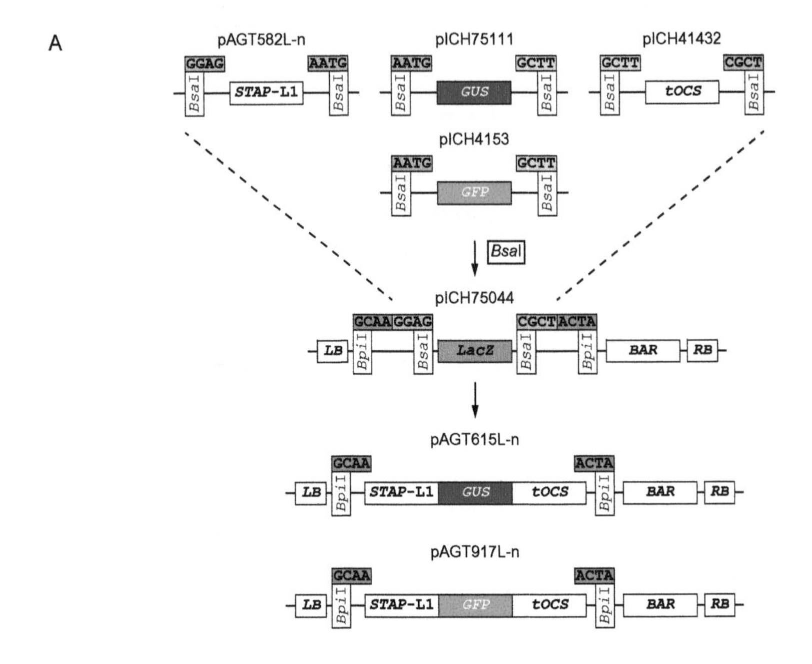

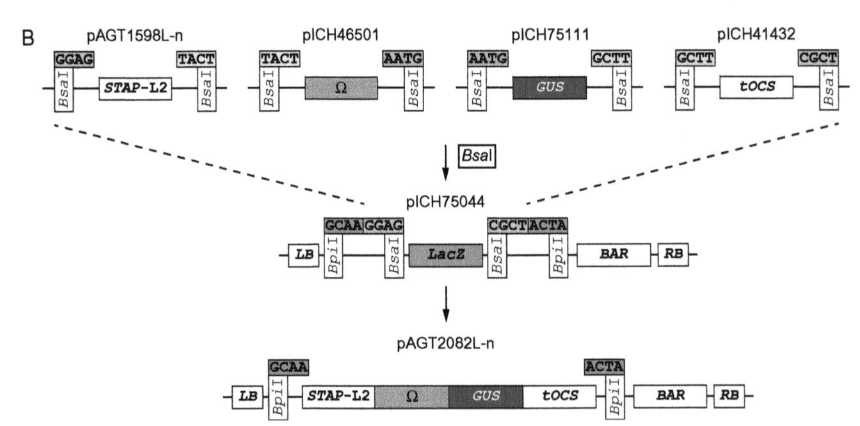

T. Schreiber and A. Tissier, Fig. 3 Cloning strategy for STAP reporters. (A) To analyze the dTALE2-mediated transcriptional induction of single STAPs several T-DNA reporter construct were generated. Level 0 modules were cloned by GG cloning with *Bsa*I into the level 1 vector pICH75044. Promoter modules of the STAP library L1 (pAGT582L-n) were combined with the ORF modules of *GUS* (pICH75111) and *GFP* (pICH4153) together with the terminator module tOCS (pICH41432) resulting in the T-DNA vectors pAGT615L-n and pAGT917L-n, respectively. (B) Promoter modules of the STAP library L2 were combined with the Ω enhancer module (pICH46501), the ORF module of *GUS* (pICH75111), and the terminator module *tOCS* (pICH41432) resulting in the T-DNA vector pAGT2082L-n.

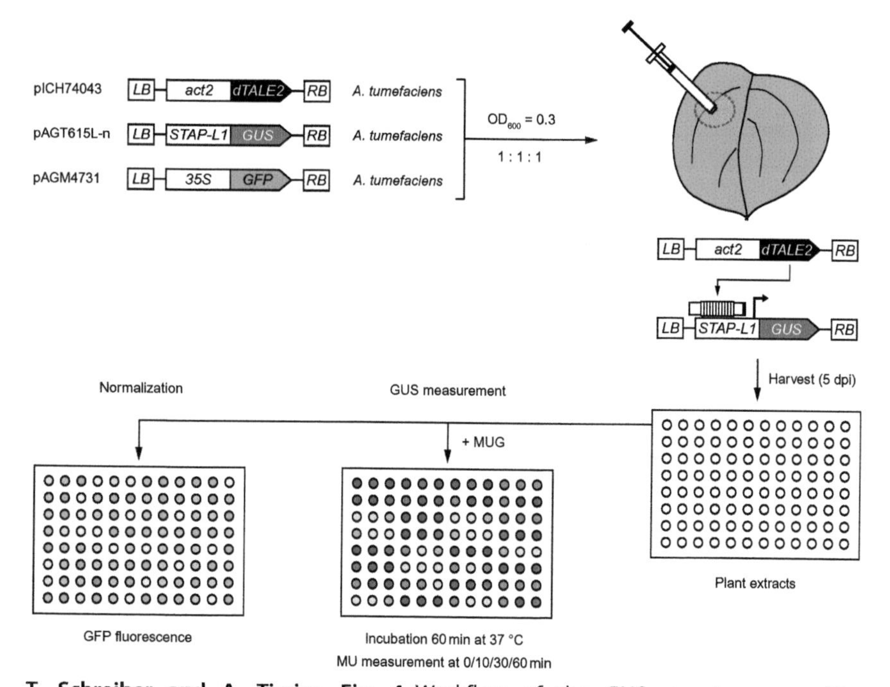

T. Schreiber and A. Tissier, Fig. 4 Workflow of the GUS reporter assay. The *Agrobacterium* strains carrying the different T-DNA constructs (pICH74043, *act2:dTALE2*; pAGT615L-n, *STAP-L1:GUS*; pAGM4731, *35S:GFP*) were resuspended in *Agrobacterium infiltration media* (AIM) and diluted to an OD_{600} of 0.3. The strains were mixed to equal amounts and inoculated in *N. benthamiana*. Five days postinoculation (5 dpi) leaf material was harvested, processed to plant extracts, and samples transferred into a 96-well plate by filtration. Plant extracts were separated for normalization by GFP measurement and measurement of GUS activity. For the measurement of GUS activity, 1 m*M* 4-methylumbelliferyl-β-D-glucuronide (MUG) was added and incubated for 60 min at 37°C. Reaction was stopped and measured at subsequent time points (0, 10, 30, and 60 min).

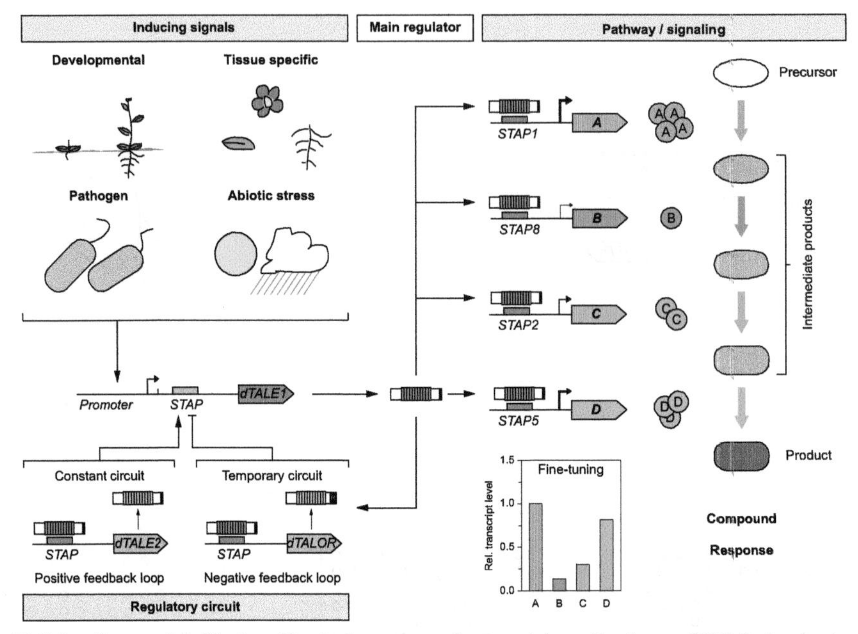

T. Schreiber and A. Tissier, Fig. 5 Overview of potential applications of STAPs *in planta*. A synthetic circuit can be initiated by different external or internal signals (developmental signals, abiotic stresses, etc.). A simple way to generate an output from these signals is to put a dTALE under the control of a signal-responsive promoter, in which the dTALE gets expressed under certain conditions or developmental stages to induce a signal-adapted reaction. Besides an induced reaction this system can also be used to induce a pathway resulting in production of a high-value compound. A STAP library can be used to induce enzymes of a pathway with different expression levels by a single activator (dTALE1). This enables the generation of efficient biosynthetic pathways by fine-tuning of intermediate enzymatic steps (eg, no accumulation of intermediate products). In some cases, inducing signals are only transient. To ensure a constant activity of a pathway one can introduce positive feedback loops. In this scenario, dTALE1 may also induce another dTALE with different DNA-binding specificity (dTALE2), which then in turn induces the dTALE1 if a dTALE2-responsible STAP is fused in front of dTALE1. Responses however, sometimes only need to occur transiently. To accomplish a transient expression one can introduce a negative feedback loop. Therefore, dTALE1 may induce a dTALE2-based transcriptional repressor, which then in turn repress the expression of dTALE1 and hence the subsequent dTALE1-induced response.

Edwards Brothers Malloy
Ann Arbor MI. USA
August 8, 2016